THE STRUCTURE OF SCIENCE
Problems in the Logic of Scientific Explanation

Ernest Nagel   COLUMBIA UNIVERSITY

THE STRUCTURE
OF SCIENCE

Problems in the Logic of
Scientific Explanation

HACKETT PUBLISHING COMPANY

INDIANAPOLIS · CAMBRIDGE

10 09 08 07 06 05 04 03   6 7 8 9 10 11 12

Library of Congress Catalog Card Number 60–15504

ISBN: 0–915144–71–9 (pbk.)
ISBN: 0–915144–72–7 (cloth)

For further information address the publisher
Hackett Publishing Company, Inc.
P.O. Box 44937
Indianapolis, Indiana 46244-0937

The paper used in this publication meets the minimum requirements of American National Standard for Information Sciences—Permanence of Paper for Printed Library Materials, ANSI Z39.48-1984.

♾

To
Edith

Preface

Science as an institutionalized art of inquiry has yielded varied fruit. Its currently best-publicized products are undoubtedly the technological skills that have been transforming traditional forms of human economy at an accelerating rate. It is also responsible for many other things not at the focus of present public attention, though some of them have been, and continue to be, frequently prized as the most precious harvest of the scientific enterprise. Foremost among these are: the achievement of generalized theoretical knowledge concerning fundamental determining conditions for the occurrence of various types of events and processes; the emancipation of men's minds from ancient superstitions in which barbarous practices and oppressive fears are often rooted; the undermining of the intellectual foundations for moral and religious dogmas, with a resultant weakening in the protective cover that the hard crust of unreasoned custom provides for the continuation of social injustices; and, more generally, the gradual development among increasing numbers of a questioning intellectual temper toward traditional beliefs, a development frequently accompanied by the adoption in domains previously closed to systematic critical thought of logical methods for assessing, on the basis of reliable data of observation, the merits of alternative assumptions concerning matters of fact or of desirable policy.

Despite the brevity of this partial list, it suffices to make evident how much the scientific enterprise has contributed to the articulation as well as to the realization of aspirations generally associated with the idea of a liberal civilization. For this reason alone, it is not astonishing that science as a way of acquiring competent intellectual and practical mastery over events should be a perennial subject for attentive study. But however this may be, the record of reflection on the nature of scientific inquiry and its significance for human life goes back to the beginnings of theoretical science in Greek antiquity; and there are few notable figures in the history of Western philosophy who have not given serious thought to problems raised by the sciences of their day.

In consequence, although the use of the term 'philosophy of science' as the name for a special branch of study is relatively recent, the name designates investigations continuous with those that have been pursued for centuries under such headings for traditional divisions of philosophy as 'logic,' 'theory of knowledge,' 'metaphysics,' and 'moral and social philosophy.' Moreover, despite the impression sometimes created by the wide currency of the term in titles given to books, courses of instruction, and learned societies that it denotes a clearly delimited discipline which deals with a group of closely interrelated questions, the philosophy of science as currently cultivated is not a well-defined area of analysis. On the contrary, contributors to the area often manifest sharply contrasting aims and methods; and the discussions commonly classified as belonging to it collectively range over most of the heterogeneous set of problems that have been the traditional concern of philosophy.

The present book, although an essay in the philosophy of science, nevertheless deals with a more integrated group of questions, and its scope is controlled by the objective of analyzing the logic of scientific inquiry and the logical structure of its intellectual products. It is primarily an examination of logical patterns exhibited in the organization of scientific knowledge as well as of the logical methods whose use (despite frequent changes in special techniques and revolutions in substantive theory) is the most enduring feature of modern science. The book accordingly ignores many issues, often discussed at length in standard works and courses on the philosophy of science, that do not seem to me relevant to its objective—for example, issues in the epistemology of sense perception, or in proposed cosmic syntheses purporting to make "intelligible" the totality of specialized scientific findings. On the other hand, I have not hesitated to consider issues that may appear to be only remotely related to the actual practice of science, if their discussion might contribute to a clarified understanding of scientific method and its fruits —for example, questions dealing with the translatability of scientific theories into statements about data of sensory observation, or with the import of the belief in universal determinism for ascriptions of moral responsibility.

The order in which problems are discussed in this book reflects in part the stress I place upon the achievement of encompassing well-grounded explanations as a major and distinctive scientific ideal. But irrespective of this stress, the study of the logic of science can be divided for the sake of convenient analysis and exposition into three principal parts. The first division is addressed to problems dealing primarily with the nature of scientific explanations: with their logical structures, their mutual relations, their functions in inquiry, and their devices for systematizing knowledge. The second division concentrates on questions concerned with the logical structure of scientific concepts: with their articulation by way of diverse techniques of definition and measurement, their linkages to data of observation, and the conditions under which they are scientifically meaningful. The third division is directed to problems dealing with the evaluation of claims to knowledge in various sciences: with the structure of probable inference, the principles employed in weighing evidence, and the validation of inductive arguments. These three partly overlapping groups of problems constitute the scope of a systematically unified study of the logic of science; nevertheless, each group of questions can be explored with only occasional reference to issues subsumed under the others. Accordingly, although the present volume is devoted mainly to questions falling into the first of the above tripartite divisions—the problems in the other two are reserved for detailed discussion in a volume in active preparation—the volume is fully self-contained; and issues central to the other divisions but requiring immediate notice in the present volume receive at least brief attention in it.

I have tried to write this book for a wider audience than that of professional students of philosophy, in the conviction that even if some questions discussed in it are perhaps of little interest to anyone else the book as a whole deals with matters which are of more than narrowly limited professional concern. I have therefore avoided highly formalized presentations of analyses or the use of the special symbolic notation of modern formal logic, however valuable a precise formalism may be for the resolution of certain technical problems. It would have been inconsistent with the central aim of the book to exclude all mention of difficult technical notions employed in special branches of science; on the other hand, the book attempts to explain such notions when they are not likely to be familiar to many readers whom I would like to reach. I have also sought to exhibit the character of scientific method in a variety of concrete domains—in the social and biological sciences as well as in physics. I have sought to do this, even though with the omission of several other special disciplines I had originally intended to canvass, partly in order to make clear to a varied audience that despite important differences there is a basic logical continuity in the operations of scientific intelligence, and partly to provide for members of such an audience a broad foundation for assessing in a judicious spirit the current tide of criticism directed (frequently on behalf of some "higher wisdom") against the works of scientific reason.

Several chapters in this volume include in considerably revised form material previously published. I want to thank the publishers of the following articles for their kind permission to use them in the present book: "The Causal Character of Modern Physical Theory," in *Freedom and Reason* (ed. by S. Baron, E. Nagel, and K. S. Pinson), The Free Press, Glencoe, Ill., 1951; "The Meaning of Reduction in the Natural Sciences," in *Science and Civilization* (ed. R. C. Stauffer), The University of Wisconsin Press, 1949, by permission of the Regents of the University of Wisconsin; "Teleological Explanation and Teleological Systems," in *Vision and Action* (ed. by S. Ratner), Rutgers University Press, 1953; "Science, With and Without Wisdom," *The Saturday Review of Literature*, 1945; "Wholes, Sums and Organic Unities," *Philosophical Studies*, 1952; "Mechanistic Explanation and Organismic Biology" and "Determinism in History," *Philosophy and Phenomenological Research*, 1951 and 1960; and "Some Issues in the Logic of Historical Analysis," *Scientific Monthly*, 1952, by permission of the American Association for the Advancement of Science.

It is an author's privilege to acknowledge the personal debts he has incurred in writing his book, and although it is not possible for me to list them all I record with pleasure my chief creditors. My interest in the philosophy of science was first aroused by my teacher, the late Morris R. Cohen, and I remain grateful to him for the direction he gave to my thinking as well as for the continuing stimulus of his teachings. Neither Rudolf Carnap nor Philipp Frank have been formally my teachers, but I have profited greatly from the numerous conversations I have had with them since 1934 on the logic of science; and I have obtained equally invaluable instruction on the methodological problems of empirical social research from the enlightening talks I have had with Paul F. Lazarsfeld for many years. I have also received much help and encouragement from other friends: from Abraham Edel, Albert Hofstadter, and Sidney Hook, with each of whom I have enjoyed high philosophical discourse since all of us were young men, and who gave me the benefit of their criticisms on various portions of the manuscript at various stages of its completion; from John C. Cooley, Paul Edwards, Herbert Feigl, Charles Frankel, John Gregg, Carl G. Hempel, Sidney Morgenbesser, Meyer Schapiro, and Patrick Suppes, who have contributed heavily to the clarification of my ideas during the many discussions I have had with them; and from my wife, to whom the volume is dedicated, who has served patiently as a touchstone for the intelligibility of most of the things said in it. I am deeply grateful to the John Simon Guggenheim Memorial Foundation, the Rockefeller Foundation, and the Center for Advanced Study in the Behavioral Sciences for the leisure to study and write which they made possible.

E. N.

South Wardsboro, Vermont
August, 1960

Contents

THE STRUCTURE OF SCIENCE
Problems in the Logic of Scientific Explanation

Introduction: Science and
Common Sense

1 Long before the beginnings of modern civilization, men acquired vast funds of information about their environment. They learned to recognize substances which nourished their bodies. They discovered the uses of fire and developed skills for transforming raw materials into shelters, clothing, and utensils. They invented arts of tilling the soil, communicating, and governing themselves. Some of them discovered that objects are moved more easily when placed on carts with wheels, that the sizes of fields are more reliably compared when standard schemes of measurement are employed, and that the seasons of the year as well as many phenomena of the heavens succeed each other with a certain regularity. John Locke's quip at Aristotle—that God was not so sparing to men as to make them merely two-legged creatures, leaving it to Aristotle to make them rational—seems obviously applicable to modern science. The acquisition of reliable knowledge concerning many aspects of the world certainly did not wait upon the advent of modern science and the self-conscious use of its methods. Indeed, in this respect, many men in every generation repeat in their own lives the history of the race: they manage to secure for themselves skills and competent information, without benefit of training in the sciences and without the calculated adoption of scientific modes of procedure.

If so much in the way of knowledge can be achieved by the shrewd exercise of native gifts and "common-sense" methods, what special excellence do the sciences possess, and what do their elaborate intellectual

1

and physical tools contribute to the acquisition of knowledge? The question requires a careful answer if a definite meaning is to be associated with the word 'science.'

The word and its linguistic variants are certainly not always employed with discrimination, and they are frequently used merely to confer an honorific distinction on something or other. Many men take pride in being "scientific" in their beliefs and in living in an "age of science." However, quite often the sole discoverable ground for their pride is a conviction that, unlike their ancestors or their neighbors, they are in possession of some alleged final truth. It is in this spirit that currently accepted theories in physics or biology are sometimes described as scientific, while all previously held but no longer accredited theories in those domains are firmly refused that label. Similarly, types of practice that are highly successful under prevailing physical and social conditions, such as certain techniques of farming or industry, are occasionally contrasted with the allegedly "unscientific" practices of other times and places. Perhaps an extreme form of the tendency to rob the term 'scientific' of all definite content is illustrated by the earnest use that advertisers sometimes make of such phrases as 'scientific haircutting,' 'scientific rug cleaning,' and even 'scientific astrology.' It will be clear, however, that in none of the above examples is a readily identifiable and differentiating characteristic of beliefs or practices associated with the word. It would certainly be ill-advised to adopt the suggestion, implicit in the first example, to limit the application of the adjective 'scientific' to beliefs that are indefeasibly true—if only because infallible guaranties of truth are lacking in most if not all areas of inquiry, so that the adoption of such a suggestion would in effect deprive the adjective of any proper use.

The words 'science' and 'scientific' are nevertheless not quite so empty of a determinate content as their frequently debased uses might indicate. For in fact the words are labels either for an identifiable, continuing enterprise of inquiry or for its intellectual products, and they are often employed to signify traits that distinguish those products from other things. In the present chapter we shall therefore survey briefly some of the ways in which "prescientific" or "common-sense" knowledge differs from the intellectual products of modern science. To be sure, no sharp line separates beliefs generally subsumed under the familiar but vague rubric of "common sense" from those cognitive claims recognized as "scientific." Nevertheless, as in the case of other words whose fields of intended application have notoriously hazy boundaries (such as the term 'democracy'), absence of precise dividing lines is not incompatible with the presence of at least a core of firm meaning for each of these words. In their more sober uses, at any rate, these words do in fact connote important and recognizable differences. It is these differences that we must attempt to identify, even if we are compelled to sharpen some of them for the sake of expository emphasis and clarity.

1. No one seriously disputes that many of the existing special sciences have grown out of the practical concerns of daily living: geometry out of problems of measuring and surveying fields, mechanics out of problems raised by the architectural and military arts, biology out of problems of human health and animal husbandry, chemistry out of problems raised by metallurgical and dyeing industries, economics out of problems of household and political management, and so on. To be sure, there have been other stimuli to the development of the sciences than those provided by problems of the practical arts; nevertheless, these latter have had, and still continue to have, important roles in the history of scientific inquiry. In any case, commentators on the nature of science who have been impressed by the historical continuity of common-sense convictions and scientific conclusions have sometimes proposed to differentiate between them by the formula that the sciences are simply "organized" or "classified" common sense.

It is undoubtedly the case that the sciences are organized bodies of knowledge and that in all of them a classification of their materials into significant types or kinds (as in biology, the classification of living things into species) is an indispensable task. It is clear, nonetheless, that the proposed formula does not adequately express the characteristic differences between science and common sense. A lecturer's notes on his travels in Africa may be very well organized for the purposes of communicating information interestingly and efficiently, without thereby converting that information into what has historically been called a science. A librarian's card catalogue represents an invaluable classification of books, but no one with a sense for the historical association of the word would say that the catalogue is a science. The obvious difficulty is that the proposed formula does not specify what *kind* of organization or classification is characteristic of the sciences.

Let us therefore turn to this question. A marked feature of much information acquired in the course of ordinary experience is that, although this information may be accurate enough within certain limits, it is seldom accompanied by any explanation of why the facts are as alleged. Thus societies which have discovered the uses of the wheel usually know nothing of frictional forces, nor of any reasons why goods loaded on vehicles with wheels are easier to move than goods dragged on the ground. Many peoples have learned the advisability of manuring their agricultural fields, but only a few have concerned themselves with the reasons for so acting. The medicinal properties of herbs like the foxglove have been recognized for centuries, though usually no account was given of the grounds for their beneficent virtues. Moreover, when "common sense" does attempt to give explanations for its facts—as when the value of the foxglove as a cardiac stimulant is explained in terms of the similarity in shape of the flower and the human heart—the explanations are frequently accepted without critical tests of their relevance to the facts. Common sense is often eligible to receive the well-known advice Lord Mansfield gave to a newly appointed governor of a colony who was unversed in the law: "There is no difficulty in deciding a case—only hear both sides patiently, then consider what you think justice requires, and decide accordingly; but never give your reasons, for your judgment will probably be right, but your reasons will certainly be wrong."

It is the desire for explanations which are at once systematic and controllable by factual evidence that generates science; and it is the organization and classification of knowledge on the basis of explanatory principles that is the distinctive goal of the sciences. More specifically, the sciences seek to discover and to formulate in general terms the conditions under which events of various sorts occur, the statements of such determining conditions being the explanations of the corresponding happenings. This goal can be achieved only by distinguishing or isolating certain properties in the subject matter studied and by ascertaining the repeatable patterns of dependence in which these properties stand to one another. In consequence, when the inquiry is successful, propositions that hitherto appeared to be quite unrelated are exhibited as linked to each other in determinate ways by virtue of their place in a system of explanations. In some cases, indeed, the inquiry can be carried to remarkable lengths. Patterns of relations may be discovered that are pervasive in vast ranges of fact, so that with the help of a small number of explanatory principles an indefinitely large number of propositions about these facts can be shown to constitute a logically unified body of knowledge. The unification sometimes takes the form of a deductive system, as in the case of demonstrative geometry or the science of mechanics. Thus a few principles, such as those formulated by Newton, suffice to show that propositions concerning the moon's motion, the behavior of the tides, the paths of projectiles, and the rise of liquids in thin tubes are intimately related, and that all these propositions can be rigorously deduced from those principles conjoined with various special assumptions of fact. In this way a systematic explanation is achieved for the diverse phenomena which the logically derived propositions report.

Not all the existing sciences present the highly integrated form of systematic explanation which the science of mechanics exhibits, though for many of the sciences—in domains of social inquiry as well as in the various divisions of natural science—the idea of such a rigorous logical systematization continues to function as an ideal. But even in those branches of departmentalized inquiry in which this ideal is not generally pursued, as in much historical research, the goal of finding explanations for facts is usually always present. Men seek to know why the thirteen American colonies rebelled from England while Canada did not, why the ancient Greeks were able to repel the Persians but succumbed to the Roman armies, or why urban and commercial activity developed in medieval Europe in the tenth century and not before. To explain, to establish some relation of dependence between propositions superficially unrelated, to exhibit systematically connections between apparently miscellaneous items of information are distinctive marks of scientific inquiry.

2. A number of further differences between common sense and scientific knowledge are almost direct consequences of the systematic character of the latter. A well-recognized feature of common sense is that, though the knowledge it claims may be accurate, it seldom is aware of the limits within which its beliefs are valid or its practices successful. A community, acting on the rule that spreading manure preserves the fertility of the soil, may in many cases continue its mode of agriculture successfully. However, it may continue to follow the rule blindly, in spite of the manifest deterioration of the soil, and it may therefore be helpless in the face of a critical problem of food supply. On the other hand, when the reasons for the efficacy of manure as a fertilizer are understood, so that the rule is connected with principles of biology and soil chemistry, the rule comes to be recognized as only of restricted validity, since the efficiency of manure is seen to depend on the persistence of conditions of which common sense is usually unaware. Few who know them are capable of withholding admiration for the sturdy independence of those farmers who, without much formal education, are equipped with an almost endless variety of skills and sound information in matters affecting their immediate environment. Nevertheless, the traditional resourcefulness of the farmer is narrowly circumscribed: he often becomes ineffective when some break occurs in the continuity of his daily round of living, for his skills are usually products of tradition and routine habit and are not informed by an understanding of the reasons for their successful operation. More generally, common-sense knowledge is most adequate in situations in which a certain number of factors remain practically unchanged. But since it is normally not recognized that this adequacy does depend on the constancy of such factors—indeed, the very existence of the pertinent factors may not be recognized—common-sense knowledge suffers from a serious incompleteness. It is the aim of systematic science to remove this incompleteness, even if it is an aim which frequently is only partially realized.

The sciences thus introduce refinements into ordinary conceptions by the very process of exhibiting the systematic connections of propositions about matters of common knowledge. Not only are familiar practices thereby shown to be explicable in terms of principles formulating relations between items in wide areas of fact; those principles also provide clues for altering and correcting habitual modes of behavior, so as to make them more effective in familiar contexts and more adaptable to novel ones. This is not to say, however, that common beliefs are necessarily mistaken, or even that they are inherently more subject to change under the pressure of experience than are the propositions of science. Indeed, the age-long and warranted stability of common-sense convictions, such as that oaks do not develop overnight from acorns or that water solidifies on sufficient cooling, compares favorably with the relatively short life span of many theories of science. The essential point to be observed is that, since common sense shows little interest in systematically explaining the facts it notes, the range of valid application of its beliefs, though in fact narrowly circumscribed, is not of serious concern to it.

3. The ease with which the plain man as well as the man of affairs entertains incompatible and even inconsistent beliefs has often been the subject for ironic commentary. Thus, men will sometimes argue for sharply increasing the quantity of money and also demand a stable currency; they will insist upon the repayment of foreign debts and also take steps to prevent the importation of foreign goods; and they will make inconsistent judgments on the effects of the foods they consume, on the size of bodies they see, on the temperature of liquids, and the violence of noises. Such conflicting judgments are often the result of an almost exclusive preoccupation with the immediate consequences and qualities of observed events. Much that passes as common-sense knowledge certainly is about the effects familiar things have upon matters that men happen to value; the relations of events to one another, independent of their incidence upon specific human concerns, are not systematically noticed and explored.

The occurrence of conflicts between judgments is one of the stimuli to the development of science. By introducing a systematic explanation of facts, by ascertaining the conditions and consequences of events, by exhibiting the logical relations of propositions to one another, the sciences strike at the sources of such conflicts. Indeed, a large number of extraordinarily able minds have traced out the logical consequences of basic principles in various sciences; and an even larger number of investigators have repeatedly checked such consequences with other propositions obtained as a result of critical observation and experiment. There is no iron-clad guaranty that, in spite of this care, serious inconsistencies in these sciences have been eliminated. On the contrary, mutually incompatible assumptions sometimes serve as the bases for inquiries in different branches of the same science. For example, in certain parts of physics atoms were at one time assumed to be perfectly elastic bodies, although in other branches of physical science perfect elasticity was not ascribed to atoms. However, such inconsistencies are sometimes only apparent ones, the impression of inconsistency arising from a failure to note that different assumptions are being employed for the solution of quite different classes of problems. Moreover, even when the inconsistencies are genuine, they are often only temporary, since incompatible assumptions may be employed only because a logically coherent theory is not yet available to do the complex job for which those assumptions were originally introduced. In any event, the flagrant inconsistencies that so frequently mark common beliefs are notably absent from those sciences in which the pursuit of unified systems of explanation has made considerable headway.

4. As has already been noted, many everyday beliefs have survived centuries of experience, in contradistinction to the relatively short life span that is so often the fate of conclusions advanced in various branches of modern science. One partial reason for this circumstance merits attention. Consider some instance of common-sense beliefs, such as that water solidifies when it is sufficiently cooled; and let us ask what is signified by the terms 'water' and 'sufficiently' in that assertion. It is a familiar fact that the word 'water,' when used by those unacquainted with modern science, generally has no clear-cut meaning. It is then frequently employed as a name for a variety of liquids despite important physicochemical differences between them, but is frequently rejected as a label for other liquids even though these latter liquids do not differ among themselves in their essential physicochemical characteristics to a greater extent than do the former fluids. Thus, the word may perhaps be used to designate the liquids falling from the sky as rain, emerging from the ground in springs, flowing in rivers and roadside ditches, and constituting the seas and oceans; but the word may be employed less frequently if at all for liquids pressed out of fruits, contained in soups and other beverages, or evacuated through the pores of the human skin. Similarly, the word 'sufficiently' when used to characterize a cooling process may sometimes signify a difference as great as that between the maximum temperature on a midsummer day and the minimum temperature of a day in midwinter; at other times, the word may signify a difference no greater than that between the noon and the twilight temperatures on a day in winter. In short, in its common-sense use for characterizing temperature changes, the word 'sufficiently' is not associated with a precise specification of their extent.

If this example can be taken as typical, the language in which common-sense knowledge is formulated and transmitted may exhibit two important kinds of indeterminacy. In the first place, the terms of ordinary speech may be quite vague, in the sense that the class of things designated by a term is not sharply and clearly demarcated from (and may in fact overlap to a considerable extent with) the class of things not so designated. Accordingly, the range of presumed validity for statements employing such terms has no determinate limits. In the second place, the terms of ordinary speech may lack a relevant degree of specificity, in the sense that the broad distinctions signified by the terms do not suffice to characterize more narrowly drawn but important differences between the things denoted by the terms. Accordingly, relations of dependence between occurrences are not formulated in a precisely determinate manner by statements containing such terms.

As a consequence of these features of ordinary speech, experimental control of common-sense beliefs is frequently difficult, since the distinction between confirming and contradicting evidence for such beliefs cannot be easily drawn. Thus, the belief that "in general" water solidifies when sufficiently cooled may answer the needs of men whose interest in the phenomenon of freezing is circumscribed by their concern to achieve the routine objectives of their daily lives, despite the fact that the language employed in codifying this belief is vague and lacks specificity. Such men may therefore see no reason for modifying their belief, even if they should note that ocean water fails to freeze although its temperature is sensibly the same as that of well water when the latter begins to solidify, or that some liquids must be cooled to a greater extent than others before changing into the solid state. If pressed to justify their belief in the face of such facts, these men may perhaps arbitrarily exclude the oceans from the class of things they denominate as water; or, alternatively, they may express renewed confidence in their belief, irrespective of the extent of cooling that may be required, on the ground that liquids classified as water do indeed solidify when cooled.

In their quest for systematic explanations, on the other hand, the sciences must mitigate the indicated indeterminacy of ordinary language by refashioning it. For example, physical chemistry is not content with the loosely formulated generalization that water solidifies if it is sufficiently cooled, for the aim of that discipline is to explain, among other things, why drinking water and milk freeze at certain temperatures although at those temperatures ocean water does not. To achieve this aim, physical chemistry must therefore introduce clear distinctions between various kinds of water and between various amounts of cooling. Several devices reduce the vagueness and increase the specificity of linguistic expressions. Counting and measuring are for many purposes the most effective of these techniques, and are perhaps the most familiar ones. Poets may sing of the infinity of stars which stud the visible heavens, but the astronomer will want to specify their exact number. The artisan in metals may be content with knowing that iron is harder than lead, but the physicist who wishes to explain this fact will require a precise measure of the difference in hardness. Accordingly, an obvious but important consequence of the precision thus introduced is that statements become capable of more thorough and critical testing by experience. Prescientific beliefs are frequently incapable of being put to definite experiential tests, simply because those beliefs may be vaguely compatible with an indeterminate class of unanalyzed facts. Scientific statements, because they are required to be in agreement with more closely specified materials of observation, face greater risks of being refuted by such data.

This difference between common and scientific knowledge is roughly analogous to differences in standards of excellence which may be set up for handling firearms. Most men would qualify as expert shots if the standard of expertness were the ability to hit the side of a barn from a distance of a hundred feet. But only a much smaller number of individuals could meet the more rigorous requirement of consistently centering their shots upon a three-inch target at twice that distance. Similarly, a prediction that the sun will be eclipsed during the autumn months is more likely to be fulfilled than a prediction that the eclipse will occur at a specific moment on a given day in the fall of the year. The first prediction will be confirmed should the eclipse take place during any one of something like one hundred days; the second prediction will be refuted if the eclipse does not occur within something like a small fraction of a minute from the time given. The latter prediction could be false without the former being so, but not conversely; and the latter prediction must therefore satisfy more rigorous standards of experiential control than are assumed for the former.

This greater determinacy of scientific language helps to make clear why so many common-sense beliefs have a stability, often lasting for many centuries, that few theories of science possess. It is more difficult to devise a theory that remains unshaken by repeated confrontation with the outcome of painstaking experimental observation, when the standards are high for the agreement that must obtain between such experimental data and the predictions derived from the theory, than when such standards are lax and the admissible experimental evidence is not required to be established by carefully controlled procedures. The more advanced sciences do in fact specify almost invariably the extent to which predictions based on a theory may deviate from the results of experiment without invalidating the theory. The limits of such permissible deviations are usually quite narrow, so that discrepancies between theory and experiment which common sense would ordinarily regard as insignificant are often judged to be fatal to the adequacy of the theory.

On the other hand, although the greater determinacy of scientific statements exposes them to greater risks of being found in error than are faced by the less precisely stated common-sense beliefs, the former have an important advantage over the latter. They have a greater capacity for incorporation into comprehensive but clearly articulated systems of explanation. When such systems are adequately confirmed by experimental data, they codify frequently unsuspected relations of dependence between many varieties of experimentally identifiable but distinct kinds of fact. In consequence, confirmatory evidence for statements belonging to such a system can often be accumulated more rapidly and in larger quantities than for statements (such as those expressing common-sense beliefs) not belonging to such a system. This is so because evidence for statements in such a system may be obtainable by observations of an extensive class of events, many of which may not be explicitly mentioned by those statements but which are nevertheless relevant sources of evidence for the statements in question, in view of the relations of dependence asserted by the system to hold between the events in that class. For example, the data of spectroscopic analysis are employed in modern physics to test assumptions concerning the chemical structure of various substances; and experiments on thermal properties of solids are used to support theories of light. In brief, by increasing the determinacy of statements and incorporating them into logically integrated systems of explanation, modern science sharpens the discriminating powers of its testing procedure and augments the sources of relevant evidence for its conclusions.

5. It has already been mentioned in passing that, while common-sense knowledge is largely concerned with the impact of events upon matters of special value to men, theoretical science is in general not so provincial. The quest for systematic explanations requires that inquiry be directed to the relations of dependence between things irrespective of their bearing upon human values. Thus, to take an extreme case, astrology is concerned with the relative positions of stars and planets in order to determine the import of such conjunctions for the destinies of men; in contrast, astronomy studies the relative positions and motions of celestial bodies without reference to the fortunes of human beings. Similarly, breeders of horses and of other animals have acquired much skill and knowledge relating to the problem of developing breeds that will implement certain human purposes; theoretical biologists, on the other hand,

are only incidentally concerned with such problems, and are interested in analyzing among other things the mechanisms of heredity and in obtaining laws of genetic development.

One important consequence of this difference in orientation between theoretical and common-sense knowledge, however, is that theoretical science deliberately neglects the immediate values of things, so that the statements of science often appear to be only tenuously relevant to the familiar events and qualities of daily life. To many people, for example, an unbridgeable chasm seems to separate electromagnetic theory, which provides a systematic account of optical phenomena, and the brilliant colors one may see at sunset; and the chemistry of colloids, which contributes to an understanding of the organization of living bodies, appears to be an equally impossible distance from the manifold traits of personality exhibited by human beings.

It must certainly be admitted that scientific statements make use of highly abstract concepts, whose pertinence to the familiar qualities which things manifest in their customary settings is by no means obvious. Nevertheless, the relevance of such statements to matters encountered in the ordinary business of life is also indisputable. It is well to bear in mind that the unusually abstract character of scientific notions, as well as their alleged "remoteness" from the traits of things found in customary experience, are inevitable concomitants of the quest for systematic and comprehensive explanations. Such explanations can be constructed only if the familiar qualities and relations of things, in terms of which individual objects and events are usually identified and differentiated, can be shown to depend for their occurrence on the presence of certain other pervasive relational or structural properties that characterize in various ways an extensive class of objects and processes. Accordingly, to achieve generality of explanation for qualitatively diverse things, those structural properties must be formulated without reference to, and in abstraction from, the individualizing qualities and relations of familiar experience. It is for the sake of achieving such generality that, for example, the temperature of bodies is defined in physics not in terms of directly felt differences in warmth, but in terms of certain abstractly formulated relations characterizing an extensive class of reversible thermal cycles.

However, although abstractness in formulation is an undoubted feature in scientific knowledge, it would be an obvious error to suppose that common-sense knowledge does not involve the use of abstract conceptions. Everyone who believes that man is a mortal creature certainly employs the abstract notions of humanity and mortality. The conceptions of science do not differ from those of common sense merely in being abstract. They differ in being formulations of pervasive structural properties, abstracted from familiar traits manifested by limited classes of things usually only under highly specialized conditions, related to matters open to direct observation only by way of complex logical and experimental procedures, and articulated with a view to developing systematic explanations for extensive ranges of diverse phenomena.

6. Implicit in the contrasts between modern science and common sense already noted is the important difference that derives from the deliberate policy of science to expose its cognitive claims to the repeated challenge of critically probative observational data, procured under carefully controlled conditions. As we had occasion to mention previously, however, this does not mean that common-sense beliefs are invariably erroneous or that they have no foundations in empirically verifiable fact. It does mean that common-sense beliefs are not subjected, as a matter of established principle, to systematic scrutiny in the light of data secured for the sake of determining the accuracy of those beliefs and the range of their validity. It also means that evidence admitted as competent in science must be obtained by procedures instituted with a view to eliminating known sources of error; and it means, furthermore, that the weight of the available evidence for any hypothesis proposed as an answer to the problem under inquiry is assessed with the help of canons of evaluation whose authority is itself based on the performance of those canons in an extensive class of inquiries. Accordingly, the quest for explanation in science is not simply a search for any *prima facie* plausible "first principles" that might account in a vague way for the familiar "facts" of conventional experience. On the contrary, it is a quest for explanatory hypotheses that are genuinely testable, because they are required to have logical consequences precise enough not to be compatible with almost every conceivable state of affairs. The hypotheses sought must therefore be subject to the possibility of rejection, which will depend on the outcome of critical procedures, integral to the scientific quest, for determining what the actual facts are.

The difference just described can be expressed by the dictum that the conclusions of science, unlike common-sense beliefs, are the products of scientific method. However, this brief formula should not be misconstrued. It must not be understood to assert, for example, that the practice of scientific method consists in following prescribed rules for making experimental discoveries or for finding satisfactory explanations for matters of established fact. There are no rules of discovery and invention in science, any more than there are such rules in the arts. Nor must the formula be construed as maintaining that the practice of scientific method consists in the use in all inquiries of some special set of techniques (such as the techniques of measurement employed in physical science), irrespective of the subject matter or the problem under investigation. Such an interpretation of the dictum is a caricature of its intent; and in any event the dictum on that interpretation is preposterous. Nor, finally, should the formula be read as claiming that the practice of scientific method effectively eliminates every form of personal bias or source of error which might otherwise impair the outcome of the inquiry, and more generally that it assures the truth of every conclusion reached by inquiries employing the method. But no such assurances can in fact be given; and no antecedently fixed set of rules can serve as automatic safeguards against unsuspected prejudices and other causes of error that might adversely affect the course of an investigation.

The practice of scientific method is the persistent critique of arguments, in the light of tried canons for judging the reliability of the procedures by which evidential data are obtained, and for assessing the probative force of the evidence on which conclusions are based. As estimated by standards prescribed by those canons, a given hypothesis may be strongly supported by stated evidence. But this fact does not guarantee the truth of the hypothesis, even if the evidential statements are admitted to be true—unless, contrary to standards usually assumed for observational data in the empirical sciences, the degree of support is that which the premises of a valid deductive argument give to its conclusion. Accordingly, the difference between the cognitive claims of science and common sense, which stems from the fact that the former are the products of scientific method, does not connote that the former are invariably true. It does imply that, while common-sense beliefs are usually accepted without a critical evaluation of the evidence available, the evidence for the conclusions of science conforms to standards such that a significant proportion of conclusions supported by similarly structured evidence remains in good agreement with additional factual data when fresh data are obtained.

Further discussion of these considerations must be postponed. However, one brief addendum is required at this point. If the conclusions of science are the products of inquiries conducted in accordance with a definite policy for obtaining and assessing evidence, the rationale for confidence in those conclusions as warranted must be based on the merits of that policy. It must be admitted that the canons for assessing evidence which define the policy have, at best, been explicitly codified only in part, and operate in the main only as intellectual habits manifested by competent investigators in the conduct of their inquiries. But despite this fact the historical record of what has been achieved by this policy in the way of dependable and systematically ordered knowledge leaves little room for serious doubt concerning the superiority of the policy over alternatives to it.

This brief survey of features that distinguish in a general way the cognitive claims and the logical method of modern science suggests a variety of questions for detailed study. The conclusions of science are the fruits of an institutionalized system of inquiry which plays an increasingly important role in the lives of men. Accordingly, the organization of that social institution, the circumstances and stages of its development and influence, and the consequences of its expansion have been repeatedly explored by sociologists, economists, historians, and moralists. However, if the nature of the scientific enterprise and its place in contemporary society are to be properly understood, the types and the articulation of scientific statements, as well as the logic by which scientific conclusions are established, also require careful analysis. This is a task—a major if not exclusive task—that the philosophy of science undertakes to execute. Three broad areas for such an analysis are in fact suggested by the survey just concluded: the logical patterns exhibited by explanations in the sciences; the construction of scientific concepts; and the validation of scientific conclusions. The chapters that follow deal largely though not exclusively with problems concerning the structure of scientific explanations.

Patterns of Scientific Explanation

2 The preceding chapter has argued that the distinctive aim of the scientific enterprise is to provide systematic and responsibly supported explanations. As we shall see, such explanations may be offered for individual occurrences, for recurring processes, or for invariable as well as statistical regularities. This task is not the sole preoccupation of the sciences, if only because much of their effort goes into ascertaining what the facts are in fresh areas of experience for which explanations may be subsequently sought. It is indeed patent that at any given time the various sciences differ in the emphasis they place upon developing systematic explanations, and also in the degree of completeness with which they achieve such explanatory systems. Nevertheless, the quest for systematic explanations is never totally absent from any of the recognized scientific disciplines. To understand the requirements for, and structure of, scientific explanations is therefore to understand a pervasive feature of the scientific enterprise. The present chapter seeks to prepare the ground for such an understanding, by noting in a preliminary way ostensibly different forms of explanations encountered in the various sciences.

I. Illustrations of Scientific Explanation

Explanations are answers to the question 'Why?' However, very little reflection is needed to reveal that the word 'why' is not unambiguous and that with varying contexts different sorts of answers are relevant responses to it. The following brief list contains examples of the use of 'why,' several of which impose certain distinctive restrictions upon admissible answers to questions put with the help of the word.

1. Why is the sum of any number of consecutive odd integers beginning with 1 always a perfect square (for example, $1 + 3 + 5 + 7 = 16 = 4^2$)? Here the "fact" to be explained (called the *explicandum*) will be assumed to be a claimant for the familiar though not transparently clear label of "necessary truth," in the sense that its denial is self-contradictory. A relevant answer to the question is therefore a demonstration which establishes not only the universal truth but also the necessity of the explicandum. The explanation will accomplish this if the steps of the demonstration conform to the formal requirements of logical proof and if, furthermore, the premises of the demonstration are themselves in some sense necessary. The premises will presumably be the postulates of arithmetic; and their necessary character will be assured if, for example, they can be construed as true in virtue of the meanings associated with the expressions occurring in their formulation.

2. Why did moisture form on the outside of the glass when it was filled with ice water yesterday? Here the fact to be explained is the occurrence of an individual event. Its explanation, in broad outlines, might run as follows: The temperature of the glass after it was filled with ice water was considerably lower than the temperature of the surrounding air; the air contained water vapor; and water vapor in air is in general precipitated as a liquid whenever the air comes into contact with a sufficiently cold surface. In this example, as in the previous one, the formal pattern of the explanation appears to be that of a deduction. Indeed, were the explanatory premises formulated more fully and carefully, the deductive form would be unmistakable. However, the explicandum in this case is not a necessary truth, and on the face of it neither are the explanatory premises. On the contrary, the premises are statements which are presumably based on pertinent observational or experimental evidence.

3. Why did a smaller percentage of Catholics commit suicide than did Protestants in European countries during the last quarter of the nineteenth century? A well-known answer to the question is that the institutional arrangements under which Catholics lived made for a greater degree of "social cohesion" than did the social organizations of Protestants, and that in general the existence of strongly knit social bonds between members of a community helps to sustain human beings during

periods of personal stress. The explicandum in this case is a historical phenomenon that is statistically described, in contrast with the individual event of the previous example; and the proposed explanation does not therefore attempt to account for any individual suicide in the period under discussion. Indeed, although the explanatory premises are stated neither precisely nor completely, it is clear that some of them have a statistical content, just as does the explicandum. But, since the premises are not fully formulated, it is not quite clear just what the logical structure of the explanation is. We shall assume, however, that the implicit premises can be made explicit and that, moreover, the explanation then exhibits a deductive pattern.

4. Why does ice float on water? The explicandum in this example is not a historical fact, whether individual or statistical, but a universal law which asserts an invariable association of certain physical traits. It is familiarly explained by exhibiting it as the logical consequence of other laws—the law that the density of ice is less than that of water, the Archimedean law that a fluid buoys up a body immersed in it with a force equal to the weight of the fluid displaced by the body, and further laws concerning the conditions under which bodies subjected to forces are in equilibrium. It is worth noting that in this case, in contrast with the two immediately preceding examples, the explanatory premises are statements of universal law.

5. Why does the addition of salt to water lower its freezing point? The explicandum in this case is once more a law, so that in this respect the present example does not differ from the preceding one. Moreover, its current explanation consists in deducing it from the principles of thermodynamics conjoined with certain assumptions about the composition of heterogeneous mixtures; and in consequence, the present example also agrees with the previous one in regard to the formal pattern of the explanation. Nevertheless, the example is included for future reference, because the explanatory premises exhibit certain *prima facie* distinctive features having considerable methodological interest. For the thermodynamical principles included among the explanatory premises in the present example are assumptions much more comprehensive than any of the laws cited in previous examples. Unlike those laws, these assumptions make use of "theoretical" notions, such as energy and entropy, that do not appear to be associated with any overtly fixed experimental procedures for identifying or measuring the physical properties those notions presumably represent. Assumptions of this sort are frequently called "theories" and are sometimes sharply distinguished from "experimental laws." However, we must postpone for later discussion the question of whether the

distinction has any merit, and if so what is its importance. For the moment, the present example simply puts on record an allegedly distinctive species of deductive explanations in science.

6. Why is it that in the progeny of inbred hybrid peas, obtained by crossing round and wrinkled parents, approximately ¾ of the peas are always round whereas the remaining ¼ are wrinkled? The explicandum is currently explained by deducing it from the general principles of the Mendelian theory of heredity, coupled with certain further assumptions about the genetic constitution of peas. It is obvious that the fact here explained is a statistical regularity, not an invariable association of attributes, which is formulated as the relative frequency of a given trait in a certain population of elements. Moreover, as becomes evident when the explanatory premises are carefully stated, some of the premises also have a statistical content, since they formulate the probability (in the sense of a relative frequency) with which parent peas transmit determinants of given genetic traits to their offspring. The present example is similar to the preceding one in illustrating a deductive pattern of explanation containing theoretical assumptions among its premises. It is unlike any previous example, however, in that the explicandum as well as some of the premises are ostensibly statistical laws, which formulate statistical rather than invariable regularities.

7. Why did Cassius plot the death of Caesar? The fact to be explained is once more an individual historical occurrence. If we can believe Plutarch, the explanation is to be found in the inbred hatred which Cassius bore toward tyrants. However, this answer is obviously incomplete without a number of further general assumptions, such as some assumption concerning the way hatred is manifested in a given culture by persons of a certain social rank. It is unlikely, however, that such assumptions, if they are to be credible, can be asserted with strict universality. If the assumption is to be in agreement with the known facts, it will at best be only a statistical generalization. For example, a credible generalization may assert that most men (or a certain percentage of men) of a certain sort in a certain kind of society will behave in a certain manner. Accordingly, since the fact to be explained in this example is an individual historical occurrence, while the crucial explanatory assumption is statistical in form, the explicandum is not a deductive consequence of the explanatory premises. On the contrary, the explicandum in this case is simply made "probable" by the latter. This is a distinguishing feature of the present example, and sets it off from the preceding ones. Furthermore, there is an important substantive difference between this and previous examples, in that the explanatory premises

in the present instance mention a psychological disposition (e.g., an emotional state or attitude) as one of the springs of an action. Accordingly, if the question 'Why?' is raised in order to obtain an answer in terms of psychological dispositions, the question is significant only if there is some warrant for assuming that such dispositions do in fact occur in the subject matter under consideration.

8. Why did Henry VIII of England seek to annul his marriage to Catherine of Aragon? A familiar explanation for this historical occurrence consists in imputing to Henry a consciously entertained objective rather than a psychological disposition, as was the case in the immediately preceding example. Thus historians often account for Henry's efforts to annul his marriage to Catherine by citing the fact that, since she bore him no son, he wished to remarry in order to obtain a male heir. Henry doubtless possessed many psychological dispositions that may have been partly responsible for his behavior toward Catherine. However, in the explanation as just stated such psychological "springs of action" for Henry's conduct are not mentioned, and his efforts at obtaining an annulment are explained as deliberate means instituted for realizing a conscious goal (or end-in-view). Accordingly, the difference between the present and the preceding example hinges on the distinction between a psychological disposition or spring of action (of which an individual may be unaware even though it controls his actions) and a consciously held end-in-view (for the sake of which an individual may adopt certain means). This distinction is commonly recognized. A man's behavior is sometimes explained in terms of springs of action, even when he has no end-in-view for that behavior. On the other hand, no explanation is regarded as satisfactory for a certain class of human actions if the explanation does not refer to some conscious goal for the realization of which those actions are instituted. In consequence, in certain contexts a requirement for the intelligibility of questions introduced by 'Why?' is that in those contexts explicitly entertained objectives can be asserted.

9. Why do human beings have lungs? The question as it stands is ambiguous, for it may be construed either as raising a problem in the historical evolution of the human species or as requesting an account of the function of lungs in the human body at its present stage of evolutionary development. It is in this latter sense that the question is here intended. When so understood, the usual answer as supplied by current physiology calls attention to the indispensability of oxygen for the combustion of food substances in the body, and to the instrumental role of the lungs in conveying oxygen from the air to the blood and so eventually to various cells of the organism. Accordingly, the explanation de-

scribes the operation of the lungs as essential for the maintenance of certain biological activities. The explanation thus exhibits a *prima facie* distinctive form. The explanation does not explicitly mention the conditions under which the complex events called "the operation of the lungs" occur. It states rather in what way the lungs, as a specially organized part of the human body, contribute to the continuance of some of the other activities of the body.

10. Why does the English language in its current form have so many words of Latin origin? The historical fact for which an explanation is requested here is a complex set of linguistic habits, manifested by men during a somewhat loosely delimited historical period in various parts of the world. It is also important to note that the question 'Why?' in the present example, unlike the questions in preceding ones, tacitly calls for an account of how a certain system has developed into its current form from some earlier stage of the system. For the system under consideration, however, we do not possess general "dynamical laws of development," such as are available in physics, for example, for the development of a rotating gaseous mass. An admissible explanation for the historical fact in question will therefore have to mention sequential changes over a period of time, and not merely a set of occurrences at some antecedent initial time. Accordingly, the standard explanation for that fact includes reference to the Norman Conquest of England, to the speech employed by the victors and vanquished before the Conquest, and to developments in England and elsewhere after the Conquest. Moreover, the explanation assumes a number of more or less vague generalizations (not always explicitly stated, and some of them undoubtedly possessing a statistical content) concerning ways in which speech habits in different linguistic communities are altered when such communities enter into stated relations with each other. In short, the explanation requested in the present example is a genetic one, whose structure is patently more complex than the structure of explanations previously illustrated. The complexity should not be attributed to the circumstance that this explicandum happens to be a fact of human behavior. A comparable complexity is exhibited by a genetic explanation for the fact that the saline content of the oceans is at present about three per cent by volume.

II.   *Four Types of Explanation*

The above list does not exhaust the types of answers that are sometimes called "explanations." It is long enough, however, to establish the important point that even answers to the limited class of questions introduced by 'Why' are not all of the same kind. Indeed, the list clearly suggests that explanations offered in various sciences in response to such questions may differ in the way explanatory assumptions are related to their explicanda, so that explanations fall into distinct logical patterns.

We shall proceed on that suggestion, and characterize what appear to be the distinct types of explanation under which the examples in the above list can be classified. However, we shall at this point not engage the issue whether what seem to be different logical patterns of explanation are in fact only imperfectly formulated variants or limiting cases of some common pattern. For the present, at any rate, we identify four major and ostensibly different patterns of explanation.

1. *The deductive model.* A type of explanation commonly encountered in the natural sciences, though not exclusively in those disciplines, has the formal structure of a deductive argument, in which the explicandum is a logically necessary consequence of the explanatory premises. Accordingly, in explanations of this type the premises state a sufficient (and sometimes, though not invariably, a necessary) condition for the truth of the explicandum. This type has been extensively studied since ancient times. It has been widely regarded as the paradigm for any "genuine" explanation, and has often been adopted as the ideal form to which all efforts at explanation should strive.

The first six examples in the above list are *prima facie* illustrations of this type. Nevertheless, there are significant differences between them that are worth reviewing. In the first example, both the explicandum and the premises are necessary truths. However, although the point will need further discussion, few if any experimental scientists today believe that their explicanda can be shown to be inherently necessary. Indeed, it is just because the propositions (whether singular or general) investigated by the empirical sciences can be denied without logical absurdity that observational evidence is required to support them. Accordingly, the justification of claims as to the necessity of propositions, as well as the explanation of why propositions are necessary, are the business of formal disciplines like logic and mathematics, and not of empirical inquiry.

In both the second and third examples, the explicandum is a historical fact. However, in the second the fact is an individual event, while in the third it is a statistical phenomenon. In both examples the premises contain at least one "lawlike" assumption that is general in form, and at least one singular statement (whether individual or statistical). On the other hand, the explanation of the statistical phenomenon is distinguished by the presence of a statistical generalization in the premises.

In the fourth, fifth, and sixth examples the explicandum is a law—in the fourth and fifth cases a strictly universal statement asserting an invariable association of certain traits, in the sixth case a statistical law. However, the law in the fourth example is explained by deducing it from assumptions each of which is an "experimental law" in the sense already indicated briefly. In the fifth and sixth examples, on the other hand, the explanatory premises include so-called "theoretical" statements; in the sixth example, with a statistical law as the explicandum, the explanatory theory itself contains assumptions of a statistical form.

The differences just noted between explanations conforming to the deductive model have been described only in a schematic way. A fuller account of them will be given subsequently. Moreover, the purely formal requirements which deductive explanations must meet do not exhaust all the conditions that satisfactory explanations of this type are frequently expected to meet; and a number of further conditions will need to be discussed. In particular, although the important role of general laws in deductive explanations has been briefly noticed, the much-debated question remains whether laws can be characterized simply as presumptively true universal statements, or whether in order to serve as a premise in a satisfactory explanation a universal statement must in addition possess a distinctive type of relational structure. Furthermore, while mention has been made of the fact that highly integrated and comprehensive systems of explanation are achieved in science through the use of so-called "theoretical" assumptions, it will be necessary to inquire more closely what are the traits that distinguish theories from other laws, what features in them account for their power to explain a wide variety of facts in a systematic manner, and what cognitive status can be assigned to them.

2. *`Probabilistic explanations.`* Many explanations in practically every scientific discipline are *prima facie* not of deductive form, since their explanatory premises do not formally imply their explicanda. Nevertheless, though the premises are logically insufficient to secure the truth of the explicandum, they are said to make the latter "probable."

Probabilistic explanations are usually encountered when the explanatory premises contain a statistical assumption about some class of elements, while the explicandum is a singular statement about a given individual member of that class. This type of explanation is illustrated by both the seventh and tenth examples in the above list, though more clearly by the seventh. When this latter is formulated somewhat more explicitly, it runs as follows: In ancient Rome the relative frequency (or probability) was high (e.g., greater than one-half) that an individual belonging to the upper strata of society and possessed by great hatred of tyranny would plot the death of men who were in a position to secure tyrannical power. Cassius was such a Roman and Caesar such a potential tyrant. Hence, though it does not follow that Cassius plotted the death of Caesar, it is highly probable that he did do so.

A few observations are in order. It is sometimes maintained that probabilistic explanations are only temporary halfway stations on the road to the deductive ideal, and do not therefore constitute a distinct type. All that need be done, so it has been suggested, is to replace the statistical assumptions in the premises of probabilistic explanations by a strictly universal statement—in the above illustration, for example, by a statement asserting an invariable association between certain carefully delimited psychosociological traits (which Cassius presumably possessed) and participation in assassination plots. But, though the suggestion is not necessarily without merit and may be a goad to further inquiry, in point of fact it is extremely difficult in many subject matters to assert with even moderate plausibility strictly universal laws that are not trivial and hence otiose. Often the best that can be established with some warrant is a statistical regularity. Accordingly, probabilistic explanations cannot be ignored, on pain of excluding from the discussion of the logic of explanation important areas of investigation.

It is essential not to confuse the question whether the premises of an explanation are known to be true, with the question whether an explanation is of the probabilistic type. It may well be that in no scientific explanation are the general assumptions contained in the premises known to be true, and that every such assumption can be asserted as only "probable." However, even if this is so, it does not abolish the distinction between deductive and probabilistic types of explanation. For the distinction is based on patent differences in the way the premises and the explicanda are related to one another, and not on any supposed differences in our knowledge of the premises.

It should be noted, finally, that it is still an unsettled question whether an explanation must contain a statistical assumption in order to be a probabilistic one, or whether nonstatistical premises may not make an explicandum "probable" in some nonstatistical sense of the word. Nor are students of the subject in general agreement as to how the relation between premises and explicanda is to be analyzed, even in those probabilistic explanations in which the premises are statistical and the explicanda are statements about some individual. These questions will receive attention later.

3. *Functional or teleological explanations.* In many contexts of inquiry—especially, though not exclusively, in biology and in the study of human affairs—explanations take the form of indicating one or more functions (or even dysfunctions) that a unit performs in maintaining or realizing certain traits of a system to which the unit belongs, or of

stating the instrumental role an action plays in bringing about some goal. Such explanations are commonly called "functional" or "teleological." It is characteristic of functional explanations that they employ such typical locutions as "in order that," "for the sake of," and the like. Moreover, in many functional explanations there is an explicit reference to some still future state or event, in terms of which the existence of a thing or the occurrence of an act is made intelligible.

It is implicit in what has just been said that two subsidiary cases of functional explanation can be distinguished. A functional explanation may be sought for a particular act, state, or thing occurring at a stated time. This case is illustrated by the eighth example in the above list. Or alternatively, a functional explanation may be given for a feature that is present in all systems of a certain kind, at whatever time such systems may exist. This case is illustrated by the ninth of the above examples. Both examples exhibit the characteristic features of functional explanations. Thus, Henry's efforts to annul his first marriage are explained by indicating that they were undertaken for the sake of obtaining a male heir in the future; and the occurrence of lungs in the human body is explained by showing that they operate in a stated manner in order to maintain a certain chemical process and thereby to assure a continuance of life for the body into the future.

What the detailed structure of functional explanations is, how they are related to nonteleological ones, and why teleological explanations are frequent in certain areas of inquiry but rare in others are questions which must be reserved for later discussion. However, two common misconceptions concerning teleological explanations require brief notice immediately.

It is a mistaken supposition that teleological explanations are intelligible only if the things and activities so explained are conscious agents or the products of such agents. Thus, in the functional explanation of lungs, no assumption is made, either explicitly or tacitly, that the lungs have any conscious ends-in-view or that they have been devised by any agent for a definite purpose. In short, the occurrence of teleological explanations in biology or elsewhere is not necessarily a sign of anthropomorphism. On the other hand, some teleological explanations patently do assume the existence of deliberate plans and conscious purposes; but such an assumption is not illegitimate when, as in the case of teleological explanations of certain aspects of human behavior, the facts warrant it.

It is also a mistake to suppose that, because teleological explanations contain references to the future in accounting for what already exists, such explanations must tacitly assume that the future acts causally on the present. Thus, in accounting for Henry's efforts at obtaining an annulment of his marriage, no assumption is made that the unrealized future state of his possessing a male heir caused him to engage in certain activities. On the contrary, the explanation of Henry's behavior is entirely compatible with the view that it was his existing desires for a certain kind of future, and not the future itself, which were causally responsible for his conduct. Similarly, no assumption is made in the functional explanation of human lungs that it is the future oxidation of foods in the body which brings lungs into existence or causes them to operate; and the explanation does not depend on denying that the operation of the lungs is causally determined by the existing constitution of the body and its environment. By giving a teleological explanation one is therefore not necessarily giving hostages to the doctrine that the future is an agent in its own realization.

4. *Genetic explanations.* One remaining kind of explanation remains to be mentioned, though it is a moot question whether it constitutes a distinctive type. Historical inquiries frequently undertake to explain why it is that a given subject of study has certain characteristics, by describing how the subject has evolved out of some earlier one. Such explanations are commonly called "genetic," and they have been given for animate as well as inanimate things, for traits of an individual as well as for characteristics of a group. The tenth example in the above list illustrates the type.

The task of genetic explanations is to set out the sequence of major events through which some earlier system has been transformed into a later one. The explanatory premises of such explanations will therefore necessarily contain a large number of singular statements about past events in the system under inquiry. Two further points about the explanatory premises of genetic explanations should be noted. The first is the obvious one that not every past event in the career of the system will be mentioned. The second is that those events which are mentioned are selected on the basis of assumptions (frequently tacit ones) as to what sorts of events are causally relevant to the development of the system. Accordingly, in addition to the singular statements the premises will also include (whether explicitly or implicitly) general assumptions about the causal dependencies of various kinds of events.

These general assumptions may be fairly precise developmental laws, for which independent inductive evidence is available. (This may happen when the system under study can be regarded for the purposes at hand as a member of a class of similar systems which undergo a similar evolution—for example, in the study of the development of biological traits of an individual member of some species. For it is then often possible to employ methods of comparative analysis to establish such developmental laws.) In other cases, the general assumptions may be only vague generalizations, perhaps statistical in content, and may contain no reference to some of the highly specific features of the subject matter under study. (This often happens when the system investigated is a relatively unique one—for example, when the development of some institution in a particular culture is investigated.) However, in neither case do the explanatory premises in familiar examples of genetic explanations state the sufficient conditions for the occurrence of the fact stated in the explicandum, though the premises often do state some of the conditions which, under the circumstances generally taken for granted, are necessary for the latter. It is therefore a reasonable conclusion that genetic explanations are by and large probabilistic. However, further consideration of the structure of genetic explanations—and more generally of historical explanations—must be postponed.

III. *Do the Sciences Explain?*

These four major types of explanation have been distinguished because they appear to correspond to real structural differences in the examples of explanation that have been surveyed, and because the classification provides a convenient framework for examining important issues in the construction of systematic explanations. The next chapter will consider some of the problems associated especially, though not exclusively, with deductive explanations.

But before turning away from the outline of explanatory patterns developed in the present chapter, a historically influential objection to the claim that the sciences do in fact explain deserves brief comment. No science (and certainly no physical science), so the objection runs, really answers questions as to *why* any event occurs, or *why* things are related in certain ways. Such questions could be answered only if we were able to show that the events which occur *must* occur and that the relations which hold between things *must* hold between them. However, the experimental methods of science can detect no absolute or logical necessity in the phenomena which are the ultimate subject matter of every empirical inquiry; and, even if the laws and theories of science are true, they are no more than logically contingent truths about the relations of concomitance or the sequential orders of phenomena. Accordingly, the questions which the sciences answer are questions as to *how* (in what manner or under what circumstances) events happen and things are related. The sciences therefore achieve what are at best only comprehensive and accurate systems of *description*, not of explanation.[1]

---

[1] "The very common idea that it is the function of Natural Science to explain physical phenomena cannot be accepted as true unless the word 'explain' is used in a

More issues are raised by this argument than can be discussed with profit at this point. In particular, the question whether laws and theories are merely formulations of relations of concomitance and sequence between phenomena needs more attention than can now be given to it. But even if this view of laws and theories is granted, it is evident that the argument hinges in some measure on a verbal issue. For the argument assumes that there is just one correct sense in which 'why' questions can be raised—namely, the sense in which the proper answer to it is a proof of the inherent necessity of a proposition. However, this is a mistaken assumption, as the above list of examples amply testifies. Accordingly, a sufficient reply to the argument when it is construed as resting on this assumption is that there are in fact well-established uses for the words 'why' and 'explanation' such that it is entirely appropriate to designate an answer to a 'why' question as an explanation, even when the answer does not supply reasons for regarding the explicandum as intrinsically necessary. Indeed, even writers who officially reject the view that the sciences can ever explain anything sometimes lapse into language which describes certain scientific discoveries as "explanations."[2]

To the extent that the argument rests exclusively on assumptions about linguistic usage, it is neither important nor interesting. In point of fact, however, the argument does have more substance. The objection it voices was originally directed against several targets. One was the surviving anthropomorphisms in physics and biology, some of which lurked in the meanings commonly associated even with technical notions such as force and energy, while others were manifest in the uncritical use of teleological categories. In this respect, the objection was a demand for intellectual house cleaning, and it stimulated a program of careful analysis of scientific ideas that remains an active one. A second target of the objection was a conception of science that was at one time widespread and that continues in variant forms to retain distinguished adherents. According to this conception, it is the business of science to

very limited sense. The notions of efficient causation, and of logical necessity, not being applicable to the world of physical phenomena, the function of Natural Science is to describe conceptually the sequences of events which are to be observed in Nature; but Natural Science cannot account for the existence of such sequences, and therefore cannot explain the phenomena in the physical worlds, in the strictest sense in which the term explanation can be used. Thus Natural Science describes, so far as it can, *how*, or in accordance with what rules, phenomena happen, but it is wholly incompetent to answer the question *why* they happen."—E. W. Hobson, *The Domain of Natural Science*, London, 1923, pp. 81-82.

[2] For example, Mach describes Galileo's analysis of equilibrium on an inclined plane in terms of the principle of the lever, as an explanation of the former. (Ernst Mach, *The Science of Mechanics*, La Salle, Ill., 1942, p. 31.)

explain phenomena on the basis of laws of nature that codify a necessary order in things, and that are therefore more than contingently true. The objection is thus a denial of the claim that laws of nature possess more than a *de facto* universality, a denial which coincides with one major conclusion of David Hume's analysis of causality. The real issue which the argument raises is not a trivial one about linguistic usage, but a substantial one about the adequacy of an essentially Humean account of scientific law. This issue will receive attention in Chapter 4.

The Deductive Pattern
of Explanation

Ever since Aristotle analyzed the structure of what he believed to be the ideal of science, the view that scientific explanations must always be rendered in the form of a logical deduction has had wide acceptance. Although the universality of the deductive pattern may be open to question, even when the pattern is projected as an ideal, it is hardly disputable that many explanations in the sciences—and indeed the most comprehensive and impressive systems of explanation—are of this form. Moreover, many explanations that ostensibly fail to realize this form can be shown to exemplify it, when the assumptions that are taken for granted in the explanations are made explicit; and such cases must count not as exceptions to the deductive model, but as illustrations of the frequent use of enthymematic arguments.[1]

We must however inquire whether, in addition to the definitional requirement that in deductive types of explanation the explicandum follow logically from the explanatory premises, satisfactory explanations of this type must fulfill further conditions. For it is evident that not every proposed explanation is acceptable merely because it embodies a deductive

[1] For example, the expansion of a piece of wire on a given occasion may be explained by citing the fact that the wire had just been heated; and it is evident that the explicandum does not follow logically from the explanatory premise as stated. However, it seems most plausible that the proposed explanation tacitly assumes additional premises—for example, that the wire is copper and that copper always expands when heated. When these additional assumptions are made explicit, the explanation does conform to the deductive model.

structure. For example, no one is likely to regard as satisfactory a proposed explanation of the fact that Jupiter has at least one satellite which cites the further fact that Jupiter has eight moons—even though the former statement follows logically from the latter. Discussions of this question go back to Greek antiquity, and a large variety of such additional conditions have been suggested. These conditions can be classified for convenience under three heads: *logical,* which specify various formal requirements for explanatory premises; *epistemic,* which stipulate in what cognitive relations one ought to stand to the premises; and *substantive,* which prescribe what sort of content (empirical or otherwise) the premises ought to have. What these labels signify will become clearer as we proceed. It would, however, be awkward and would involve needless repetition were each type of condition always discussed separately; and such rigidly compartmentalized analysis will therefore not be attempted. Nevertheless, most of the logical conditions that need to be noted will be considered in the present chapter.

I. *Explanations of Individual Events*

Let us begin with an example of a deductive explanation in which the explicandum is the occurrence of some individual event. Consider the illustration, mentioned in the previous chapter, of moisture forming on the surface of a glass on a certain day. The explanation, stated more carefully though also more pedantically than before, can be set out as follows:

> Whenever the temperature of any volume of air containing water vapor is reduced below the point where the density of vapor in the air is greater than the saturation density of water vapor in air at that temperature, the vapor in the air condenses into liquid water at those places where the temperature of the air has dropped below this saturation point.
> The volume of air surrounding the glass yesterday contained water vapor.
> The temperature of the layer of air immediately adjacent to the glass was reduced when ice water was poured into the glass.
> The actual density of vapor in this layer of air when its temperature dropped was greater than the saturation density at the new temperature.
> Accordingly, the vapor in this adjacent layer of air was condensed on the glass surface as liquid water—in short, moisture was formed on the glass.

The first point to note in this example is that the premises contain a statement that is universal in form and that asserts an invariable connection of certain properties. In other examples, more than one such universal law may occur in the premises.[2] If now we generalize from

---

[2] In point of fact, even in the present example other laws are tacitly assumed. One such law is that for each temperature air has a definite saturation density. Other

this example, it appears that at least one of the premises in a deductive explanation of a singular explicandum must be a universal law, and one moreover that is not a sleeping partner but plays an essential role in the derivation of the explicandum.[8] It is evident that this requirement is sufficient to exclude as a bona fide case of explanation the deduction, mentioned earlier, of the fact that Jupiter has at least one satellite from the fact that the planet has eight of them.

But in addition to a universal law, the above premises also contain a number of singular or instantial statements, which assert that certain events have occurred at indicated times and places or that given objects have definite properties. Such singular statements will be referred to as "statements of initial conditions" (or more briefly as "initial conditions"). More generally, initial conditions constitute the special cir-

laws that easily escape notice because they are so familiar lurk in the characterization of various things as water, glass, and so on. These latter laws in effect affirm that there are distinct kinds of substances, each of which exhibits certain fixed concatenations of traits and modes of behavior. For example, the statement that something is water implicitly asserts that a number of properties (a certain state of aggregation, a certain color, a certain freezing and boiling point, certain affinities for entering into chemical reactions with other kinds of substances, etc.) are uniformly associated with each other. The discovery and classification of kinds is an early but indispensable stage in the development of systematic knowledge; and all the sciences, including physics and chemistry, assume as well as continue to refine and modify distinctions with respect to kinds that have been initially recognized in common experience. Indeed, the development of comprehensive theoretical systems seems to be possible only after a preliminary classification of kinds has been achieved, and the history of science repeatedly confirms the view that the noting and mutual ordering of various kinds—a stage of inquiry often called "natural history"—is a prerequisite for the discovery of more commonly recognized types of laws and for the construction of far-reaching theories. Modern physics and chemistry did not come into being until after such preliminary classifications of kinds (whose beginnings are lost in primitive antiquity) were accomplished; traditional botany and zoology consist largely of specifications and subordinations of kinds; and some of the social sciences are still struggling to achieve usable and reliable formulations of kinds of human beings and of social institutions. Recognition of different kinds goes hand in hand with subordination (or inclusion) of one kind in another kind. Thus, chemistry not only distinguishes between the kinds copper and sulfur but also between metals and nonmetals; it includes copper in the metals and sulfur in the nonmetals. Similarly, biology includes the kinds or species tiger and lion in the common genus cat, the kind cat in the more inclusive order of carnivore, the carnivore in the class mammal, and so on. When a system of inclusion between kinds is achieved, it is possible to explain (even if only in a crude fashion) why some individual thing is a member of a specified kind by showing that the individual is a member of a subordinate kind (for example, the family pet is a mammal because it is a cat and cats are mammals). Such explanations are obviously far removed from the sort of explanations to which the modern theoretical sciences have accustomed us; nevertheless, they are early steps on the road which leads to the latter.

[8] This proviso is introduced in order to eliminate trivial exceptions. Thus, though "Brown is older than Smith" is deducible from the two premises "Smith is younger than Brown" and "All mammals are vertebrates," it will not count as an explanation even though the premises include a general law, simply because the second premise is not required for the deduction.

cumstances to which the laws included in the explanatory premises are applied. However, it is not possible to state in general terms which circumstances are to be selected to serve as appropriate initial conditions, for the answer to the question depends on the specific content of the laws used as well as on the special problems for whose solution the laws are invoked.

The indispensability of initial conditions for the deductive explanation of individual occurrences is obvious as a point in formal logic. For it is logically impossible to deduce a statement instantial in form from statements that have the form of a universal conditional. (For example, it is impossible to derive an instantial statement of the form 'x is B' from a universal conditional of the form 'For any x, if x is A then x is B.') But obvious though the point may be, it is an important one that is frequently neglected in discussions of scientific procedure. Its neglect is at least partly responsible for the cavalier way in which broad generalizations are sometimes used to account for detailed matters of fact (especially in the study of human affairs) and for the low esteem which observers sometimes have for painstaking investigations of what are the actual facts. It is, however, often difficult to make concrete uses of laws and theories, simply because the specific initial conditions for their application are inaccessible and therefore unknown. And, conversely, mistaken explanations and false predictions are frequently proposed because the general assumptions that are employed, though sound enough in themselves, are applied to situations which do not constitute appropriate initial conditions for those assumptions. Though laws of one sort or another are indispensable in scientific explanations of the actual course of events, what actually transpires cannot be explained exclusively by reference to laws. In the pursuit of scientific understanding, as in the settlement of legal disputes, general principles alone do not determine any individual case.

Accordingly, a deductive scientific explanation, whose explicandum is the occurrence of some event or the possession of some property by a given object, must satisfy two logical conditions. The premises must contain at least one universal law, whose inclusion in the premises is essential for the deduction of the explicandum. And the premises must also contain a suitable number of initial conditions.[4]

---

[4] Although the explanation of a singular fact requires the inclusion in the premises of both statements of law and statements of initial conditions, inquiries may differ according as they are directed toward finding and establishing one type of premise rather than another. Thus, we may note the occurrence of some event, and then seek to explain it by discovering some other event which, on the basis of an already established law, is assumed to be the condition for the occurrence of the given event. For example, if a tire on an automobile goes flat we may initiate a search for a puncture, on the general assumption that flat tires are the consequences of punctured tubes. On the other hand, we may note the occurrence of two or more events, sus-

II. *Explanations of Laws*

Treatises devoted to expounding systematically some branch of deductively organized science usually contain no explanations of individual occurrences and particular facts; and when they do so it is often merely by way of illustrating the uses of laws and theories. In the more advanced physical sciences, at any rate, the paramount concern is with the explanation of laws and, in consequence, with the systematic interrelations of laws.

All explanations of laws seem to be of the deductive type,[5] and we must examine what special features characterize them. We shall consider first the explanation of universal laws. Moreover, we shall ignore for the present not only statistical laws but also the distinction mentioned earlier between explanations in which all the premises are "experimental laws" and those in which the premises include "theoretical" assumptions. Let us therefore turn to the example, cited in the preceding chapter, of the explanation of the law that ice floats in water. It would be wearisome, however, to spell out in precise detail the rigorous deduction of this law from the premises that physicists usually assume when they explain it. The hints given earlier as to the identity of these premises will suffice for our purposes.[6]

pect them to be significantly related, and attempt to discover the laws which formulate the specific modes of dependence between events of that character. Thus we may happen to note that a person's pulse rate increases after he engages in some vigorous exercise; and, if we suspect that pulse rate is in some way contingent on exercise, we may institute inquiry into the precise mode of connection between the activities so as to yield a general formula for their relation of dependence. Again, in the attempt to explain some events, inquiry may be directed to discovering both types of suitable explanatory premises. For example, we may know no law relevant to the appearance of a certain cancerous growth, and may also be ignorant concerning the specific events upon which such growth is contingent. We may therefore seek to discover both the particular circumstances that initiated the cancer and the laws connecting such circumstances with cancerous growths.

[5] This does not mean, of course, that laws are always *established* by deductive means alone. Most laws are in fact shown to be warranted by adducing observational evidence for them.

[6] A first approximation to such a deduction is the following: The buoyant force of a liquid upon a body immersed in it is in a direction perpendicular to the surface of the liquid, and is equal but oppositely directed to the weight of the liquid displaced by the body. [Therefore, the buoyant force of water on ice immersed in it is in a direction perpendicular to the surface of the water, and is equal to the weight of the water displaced by the ice.]

A body is in equilibrium if, and only if, the vectorial sum of the forces acting on it is zero. [Therefore, ice immersed in water is in equilibrium if, and only if, the vectorial sum of the forces acting on the ice is zero.]

The vectorial sum of the forces acting on a body immersed in a liquid, in a direction parallel to the surface of the liquid, is zero.

Every force is the vectorial sum of two forces (called the "components" of the given force) whose directions are at right angles to each other. [Therefore, ice im-

Three points are immediately evident in this explanation: all the premises are universal statements; there is more than one premise, each of which is essential in the derivation of the explicandum;[7] and the premises, taken singly or conjointly, do not follow logically from the explicandum. The first point needs little comment, for it is logically inevitable, since the explicandum is itself a universal law. Accordingly, the introduction of initial conditions into the premises would be gratuitous in the explanation of universal laws.

But the second point does raise the issue whether the presence of more than one universal law in the premises is just a peculiarity of the example used or whether it is an essential mark of all acceptable explanations. The question cannot be decided with finality, since there is no precise criterion for distinguishing between explanations that are satisfactory and those that are not. It is pertinent to ask, nevertheless, whether the deduction of a universal law from a single premise would normally be regarded as an explanation of the former. To fix our ideas, consider Archimedes' law, which says that the buoyant force of a liquid upon a body immersed in it is equal to the weight of the liquid displaced by that body. From this it follows as a special case that the buoyant force of water upon ice immersed in it is equal to the weight of the water displaced by the ice.[8] It seems, however, unlikely that most physicists

mersed in water is in equilibrium if, and only if, the vectorial sum of the forces acting on the ice in a direction *perpendicular* to the surface of the water is zero. Therefore, also, if the only forces acting on ice immersed in water are the buoyant force of the water and the force of the weight of the ice, ice immersed in water is in equilibrium if, and only if, the buoyant force of the water is equal, but oppositely directed, to the total weight of the ice.]

The density of water is greater than the density of ice. [Therefore, the weight of a given volume of water is greater than the weight of an equal volume of ice.]

Therefore, if the only forces acting on ice immersed in water are the buoyant forces of the water and its own weight, ice immersed in water is in equilibrium if, and only if, a portion of the ice is not submerged, and the buoyant force of the water is equal and opposite to the weight of the water displaced by the submerged portion of the ice. In short, ice immersed in water (and under the action of none but the "normal" forces) is in equilibrium if, and only if, it floats.

[7] It is always possible to obtain just one premise, by forming the conjunction of several premises. What is intended in the text is that if there were only a single conjunctive premise it would be equivalent to a class of logically independent premises in which the class would contain more than one member.

[8] The deduction is in effect performed by substituting particular values for the "variables" implicit in the formulation of Archimedes' law. The schematic form of the deduction is as follows:

For all properties P which are in K_1 and for all properties Q which are in K_2, all P's are Q's.

A is in K_1, and B is in K_2, *ex vi terminorum*.

Hence all A's are B's.

The deduction is quite analogous to the derivation of Boyle's law which says that

would say that this special law has thus been explained; and certainly few people would "experience" this deduction of the special law as an explanation. If this example can be taken as typical, and if these conjectures as to how scientists would respond to it are sound, it seems a reasonable logical requirement for the explanation of laws that the explanatory assumptions contain at least two formally independent premises.

A further consideration speaks in favor of this requirement, though the point carries little independent weight. We often reserve the word "explanation" in discussing laws for one of two cases. In the first of these, the "phenomenon" formulated by the law is shown to be the resultant of several independent factors that enter into some special set of relations. In the second case, the invariable association between the traits asserted by the law is shown to be the product of two or more associations, where the latter hold between the traits mentioned in the law and various other traits that are intermediate links in a chain or network. What is intended by these alternatives will perhaps be clearer from the following schematic illustrations. Let us assume that a universal law has the form of a simple universal conditional: 'For any x, if x is A then x is B' (or 'All A's are B's'), where 'A' and 'B' designate definite properties. Suppose that the property A occurs if and only if the properties A_1 and A_2 jointly occur; and suppose, similarly, that B occurs if and only if B_1 and B_2 occur jointly. Assume, furthermore, that all A_1's are B_1's and all A_2's are B_2's. It then follows that all A's are B's, so that this law is now explained. This schematism illustrates the first of the above alternatives. A concrete example is the explanation of the law that ice floats in water, since the behavior of ice in water is exhibited as the resultant of several independent forces acting on the immersed body. However, the actual logical structure of this explanation is far more complex than the one portrayed by the above simple schematism.

A schematic illustration for the second alternative is provided by an explanation of a law having the form 'All A's are B's,' when it is deduced from two laws having the forms, respectively, 'All A's are C's' and 'All C's are B's.' A concrete example for this case is the explanation of the law "When gases containing water vapor are sufficiently expanded without changing their heat content, the vapor condenses," when it is deduced from the two laws "When gases expand without change in heat content, their temperature falls" and "When the temperature of a gas containing water vapor is lowered, the saturation density of the vapor is decreased."

It will be evident that explanations subsumable under either of these

for any ideal gas the pressure of the gas times its volume is constant when the temperature of the gas is constant from the Boyle-Charles' law, which says that for any ideal gas the pressure of the gas times its volume is proportional to its temperature.

alternative schemata employ at least two premises. But whether or not we adopt the requirement that at least two premises must be present in a satisfactory explanation, we can be reasonably certain that we will not find in the sciences many explanations that violate it.

The third point noted above concerning the ice example—that the explicandum does not logically imply the premises—is less debatable as a general requirement for explanation. For were this condition not satisfied, the conjunction of the premises would be logically equivalent to the explicandum; and in that event the premises would simply restate the law for which the explanation is proposed. Take, for example, the law that the time required for a freely falling body to cover a given distance is proportional to the square root of the magnitude of that distance. This law follows logically from the law that the distance covered by a freely falling body is proportional to the square of the duration of the fall. However, no one is likely to call this an explanation of the former law, since the premise is just a mathematically equivalent transformation of the explicandum. (This example violates the requirement that an explanation must have more than one premise. Examples which do not violate this condition but in which the premises and explicandum are nevertheless logically equivalent—e.g., the Newtonian formulation of mechanics, familiar to beginning students of physics, and the less familiar because mathematically less elementary but more general formulation of the theory given by the eighteenth-century theoretical physicist Joseph Lagrange—are, however, too complex to state in detail.) Were anyone to do so, he might just as well take the explicandum as an explanation of itself.

It seems clear, therefore, that we expect the explanatory premises in a satisfactory explanation to assert something more than what is asserted by the explicandum. Stated more fully, our expectation is that at least one of the premises in the explanation of a given law will meet the following requirement: when conjoined with suitable additional assumptions, the premise should be capable of explaining other laws than the given one; on the other hand, it should not in turn be possible to explain the premise with the help of the given law, even when those additional assumptions are adjoined to the law. If none of the premises in an explanation satisfied this requirement, two undesirable consequences would follow: it would be impossible to obtain evidence for the premises other than the evidence supplied by the explicandum in question; and the explanation would not significantly advance the organization of the subject matter into a system, since except in isolated instances the known facts as well as those yet to be discovered would remain unrelated.

The requirement that the premises must not be equivalent to the explicandum is sufficient to eliminate many pseudo-explanations, in which the premises simply rebaptize the facts to be explained by coining new names for them. The classical example of such pseudo-explanations is the butt of Molière's satire in which he ridicules those who explain the fact that opium induces sleep by invoking the dictum that opium possesses a dormative virtue. A less obvious illustration, sometimes found in popular expositions of science, is the explanation of the law that the velocity of a body remains constant unless the body is acted on by an unbalanced external force, because all bodies possess an inherent force of inertia. This is a pseudo-explanation, since the word "inertia" is just another label for the fact stated in the law.

III. *Generality in Explanations*

There is, however, an additional requirement for satisfactory explanations of laws, closely related to the one just considered, that has often been proposed.[9] According to this requirement, at least one of the premises must be "more general" than the law being explained. Thus, Archimedes' law (which appears in the premises of the floating ice example) is said to be more general than the law that ice floats in water, because the former asserts something of all liquids and not only of water, and of all bodies immersed in liquids and not only of ice. Analogously, the law of the lever is held to be more general than are the laws dealing with the motions of vertebrates; more inclusively, though perhaps in a looser sense, the laws of physics are frequently declared to be more general than the laws of biology.

However, although the sense of "more general" may be clear enough in particular examples of its use, it is not easy to give a precise explication of the notion. We must nevertheless attempt such an explication, and note some of the difficulties that emerge. When it is said that one statement S_1 is more general than a second statement S_2, it is presumably not intended that S_1 must logically imply S_2; for such an implication does not obtain between Archimedes' law and the law that ice floats in water, despite the fact that the former is held to be more general than the latter. Moreover, it is plausible to construe the meaning of the expression "more general" in such a way that S_1 is said to be more general than S_2 not merely because S_1 logically implies S_2. For example, the statement 'All planets move on elliptic orbits' logically implies 'All planets move on orbits which are conic sections,' but presumably the first of these is not more general than the second. Accordingly, for S_1 to be more

[9] Cf. John Stuart Mill, *A System of Logic*, London, 1879, Book 3, Chap. 12, Sec. 4; Norman R. Campbell, *Physics, the Elements*, Cambridge, England, 1920, pp. 114ff.; Karl Popper, *Logik der Forschung*, Vienna, 1935, p. 75.

general than S_2, it appears to be neither necessary nor sufficient that S_1 logically imply S_2.

If we limit ourselves to a special class of statements which might be comparable in respect to their relative "generality," an obvious way of defining this relation is the following.[10] Consider only laws capable of being stated as universal conditionals of the simplest form. Let S_1 be a statement of the form 'For any x, if x is A then x is B' (or, in more customary mode of expression, of the form 'All A is B'), and S_2 of the form 'All C is D.' Then S_1 will be said to be more general than S_2 if, and only if, 'All C is A' is *logically* true but its converse 'All A is C' is not. Furthermore, S_1 will be said to be as general as S_2 if, and only if, 'All A is C' and 'All C is A' are both logically true. If neither statement having one of the latter two forms is logically true, then S_1 and S_2 will be said to be not comparable in respect to their generality. For example, the law that all objects immersed in liquids are acted on by a buoyant force equal to the weight of the liquid displaced by the object (Archimedes' law) is more general, on the basis of this definition, than is the law that ice immersed in water floats. For the statement 'Ice immersed in water is an object immersed in a liquid' is true by virtue of the meanings associated with its terms, while its converse clearly is not.

Although at first sight this definition appears to provide a satisfactory explication of what is presumably intended when one statement is said to be more general than another, it nevertheless leads to difficulties. For the requirement that two logically equivalent statements should be equally general seems reasonable, so that, if S_1 is more general than S_2, and S_2 is logically equivalent to a third statement S_3, then S_1 is also more general than S_3. However, this requirement is not fulfilled when 'more general' is understood in agreement with the proposed definition. Thus, suppose that 'All A is B' is more general than 'All C is D' (so that 'All C is A' but not its converse is logically true). But 'All non-B is non-A' is logically equivalent to 'All A is B,' and consistent with the suggested requirement it ought therefore to be more general than 'All C is D.' For this to be the case on the basis of the proposed definition, 'All C is non-B' would have to be logically true, although in point of fact this would usually not happen. For example, 'All living organisms are mortal' is more general on the proposed definition than 'All human beings are mortal' (because 'All human beings are living organisms' though not its converse is a logical truth); and 'All living organisms are mortal' is also logically equivalent to 'All non-mortals are non-(living organisms).' But since 'All human beings are non-mortals' is patently not a logical truth, the statement 'All non-mortals are non-(living organisms)' is not more

[10] Popper, *ibid.*

general, when judged in accordance with the proposed definition, than 'All human beings are mortal.'[11]

These difficulties are not necessarily fatal to the proposed explication of the notion of greater generality. But to avoid them one must drop the seemingly plausible requirements that logically equivalent statements must be equally general, and adopt the position that the comparative generality of laws is relative to the way they are formulated. It might be objected, however, that such a course opens the door to unlimited arbitrariness in classifying laws according to their generality, since for a given statement there are an indefinite number of logical equivalents differing in their formulation. Nevertheless, the arbitrariness may not be as serious as it looks at first sight. For the actual formulation of a law frequently indicates what is the range of things that are the subjects of predication in given contexts, where this identification of the intended scope of the law is controlled by the nature of the particular inquiry. But there is nothing especially arbitrary in this, other than the arbitrariness inherent in dealing with one set of problems rather than with another set. Accordingly, insofar as the subject term in the statement of a law indicates the intended scope of the law in a concrete context (or class of contexts) of its use, the assertion that one law is more general than another is not fatally arbitrary—even if in some other context a different comparative judgment is required. For example, the law that ice floats in water is commonly so used that its range of application is the indefinitely large class of instances of ice which are (or have been or will be) immersed in water. The law is rarely if ever used so that its range of application is taken to be the miscellaneous collection of things which do not float in water (whether in the past, present, or future). Indeed, it is a plausible claim that, were the law used in this latter way in some context, its customary formulation would in that context be appropriately modified. At any rate, there appears to be a tacit reference to contexts of use in the actual formulations of laws. But if this is so, the proposed explication of the notion of greater generality is not hopelessly defective.

However, since the explication discussed thus far does not capture a

---

[11] Additional difficulties of a similar nature can be devised, by making use of other equivalences that are shown to hold in formal logic. For example, 'All A is B' is more general than 'All AE is B,' since 'All AE is A' is a logical truth while 'All A is AE' is not. However, 'All AE is B' is logically equivalent to 'All A is either B or not–E.' But 'All A is B' is not more general than 'All A is either B or not–E,' despite the fact that it is more general than a statement logically equivalent with the latter. These difficulties cannot be obviated by modifying the requirement in the initial exposition of the necessary and sufficient conditions for greater generality (according to which 'All C is A' but not its converse must be a *logical* truth) into the weaker condition that 'All C is A' but not its converse be only contingently (or factually) true.

more inclusive if vague sense of "more general," the matter deserves brief attention. This sense is illustrated when it is said that physics is a more general science than is biology, or more particularly when the law of the lever is declared to be more general than, say, the law that blue-eyed human parents have only blue-eyed children. What is sometimes meant by such statements is perhaps that biological phenomena can be explained on the basis of the laws of physics, though not conversely. But quite irrespective of the truth of this claim, it does not convey the sense always intended by the illustrative statements, for it is doubtful whether anyone ever has maintained that the law of the lever can explain any law of human heredity. The sense more frequently associated with such statements is perhaps more like this: The law of the lever (and, more inclusively, the science of physics) formulates certain characteristics of things irrespective of whether those things are animate or inanimate. On the other hand, the law about eye color (and, more inclusively, the science of biology) asserts something about traits that are exhibited only by a special class of systems, some (though not necessarily all) of which also manifest those characteristics formulated by the law of the lever. The law of the lever thus abstracts from many features of things that are considered by the biological law, and the descriptive expressions occurring in the law of the lever are therefore predicable of a more inclusive class of systems than are the descriptive expressions occurring in the biological law.

Let us attempt a formally more precise account of this interpretation of a sense of "more general." Let L_1 be a law (or a set of laws and theories constituting some special science such as physics), and let 'P_1,' 'P_2,' . . . , 'P_n,' be a set of "primitive" predicates in terms of which the predicates occurring in L_1 are in some sense definable. (For the sake of simplicity, and without any essential loss in generality of statement, we shall assume that the predicates are all adjectives or "one-place" predicates such as 'rigid' or 'heavy,' and include no relational expressions such as 'longer than' or 'ancestor of.' Accordingly, the predicates can be used to construct statements of the form 'x is rigid' containing but one individual name.) Similarly, let 'Q_1,' 'Q_2,' . . . , 'Q_s,' be the corresponding set of primitives for a law L_2. Finally, let K be a class of objects, each of which can be significantly (or meaningfully) characterized, whether truly or falsely, by the predicates of either set. Thus, if 'heavy' is a predicate belonging to the first set and 'mammalian' a predicate of the second set, K will contain only such elements (e.g., rocks, tables, animals) of each of which it is meaningful (though perhaps false) to say that it is heavy and mammalian. We shall also say that an object in K satisfies a law L "non-vacuously" only if the object actually possesses the various traits mentioned in the law and, moreover, the traits do stand to each other in the relations asserted by the law. Objects not possessing all the traits mentioned in L, so that they cannot be counted as disconfirming or counter-instances for L, will be said to satisfy the law "vacuously." For example, a system consisting of a heavy object suspended by a string with negligible weight satisfies non-vacuously the law for the period of a simple pendulum. On the other hand, this law is satisfied only vacuously by a system consisting of a book resting on a table, because, although the law would normally not be said to be falsified by this system, the system does not in fact possess the traits whose relations the law formulates—in short, the system is not a simple pendulum.

We now assume the following conditions: (1) Some (and perhaps all) of the predicates in the first set occur in the second, but some predicates in the second set do not belong to the first set. (2) Every object in K has at least one P-property, that is, a property designated by a predicate in the first set. (3) There is a non-empty subclass A of objects in K possessing only P-properties. (4) There is a non-empty subclass A of objects in K each of which possesses at least one Q-property that is not a P-property. (As a consequence of these stipulations, the range of objects to which one or the other of the first set of predicates actually applies is more inclusive than is the corresponding range for the second set.) (5) There is a non-empty (but not necessarily proper) subclass B of objects in K each of which satisfies L_1 non-vacuously, and such that some objects in B belong to A while others belong to $\overline{A}$. (Accordingly, when L_1 is satisfied non-vacuously, it holds irrespective of whether or not an object possesses only P-properties.) (6) There is a non-empty subclass C of objects in $\overline{A}$ for which L_2 holds non-vacuously, and such that some (and perhaps all) of the objects in C also belong to B. (Hence L_2, unlike L_1, is satisfied non-vacuously only by objects possessing some Q-property that is not a P-property. It is not excluded, however, that L_2 holds non-vacuously only for those objects for which L_1 also holds non-vacuously.) When these six conditions are fulfilled, L_1 may be said to be more general in K than is L_2 (in the more extended sense of "more general" now under discussion). If in the sixth condition the stronger requirement is introduced that C must be completely included in B, the present sense of "more general" is specialized to approximate the narrower sense of "more general" previously discussed.

This formal account of an inclusive meaning of "more general" requires elaboration in several directions in order to be entirely satisfactory. For example, the nature of the "definitions" assumed for the predicates in L_1 and L_2 needs discussion; the sense in which the L's are supposed to "hold" for objects requires clarification; and restrictions need to be imposed on the types of objects which may be members of K, as well as upon the distribution of *P*-properties among them. But these problems cannot be pursued any further. However, enough has been said for purposes of the present discussion to indicate that at least two fairly clear senses of "more general" can be distinguished, and that universal statements are frequently comparable in respect to their relative generality, whether in the narrower or broader sense of the phrase. The reason for dwelling on this point is that the premises of satisfactory explanations do appear to be more general than the explicanda. This greater generality of the explanatory premises is of considerable importance because this feature contributes to the achievement of inclusive systems of explanation. We shall presently examine one important device by which universal statements in some of the sciences acquire a comprehensive generality.

IV.   *Epistemic Requirements for Explanations*

The requirements for explanations considered thus far have been almost exclusively logical conditions. But it is obvious that other requirements must also be recognized. If, for example, an initial condition in a proposed explanation for the occurrence of an individual event were known to be false, the proposal would be immediately rejected as unsatisfactory. Let us therefore turn briefly to some epistemic requirements for adequate explanations.

In discussing this question, Aristotle maintained that the premises in a deductive explanation must among other things be true, that they must be known to be true, and that they must be "better known" than the explicandum.[12] We shall examine these conditions in turn and discuss some related ones as well.

1.   Any evaluation of the suggestion that the premises of an explanation must be true is complicated by an important circumstance. There frequently occur among the explicit premises of scientific explanations universal statements which are part of some comprehensive scientific theory. However, competent thinkers are divided on the issue whether such statements (and indeed, whether any scientific theory) can be appropriately characterized as either true or false. Accordingly, anyone who subscribes to the view that these characterizations are misapplied when used in connection with such statements will automatically reject the requirement that the explicit premises in a satisfactory explanation must be true. The rejection of this requirement thus hinges on the way the issue mentioned is resolved. The issue will occupy us later. For the present, however, we shall assume that every statement which may appear as a premise in an explanation is either true or false.

[12] *Posterior Analytics*, Book 1, Chap. 2.

If this assumption is made, the requirement that the premises in a satisfactory explanation must be true seems inescapable. It is always relatively easy to invent an arbitrary set of premises which satisfy the logical conditions for deductive explanations; and unless further restrictions were placed on the premises, only a moderate logical and mathematical ability would be required for explaining any fact in the universe without leaving one's armchair. In point of fact, however, all such arbitrarily constructed explanations would be dismissed as inadequate if any of the premises were known to be false. The truth of the premises is undoubtedly a desirable condition for satisfactory explanations.

2. However, this requirement does not carry us far in judging the worth of a proposed explanation if we are not in a position to say whether or not the premises are false. The Aristotelian requirement that the premises must be *known* to be true thus provides an apparently effective criterion for eliminating many proposed explanations as unsatisfactory. But this requirement is much too strong. Were it adopted, few if any of the explanations given by modern science could be accepted as satisfactory. For in point of fact, we do *not* know whether the unrestrictedly universal premises assumed in the explanations of the empirical sciences are indeed true; and, were the requirement adopted, most of the widely accepted explanations in current science would have to be rejected as unsatisfactory. This is in effect a *reductio ad absurdum* of the requirement. In practice it would lead simply to the introduction of another term, perhaps freshly coined for the purpose, for distinguishing those explanations that are judged to have merit by the scientific community—despite their nominal "unsatisfactory" character under the requirement—from those explanations that are judged differently. There is therefore no point in adopting the strong Aristotelian requirements for the adequacy of explanations.

Nevertheless, a stipulation of some kind, though weaker than this Aristotelian one, is needed concerning the cognitive status of explanatory premises. A reasonable candidate for such a weaker condition is the requirement that the explanatory premises be compatible with established empirical facts and be in addition "adequately supported" (or made "probable") by evidence based on data other than the observational data upon which the acceptance of the explicandum is based. The first part of this requirement is simply the demand that there should be no grounds for regarding the premises as false. The second part seeks not only to exclude so-called *ad hoc* premises for which there is no evidence whatsoever. It also seeks, among other things, to eliminate explanations that are in a sense circular and therefore trivial because one or more of the premises is established (and perhaps can be established) only by way of the evidence used to establish the explicandum. Suppose, for example, that we undertook to explain the explosive noises known as static which issued from a radio on a given day; and suppose that one of the explanatory premises stated the initial condition that on this day there were violent magnetic storms on the sun. However, if the sole evidence for the occurrence of these storms were the static on the radio, the explanation would suffer from a species of circularity and would generally be regarded as defective. In this example, however, evidence for the instantial premise could in point of fact be obtained independently of the noises issuing from the radio. The explanation would have doubtful merit if no such independent evidence could be given.[18]

This weaker condition concerning the cognitive status of premises in explanations is undoubtedly vague. For at present no precise and generally accepted standard is available for judging whether an assumption is indeed "adequately supported" by given evidence. Despite this vagueness, however, those who acquire competence in some field of inquiry often are in good agreement as to the adequacy of the supporting evidence for a definite assumption. In actual practice, at any rate, the use of the weaker condition results in a fair consensus concerning the worth of a proposed explanation. The objection may nevertheless be raised against this condition that, since the evidence for a supposed universal law does not remain constant in time, an explanation that includes the law in its premises and that is satisfactory at one time may cease to be satisfactory when unfavorable evidence for the law is discovered. But the objection is not a disturbing one, unless the dubious assumption is made that in judging an explanation to be satisfactory a timeless property is being predicated of the explanation. It does not seem unreasonable, therefore, to adopt the condition as an epistemic requirement for adequate explanations.

---

[18] Essentially the same point is made more formally by C. G. Hempel and Paul Oppenheim, "Studies in the Logic of Explanation," *Philosophy of Science*, Vol. 15 (1948), pp. 135-78. These authors argue that, unless the restriction mentioned in the text is adopted, any instantial explicandum can be "explained" with the help of any arbitrarily selected universal premise and an appropriately constructed "initial condition." Thus, let E be any explicandum; L the law that, for any x, if x is A then x is B; and C the initial condition which says that either a given individual i is A but not B, or E. Then E follows logically from the premises L and C. For from L we can obtain the consequence that it is not the case that the individual i is A but not B; and if we combine this with C, E is obtained. However, if we ask how one can establish C, it is clear that the only way of doing this, on the assumption that L is true, is to reason as follows: E is true, by hypothesis; hence either E is true or the individual i is A but not B. Accordingly, C can be established only by way of first establishing E. Hempel and Oppenheim therefore propose the condition that the truth of the law L must not imply that every class of true evidential statements from which C is deducible will also yield E—or, alternately, that there must be at least one class of evidential statements such that the initial condition C is deducible from it but neither E nor the denial of L is deducible from it. Cf. especially pp. 162-63.

3. The Aristotelian requirement, that the premises in a scientific explanation must be "better known" than the explicandum, is intimately related to Aristotle's conception of what constitutes the proper object of scientific knowledge; the requirement was intended by him to apply exclusively to the explanation of scientific laws. On this conception genuine scientific knowledge is possible only of what cannot be otherwise than it is. Accordingly, there can be no scientific knowledge of particular events; and universal laws concerning some area of nature, when they are not immediately recognized as inherently "necessary," must be explained by exhibiting them as the consequences of the "first principles" for that area which can be directly grasped as possessing such necessity. These first principles are therefore the ultimate premises in scientific explanations; and they are "better known" than any of the explicanda because their necessity is intrinsic and transparent to the intellect. The branch of knowledge that undoubtedly served as the model for this conception of science is demonstrative geometry. For according to the view widely held about geometry until fairly recently, each of its theorems states what must be the case universally; and while neither this necessity nor this universality may be immediately evident, both are established when a theorem is deduced from the more general axioms or first principles whose universality is "self-evident." In maintaining that the premises in an explanation must be "better known" than the explicandum, Aristotle was thus simply making explicit his conception of the nature of science.

This conception is true of nothing that can be identified as part of the asserted content of modern empirical science. Accordingly, Aristotle's requirement that the explanatory premises be better known than the explicandum is entirely irrelevant as a condition for anything that would today be regarded as an adequate scientific explanation. On the other hand, various psychologized versions of the Aristotelian requirement have enjoyed wide currency and have frequently been advanced by distinguished men of science as essential conditions for satisfactory explanations. The substance of these suggested conditions is that, since what normally requires explanation is something strange and unexpected, an explanation will yield genuine intellectual satisfaction only if it makes what is unfamiliar intelligible in terms of what is familiar. For example, an eminent contemporary physicist maintains that an "Explanation consists merely in analyzing our complicated systems into simpler systems in such a way that we recognize in the complicated system the interplay of elements already so familiar to us that we accept them as not needing explanation." [14] And he argues that, since current quantum theory does not show how the physical systems falling into its province are the re-

[14] P. W. Bridgman, *The Nature of Physical Theory*, Princeton, 1936, p. 63.

sultants of familiar modes of action between familiar kinds of constituents, the theory does not give us a sense of explaining anything, despite its admittedly remarkable systematizing achievements. Similar views have been expressed by many other thinkers, in the natural as well as in the social sciences.

It would be flying into the teeth of the evidence to deny that important developments in the history of science have been controlled by the desire to explain new domains of fact in terms of something already familiar. One need only recall the persistent use of familiar mechanical models in constructing explanations for the phenomena of heat, light, electricity, and even human behavior in order to recognize the influence of this conception of explanation. Nevertheless, explanations are not invariably judged to be unsatisfactory unless they effect a reduction of the familiar to the unfamiliar. When the bleaching effect of sunlight on colored materials is explained in terms of physical and chemical assumptions about the composition of light and of colored substances, the explanation is not rejected as unsatisfactory, even though it is the familiar which is being accounted for in terms of what to most men is quite unfamiliar. Moreover, the conception of explanation under discussion is in patent disaccord with the fact that throughout the history of science explanatory hypotheses have frequently been introduced which postulate modes of interrelation between assumed elements, where the interrelations and elements are initially strange and occasionally even seemingly paradoxical.

Nevertheless, two brief points should be noted. If an explanation satisfies the epistemic condition previously discussed, then, even though its explanatory premises may at one time have been unfamiliar, they must have finally achieved the status of assumptions well-supported by evidence. Accordingly, even if the explanation does not reduce the unfamiliar to what was initially familiar, it is an acceptable explanation because the premises are firmly grounded in evidence that has ceased to be unfamiliar to some part of the scientific community. In the second place, though explanatory premises may make use of quite unfamiliar ideas, such ideas often exhibit important analogies with notions employed in connection with already familiar subject matters. The analogies help to assimilate the new to the old, and prevent novel explanatory premises from being radically unfamiliar. But we must postpone for a later chapter a fuller discussion of the role of analogy in the development of comprehensive systems of explanation.

The Logical Character
of Scientific Laws

4 The requirements for adequate explanations considered thus far have been discussed with only incidental reference to the nature of the relations asserted by scientific laws or theories. It has been tacitly assumed that laws have the form of generalized conditionals, in the simplest case represented by the schema 'For any x, if x is A then x is B' (or alternatively by 'All A is B').[1] However, it is by no means the case that any true

[1] The assumption that this simple schema is an adequate representation of the logical form of scientific laws has been repeatedly made in previous chapters, and will frequently be made throughout the volume. However, this assumption is adopted in the main for the sake of avoiding complexities that would arise were less simple but more realistic schema recognized—complexities that are largely irrelevant to the chief points under discussion. There undoubtedly are many scientific laws which do exhibit the simple formal structure mentioned above. Nevertheless, there are also many laws whose logical form is more complicated—a fact of considerable importance in analyzing the rationale of inductive and verificatory procedures in science, though only of subsidiary interest in the present context of discussion.

One type of complexity in the formal structure of laws is illustrated by the following two examples. The content of the law that copper expands if heated is made more explicit by restating it as: 'For any x and for any y, if x is copper and x is heated at time y, then x expands at time y.' As in other conditionals (or "if-then" formulations), the clause introduced by 'if' is known as the "antecedent," and the clause introduced by 'then' as the "consequent." The present example also contains as "prefixes" the two expressions 'for every x' and 'for every y' (technically known as "universal quantifiers"), unlike the simple schema in the text, which contains only one universal quantifier. Again, the so-called "law of biogenesis" that all life comes from pre-existing life can be rendered by: 'For any x, there is a y, such that if x is a living organism then y is a parent of x.' In this case, the statement contains not only the universal quantifier 'for every x,' but also the expression 'there is a y' (called

statement having this form is invariably counted among the laws of nature. In any event, despite the fact that proposed explanations may conform to the requirements already mentioned, they are frequently rejected as unsatisfactory on at least two grounds: although the universal premises of an explanation may be acknowledged to be true, they are said for one reason or another not to be genuine "laws"; and, although the universal premises may be admitted to the status of scientific law, they allegedly fail to meet some further conditions, such as that of being "causal" laws.

Suppose, for example, that, in answer to the question why a certain screw *s* is rusty, one is told that all the screws in Smith's current car are rusty and that *s* is a screw in Smith's car. Such an explanation is likely to be rejected as quite unsatisfactory, on the ground that the universal premise is not even a law of nature, to say nothing of its not being a causal law. A *prima facie* distinction between "lawlike" universal statements (i.e., statements which, if true, qualify for the designation "law of nature") and universal statements that are not lawlike thus underlies the objection to the proposed explanation.

On the other hand, a proposed explanation of the fact that a given bird *b* is black, because all crows are black and *b* is a crow, is sometimes put aside as inadequate on the ground that even if the universal premise is assumed to be a law of nature it does not "really" explain why *b* is black. Now on one interpretation of this objection it undoubtedly confounds two different things: an explanation of the fact that *b* is black, as distinct from an explanation of the assumed law that all crows are black. Accordingly, a decisive rejoinder to the objection might very well be that, while the explanation does not explain why all crows are black, it does explain why *b* is black: for the explanation shows at the very least that the color of *b*'s plumage is not an idiosyncrasy of *b* but is a trait *b* shares with every other bird which, like itself, is a crow. Nevertheless, the objection can also be understood as an expression of dissatisfaction with the proposed explanation of *b*'s black plumage, on the ground that the assumed law does not give a causal account of the bird's coloring.

These examples, which illustrate a widespread if tacit acceptance of conditions for satisfactory explanations additional to those already discussed, thus invite a consideration of some of the features that sup-

---

an "existential quantifier"). This statement thus contains more than one quantifier, and these moreover of an unlike (or "mixed") kind. A large fraction of quantitative laws, especially in theoretical physics, contain several quantifiers, often of a mixed kind. However, it seems unlikely that any statement would normally be counted as a law if it did not contain at least one universal quantifier, usually as the initial prefix. It is for this reason that the simplifying assumption adopted in the text does not appear to be a fatal oversimplification.

posedly distinguish natural laws from other universal conditionals, and causal laws from noncausal ones. We must examine several far-flung issues generated by these distinctions.

I. *Accidental and Nomic Universality*

The label 'law of nature' (or similar labels such as 'scientific law,' 'natural law,' or simply 'law') is not a technical term defined in any empirical science; and it is often used, especially in common discourse, with a strong honorific intent but without a precise import. There undoubtedly are many statements that are unhesitatingly characterized as 'laws' by most members of the scientific community, just as there is an even larger class of statements to which the label is rarely if ever applied. On the other hand, scientists disagree about the eligibility of many statements for the title of 'law of nature,' and the opinion of even one individual will often fluctuate on whether a given statement is to count as a law. This is patently the case for various theoretical statements, to which reference was made in the previous chapter, which are sometimes construed to be at bottom only procedural rules and therefore neither true nor false, although viewed by others as examples par excellence of laws of nature. Divergent opinions also exist as to whether statements of regularities containing any reference to particular individuals (or groups of such individuals) deserve the label of 'law.' For example, some writers have disputed the propriety of the designation for the statement that the planets move on elliptic orbits around the sun, since the statement mentions a particular body. Similar disagreements occur over the use of the label for statements of statistical regularities; and doubts have been expressed whether any formulation of uniformities in human social behavior (e.g., those studied in economics or linguistics) can properly be called 'laws.' The term 'law of nature' is undoubtedly vague. In consequence, any explication of its meaning which proposes a sharp demarcation between lawlike and nonlawlike statements is bound to be arbitrary.

There is therefore more than an appearance of futility in the recurring attempts to define with great logical precision what is a law of nature—attempts often based on the tacit premise that a statement is a law in virtue of its possessing an inherent "essence" which the definition must articulate. For not only is the term 'law' vague in its current usage, but its historical meaning has undergone many changes. We are certainly free to designate as a law of nature any statement we please. There is often little consistency in the way we apply the label, and whether or not a statement is *called* a law makes little difference in the way in which the statement may be used in scientific inquiry. Nevertheless, members of the scientific community agree fairly well on the applicability of the term for a considerable though vaguely delimited class of universal statements. Accordingly, there is some basis for the conjecture that the predication of the label, at least in those cases where the consensus is unmistakable, is controlled by a felt difference in the "objective" status and function of that class of statements. It would indeed be futile to attempt an ironclad and rigorously exclusive definition of 'natural law.' It is not unreasonable to indicate some of the more prominent grounds upon which a numerous class of statements is commonly assigned a special status.

The *prima facie* difference between lawlike and non-lawlike universal conditionals can be brought out in several ways. One effective way depends on first recalling in what manner modern formal logic construes statements that have the form of universal conditionals. Two points must be noted in this connection. Such statements are interpreted in modern logic to assert merely this: any individual fulfilling the conditions described in the antecedent clause of the conditional also fulfills, *as a matter of contingent fact,* the conditions described in the consequent clause. For example, in this interpretation the statement 'All crows are black' (which is usually transcribed to read 'For any x, if x is a crow then x is black') merely says that any individual thing which happens to exist whether in the past, present, or future and which satisfies the conditions for being a crow is in point of fact also black. Accordingly, the sense assigned to the statement by this interpretation is also conveyed by the equivalent assertions that there never was a crow that was not black, there is no such crow at present, and there never will be such a crow. Universal conditionals construed in this way, so that they assert only matter-of-fact connections, are sometimes said to formulate only a "constant conjunction" of traits and to express "accidental" or *de facto* universality.

The second point to be noted in this interpretation is an immediate consequence of the first. On this interpretation a universal conditional is true, provided that there are no things (in the omnitemporal sense of 'are') which satisfy the conditions stated in the antecedent clause. Thus, if there are no unicorns, then all unicorns are black; but also, if there are no unicorns, then all unicorns are red.[2] Accordingly, on the construc-

---

[2] This will be evident from the following: If there is no x such that x is a unicorn, then clearly there is no x such that x is a unicorn that is not black. But on the standard interpretation of the universal conditional, this latter statement immediately yields the conclusion that for any x, if x is a unicorn then x is black. Accordingly, if there are no unicorns then all unicorns are black.

It can also be shown that a universal conditional is true no matter what its antecedent clause may be, provided that everything of which the consequent clause can be significantly predicated satisfies the consequent clause. But we shall ignore any difficulties generated by this feature of universal conditionals.

tion placed upon it in formal logic, a *de facto* universal conditional is true, irrespective of the content of its consequent clause, if in point of fact there happens to be nothing which satisfies its antecedent clause. Such a universal conditional is said to be "vacuously" true (or "vacuously satisfied").

Does a law of nature assert no more than accidental universality? The answer commonly given is in the negative. For a law is often held to express a "stronger" connection between antecedent and consequent conditions than just a matter-of-fact concomitance. Indeed, the connection is frequently said to involve some element of "necessity," though this alleged necessity is variously conceived and is described by such qualifying adjectives as 'logical,' 'causal,' 'physical,' or 'real.'[8] The contention is that to say that 'Copper always expands on heating' is a law of nature is to claim more than that there never has been and never will be a piece of heated copper that does not expand. To claim for that statement the status of a law is to assert, for example, not merely that there does not happen to exist such a piece of copper, but that it is "physically impossible" for such a piece of copper to exist. When the statement is assumed to be a law of nature, it is thus construed to assert that heating any piece of copper "physically necessitates" its expansion. Universal conditionals understood in this way are frequently described as "universals of law" or "nomological universals," and as expressing a "nomic" universality.

The distinction between accidental and nomic universality can be brought out in another way. Suppose that a piece of copper c which has never been heated is called to our attention, and is then destroyed so that it will never be heated. Suppose, further, that after the work of destruction is over we are asked whether c would have expanded had it been heated, and that we reply in the affirmative. And suppose, finally, that we are pressed for a reason for this answer. What reason can be advanced? A reason that would generally be accepted as cogent is that the natural law 'All copper when heated expands' warrants the contrary-to-fact conditional 'If c had been heated, it would have expanded.' Indeed, most people are likely to go further, and maintain that the nomological universal warrants the subjunctive conditional 'For any x, if x were copper and were heated, then x would expand.'

Laws of nature are in fact commonly used to justify subjunctive and contrary-to-fact conditionals, and such use is characteristic of all nomological universals. Moreover, this function of universals of law also suggests that the mere fact that nothing happens to exist (in the omnitem-

[8] Cf. A. C. Ewing, *Idealism*, London, 1934, p. 167; C. I. Lewis, *An Analysis of Knowledge and Valuation*, La Salle, Ill., 1946, p. 228; Arthur W. Burks, "The Logic of Causal Propositions," *Mind*, Vol. 60 (1951), pp. 363-82.

poral sense) which satisfies the antecedent clause of a nomological conditional is not sufficient to establish its truth. Thus, the assumption that the universe contains no bodies which are under the action of no external forces suffices to establish neither the subjunctive conditional that if there were such bodies their velocities would remain constant, nor the nomological universal that every body which is under the action of no external forces does not maintain a constant velocity.

On the other hand, the patently accidental universal 'All the screws in Smith's current car are rusty' does not justify the subjunctive conditional 'For any x, if x were a screw in Smith's current car x would be rusty.' [4] Certainly no one is likely to maintain on the strength of this *de facto* universal that, if a particular brass screw now resting on a dealer's shelf were inserted into Smith's car, that screw would be rusty. This *prima facie* difference between accidental and nomic universality can be briefly summarized by the formula: A universal of law "supports" a subjunctive conditional, while an accidental universal does not.

II. *Are Laws Logically Necessary?*

No one seriously disputes the claim that a distinction something like the one baptized by the labels 'accidental' and 'nomic' universality is recognized in common speech and in practical action. The question in dispute is whether the *prima facie* differences that have been noted require the acceptance of the "necessity" associated with universals of law as something "ultimate" or whether nomic universality can be explicated in terms of notions that are less opaque. If this necessity is interpreted, as it has been, as a form of *logical* necessity, the meaning of 'necessary' in this sense is quite transparent; and indeed a systematic and generally accepted analysis of such necessity is provided by logical theory. Accordingly, though the view that nomological universals are logically necessary faces grave difficulties, as will be noted in a moment, the view has at least the merit of clarity. On the other hand, those who maintain that the necessity of universals of law is *sui generis* and at bottom not further analyzable postulate a property whose nature is essentially obscure. This obscurity is only named and not lightened by such labels as 'physical necessity' or 'real necessity.' Moreover, since it is generally supposed that this allegedly special type of necessity can be recognized only by some "intuitive apprehension," predicating such necessity

---

[4] This subjunctive conditional is not to be construed as saying that if any screw were *identical with* one of the screws in Smith's car it would be rusty. The latter subjunctive conditional is clearly true if indeed all the screws in Smith's current car are rusty. The subjunctive conditional in the text is to be understood as saying that for any object x—whether or not it is identical with one of the screws now in Smith's car—if x were a screw in that car it would be rusty.

(whether of statements or of relations between events) is subject to all the vagaries of intuitive judgments. To be sure, the necessity that ostensibly characterizes universals of law may indeed be unique and not analyzable, but for the reasons noted it seems advisable to accept this conclusion only as a last resort.

The view that universals of law in general and causal laws in particular state a logical necessity has been frequently advanced. However, those who adopt this position usually do not claim that the logical necessity of nomological universals can in fact be established in every case. They contend only that genuine nomologicals are logically necessary, and could "in principle" be shown to be such, even though for the most part a demonstration of the necessity is lacking. For example, in discussing the nature of causality a contemporary writer maintains that "The cause logically entails the effect in such a way that it would be in principle possible, with sufficient insight, to see what kind of effect must follow from examination of the cause alone without having learnt by previous experience what were the effects of similar causes." [5] In some cases such a view is based on an allegedly direct perception of the logical necessity of at least a few nomological universals, and on the assumption that all other nomologicals must therefore also share this characteristic; in other cases, the view is adopted because the validity of scientific induction is held to depend on it; [6] and at least one proponent of the position has frankly admitted that the most impressive arguments for it are the objections to any alternative view. [7]

The difficulties confronting the position are nevertheless formidable. In the first place, none of the statements generally labeled as laws in the various positive sciences are in point of fact logically necessary, since their formal denials are demonstrably not self-contradictory. Accordingly, proponents of the view under discussion must either reject all these statements as not cases of "genuine" laws (and so maintain that no laws have yet been discovered in any empirical science), or reject the proofs that these statements are not logically necessary (and so challenge the validity of established techniques of logical proof). Neither horn of the dilemma is inviting. In the second place, if laws of nature are logically necessary, the positive sciences are engaged in an incongruous performance whenever they seek experimental and observational evidence for a

[5] A. C. Ewing, "Mechanical and Teleological Causation," *Aristotelian Society,* Suppl. Vol. 14 (1935), p. 66. Cf. also G. F. Stout: "If we had a sufficiently comprehensive and accurate knowledge of what really takes place, we should see how and why the effect follows from the cause with logical necessity."—*Aristotelian Society,* Suppl. Vol. 14 (1935), p. 46.

[6] A. C. Ewing, "Mechanical and Teleological Causation," *Aristotelian Society,* Suppl. Vol. 14 (1935), p. 77.

[7] C. D. Broad, *Aristotelian Society,* Suppl. Vol. 14 (1935), p. 94.

supposed law. The procedure appropriate for establishing a statement as logically necessary is that of constructing a demonstrative proof in the manner of mathematics, and not that of experimentation. No one today knows whether Goldbach's conjecture (that every even number is the sum of two primes) is logically necessary; but also no one who understands the problem will try to establish the conjecture as logically necessary by performing physical experiments. It is, however, fantastic to suggest that when the truth of an alleged physical law, for example about light, is in doubt the physicist ought to proceed as a mathematician does. And finally, despite the fact that the statements believed to be laws of nature are not known to be logically necessary, those statements successfully play the roles in science that are assigned to them. It is therefore gratuitous to maintain that, unless those statements were logically necessary, they could not do the tasks which they manifestly do perform. The statement known as Archimedes' law of buoyancy, for example, enables us to explain and predict a large class of phenomena, even though there are excellent reasons for believing that the law is not logically necessary. However, the assumption that the law really must be necessary does not follow from the fact that it is successfully used to explain and predict. Accordingly, the assumption postulates a characteristic that plays no identifiable part in the actual use made of the law.

It is, nevertheless, not difficult to understand why laws of nature sometimes appear to be logically necessary. For a given *sentence* may be associated with quite different meanings, so that while it is used in one context to express a logically contingent truth, in another context the *same sentence* may state something that is logically necessary. There was a time, for example, when copper was identified on the basis of properties that included none of the electrical properties of the substance. After electricity was discovered, the sentence 'Copper is a good electrical conductor' was asserted on experimental grounds as a law of nature. Eventually, however, high conductivity was absorbed into the defining properties of copper, so that the sentence 'Copper is a good electrical conductor' acquired a new use and meaning. In its new use, the sentence no longer expressed merely a logically contingent truth, as it did before, but served to convey a logically necessary truth. There is undoubtedly no sharp line separating those contexts in which copper is identified without reference to properties of conductivity from those contexts in which high conductivity is taken to belong to the "nature" of copper. In consequence, the status of what is being asserted by the sentence 'Copper is a good electrical conductor' is not always clear, so that the logical character of the assertion it is used to make in one context can be easily confounded with the character of the assertion made by it in some other context.[8] Such varying usage for the same sentence helps to explain why the view that laws of nature are logically necessary seems so plausible to many thinkers. It suggests a source for the conviction, apparent in declarations such as the following, that any alternative to this view is absurd: "I can attach no meaning to a causation in which the effect is not necessarily determined, and I can attach no meaning to a necessary determination which would leave it perfectly possible for the necessarily determined event to be different without contradicting either its own nature or the nature of that which determines it."[9] But in any event, the shifts in meaning to which sentences are subject as a consequence of advances in knowledge are an important feature in the development of comprehensive systems of explanation. It is a feature that will receive further attention in subsequent chapters.

The issue as to the nature of the ostensible necessity of nomological universals has occupied many thinkers since Hume proposed an analysis of causal statements in terms of constant conjunctions and *de facto* uniformities. Ignoring important details in Hume's account of the spatiotemporal relations of events which are said to be causally connected, the substance of the Humean position is briefly as follows. The objective content of the statement that a given event c is the cause of another event e, is simply that c is an instance of a property C, e is an instance of a property E (these properties may be quite complex), and any C is

[8] Another example may help to make the point more clearly. Consider the law of the lever in the form that if equal weights are placed at the extremities of a homogeneous rigid bar suspended at its midpoint, the lever is in equilibrium; and suppose that none of the expressions used in the statement of the law are defined in ways which involve assumptions about the behavior of levers. On this supposition, the statement is clearly an empirical law, and is not a logically necessary statement. On the other hand, suppose that two bodies are defined to be equal in weight if, when they are placed at the extremities of the equal arms of levers, the levers are in equilibrium. In contexts in which such a definition of "equality in weight" is employed, the above sentence about levers cannot be denied without self-contradiction, so that it does not express an empirical law for which experimental evidence is relevant but states a logically necessary truth. Sentences which appear to state laws but which are in fact employed as definitions, are commonly called "conventions"; and the role of such conventions and their articulation with laws will be discussed at greater length later on.

[9] A. C. Ewing, reference cited in footnote 5. It is only by an ellipsis that effects are said to be inferred from causes, since from the statement that an alleged cause has occurred the statement about the occurrence of a corresponding effect does not in fact follow logically. To infer the statement about the effect, the statement about the cause must be supplemented with a general law. Thus, the statement that a given billiard ball collides with a second ball does not logically entail any statement about the subsequent behavior of the second ball. Such a further statement can be derived only if some law (e.g., concerning the conservation of momentum) is added to the initial statement. The thesis that statements about causes logically imply statements about effects thus confounds the relation of logical necessity that holds between a set of explanatory premises and the explanandum, with the contingent relation affirmed by laws contained in these premises.

as a matter of fact also *E*. On this analysis, the "necessity" allegedly characterizing the relation of *c* to *e* does not reside in the objective relations of the events themselves. The necessity has its locus elsewhere—according to Hume, in certain habits of expectation that have been developed as a consequence of the uniform but *de facto* conjunctions of *C* and *E*.

The Humean account of causal necessity has been repeatedly criticized, partly on the ground that it rests on a dubious psychology; and the merits of criticisms of this sort are now generally acknowledged. However, Hume's psychological preconceptions are not essential to his central thesis—namely, that universals of law can be explicated without employing irreducible modal notions like "physical necessity" or "physical possibility." Accordingly, the burden of much of the current criticism of the Humean analysis is that the use of such modal categories is unavoidable in any adequate analysis of nomic universality. The issue remains unsettled and continues to be debated; and some of the problems connected with it have come to be discussed on a highly technical level. It will not be profitable to examine most of these technical details,[10] and only the outlines of an essentially Humean interpretation of nomic universality will be developed.

III. *The Nature of Nomic Universality*

With this end in view, let us consider whether, if by imposing a number of logical and epistemic requirements upon universal conditionals (interpreted in the manner of modern formal logic, as explained above), conditionals satisfying them can plausibly be regarded as lawlike statements. It will be helpful to begin by comparing a patently accidental universal ('All the screws in Smith's current car are rusty,' or in more expanded form 'For every x, if x is a screw in Smith's car during the time period a, then x is rusty during a,' where a designates some definite time period), with a commonly acknowledged example of a universal of law ('All copper expands on heating' or, more explicitly, 'For any x and for any t, if x is copper and x is heated at time t, then x expands at time t').

[10] Some of these technical details are relevant only on an assumption that does not appear to be reasonable. The implicit assumption is that, short of adopting some modal notions as ultimate, if an adequate explication of nomic universality is to be obtained each universal law must be treated as a unit and shown to be translatable into a properly constructed *de facto* universal also treated as a complete unit. But there surely is an alternative to this assumption: namely, the explication of nomological universals by indicating some of the logical and epistemic conditions under which *de facto* universals are accepted as universals of law. Moreover, some of the technical details are generated by the aim to exclude every possible "queer" case that could theoretically arise, even though they rarely if ever arise in scientific practice.

1. Perhaps the first thing that will strike us is that the accidental universal contains designations for a particular individual object and for a definite date or temporal period, while the nomological universal does not. Is this difference the decisive one? Not if we wish to count among the laws of nature a number of statements frequently so classified—for example, the Keplerian laws of planetary motion, or even the statement that the velocity of light in vacuum is 300,000 kilometers per second. For the Keplerian laws mention the sun (the first of the three laws, for instance, asserts that the planets move on elliptic orbits with the sun at one focus of each ellipse); and the law about the velocity of light tacitly mentions the earth, since the units of length and time used are defined by reference to the size of the earth and the periodicity of its rotation. But although we can exclude such statements from the class of laws, it would be highly arbitrary to do so. Moreover, the refusal to count such statements as laws would lead to the conclusion that there are few if any laws, should there be merit in the suggestion (discussed more fully in Chapter 11) that the relations of dependence codified as laws undergo evolutionary changes. According to the suggestion, different cosmic epochs are characterized by different regularities in nature, so that every statement properly formulating a regularity must contain a designation for some specific temporal period. However, no statement containing such a designation could be counted as a law by those who find the occurrence of a proper name in a statement to be incompatible with the statement's being a nomological universal.

A way of outflanking this difficulty has been proposed in several recent discussions of lawlike statements. In the first place, a distinction is drawn between predicates which are "purely qualitative" and those which are not, where a predicate is said to be purely qualitative if "a statement of its meaning does not require reference to any *particular* object or spatiotemporal location." [11] Thus, 'copper' and 'greater current strength' are examples of purely qualitative predicates, while 'lunar' and 'larger than the sun' are not. In the second place, a distinction is introduced between "fundamental" and "derivative" lawlike statements. Ignoring fine points, a universal conditional is said to be fundamental if it contains no individual names (or "individual constants") and all its predicates are purely qualitative; a universal conditional is said to be derivative if it is a logical consequence of some set of fundamental lawlike statements; and finally, a universal conditional is said to be lawlike if it is either fundamental or derivative. Accordingly, the Keplerian statements can be counted among the laws of nature if they are the logical

[11] Carl G. Hempel and Paul Oppenheim, "Studies in the Logic of Explanation," *Philosophy of Science*, Vol. 15 (1948), p. 156.

consequences of presumably true fundamental laws, such as Newtonian theory.

On the face of it, this proposed explication is most attractive, and reflects an undoubted tendency in current theoretical physics to formulate its basic assumptions exclusively in terms of qualitative predicates. The proposal nevertheless runs into two unresolved difficulties. In the first place, it just happens that universal conditionals containing predicates that are not purely qualitative are sometimes called laws, even if they are not known to follow logically from some set of fundamental laws. This was the case, for example, for the Keplerian laws before the time of Newton; and if we label as "law" (as some do) the statement that the planets all revolve around the sun in the same direction, this is the case today for this law. But in the second place, it is far from certain that such statements as Kepler's are in fact logically derivable even today from fundamental laws *alone* (as is required by the proposal under discussion, if those statements are to be classified as laws). There appears to be no way of deducing the Keplerian laws from Newtonian mechanics and gravitational theory, *merely* by substituting constant terms for variables occurring in the latter and *without* using additional premises whose predicates are not purely qualitative. And if this is so, the proposed explication would rule out from the class of lawlike statements an indefinite number of statements that are commonly said to be laws.[12] In effect, therefore, the proposed explication is far too restrictive, and fails to do justice to some of the important reasons for characterizing a statement as a law of nature.

Let us therefore compare that paradigm of accidental universality, 'For any x, if x is a screw in Smith's car during the time period a, then x is rusty during a,' with the first Keplerian law 'All planets move on elliptic orbits with the sun at one focus of each ellipse' (or, in comparable logical form, 'For any x and for any sufficiently long time interval t, if x is a planet, then x moves on an elliptic orbit during t and the sun is at one focus of this ellipse'). Both statements contain names of individuals and

[12] On the other hand, if one relaxes the requirement that all the premises from which a derivative law is to be deduced must be fundamental, such patently unlawlike statements as the one about the screws in Smith's car will have to be counted as laws. Thus this statement follows from the presumably fundamental law that all iron screws exposed to oxygen rust, conjoined with the additional premises that all the screws in Smith's current car are iron and have been exposed to oxygen.

It is indeed possible to deduce from Newtonian theory that a body which is under the action of an inverse-square law will move on an orbit that is a conic section with its focus as the origin of the central force. But in order to derive the further conclusion that the conic is an ellipse, additional premises appear to be unavoidable—premises which state the relative masses and the relative velocities of the planets and the sun. This circumstance is one reason for doubting that Kepler's laws are deducible from premises containing only fundamental laws.

predicates that are not purely qualitative. Nevertheless, there is a difference between them. In the accidental universal, the objects of which the predicate 'rusty during the time period a' is affirmed (let us call the class of such objects the "scope of predication" of the universal) are severely restricted to things that fall into a specific spatiotemporal region. In the lawlike statement, the scope of predication of the somewhat complex predicate 'moving on an elliptic orbit during the time interval t and the sun is at one focus of this ellipse' is not restricted in this way: the planets and their orbits are not required to be located in a fixed volume of space or a given interval of time. For convenience, let us call a universal whose scope of predication is not restricted to objects falling into a fixed spatial region or a particular period of time an "unrestricted universal." It is plausible to require lawlike statements to be unrestricted universals.

It must be noted, however, that whether or not a universal conditional is unrestricted cannot invariably be decided on the basis of the purely grammatical (or syntactical) structure of the sentence used to state the conditional, even if grammatical structure is often a reasonably safe guide. For example, one might coin the word 'scarscrew' to replace the expression 'screw in Smith's car during the period a,' and then render the accidental universal by 'All scarscrews are rusty.' But the syntactical structure of this new sentence does not reveal that its scope of predication is restricted to objects satisfying a given condition during only a limited period. Accordingly, familiarity must be assumed with the use or meaning of the expressions occurring in a sentence, in deciding whether the statement conveyed by the sentence is unrestrictedly universal. It must also be noted that, though a universal conditional is unrestricted, its scope of predication may actually be finite. On the other hand, though the scope is finite, the fact that it is finite must not be inferrible from the term in the universal conditional which formulates the scope of predication, and must therefore be established on the basis of independent empirical evidence. For example, though the number of known planets is finite, and though we have some evidence for believing that the number of times the planets revolve around the sun (whether in the past or distant future) is also finite, these are facts which cannot be deduced from Kepler's first law.

2.   But though unrestricted universality is often taken as a necessary condition for a statement to be a law, it is not a sufficient one. An unrestricted universal conditional may be true, simply because it is vacuously true (i.e., nothing whatever satisfies its antecedent clause). But if such a conditional is accepted for this reason alone, it is unlikely that anyone will number it among the laws of nature. For example, if we

assume (as we have good reason to) that there are no unicorns, the rules of logic require us also to accept as true that all unicorns are fleet of foot. Despite this, however, even those familiar with formal logic will hesitate to classify this latter statement as a law of nature—especially since logic also requires us to accept as true, on the basis of the very same initial assumption, that all unicorns are slow runners. Most people would in fact regard it as at best a mild joke were a universal conditional labeled a law because it is vacuously true. The reason for this lies in good part in the uses which are normally made of laws: to explain phenomena and other laws, to predict events, and in general to serve as instruments for making inferences in inquiry. But if a universal conditional is accepted on the ground that it is vacuously true, there is nothing to which it can be applied, so that it cannot perform the inferential functions which laws are expected to perform.

It may therefore seem plausible that a universal conditional is not to be called a law unless we know that there is at least one object which satisfies its antecedent. However, this requirement is too restrictive, for we are not always in the position to know this much, even though we are prepared to call a statement a law. For example, we may not know whether there are in fact any pieces of copper wire at minus 270° C temperature, and yet be willing to classify as a law the statement that all copper wire at minus 270° C temperature is a good conductor of electricity. But if we do accept the statement as a law, on what evidence do we accept it? By hypothesis, we have no direct evidence for it, since we have assumed that we do not know whether there is any copper wire at near absolute zero temperatures, and have therefore not performed any experiments on such wire. The evidence must accordingly be indirect: the statement is accepted as a law, presumably because it is a consequence of some *other* assumed laws for which there is evidence of some kind. For example, the statement is a consequence of the ostensible law that all copper is a good electric conductor, for which there is considerable evidence. Accordingly, we can formulate as follows an additional requirement implicit in classifying an unrestricted universal as a law of nature: the vacuous truth of an unrestricted universal is not sufficient for counting it a law; it counts as a law only if there is a set of other assumed laws from which the universal is logically derivable.

Unrestricted universals whose antecedent clauses are believed to be satisfied by nothing in the universe thus acquire their status as laws, because they are part of a system of deductively related laws and are supported by the empirical evidence—often comprehensive and of a wide variety—which supports the system. It is pertinent to ask, nevertheless, why, even though a universal statement has such support, it should be classified as a law if it is also alleged to be vacuously true. Now there are two possible reasons for such an allegation. One is that no instances satisfying the antecedent clause of the law have been found, despite persistent search for such instances. Although such negative evidence may sometimes be impressive, it is frequently not very conclusive, since such instances may after all occur in overlooked places or under special circumstances. The law may then be employed for calculating the logical consequences of the supposition that there are in fact positive instances in some unexplored regions or under imagined conditions. The calculation may thus suggest how the area of further search for positive instances can be narrowed, or what experimental manipulations should be undertaken for generating such instances. The second and usually more decisive reason for believing a law to be vacuously true is a proof to the effect that the assumed existence of any positive instances for the law is logically incompatible with other laws belonging to the system. The vacuously true law may then indeed be otiose, and may represent so much dead lumber, because it serves no inferential function. On the other hand, if the laws used to establish this vacuous truth are themselves suspect, the ostensibly vacuously true law may be used as a basis for obtaining further critical evidence for these laws. There are doubtless other imagined uses for vacuously true laws. The point is that unless they do have some use, they are not likely to be included in codified bodies of knowledge.

One further question requires brief notice in this connection. It is frequently maintained that some of the laws in physics (and in other disciplines as well, for example in economics), accepted as at least temporarily ultimate, are known to be vacuously true. In consequence, the present account seems not to be adequate, since unrestricted universals are called "laws" despite the fact that they are not derived from other laws. A familiar example of such an ultimate vacuously true law is Newton's first law of motion, according to which a body under the action of no external forces maintains a constant velocity; and the familiar claim is that there are in fact no such bodies, since the assumption that there are is incompatible with Newtonian gravitation theory. Little will be said at this place about the example, since it will receive considerable attention in a later chapter. But two points can be made quickly. Even if the claim is granted that the Newtonian law is vacuously true, it is not for this reason that it is accepted as a law. Why then is it accepted? Waiving the question as to how the Newtonian statement is to be interpreted (e.g., whether or not it is in effect a definitional statement of what it is to be a body under the action of no external forces), and waiving also the question whether or not it is deducible from some other assumed law (e.g., Newton's second law of motion), an examination of its use shows that when the motions of bodies are analyzed in terms of the vectorial components of the motions, the velocities of bodies are constant in directions along which there are no effective forces acting on the bodies. In short, it is a gross oversimplification to claim that the law is vacuously true; for the law is an element in a system of laws for which there are certainly confirmatory instances. More generally, if an "ultimate" law were vacuously satisfied, it would be difficult to understand what use it would have in the system of which it is a part.

3.   It is plausible to assume that candidates for the title of "law of nature" must satisfy another condition, one suggested by the considerations just mentioned. Quite apart from the fact that the paradigm accidental universal about the rusty screws in Smith's current car is not unrestrictedly universal, it exhibits a further feature. That universal conditional (let us refer to it as S) can be construed as a compendious way of asserting a finite conjunction of statements, each conjunct being a statement about a particular screw in a finite class of screws. Thus, S is equivalent to the conjunction: 'If s_1 is a screw in Smith's car during period a then s_1 is rusty during a, and if s_2 is a screw in Smith's car during period a then s_2 is rusty during a, and . . . , and if s_n is a screw in Smith's car during period a then s_n is rusty during period a,' where n is some finite number. S can therefore be established by establishing the truth of a finite number of statements of the form: 's_i is a screw in Smith's car during period a and s_i is rusty during period a.'

Accordingly, if we accept S, we do so because we have examined some fixed number of screws which we have reason to believe exhaust the scope of predication of S. If we had grounds for suspecting that the examined screws did not exhaust the lot in Smith's car, but that there are an indefinite number of further screws in the car which have not been examined, we would not be in the position to assert S as true. For in asserting S we are in effect saying that each of the examined screws is rusty, and that the examined screws are all the screws there are in Smith's car. It is important, however, not to misunderstand what is the point being stressed. In the first place, S might be accepted as true, not because each screw in Smith's car has been found to be rusty, but because S is deduced from some other assumptions. For example, we might deduce S from the premises that all the screws in Smith's current car are iron, that they have been exposed to free oxygen, and that iron always rusts in the presence of oxygen. But even in this case the acceptance of S depends on our having established a fixed number of statements having the form 's_i is an iron screw in Smith's car and it has been exposed to oxygen,' where the examined screws exhaust the scope of application of S. In the second place, S might be accepted on the ground that we have examined only a presumably "fair sample" of screws in Smith's car, and have inferred the character of the unexamined screws from the observed character of the screws in the sample. But here too the presumption of the inference is that the screws in the sample come from a class of screws that is complete and will not be augmented. For example, we assume that no one will remove a screw from the car and replace it by another, or that no one will drill a fresh hole in the car in order to insert a new screw. If we accept S as true on the basis of what we find in the sample, we do so in part because we assume that the sample has been obtained from a population of screws that will neither increase nor be altered during the period mentioned in S.

On the other hand, no analogous assumption appears to be made concerning the evidence on which statements called laws are accepted. Thus, although the law that iron rusts in the presence of free oxygen was at one time based exclusively on evidence drawn from an examination of a finite number of iron objects which had been exposed to oxygen, that evidence was not assumed to exhaust the scope of predication of the law. However, if there had been reason to suppose that this finite number of objects exhausted the class of iron objects exposed to oxygen which have ever existed or will ever exist in the future, it is doubtful whether the universal conditional would be called a law. On the contrary, if the observed cases were believed to exhaust the scope of application of the conditional, it is more likely that the statement would be classified simply as a historical report. In calling a statement a law, we are apparently asserting at least tacitly that as far as we know the examined instances of the statement do not form the exhaustive class of its instances. Accordingly, for an unrestricted universal to be called a law it is a plausible requirement that the evidence for it is not known to coincide with its scope of predication and that, moreover, its scope is not known to be closed to any further augmentation.

The rationale for this requirement is again to be found in the inferential uses to which statements called laws are normally put. The primary function of such statements is to explain and to predict. But if a statement asserts in effect no more than what is asserted by the evidence for it, we are being slightly absurd when we employ the statement for explaining or predicting anything included in this evidence, and we are being inconsistent when we use it for explaining or predicting anything not included in that evidence. To call a statement a law is therefore to say more than that it is a presumably true unrestricted universal. To call a statement a law is to assign a certain function to it, and thereby to say in effect that the evidence on which it is based is assumed not to constitute the total scope of its predication.

This requirement appears to be sufficient for denying the title of "law" to a certain class of manufactured statements that would normally not be so classified but which ostensibly satisfy the requirements previously discussed. Consider the statement 'All men who are the first to see a living human retina contribute to the establishment of the principle of conservation of energy.' Let us assume that the statement is not vacuously true and that it is an unrestricted universal, so that it can be transcribed as 'For any *x* and any *t*, if *x* is the man who sees a living human retina at time *t* and no man sees a living human retina at any time before *t*, then *x* contributes to the establishment of the principle of conservation of energy.' [18] Everyone who recalls the history of science will recognize the reference to Helmholtz, who was both the first to see a living human retina and also a founder of the conservation principle. Accordingly, the above statement is true, and by hypothesis satisfies the requirement of unrestricted universality. Nevertheless, it is plausible to assume that most people would be loath to call it a law. The reason for this conjectured reluctance becomes clear when we examine what evidence is needed to establish the statement. To establish it as true it is sufficient to show that Helmholtz was indeed the first human being to see a living human retina, and that he did contribute to founding the conservation principle. However, if Helmholtz was such a person, then, in the nature of the case, *logically* there cannot be another human being who satisfies the conditions described in the antecedent clause of the above statement. In brief, we *know* in this case that the evidence upon which the statement is accepted coincides with the scope of its predication. The statement is useless for explaining or predicting anything not included in the evidence, and is therefore not given the status of a law of nature.

4. One further point concerning statements commonly designated as laws requires to be noted, though it is difficult in this connection to formulate anything like a "requirement" that lawlike statements must invariably satisfy. The point bears on the standing which laws have in the corpus of our knowledge, and on the cognitive attitude we often manifest toward them.

The evidence on the strength of which a statement *L* is called a law can be distinguished as either "direct" or "indirect." (a) It may be "direct" evidence, in the familiar sense that it consists of instances falling into the scope of predication of *L*, where all the examined instances possess the property predicated by *L*. For example, direct evidence for the law that copper expands on heating is provided by lengths of copper wire which expand on heating. (b) The evidence for *L* may be "indirect" in two senses. It may happen that *L* is jointly derivable with other laws

[18] Hans Reichenbach, *Nomological Statements and Admissible Operations,* Amsterdam, 1954, p. 35.

L_1, L_2, etc., from some more general law (or laws) M, so that the direct evidence for these other laws counts as (indirect) evidence for L. For example, the law that the period of a simple pendulum is proportional to the square root of its length, and the law that the distance traversed by a freely falling body is proportional to the square of the time of fall, are jointly derivable from the assumptions of Newtonian mechanics. It is customary to count the direct confirming evidence for the first of these laws as confirmatory evidence, although only as "indirectly" confirmatory evidence, for the second law. However, the evidence for L may be "indirect" in the somewhat different sense that L can be combined with a variety of special assumptions to yield other laws each possessing a distinctive scope of predication, so that the direct evidence for these derivative laws counts as "indirect" evidence for L. For example, when the Newtonian laws of motion are conjoined with various special assumptions, the Keplerian laws, the law for the period of a pendulum, the law for freely falling bodies, and the laws concerning the shapes of rotating masses can all be deduced. Accordingly, the direct evidence for these derived laws serves as indirect evidence for the Newtonian laws.

Suppose now that, while some of the evidence for L is direct, there is also considerable indirect evidence for L (in either sense of "indirect"). But suppose also that some apparent exceptions to L are encountered. We may nevertheless be most reluctant to abandon L despite these exceptions, and for at least two reasons. In the first place, the combined direct and indirect confirmatory evidence for L may outweigh the apparently negative evidence. In the second place, in virtue of its relations to other laws and to the evidence for these latter, L does not stand alone, but its fate affects the fate of the system of laws to which L belongs. In consequence, the rejection of L would require a serious reorganization of certain parts of our knowledge. However, such a reorganization may not be feasible because no suitable replacement is momentarily available for the hitherto adequate system; and a reorganization may perhaps be avoided by reinterpreting the apparent exceptions to L, so that these latter are construed as not "genuine" exceptions after all. In that event, both L and the system to which it belongs can be "saved," despite the ostensible negative evidence for the law. This point is illustrated when an apparent failure of a law is construed as the result of careless observation or of inexpertness in conducting an experiment. But it can be illustrated by more impressive examples. Thus, the law (or principle) of the conservation of energy was seriously challenged by experiments on beta-ray decay whose outcome could not be denied. Nevertheless, the law was not abandoned, and the existence of a new kind of entity (called a "neutrino") was assumed in order to bring the law into concordance with experimental data. The rationale for this assumption is that the rejection of the conservation law would deprive a large part of our physical knowledge of its systematic coherence. On the other hand, the law (or principle) of the conservation of parity in quantum mechanics (which asserts that, for example, in certain types of interactions atomic nuclei oriented in one direction emit beta-particles with the same intensity as do nuclei oriented in the opposite direction) has recently been rejected, even though at first only relatively few experiments indicated that the law did not hold in general. This marked difference in the fates of the energy and parity laws is an index of the different positions these assumptions occupy at a given time in the system of physical knowledge, and of the greater intellectual havoc that would ensue at that stage from abandoning the former assumption than is involved in rejecting the latter.

More generally, we are usually quite prepared to abandon an assumed law for which the evidence is exclusively direct evidence as soon as *prima facie* exceptions to it are discovered. Indeed, there is often a strong disinclination to call a universal conditional *L* a "law of nature," despite the fact that it satisfies the various conditions already discussed, if the only available evidence for *L* is direct evidence. The refusal to call such an *L* a "law" is the more likely if, on the assumption that *L* has the form 'All *A* is *B*,' there is a class of things *C* that are not *A* which resemble things that are *A* in some respects deemed "important," such that while some members of *C* have the property *B*, nevertheless *B* does not invariably characterize the members of *C*. For example, although all the available evidence confirms the universal statement that all ravens are black, there appears to be no indirect evidence for it. However, even if the statement is accepted as a "law," those who do so will probably not hesitate to reject it as false and so to withdraw the label should a bird be found that is ostensibly a raven but has white plumage. Moreover, the color of plumage is known to be a variable characteristic of birds in general; and in fact species of birds similar to ravens in biologically significant respects, but lacking a completely black plumage, have been discovered. Accordingly, in the absence of known laws in terms of which the black color of ravens can be explained, with the consequent absence of a comprehensive variety of indirect evidence for the statement that all ravens are black, our attitude to this statement is less firmly settled than it is toward statements called laws for which such indirect evidence is available.

Such differences in our readiness to abandon a universal conditional in the face of apparently contradictory evidence is sometimes reflected in the ways we employ laws in scientific inference. Up to this point we have assumed that laws are used as premises *from which* consequences are derived in accordance with the rules of formal logic. But when a law is regarded as well-established and as occupying a firm position in the body of our knowledge, the law may itself come to be used as an empirical principle *in accordance with which* inferences are drawn. This difference between premises and rules of inference can be illustrated from elementary syllogistic reasoning. The conclusion that a given piece of wire *a* is a good electric conductor can be derived from the two premises that *a* is copper and all copper is a good electrical conductor, in accordance with the rule of formal logic known as the *dictum de omni*. However, that same conclusion can also be obtained from the single premise that *a* is copper, if we accept as a principle of inference the rule that a statement of the form '*x* is a good electrical conductor' is derivable from a statement of the form '*x* is copper.'

On the face of it, this difference is only a technical one; and from a purely formal standpoint it is always possible to eliminate a universal premise without invalidating a deductive argument, provided we adopt a suitable rule of inference to replace that premise. Nevertheless, this technical maneuver is usually employed in practice only when the universal premise has the status of a law which we are not prepared to abandon merely because occasionally there are apparent exceptions to it. For when such a premise is replaced by a rule of inference, we are along the road to transforming the meanings of some of the terms employed in the premise, so that its empirical content is gradually absorbed into the meanings of those terms. Thus, in the above example, the statement that copper is a good electrical conductor is assumed to be a factual one, in the sense that possession of high conductivity is not one of the defining traits of what it is to be copper, so that empirical evidence is required for establishing the statement. On the other hand, when that statement is replaced by a rule of inference, electrical conductivity tends to be taken as a more or less "essential" trait of copper, so that in the end no thing may come to be classified as copper unless the object is a good conductor. As has already been noted, this tendency helps to account for the view that genuine laws express relations of logical necessity. But in any event, when this tendency has run its full course, the discovery of a poorly conducting substance that is in other respects like copper would require a reclassification of substances with a corresponding revision in the meanings associated with such terms as 'copper.' That is why the transformation of an ostensibly empirical law into a rule of inference occurs usually only when the law is assumed to be so well-established that quite overwhelming evidence is needed for dislocating it. Accordingly, although to call a universal conditional a law it is not required that we must be disposed to reinterpret apparently negative evidence so as to retain the statement as an integral part of our knowledge, many statements are classified as laws in part because we do have such an attitude toward them.

IV. Contrary-to-fact Universals

There are thus four types of considerations which seem relevant in classifying statements as laws of nature: (1) syntactical considerations relating to the form of lawlike statements; (2) the logical relations of statements to other statements in a system of explanations; (3) the functions assigned to lawlike statements in scientific inquiry; and (4) the cognitive attitudes manifested toward a statement because of the nature of the available evidence. These considerations overlap in part, since, for example, the logical position of a statement in a system is related to the role the statement can play in inquiry, as well as to the kind of evidence that can be obtained for it. Moreover, the conditions mentioned in these considerations are not asserted to be sufficient (or perhaps, in some cases, even necessary) for affixing the label "law of nature" to statements. Undoubtedly statements can be manufactured which satisfy these conditions but which would ordinarily not be called laws, just as statements sometimes called laws may be found which fail to satisfy one or more of these conditions. For reasons already stated, this is inevitable, for a precise explication of the meaning of "law of nature" which will be in agreement with every use of this vague expression is not possible. Nevertheless, statements satisfying these conditions appear to escape the objections raised by critics of a Humean analysis of nomic universality. This claim requires some defense; and something must also be said about the related problem of the logical status of contrary-to-fact conditionals.

1. Perhaps the most impressive current criticism of Humean analyses of nomic universality is the argument that *de facto* universals cannot support subjunctive conditionals. Suppose we know that there never has been a raven that was not black, that there is at present no raven that is not black, and that there never will be a raven that will not be black. We are then warranted in asserting as true the unrestricted accidental universal S: 'All ravens are black.' It has been argued, however, that S does not express what we would usually call a law of nature.[14] For suppose that in point of fact no raven has ever lived or will live in polar

---

[14] William Kneale, "Natural Laws and Contrary-to-Fact Conditionals," *Analysis*, Vol. 10 (1950), p. 123. Cf. also William Kneale, *Probability and Induction*, Oxford, 1949, p. 75. The impetus to much recent Anglo-American discussion of nomological universals and subjunctive as well as "contrary-to-fact" (or "counterfactual") conditionals was given by Roderick M. Chisholm, "The Contrary-to-fact Conditional," *Mind*, Vol. 55 (1946), pp. 289-307, and Nelson Goodman, "The Problem of Counterfactual Conditionals," *Journal of Philosophy*, Vol. 44 (1947), pp. 113-28, the latter also reprinted in Nelson Goodman, *Fact, Fiction, and Forecast*, Cambridge, Mass., 1955.

regions. But suppose further that we do not know whether or not dwelling in polar regions affects the color of ravens, so that as far as we know the progeny of ravens that might migrate into such regions may grow white feathers. Accordingly, though S is true, this truth may be only a consequence of the "historical accident" that no ravens ever live in polar regions. In consequence, the accidental universal S does not support the subjunctive conditional that if inhabitants of polar regions were ravens they would be black; and since a law of nature must, by hypothesis, support such conditionals, S cannot count as a law. In short, unrestricted universality does not explicate what we mean by nomic universality.

But though the argument may establish this latter point, it does not follow that S is not a law of nature because it fails to express an irreducible nomic necessity. For despite its assumed truth, S may be denied the status of law for at least two reasons, neither of which has anything to do with questions of such necessity. In the first place, the evidence for S may coincide with S's scope of predication, so that to anyone familiar with that evidence S cannot perform the functions which statements classified as laws are expected to perform. In the second place, though the evidence for S is by hypothesis logically sufficient to establish S as true, the evidence may be exclusively direct evidence; and one may refuse to label S as a law, on the ground that only statements for which indirect evidence is available (so that statements must occupy a certain logical position in the corpus of our knowledge) can claim title to the label.

But another consideration is no less relevant in this connection. The failure of S to support the subjunctive conditional mentioned above is a consequence of the fact that S is asserted to be true within a context of assumptions which themselves make dubious the subjunctive conditional. For example, S is asserted in the knowledge that no ravens inhabit polar regions. But it has already been suggested that we know enough about birds to know that the color of their plumage is not invariant for every species of birds. And though we do not know at present the precise factors upon which the color of plumage depends, we do have grounds for believing that the color depends at least in part on the genetic constitution of birds; and we also know that this constitution can be influenced by the presence of certain factors (e.g., high-energy radiations) which may be present in special environments. Accordingly, S does not support the cited subjunctive conditional, not because S is incapable of supporting *any* such conditional, but because the total knowledge at our command (and not only the evidence for S itself) does not warrant *this particular* conditional. It may be plausible to suppose that S does validate the subjunctive conditional that were any inhabitant of polar regions a raven not exposed to X-ray radiations, that raven would be black.

The point to be noted, therefore, is that whether or not S supports a given subjunctive conditional depends not only on the truth of S but also on other knowledge we may possess—in effect on the state of scientific inquiry. To see the point more clearly, let us apply the criticism under discussion to a statement generally counted as a law of nature. Suppose there are (omnitemporally) no physical objects that do not attract each other inversely as the square of their distances from each other. We are then entitled to assert as true the unrestricted universal S': 'All physical bodies attract each other inversely as the square of the distance between them.' But suppose also that the dimensions of the universe are finite, and that no physical bodies are ever separated by a distance greater than, say, 50 trillion light-years. Does S' support the subjunctive conditional that if there were physical bodies at distances from each other greater than 50 trillion light-years, they would attract each other inversely as the square of the distance between them? According to the argument under consideration, the answer must presumably be no. But is this answer really plausible? Is it not more reasonable to say that no answer is possible, either in the affirmative or negative— unless indeed some further assumptions are made? For in the absence of such additional assumptions, how can one adjudicate any answer that might be given? On the other hand, if such further assumptions were made—for example, if we assume that the force of gravity is independent of the total mass of the universe—it is not inconceivable that the correct answer may be an affirmative one.

In sum, therefore, the criticism under discussion does not undermine the Humean analysis of nomic universality. The criticism does bring into clearer light, however, the important point that a statement is usually classified as a law of nature because the statement occupies a distinctive position in the system of explanations in some area of knowledge, and because the statement is supported by evidence satisfying certain specifications.

2. When planning for the future or reflecting on the past, we frequently carry on our deliberations by making assumptions that are contrary to the known facts. The results of our reflections are then often formulated as contrary-to-fact conditionals (or "counterfactuals"), having the forms 'If *a* were P, then *b* would be Q,' or 'If *a* had been P, then *b* would have been (or would be) Q.' For example, a physicist designing an experiment may at some point in his calculations assert the counterfactual C: 'If the length of pendulum *a* were shortened to one-fourth its present length, its period would be half its present period.' Similarly, in at-

tempting to account for the failure of some previous experiment, a physicist can be imagined to assert the counterfactual C': 'If the length of pendulum a had been shortened to one-fourth its actual length, its period would have been half its actual period.' In both conditionals, the antecedent and consequent clauses describe suppositions presumably known to be false.

What has come to be called the "problem of counterfactuals" is the problem of making explicit the logical structure of such statements and of analyzing the grounds upon which their truth or falsity may be decided. The problem is closely related to that of explicating the notion of nomic universality. For a counterfactual cannot be translated in a straightforward way into a conjunction of statements in the indicative mood, using only the standard non-modal connectives of formal logic. For example, the counterfactual C' tacitly asserts that the length of pendulum a was not in fact shortened to one-fourth its actual length. However, C' is not rendered by the statement: 'The length of a was not shortened to a fourth of its actual length and if the length of a was shortened to one-fourth of its present length then its period was half its present period.' The proposed translation is unsatisfactory, because, since the antecedent clause of the indicative conditional is false, it follows by the rules of formal logic that if the length of a was shortened to a fourth of its present length, its period was *not* half its present period —a conclusion certainly not acceptable to anyone who asserts C'.[15] In consequence, critics of Humean analyses of nomic universality have argued that a distinctive type of nonlogical necessity is involved not only in universals of law but also in contrary-to-fact conditionals.

The content of counterfactuals can nevertheless be plausibly explicated without recourse to any unanalyzable modal notions. For what the physicist who asserts C' is saying can be rendered more clearly though more circuitously as follows. The statement 'The period of the pendulum a was half its present period' *follows logically* from the supposition 'The length of a was one-fourth its present length,' when this supposition is conjoined with the law that the period of a simple pendulum is proportional to the square root of its length, together with a number of further assumptions about initial conditions for the law (e.g., that a is a simple pendulum, that air resistance is negligible). Moreover, though the supposition and the statement deduced from it with the help of the assumptions mentioned are admittedly both false, their falsity is not included among the premises of the deduction. Accordingly, it does *not*

---

[15] This conclusion follows because of the logical rule governing the use of the connective "if-then." According to this rule both a statement of the form 'If S_1 then S_2,' and the statement of the form 'If S_1 then not S_2,' are true on the hypothesis that S_1 is false, no matter what S_2 may be.

follow from those premises that if *a*'s length was a fourth of its present length then *a*'s period was a half of its present period. In short, the counterfactual *C'* is thus asserted within some context of assumptions and special suppositions; and when these are laid bare, the introduction of modal categories other than those of formal logic is entirely gratuitous. More generally, a counterfactual can be interpreted as an implicit *metalinguistic* statement (i.e., a statement about *other* statements, and in particular about the logical relations of these other statements) asserting that the indicative form of its consequent clause follows logically from the indicative form of its antecedent clause, when the latter is conjoined with some law and the requisite initial conditions for the law.[16]

In consequence, disputes as to whether or not a given counterfactual is true can be settled only when the assumptions and suppositions on which it is based are made explicit. A counterfactual which is unquestionably true on one set of such premises may be false on another set, and may have no determinate truth-value on some third set. Thus, a physicist might reject *C'* in favor of the counterfactual 'If the length of pendulum *a* had been shortened to a fourth of its present length, the period of *a* would have been significantly more than half its present period.' He would be warranted in doing so if he is assuming, for example, that the arc of vibration of the shortened pendulum is more than 60° and if he also is assuming a modified form of the law for the periods of pendulums stated above (which is asserted only for pendulums with quite small arcs of vibration). Again, a tyro in experimental design may declare *C'* to be true, though he assumes among other things not only that the circular bob of the pendulum is three inches in diameter, but also that the apparatus enclosing the pendulum has an opening just a hairsbreadth wider than three inches at the place where the bob of the shortened pendulum has its center. It is obvious, however, that *C'* is now false because under the stated assumptions the shortened pendulum does not vibrate at all.

The various assumptions under which a counterfactual is asserted are not stated in the counterfactual itself. The evaluation of the validity of a counterfactual may therefore be quite difficult—sometimes because we do not know the assumptions under which it is asserted or because we are not clear in our minds what tacit assumptions we are making,

---

[16] Although the position adopted in the text has been reached independently, its present formulation is indebted to the views expressed in Henry Hiz, "On the Inferential Sense of Contrary-to-Fact Conditionals," *Journal of Philosophy*, Vol. 48 (1951), pp. 586-87; Julius R. Weinberg, "Contrary-to-Fact Conditionals," *Journal of Philosophy*, Vol. 48 (1951), pp. 17-22; Roderick M. Chisholm, "Law Statements and Counterfactual Inference," *Analysis*, Vol. 15 (1955), pp. 97-105; and John C. Cooley, "Professor Goodman's 'Fact, Fiction, and Forecast,'" *Journal of Philosophy*, Vol. 54 (1957), pp. 293-311.

and sometimes because we simply lack the skill to assess the logical import even of the assumptions that we make explicit. Such difficulties frequently confront us, especially in connection with counterfactuals asserted in the course of everyday affairs or even in the writings of historians. Consider, for example, the counterfactual 'If the Versailles Treaty had not imposed burdensome indemnities on Germany, Hitler would not have come into power.' This assertion has been a controversial one, not only because those participating in the discussion of it adopt different explicit assumptions, but also because much of the dispute has been conducted on the basis of implicit premises that no one has fully brought into light. In any event, it is certainly not possible to construct a general formula which will prescribe just what must be included in the assumptions upon which a counterfactual can be adequately grounded. Attempts to construct such a formula have been uniformly unsuccessful; and those who see the problem of counterfactuals as that of constructing such a formula are destined to grapple with an insoluble problem.

V. *Causal Laws*

Something must finally be said about causal laws. It would be an ungrateful and pointless task to canvass even partially the variety of senses that have been attached to the word 'cause'—varying from the ancient legal associations of the word, through the popular conception of causes as efficient agents, to the more sophisticated modern notions of cause as invariable functional dependence. The fact that the term has this wide spectrum of uses immediately rules out the possibility that there is just one correct and privileged explication for it. It is nevertheless both possible and useful to identify one fairly definite meaning associated with the word in many areas of science as well as in ordinary discourse, with a view to obtaining from this perspective a rough classification of laws that serve as premises in explanations. On the other hand, it would be a mistake to suppose that, because in one meaning of the word the notion of cause plays an important role in some field of inquiry, the notion is indispensable in all other fields—just as it would be an error to maintain that, because this notion is useless in certain parts of science, it cannot have a legitimate role in other divisions of scientific study.

The sense of 'cause' we wish to identify is illustrated by the following example. An electric spark is passed through a mixture of hydrogen and oxygen gas; the explosion that follows the passage of the spark is accompanied by the disappearance of the gases and the condensation of water vapor. The disappearance of the gases and the formation of water in this experiment are commonly said to be the effects that are caused by the spark. Moreover, the generalization based on such experiments (e.g., 'Whenever a spark passes through a mixture of hydrogen and oxygen gas, the gases disappear and water is formed') is called a "causal law."

The law is said to be a causal one apparently because the relation it formulates between the events mentioned supposedly satisfies four conditions. In the first place, the relation is an invariable or uniform one, in the sense that whenever the alleged cause occurs so does the alleged effect. There is, moreover, the common tacit assumption that the cause constitutes both a necessary and a sufficient condition for the occurrence of the effect. In point of fact, however, most of the causal imputations made in everyday affairs, as well as most of the causal laws frequently mentioned, do not state the sufficient conditions for the occurrence of the effect. Thus, we often say that striking a match is the cause of its bursting into flame, and tacitly assume that other conditions without which the effect would not occur (e.g., presence of oxygen, a dry match) are also present. The event frequently picked out as the cause is normally an event that completes the set of sufficient conditions for the occurrence of the effect, and that is regarded for various reasons as being "important." In the second place, the relation holds between events that are spatially contiguous, in the sense that the spark and the formation of water occur in approximately the same spatial region. Accordingly, when events spatially remote from each other are alleged to be causally related, it is tacitly assumed that these events are but termini in a cause-and-effect chain of events, where the linking events are spatially contiguous. In the third place, the relation has a temporal character, in the sense that the event said to be the cause precedes the effect and is also "continuous" with the latter. In consequence, when events separated by a temporal interval are said to be causally related, they are also assumed to be connected by a series of temporally adjacent and causally related events. And finally, the relation is asymmetrical, in the sense that the passage of the spark through the mixture of gases is the cause of their transformation into water, but the formation of the water is not the cause of the passage of the spark.

The ideas in terms of which this notion of cause are stated have been frequently criticized as being vague; and telling objections have been made in particular against the common-sense conceptions of spatial and temporal continuity, on the ground that they contain a nest of confusions. It is undoubtedly true, moreover, that in some of the advanced sciences such as mathematical physics this notion is quite superfluous; and it is even debatable whether the four conditions just mentioned are in fact fulfilled in alleged illustrations of this notion of cause (such as the above example), when the illustrations are analyzed in terms of modern physical theories. Nevertheless, however inadequate this notion of cause may be for the purposes of theoretical physics, it continues to play a role in many other branches of inquiry. It is a notion that is firmly embodied in the language we employ, even when abstract physical theories are used in the laboratory as well as in practical affairs for obtaining various results through the manipulation of appropriate instrumentalities. Indeed, it is because some things can be manipulated so as to yield other things, but not conversely, that causal language is a legitimate and convenient way of describing the relations of many events.

On the other hand, not all laws of nature are causal in the indicated sense of this term. A brief survey of types of laws that are used as explanatory premises in various sciences will make this evident.

1. As has already been mentioned, a basic and pervasive type of law is involved in the assumption that there are "natural kinds" or "substances." Let us understand by a "determinable" a property such as color or density, which has a number of specific or "determinate" forms. Thus, among the determinate forms of the determinable color are red, blue, green, yellow, etc.; among the determinate forms of the determinable density are the density with magnitude 0.06 (when measured in some standard fashion), the density with magnitude 2, the density with magnitude 12, etc. The determinate forms of a given determinable thus constitute a "related family" of properties such that every individual of which the determinable property can be significantly predicated must, of logical necessity, have one and only one of the determinate forms of the determinable.[17] A law of the type under consideration (e.g., 'There is the substance rock salt') then asserts that there are objects of various kinds, such that every object of a given kind is characterized by determinate forms of a set of determinable properties, and such that objects belonging to different kinds will differ in at least one (but usually more than one) determinate form of a common determinable. For example, to say that a given object *a* is rock salt is to say that there is a set of determinable properties (crystalline structure, color, melting point, hardness, etc.) such that under standard conditions *a* has a determinate form of each of these determinables (*a* has cubical crystals, it is colorless, it has a density of 2.163, a melting point of 804° C, the degree of hardness 2 on Mohs' scale, etc.). Moreover, *a* differs from an object belonging to a different kind, for example talc, in at least one (and in fact in a great many) determinate forms of these determinables. Accordingly, laws of this type assert that there is an invariable concomitance of determinate

[17] For this terminology, cf. W. E. Johnson, *Logic*, Vol. 1, Cambridge, England, 1921, Chapter 11; and Rudolf Carnap, *Logical Foundations of Probability*, Chicago, 1950, Vol. 1, p. 75.

properties in every object that is of a certain kind. It will be clear, however, that laws of this type are not causal laws—they do not assert, for example, that the density of rock salt precedes (or follows) its degree of hardness.

2. A second type of law asserts an invariable sequential order of dependence among events or properties. Two subordinate types can be distinguished. One of these is the class of causal laws, such as the law about the effect of a spark in a mixture of hydrogen and oxygen, or the law that stones thrown into water produce a series of expanding concentric ripples. A second subordinate type is the class of "developmental" (or "historical") laws, such as the law 'The formation of lungs in the human embryo never precedes the formation of the circulatory system' or the law 'Consumption of alcohol is always followed by a dilation of the blood vessels.' Both subordinate types are frequent in areas of study in which quantitative methods have not been extensively introduced, although as the examples indicate such laws are encountered elsewhere as well. Developmental laws can be construed to have the form 'If x has the property P at time t, then x has the property Q at time t' later than t.' They are commonly not regarded as causal laws, apparently for two reasons. In the first place, though developmental laws may state a necessary condition for the occurrence of some event (or complex of events), they do not state the sufficient conditions. Indeed, we usually have only the vaguest sort of knowledge as to what these sufficient conditions are. In the second place, developmental laws generally state relations of sequential order between events separated by a temporal interval of some duration. In consequence, such laws are sometimes regarded as representing only an incomplete analysis of the facts, on the ground that, since something may intervene after the earlier event to prevent the realization of the later one, the sequential order of events is not likely to be invariable. Nevertheless, whatever may be the limitations of developmental laws and however desirable it may be to supplement them by laws of another sort, both causal and developmental laws are extensively used in the explanatory systems of current science.

3. A third type of law, common in the biological and social sciences as well as in physics, asserts invariable statistical (or probabilistic) relations between events or properties. One example of such a law is: 'If a geometrically and physically symmetrical cube is repeatedly tossed, the probability (or relative frequency) that the cube will come to rest with a given face uppermost is $\frac{1}{6}$'; other examples have been previously mentioned. Statistical laws do not assert that the occurrence of one event is *invariably* accompanied by the occurrence of some other event.

They assert only that, in a sufficiently long series of trials, the occurrence of one event is accompanied by the occurrence of a second event with an *invariable relative frequency*. Such laws are manifestly not causal, though they are not incompatible with a causal account of the facts they formulate. Indeed, the above statistical law about the behavior of a cube can be deduced from laws that are sometimes said to be causal ones, if suitable assumptions are made about the statistical distribution of initial conditions for the application of those causal laws. On the other hand, there are statistical laws even in physics for which at present no causal explanations are known. Moreover, even if one assumes that "in principle" all statistical laws are the consequences of some underlying "causal order," there are areas of inquiry—in physics as well as in the biological and social sciences —in which the explanation of many phenomena in terms of strictly universal causal laws is not likely to be feasible practically. It is a reasonable presumption that however much our knowledge may increase, statistical laws will continue to be used as the proximate premises for the explanation and prediction of many phenomena.

4. A fourth type of law, characteristic of modern physical science, asserts a relation of functional dependence (in the mathematical sense of "function") between two or more variable magnitudes associated with stated properties or processes. Two subtypes can be distinguished.

a. In the first place, there are numerical laws stating an interdependence between magnitudes such that a variation in any of them is concurrent with variations in the others. An example of such a law is the Boyle-Charles' law for ideal gases, that $pV = aT$, where p is the pressure of the gas, V its volume, T its absolute temperature, and a a constant that depends on the mass and the nature of the gas under consideration. This is not a causal law. It does not assert, for example, that a change in the temperature is followed (or preceded) by some change in the volume or in the pressure; it asserts only that a change in T is concurrent with changes in p or V or in both. Accordingly, the relation stated by the law must be distinguished from the sequential order of the events that may occur when the law is being tested or used for making predictions. For example, in testing the law in a laboratory, one may diminish the volume of an ideal gas in such a way that its temperature remains constant, and then note that its pressure increases. But the law says nothing about the order in which these magnitudes may be varied, nor about the temporal sequence in which the changes may be observed. Laws of this subtype can nevertheless be used for predictive as well as explanatory purposes. For example, if in the case of a suitably "isolated" system the magnitudes mentioned in such a law satisfy the indicated relation between them at one instant, they will satisfy this relation at some future instant, even though the magnitudes may have undergone some change in the interim.

b. A second subtype consists of numerical laws asserting in what manner a magnitude varies with the time, and more generally how a change in a magnitude per unit of time is related to other magnitudes (in some cases, though not always, to temporal durations). Galileo's law for freely falling bodies in a vacuum is one illustration of such a law. It says that the distance d traversed by a freely falling body is equal to $gt^2/2$, where g is constant and t is the duration of the fall. An equivalent way of stating Galileo's law is to say that the change in the distance per unit time of a freely falling body is equal to gt. In this formulation, it is evident that a time-rate of change in one magnitude is related to a temporal interval. Another example of a law belonging to this subtype is the law for the velocity of the bob of a simple pendulum along the path of its motion. The law says that, if v_0 is the velocity of the bob at the lowest point of its motion, h the height of the bob above the horizontal line through this point, and k a constant, then at any point along the arc of its motion the bob has a velocity v such that $v^2 = v_0^2 - kh^2$. Since the velocity v is the change in distance per unit of time, the law thus says that the change in the distance of the bob along its path per unit of time is a certain mathematical function of its velocity at the lowest point of its swing and of its altitude. In this case, the time-rate of change in one magnitude is not given as a function of the time. Laws that belong to this subtype are often called "dynamical laws" because they formulate the structure of a temporal process and are generally explained on the assumption that a "force" is acting on the system under consideration. Such laws are sometimes assimilated in causal laws, although in fact they are not causal in the specific sense distinguished earlier in this section. For the relation of dependence between the variables mentioned in the law is symmetrical, so that a state of the system at a given time is determined as completely by a later state as by an earlier one. Thus, if we know the velocity of the bob of a simple pendulum at any given instant, then provided there is no external interference with the system, the above law enables us to calculate the velocity at any other time, whether it is earlier or later than the given instant.

The preceding classification of laws is not proposed as an exhaustive one; and in any event, later chapters will discuss more fully the structures of certain types of law. The classification does indicate, however, that not all the laws recognized in the sciences are of one type, and that a scientific explanation is often regarded as satisfactory even though the laws cited in the premises are not "causal" in any customary sense.

Experimental Laws
and Theories

5 Scientific thought takes its ultimate point of departure from problems suggested by observing things and events encountered in common experience; it aims to understand these observable things by discovering some systematic order in them; and its final test for the laws that serve as instruments of explanation and prediction is their concordance with such observations. Indeed, many laws in the sciences formulate relations between things or features of things that are commonly said to be themselves observable, whether with the unaided senses or with the help of special instruments of observation. The law that when water in an open container is heated it eventually evaporates is a law of this kind; and so is the law that lead melts at 327° C, as well as the law that the period of a simple pendulum is proportional to the square root of its length.

However, not all scientific laws are of this kind. On the contrary, many laws employed in some of the most impressively comprehensive explanatory systems of the physical sciences are notoriously not about matters that would ordinarily be characterized as "observable," even when the word "observable" is used as broadly as in the examples of the preceding paragraph. Thus, when the evaporation of heated water is explained in terms of assumptions about the molecular constitution of water, laws of this latter sort appear among the explanatory premises. Although we may have good observational evidence for these assumptions, neither molecules nor their motions are capable of being observed in the sense in which, for example, the temperature of boiling water or of melting lead is said to be observable.

Let us baptize the *prima facie* difference between laws just noted by calling those of the first sort "experimental laws" and those of the second sort "theoretical laws" (or simply "theories"). In consonance with this terminological stipulation and the distinction covered by it, the law that the pressure of an ideal gas whose temperature is constant varies inversely with its volume, the law that the weight of oxygen combining with hydrogen to form water is (approximately) eight times the weight of the hydrogen, and the law that children of blue-eyed human parents are blue-eyed, are all classified as experimental laws. On the other hand, the set of assumptions asserting different chemical elements to be composed of different kinds of atoms which remain undivided in chemical transformations, and the set of assumptions asserting chromosomes to be composed of different kinds of genes which are associated with the hereditary traits of organisms, are all classified as theories.[1]

These labels are not free from misleading associations. However, the terminology is firmly established in the literature for characterizing the intended distinction between kinds of laws; and in any event, no better labels are available. Two brief reminders may help to prevent misconstruing those labels. When a statement (e.g., 'All whales suckle their young') is classified as an experimental law, it is not to be construed as asserting that the law is based on laboratory experiments or that the law happens to be one for which there is thus far no explanation. The rubric "experimental law" signifies simply that a statement so characterized formulates a relation between things (or traits of things) that are observable in the admittedly loose sense of "observable" illustrated by the above examples, and that the law can be validated (even if only with some "degree of probability") by controlled observation of the things mentioned in the law. Again, when the set of assumptions about the molecular constitution of liquids is called a theory, it is not to be understood as asserting those assumptions to be entirely speculative and unsupported by any cogent evidence. What is intended by this characterization is simply that those assumptions employ terms like 'molecule,' which ostensibly designate nothing observable (in the sense indicated previously) and that the assumptions cannot be confirmed by experiments or observations of the things to which such terms ostensibly refer.

Nevertheless, although the distinction between experimental laws and theories is frequently made and appears to be at least initially plausible in the light of some of the examples used to illustrate it, it generates problems of considerable importance that cannot be ignored. Granted

[1] The present chapter leans heavily on the discussion in Norman R. Campbell's *Physics, the Elements,* Cambridge, England, 1920, especially Chapter 6. Campbell's unfinished treatise has not received the recognition that its generally admirable analyses so eminently merit.

its initial plausibility, is the distinction solidly grounded in clearly identifiable differences between two kinds of scientific laws? Moreover, even if some indisputable basis for the distinction can be specified, is the distinction as sharp as is sometimes alleged or is it only one of degree? In any event, and admitting as hardly deniable that those assumptions called "theories" yield explanatory and predictive systems far more comprehensive than are the systems whose premises are characterized as "experimental laws," what distinctive features do theories possess that account for this difference? These are the questions to which the present chapter is devoted.

I. *Grounds for the Distinction*

The above account of the distinction between experimental laws and theories is based on the contention that laws subsumed under the first of these labels, unlike the laws falling under the second one, formulate relations between observable (or experimentally determinable) traits of some subject matter. In consequence, the distinction suffers from all the notorious unclarities that attach to the word "observable." Indeed, there is a sense of the word in which none of the familiar sciences (with the possible exception of some branches of psychology) assert laws stating relations between observable things, just as there is a sense of the word in which even those assumptions called "theories" deal with observable matters.

It would certainly be a mistaken claim that scientific statements commonly cited as typical illustrations of experimental laws assert relations between data allegedly apprehended directly or noninferentially through various sense organs, that is, between the so-called "sense data" of epistemological discussion. Even if we waive the familiar difficulties concerning the possibility of identifying "pure" (i.e., noninferentially categorized) sense data, it is patent that sense data occur at best only intermittently and in patterns of sequential and concomitant order which can be formulated only with greatest difficulty (if at all) by universal laws. But however this may be, none of the customary examples of experimental laws are in fact about sense data, since they employ notions and involve assumptions that go far beyond anything directly given to sense. Consider, for example, the experimental law that the velocity of sound is greater in less dense gases than in more dense ones. This law obviously assumes that there is a state of aggregation of matter known as "gas" which is to be distinguished from other states of aggregation such as liquid and solid; that gases have different densities under determinate conditions, so that under specified conditions the ratio of the weight of a gas to its volume remains constant; that the instruments for measuring weights and volumes, distances and times, exhibit certain regularities which can be codified in definite laws, such as laws about mechanical, thermal, and optical properties of various kinds of ma-terials; and so on. It is clear, therefore, that the very meanings of the terms occurring in the law (for example, the term 'density'), and in consequence the meaning of the law itself, tacitly assume a congeries of other laws. Moreover, additional assumptions become evident when we consider what is done when evidence is adduced in support of the law. For example, in measuring the velocity of sound in a given gas, different numerical values are in general obtained when the measure-ment is repeated. Accordingly, if a definite numerical value is to be assigned to the velocity, these different numbers must be "averaged" in some fashion, usually in accordance with an assumed law of experimental error. In short, the law about the velocity of sound in gases does not formulate relations between immediate data of sense. It deals with things which can be identified only by way of procedures involving fairly com-plicated chains of inference and a variety of general assumptions.

On the other hand, though the commonly cited examples of theories are statements about things that in an obvious sense are unobservable, it is frequently possible to determine indirectly, by way of inferences drawn from experimental data in accordance with certain rules, impor-tant characteristics of what is ostensibly not observable. On the face of it, therefore, experimental laws and theories do not appear to differ radically in respect to the "observable" (or experimentally determinable) status of their respective subject matters. For example, the molecules assumed by the kinetic theory of matter as the constituents of gases are indeed not observable, in the sense in which a bit of apparatus in the laboratory or even the nucleus of a living cell viewed through a micro-scope is observable. Nevertheless, the number of molecules per unit volume of a gas, as well as their average velocities and masses, can be calculated from magnitudes ascertained by experiment; and there is no logical absurdity in the supposition that eventually all the terms of the theory which refer to unobservable matters (such as the positions of the molecules at a given time) may be associated in an analogous fash-ion with experimental data. Similarly, though the alpha particles postu-lated by contemporary electronic theories of atomic structure are not observable in the sense in which the other side of the moon is in prin-ciple observable, their ostensible tracks in Wilson cloud chambers are certainly visible.

It is pertinent to note in this connection, moreover, that reports of what are commonly regarded as experimental observations are frequently couched in the language of what is admittedly some theory. For ex-ample, experiments on beams of light passing from a given medium to a denser one show that the index of refraction varies with the source of the beam. Thus, a beam issuing from the red end of the solar spectrum has a different index of refraction than has a beam coming from the violet end. However, the experimental law based on such experiments is not formulated in unquestionably observational terms (e.g., in terms of the visible colors of light beams) but in terms of the relation between the index of refraction of a light ray and its wave frequency. The ideas of the wave theory of light are thus absorbed into the statement of the presumably experimental law. More generally, many statements of allegedly experimental laws not only take for granted other allegedly experimental laws, but apparently include as part of 'their meaning assumptions that are admittedly theoretical.

For these several reasons many students of the subject have concluded that the labels 'experimental law' and 'theory' do not signify laws fundamentally different in kind, but connote what is at best only differences in degree. In the view of these students, the distinction has therefore little if any methodological importance.

It is doubtful whether a rigorously precise sense can be usefully assigned to the word "observable"; and to the extent that the distinction between experimental laws and theories is based on a contrast between what is observable and what is not, the distinction is patently not a sharp one. In any event, no precise criterion for distinguishing between experimental laws and theories is available, and none will be proposed here. It nevertheless does not follow that the distinction is spurious because it is vague, any more than it follows that there is no difference between the front and the back of a man's head just because there is no exact line separating the two. There are in fact several well-marked features that differentiate laws we shall continue to call "experimental" from other general assumptions we designate as "theories"; and we proceed to examine those features. Despite the admitted vagueness of the distinction under discussion, we shall see that it is an important one.

1. Perhaps the most striking single feature setting off experimental laws from theories is that each "descriptive" (i.e., nonlogical) constant term in the former, but in general not each such term in the latter, is associated with at least one overt procedure for predicating the term of some observationally identifiable trait when certain specified circumstances are realized. The procedure associated with a term in an experimental law thus fixes a definite, even if only a partial, meaning for the term. In consequence, an experimental law, unlike a theoretical statement, invariably possesses a determinate empirical content which in principle can always be controlled by observational evidence obtained by those procedures. The previously cited law concerning the velocity

of sound in gases illustrates these points clearly. There are established procedures for ascertaining the density of a gas and for measuring the velocity of sound in gases; and these procedures determine the senses in which the corresponding terms in the law are to be construed. The law can therefore be tested in the light of data acquired by way of those procedures.

Accordingly, each descriptive term in an experimental law L has a meaning that is fixed by an overt observational or laboratory procedure. Moreover, if L is assumed to have a genuine empirical content (in contrast with a statement that in effect merely defines some term occurring in it), the procedures associated with the terms in L can in general be instituted without tacitly employing L. Thus, the density of a gas, as well as the velocity of sound in a gas, can be ascertained by means of procedures that make no use of the law concerning the dependence of the velocity of sound in a gas on the density of the gas. In consequence, though the operative meaning of a given term P may be augmented because of the relations P is asserted by L to have to other terms in the law, in general P has a determinate meaning independent of its occurrence in L and distinguishable from any additional meaning the term may acquire because of its occurrence in L. It is therefore possible to obtain direct evidence for an experimental law (i.e., evidence based on an examination of instances falling into the scope of predication of the law), provided that difficulties stemming from current limitations of experimental technology do not stand in the way.

It is frequently the case, however, that more than one overt procedure is available for applying a term in an experimental law to a concrete subject matter. This is generally the case when a term enters into more than one experimental law. For example, the term 'electric current' enters into at least three distinct experimental laws that relate electric currents to magnetic, chemical, and thermal phenomena, respectively. Accordingly, the strength of an electric current can be measured by the deflection of a magnetized needle, by the quantity of some element like silver which is deposited from a solution in a given time, or by the rise in temperature of a standard substance during a given temporal interval. But the tacit assumption underlying the use of such diverse procedures is that they yield concordant results. Thus, two currents that are of equal strength as determined by one procedure are of equal strength (at least approximately) as judged by the other procedures. Moreover, when several such overt procedures are available for a term in an experimental law, it is often the case in many areas of science that one procedure is selected as the standard for "defining" the term and for measuring the property it designates.

In contrast with what holds uniformly for the descriptive terms in experimental laws, the meanings of many if not all descriptive terms occurring in theories are not specified by such overt experimental procedures. To be sure, theories are frequently constructed in analogy to some familiar materials, so that most theoretical terms are associated with conceptions and images derived from the generating analogies. Nevertheless, the operative meanings of most theoretical terms either are defined only implicitly by the theoretical postulates into which the terms enter, or are fixed only indirectly in the light of the eventual uses to which the theory may be put. Thus, although the theoretical terms 'electron,' 'neutrino,' or 'gene' may be construed as "particles" possessing some (though not necessarily all) of the properties which characterize small bits of matter, there are no overt procedures for applying those terms to experimentally identifiable instances of the terms. These points will be amplified presently. For the moment we simply note the important consequence that, since the basic terms of a theory are in general not associated with definite experimental procedures for applying them, instances falling into the ostensible scope of predication of a theory cannot be observationally identified, so that a theory (unlike an experimental law) cannot be put to a direct experimental test.

2. An immediate corollary to the difference between experimental laws and theories just discussed is that while the former could, in principle, be proposed and asserted as inductive generalizations based on relations found to hold in observed data, this can never be the case for the latter. For example, Boyle based the law named after him on observations obtained by studying the changes in the volumes of gases at constant temperature when the pressures were varied; and he asserted the inverse variation of pressure and volume to hold in general, on the assumption that what was true in the observed samples is true universally. No doubt it is often possible to support an experimental law not only by direct confirmatory data but also by indirect evidence; this latter type of support can frequently be supplied when the experimental law is incorporated into an inclusive system of laws. Thus Galileo's law for freely falling bodies can be directly confirmed by data on the distances traversed by such bodies during various times; the law can also be indirectly confirmed by experiments on the periods of simple pendulums, since Galileo's law and the law for simple pendulums are intimately related by virtue of their inclusion in the system of Newtonian mechanics. It is equally undeniable that some experimental laws (e.g., the law concerning the conical refraction of light in biaxial crystals) have been first suggested by theoretical considerations, and only subsequently confirmed by direct experiment. The crucial point remains, nevertheless, that an experimental law is not held to be established until direct experimental evidence for it becomes available.

In the nature of the case, however, a theory cannot be an empirical generalization from observational data, since in general there are no experimentally identifiable instances that fall into the ostensible scope of predication of the theory. Distinguished scientists have repeatedly claimed that theories are "free creations of the mind." Such claims obviously do not mean that theories may not be *suggested* by observational materials or that theories do not require support from observational evidence. What such claims do rightly assert is that the basic terms of a theory need not possess meanings which are fixed by definite experimental procedures, and that a theory may be adequate and fruitful despite the fact that the evidence for it is necessarily indirect. Indeed, there are theories in the history of modern science that were accepted by many scientists at a time when no fresh experimental confirmation was available for those explanatory assumptions. The sole ground for accepting them at that time was the fact that they could explain experimental laws assumed to have been established by previously accumulated observational data. This was at one time the status of the Copernican theory of the solar system, the corpuscular theory of light, the atomic theory in chemistry, and the kinetic theory of gases.[2]

Accordingly, even when an experimental law is explained by a given theory and is thus incorporated into the framework of the latter's ideas (in a manner to be discussed presently), two characteristics continue to hold for the law. It retains a meaning that can be formulated independently of the theory; and it is based on observational evidence that may enable the law to survive the eventual demise of the theory. Thus, Wien's displacement law (which asserts that the wave length corresponding to the position of the maximum energy in the spectrum of the radiation emitted by a black body is inversely proportional to the absolute temperature of the radiating body) was not rejected when classical electrodynamics which explained the law was modified through the introduction of Planck's quantum hypothesis. Nor was Balmer's law abandoned (according to which the wave frequencies corresponding to the lines in the spectrum of hydrogen and other elements are terms in a series conforming to a simple numerical formula) when the Bohr theory of

[2] When Sir Arthur Eddington published his book on the general theory of relativity in 1923, he noted that the widespread interest in the theory was due to the experimental verification of certain minute deviations from Newtonian laws that were predicted by relativity theory. But he added "To those who are still hesitating and reluctant to leave the old faith, these deviations will remain the chief centre of interest; but for those who have caught the spirit of the new ideas the observational predictions form only a minor part of the subject. It is claimed for the theory that it leads to an understanding of the world of physics clearer and more penetrating than that previously attained, and it has been my aim to develop the theory in a form which throws most light on the origin and significance of the great laws of physics."—A. S. Eddington, *The Mathematical Theory of Relativity*, Cambridge, England, 1924, p. v.

the atom which explained the law was replaced by the "new quantum mechanics." Such facts indicate that an experimental law has, so to speak, a life of its own, not contingent on the continued life of any particular theory that may explain the law. Despite what appears to be the complete absorption of an experimental law into a given theory, so that the special technical language of the theory may even be employed in stating the law, the law must be intelligible (and must be capable of being established) without reference to the meanings associated with it because of its being explained by that theory. Indeed, were this not the case for the laws which a given theory purportedly explains, there would be nothing for the theory to explain. At the very least, therefore, and on pain of a fatal circularity, even if the terms occurring in an experimental law have meanings derived in part from some other theory, the terms must have determinate senses statable (albeit only partially) independently of the particular theory adopted for explaining the law.

On the other hand, theoretical notions cannot be understood apart from the particular theory that implicitly defines them. This follows from the circumstance that, although the theoretical terms are not assigned a unique set of determinate senses by the postulates of a theory, the permissible senses are limited to those satisfying the structure of interrelations into which the postulates place the terms. Accordingly, when the fundamental postulates of a theory are altered, the meanings of its basic terms are also changed, even if (as often happens) the same linguistic expressions continue to be employed in the modified theory as in the original one. The new theory will presumably continue to explain all the experimental laws that the earlier theory could explain, in addition to explaining experimental laws for which the earlier theory could not account. But in consequence of the changed theoretical content of the new theory, the observationally identifiable regularities that are formulated by experimental laws and explained by both the original and the modified theory receive what are in fact different theoretical interpretations.

These points deserve a fuller illustration. For this purpose, let us consider Millikan's famous oil-drop experiment. The experiment (first performed in 1911 and repeated a number of times with improved techniques) was conducted within the framework of a theory that postulated the existence of unobservable particles called "electrons." Electrons were assumed to possess the usual complement of traits characterizing particles (such as definite spatial positions at given times, definite velocities at those times, and masses), and in addition to be the carriers of an elementary electric charge. The aim of Millikan's experiment was to determine the numerical value e of the elementary charge. In essence, the experiment consists in comparing the velocity of a fine oil drop when it moves between two horizontal metal plates under the action of gravity alone, and its velocity when the oil drop (as a consequence of a charge having been induced on it by electric charges placed on the plates) moves under the action of both gravitational and electrostatic forces. The experiment shows that, when the amount of charge placed on the plates is varied, the velocity of the oil drop also varies. By using established experimental laws, however, it is possible to calculate the magnitudes of the induced charges on the drop which will account for the observed differences in its motion. Millikan found that within the limits of experimental error the charges on the drop are always integral multiples of an elementary charge e (4.77×10^{-10} electrostatic units); and he therefore concluded that e is the minimal electric charge, which he identified with the charge on the electron.

It is important to note, however, that we have described the oil-drop experiment (even if only in outline) without reference to electrons. A more detailed account of the experiment could be carried through in a similar fashion. Evidently, therefore, the experiment can be performed and its procedure communicated without assuming the theory of electrons. The electron theory did indeed suggest the experiment, and offers an illuminating and fruitful interpretation of its findings. However, the electron theory has undergone important modifications since Millikan first performed the experiment; and it is entirely conceivable (though at present not likely) that the electron theory may some day be completely abandoned. Nevertheless, the truth of the experimental law that Millikan helped to establish (namely, that all electric charges are integral multiples of a certain elementary charge) is not contingent upon the fate of the theory; and, provided the direct observational evidence for it continues to confirm the law, it may outlive a long series of theories that may be accepted in the future as explanations for it. On the other hand, what it is to be an electron is stated by a theory in which the word 'electron' occurs; and when the theory is altered, the meaning of the word undergoes a modification. In particular, though the same word 'electron' is used in pre-quantum theories of the electronic constitution of matter, in the Bohr theory, and in post-Bohr theories, the meaning of the word is not the same in all these theories. Accordingly, the facts established by the oil-drop experiment receive different interpretations from these various theories, even though in each case the facts are formulated by saying that the elementary charge determined by experiment is the charge "on the electron."

3.   Another conspicuous difference between experimental laws and theories is worth passing mention. An experimental law is without exception formulated in a single statement; a theory is almost without ex-

ception a system of several related statements. But this obvious difference is only an indication of something more impressive and significant: the greater generality of theories and their relatively more inclusive explanatory power. As has already been noted, experimental laws can be used to explain and predict the occurrence of individual events, as well as to explain other experimental laws. However, the materials experimental laws can explain are in certain readily identifiable respects qualitatively similar, and constitute a fairly definite class of things. For example, the Archimedean law concerning the buoyant force of liquids makes it possible to explain a variety of other experimental laws: the law that ice floats in water, the law that a solid lead sphere sinks in water but a hollow lead sphere of suitable thickness floats in it, or the law that whatever floats in oil also floats in water. But despite the differences between the materials falling into the scope of these laws, the laws all deal with phenomena of flotation. The range of things that can be explained by the Archimedean law is thus a fairly narrow one. Other experimental laws share this characteristic. This is indeed inevitable, since the terms occurring in an experimental law are associated with a small number of definite overt procedures that fix the meanings and the ranges of application of those terms.

On the other hand, many of the outstanding theories in the sciences are capable of explaining a much wider variety of experimental laws and can thus deal with an extensive range of materials that are qualitatively strikingly dissimilar. This feature of theories is related both to the fact that theoretical notions are not tied down to definite observational materials by way of a fixed set of experimental procedures, and to the fact that because of the complex symbolic structure of theories more degrees of freedom are available in extending a theory to many diverse areas. We have already noted the success of Newtonian theory in explaining the laws of planetary motion, of freely falling bodies, of tidal action, of the shapes of rotating masses; and to these we can add the laws dealing with the buoyancy of liquids and gases, with the phenomena of capillarity, with the thermal properties of gases, and much else. Similarly, contemporary quantum theory can explain the experimental laws of spectral phenomena, of thermal properties of solids and gases, of radioactivity, of chemical interactions, and of many other phenomena.

It is indeed one of the important functions of a theory to exhibit systematic connections between experimental laws about qualitatively disparate subject matters. In this respect, the theories of the natural sciences, particularly of physics, are especially noteworthy, although even in physics not all theories are of equal rank in achieving this objective. However, the explanation of already established experimental laws is not the only function theories are expected to perform. Another role played by theories which differentiates them from experimental laws is to provide suggestions for fresh experimental laws. For example, the electron theory with its assumption of electrons as carriers of an elementary charge suggested the problem of whether the magnitude of the charge can be ascertained by experiment. It is unlikely that Millikan (or anyone else) would have devised the oil-drop experiment if some atomistic theory of electricity had not first suggested a question that seemed important in the light of the theory and that the experiment was intended to settle. Thus, apparently no one has undertaken to decide by experimental means whether measurable quantities of heat are all integral multiples of an elementary "heat quantum." It is at least plausible to suppose that no such experiments have been performed because no theories of heat have been developed that assume the existence of heat quanta, so that in consequence experimental inquiry into such a hypothesis has not seemed to be a significant undertaking.

II.   *Three Major Components in Theories*

A reasonably good case can therefore be made for distinguishing experimental laws from theories, even if the distinction is not a precise one. We shall in any event adopt the distinction in large measure for reasons just surveyed, but in part also because it permits us to segregate under a convenient rubric important problems that pertain primarily to explanatory hypotheses having the generic characteristics of those we are calling "theories." We shall now look more closely at the articulation of theories, and examine in what manner they are related to matters that are usually regarded in scientific practice as objects of observation and experiment.

For the purpose of analysis, it will be useful to distinguish three components in a theory: (1) an abstract calculus that is the logical skeleton of the explanatory system, and that "implicitly defines" the basic notions of the system; (2) a set of rules that in effect assign an empirical content to the abstract calculus by relating it to the concrete materials of observation and experiment; and (3) an interpretation or model for the abstract calculus, which supplies some flesh for the skeletal structure in terms of more or less familiar conceptual or visualizable materials. We will develop these distinctions in the order just mentioned. However, they are rarely given explicit formulation in actual scientific practice, nor do they correspond to actual stages in the construction of theoretical explanations. The order of exposition here adopted must therefore not be assumed to reflect the temporal order in which theories are generated in the minds of individual scientists.

1. A scientific theory (such as the kinetic theory of gases) is often suggested by materials of familiar experience or by certain features noted in other theories. Theories are in fact usually so formulated that various more or less visualizable notions are associated with the nonlogical expressions occurring in them, that is, with "descriptive" or "subject matter" terms like 'molecule' or 'velocity,' which, unlike logical particles such as 'if-then' and 'every,' do not belong to the vocabulary of formal logic but are specific to discourse about some special subject matter. Nevertheless, the nonlogical terms of a theory can always be dissociated from the concepts and images that normally accompany them by ignoring these latter, so that attention is directed exclusively to the logical relations in which the terms stand to one another. When this is done, and when a theory is carefully codified so that it acquires the form of a deductive system (a task which, though often difficult in practice, is realizable in principle), the fundamental assumptions of the theory formulate nothing but an abstract relational structure. In this perspective, accordingly, the fundamental assumptions of a theory constitute a set of abstract or uninterpreted postulates, whose constituent nonlogical terms have no meanings other than those accruing to them by virtue of their place in the postulates, so that the basic terms of the theory are "implicitly defined" by the postulates of the theory. Moreover, insofar as the basic theoretical terms are only implicitly defined by the postulates of the theory, the postulates assert nothing, since they are statement-forms rather than statements (that is, they are expressions having the form of statements without being statements), and can be explored only with the view to deriving from them other statement-forms in conformity with the rules of logical deduction. In short, a fully articulated scientific theory has embedded in it an abstract calculus that constitutes the skeletal structure of the theory.

Some illustrations will help make clear what is meant by saying that the postulates of a theory implicitly define the terms occurring in them. A familiar example of an abstract calculus is demonstrative Euclidean geometry developed in a postulational manner. The postulates of the system are frequently stated with the expressions 'point,' 'line,' 'plane,' 'lies between,' 'congruent with,' and several others as the basic terms. Although these expressions are commonly used to characterize familiar spatial configurations and relations and are therefore generally employed with connotations associated with our spatial experience, such connotations are irrelevant to the deductive elaboration of the postulates and are best ignored. Indeed, in order to prevent the familiar although vague meanings of those expressions from compromising the rigor of proofs in the system, the postulates of demonstrative geometry are often formulated by using what are in effect predicate variables like 'P' and

'*L*,' instead of the more suggestive but also more distracting descriptive predicates 'point' and 'line.' But in any event, to the questions "What is a point?" and "What is a line?" (or analogously, "What sorts of things are *P* and *L*?"), the only answer that can be supplied within a postulational treatment of geometry is that points and lines are any things that satisfy the conditions stated in the postulates. It is in this sense that the words 'point' and 'line' are implicitly defined by the postulates.

Similarly, the assumptions that formulate a physical theory such as the kinetic theory of gases provide only an implicit definition for terms like 'molecule' or 'kinetic energy of molecules.' For the assumptions state only the structure of relations into which these terms enter, and thereby stipulate the formal conditions to be satisfied by anything for which those terms can become labels. To be sure, these terms are commonly associated with a set of intuitively satisfying images and familiar notions. In consequence, the terms have a suggestive power that makes them appear meaningful independent of the postulates in which they occur. Nevertheless, what it is to be a molecule, for example, is prescribed by the assumptions of the theory. Indeed, there is no way of ascertaining what is the "nature" of molecules except by examining the postulates of the molecular theory. It is in any event the notion of 'molecule' as implicitly defined by the postulates that does the work expected of the theory.

Because of their importance, it is desirable to illustrate more fully the character of implicit definitions. However, the geometrical calculus is far too complex to be presented here in detail; and the complexity of the calculi embedded in any of the major scientific theories is even greater. A rather simple example of implicit definitions is provided by the following set of abstract postulates. In addition to the terminology of arithmetic, the postulates employ the language of the calculus of classes. If *A* and *B* are any two classes, their logical sum $A \lor B$ is the class whose members belong either to *A* or to *B* or to both, while their logical product $A . B$ is the class whose members belong to both *A* and *B*; the complement $-A$ of a class *A* is the class of those elements that do not belong to *A*; and the null-class $\land$ is the class containing no members. The system has four postulates:

1. *K* is a class, and *F* is the class of subclasses of *K* such that if *A* is any member of *F* so is $-A$; and if *B* is also a member of *F* so are $A \lor B$ and $A . B$. (In technical language, *F* is called the "field of classes on *K*.")
2. For any *A* in *F*, there is a real number $p(A)$ associated with *A* such that $p(A) \geqq 0$.
3. $p(K) = 1$.
4. If *A* and *B* are in *F* and $A . B = \land$, then $p(A \lor B) = p(A) + p(B)$.

It is possible to derive a large number of theorems from this set, for example, the theorem that for any A in F, $0 \leqq p(A) \leqq 1$, or the theorem that for any A and B in F, $p(A \vee B) = p(A) + p(B) - p(A \cdot B)$. However, our present interest is not in the theorems, but in the implicit definitions that the postulates (and in consequence also the theorems) supply. The postulates leave completely undisclosed what specific classes are under discussion and what is the significance of the number p associated with each class. Nevertheless, the postulates impose certain conditions upon any set of classes and any set of associated numbers, if these latter are to satisfy the postulates. In particular, though the postulates do not state what definite properties of classes are measured by the associated numbers p, the numbers must lie in the interval from 0 to 1; moreover, the numbers must be so assigned that the number coupled with the logical sum of two classes is never less than the number coupled with either one of the summands. Accordingly, the property of classes that the numbers p measure is only implicitly defined. The postulates specify only the structure of systems of classes and associated numbers, not the substantive character of any particular system.

2. It is clear, however, that if a theory is to explain experimental laws, it is not sufficient that its terms be only implicitly defined. Unless something further is added to indicate how its implicitly defined terms are related to ideas occurring in experimental laws, a theory cannot be significantly affirmed or denied and in any case is scientifically useless. It obviously makes no sense to ask whether, for example, the above set of abstract postulates is true or false, or even whether $p(A)$ has a given value—say ½. For the postulates as stated do not reveal what is the subject matter, if any, for which they are supposed to hold, or what property of classes is supposed to be measured by the associated numbers. Similarly, the postulates of the kinetic theory of gases do not provide any hint as to what experimentally determinable matters its implicitly defined terms are supposed to signify—even when the term 'molecule,' for example, is taken to signify an imperceptible particle. If the theory is to be used as an instrument of explanation and prediction, it must somehow be linked with observable materials.

The indispensability of such linkages has been repeatedly stressed in recent literature, and a variety of labels have been coined for them: coordinating definitions, operational definitions, semantical rules, correspondence rules, epistemic correlations, and rules of interpretation.[3]

[3] Cf. Hans Reichenbach, *Philosophie der Raum-Zeit Lehre*, Berlin, 1928, pp. 23ff., and *The Rise of Scientific Philosophy*, Berkeley, Calif., 1951, Chap. 8; P. W. Bridgman, *The Logic of Modern Physics*, New York, 1927, Chap. 1, and *Reflections of a Physicist*, New York, 1950, Chap. 1; Rudolf Carnap, *Foundations of Logic and*

The ways in which theoretical notions are related to observational procedures are often quite complex, and there appears to be no single schema which adequately represents all of them. An example will nevertheless help bring out some important features of such correspondence rules.

The Bohr theory of the atom was devised in order to explain, among other things, experimental laws about the line spectra of various chemical elements. In brief outline the theory postulates the following. It assumes that there are atoms, each of which is composed of a relatively heavy nucleus carrying a positive electric charge and a number of negatively charged electrons with smaller mass moving in approximately elliptic orbits with the nucleus at one of the foci. The number of electrons circulating around the nucleus varies with the chemical elements. The theory further assumes that there are only a discrete set of permissible orbits for the electrons, and that the diameters of the orbits are proportional to h^2n^2, where h is Planck's constant (the value of the indivisible quantum of energy postulated in Max Planck's theory of radiation) and n is an integer. Moreover, the electromagnetic energy of an electron in an orbit depends on the diameter of the orbit. However, as long as an electron remains in any one orbit, its energy is constant and the atom emits no radiation. On the other hand, an electron may "jump" from an orbit with a higher energy level to an orbit with a lower energy level; and when it does so, the atom emits an electromagnetic radiation, whose wave length is a function of these energy differences. Bohr's theory is an eclectic fusion of Planck's quantum hypothesis and ideas borrowed from classical electrodynamic theory; and it has now been replaced by a more satisfactory theory. Nevertheless, the theory was successful in explaining a number of experimental laws of spectroscopy and for a time was a fertile guide to the discovery of new laws.

But how is the Bohr theory brought into relation with what can be ob-

---

Mathematics, International Encyclopedia of Unified Science, Vol. I, No. 3, Chicago, 1955, Chap. 3; Henry Margenau, *The Nature of Physical Reality,* New York, 1950, pp. 60ff.; F. S. C. Northrop, *The Logic of the Sciences and the Humanities,* New York, 1947, Chap. 7.

Eddington's claim that Einstein's general theory of relativity is a self-contained "closed system," all of whose concepts are cyclically defined in terms of each other, is vitiated by his general failure to distinguish between theories and experimental laws, and by his rather cavalier recognition of the need for such linkages if a theory is to have not only a formal significance but also a relevance for experimental subject matter. Eddington recognizes this need in his reference to "consciousness" as the point at which theory comes into contact with its subject matter. However, this reference is misleading, since it is not to anything "in consciousness" that theoretical notions are linked, but to characteristics of subject matter that are experimentally identifiable. See A. S. Eddington, "The Domain of Natural Science," in *Science, Religion and Reality* (ed. by Joseph Needham), London, 1925, pp. 203ff., and *The Nature of the Physical World,* New York, 1928, Chap. 12.

served in the laboratory? On the face of it, the electrons, their circulation in orbits, their jumps from orbits to orbits, and so on, are all conceptions that do not apply to anything manifestly observable. Connections must therefore be introduced between such theoretical notions and what can be identified by way of laboratory procedures. In point of fact, connections of this sort are instituted somewhat as follows. On the basis of the electromagnetic theory of light, a line in the spectrum of an element is associated with an electromagnetic wave whose length can be calculated, in accordance with the assumptions of the theory, from experimental data on the position of the spectral line. On the other hand, the Bohr theory associates the wave length of a light ray emitted by an atom with the jump of an electron from one of its permissible orbits to another such orbit. In consequence, the *theoretical* notion of an electron jump is linked to the *experimental* notion of a spectral line. Once this and other similar correspondences are introduced, the experimental laws concerning the series of lines occurring in the spectrum of an element can be deduced from the theoretical assumptions about the transitions of electrons from their permissible orbits.

3. This example of a rule of correspondence also illustrates what is meant by an interpretation or model for a theory. The Bohr theory is usually not presented as an abstract set of postulates, augmented by an appropriate number of rules of correspondence for the uninterpreted nonlogical terms implicitly defined by the postulates. It is customarily expounded, as in the above sketch, by way of relatively familiar notions, so that instead of being statement-forms the postulates of the theory appear to be statements, at least part of whose content can be visually imagined. Such a presentation is adopted, among other reasons, because it can be understood with greater ease than can an inevitably longer and more complicated purely formal exposition. But in any event, in such an exposition the postulates of the theory are embedded in a model or interpretation.

It should nevertheless be clear, despite the use of a model for stating a theory, that the fundamental assumptions of the theory provide only implicit definitions for the theoretical notions employed in them. For example, an electron according to the Bohr theory is just that sort of an "entity" which, though electrically charged and in motion, produces no electromagnetic effects as long as it remains on any one orbit. Moreover, the presentation of a theory by way of a model does not make any less imperative the need for rules of correspondence for linking the theory to experimental concepts. Although models for theories have important functions in scientific inquiry, as will be shown in the following chapter, models are not substitutes for rules of correspondence. The distinction between a model (or interpretation) for a theory and rules of correspondence for terms of the theory is a crucial one, and we must therefore discuss it further.

To fix our ideas, let us provide a model for the abstract postulates stated above for the classes K and F. Assume that there are exactly ten molecules in a certain gene G of some biological organism, and that their total mass is m grams; and call the ratio of the mass of any molecule (or set of molecules) to m 'the relative mass' (or more briefly 'the r') of the molecules (or set of molecules). For the variable letter 'K' in the postulates we now substitute the expression 'the molecules in the gene G,' for which we shall use the letter 'G' as an abbreviation; and for the letter 'F' we substitute the expression 'the set of all subclasses of the molecules in the gene G,' for which we use 'S' as an abbreviation. Counting in the null (or empty) class, S evidently contains 1024 members. Finally, for the expression '$p(A)$' in the abstract postulates we substitute the phrase 'the relative mass of A' [or in abbreviated form, '$r(A)$']. With these substitutions the abstract set of postulates is converted into a set of true statements about G, S, and r. For example, the final postulate then reads: If A and B are in S and $A . B = \wedge$, then $r(A \vee B) = r(A) + r(B)$—that is, for any two sets of molecules A and B in S having no molecules in common, the relative mass of the molecules contained either in A or in B is equal to the relative mass of the molecules in A plus the relative mass of the molecules in B. These statements (or alternatively, the *system* of "things" G, S, and r, rather than the statements) constitute what we shall understand by a "model" for the postulates.

This account of what is meant by a model can easily be generalized.[4] But the example itself suffices to bring out some useful points. On the assumption that every expression employed in formulating a model is in some sense "meaningful," a theory provided with a model is completely interpreted—in the sense that every sentence occurring in the theory is then a meaningful statement. However, though a model can be extraordinarily valuable in suggesting fresh lines of inquiry that might never occur to us when a theory is presented in a completely abstract form, presenting a theory in terms of a model runs the risk that adventitious features of a model may mislead us concerning the actual content

[4] In outline, the general formulation is as follows. Let P be a set of postulates; let P° be a set of statements obtained by substituting for each predicate variable in P some predicate that is significant for a given class of elements K; and finally, let P° consist only of true statements about the elements in K. By a model for P we understand the statements P°, or alternately the system of elements K characterized by the properties and relations that are designated by the predicates of P°. For a precise account of the notions of 'interpretation' and 'model,' see Rudolf Carnap, *Introduction to Semantics*, Cambridge, Mass., 1942, pp. 202ff.; Patrick Suppes, *Introduction to Logic*, Princeton, 1957, pp. 64ff.; Alfred Tarski, *Logic, Semantics, Metamathematics*, Oxford, 1956, Chap. 12.

of the theory. For a theory may receive alternative interpretations by way of different models; and the models may differ not only in the subject matter from which they are drawn but also in important structural properties. (For example, a structurally different model is obtained for the above postulates if the gene *G* is assumed to contain 100 instead of 10 molecules. A still different model is supplied for these postulates by the probability relations between classes of events.) Finally, and this is the central point we are concerned with making in the present context, though a theory is presented in terms of a model, it by no means follows that the theory is thereby automatically linked to experimental concepts and observational procedures. Whether a theory is so linked depends on the character of the model employed. Thus, the above statement of the molecular model for a set of postulates provides no rules for coordinating any of its nonlogical expressions (such as the expression 'the relative mass of a set of molecules in the gene *G*') with experimentally significant notions. Although a model for the postulates is specified, no rules of correspondence are given. In short, therefore, stating a model for a theory so that all its descriptive terms receive an interpretation is in general not sufficient for deducing any experimental law from the theory.

III. *Rules of Correspondence*

We must now call attention to certain features of rules of correspondence that have thus far not been explicitly mentioned.

1. The above example of a rule of correspondence for the Bohr theory of the atom provides a convenient point of departure for noting one such feature. It will be evident that the rule cited in the example does not provide an *explicit* definition of any theoretical notion in the Bohr theory, in terms of predicates used to characterize matters normally said to be observable. The example thus suggests that in general rules of correspondence do not supply such definitions.

Let us make clearer what is involved in this suggestion. When an expression is said to be "explicitly defined," the expression may always be eliminated from any context in which it occurs, since it can be replaced by the defining expression without altering the sense of the context. Thus, the expression '*x* is a triangle' is explicitly defined by the expression '*x* is a closed plane figure bounded by three straight line segments.' The former (or defined) expression can therefore be eliminated from any context in favor of the latter (or defining) expression; for example, the statement 'The area of a triangle is equal to one-half the product of its base and altitude,' can be replaced by the logically equivalent statement 'The area of a closed plane figure bounded by three straight line segments is equal to one-half of the product of its base and altitude.' On the other hand, the theoretical expression in the Bohr theory '*x* is the wave length of the radiation emitted when an electron jumps from the next-to-the-smallest to the smallest permissible orbit of the hydrogen atom' is not being explicitly defined when it is coordinated with an expression having approximately the form of '*y* is the line occurring at a certain position in the spectrum of hydrogen.' It is indeed patent that the two expressions have quite different connotations. Accordingly, although the rule of correspondence establishes a definite connection between the two expressions, the former cannot be replaced by the latter in such statements as 'Transitions of electrons from their next-to-the-smallest to their smallest permissible orbits occur in about ten per cent of hydrogen atoms.' Were the indicated replacement attempted, the result would in fact be nonsense.

No infrangibly conclusive proof is available, and perhaps no such proof is possible, that the theoretical notions employed in current science cannot be explicitly defined in terms of experimental ideas. The issue here broached will be discussed more fully in the next chapter. It is pertinent to observe, however, that no one has yet successfully constructed such definitions. Moreover, there are good reasons for believing that the rules of correspondence in actual use do not constitute explicit definitions for theoretical notions in terms of experimental concepts.

One of these reasons has already been noted. When a theory is formulated by way of a model, the language used in stating the model usually has connotations that the language of experimental procedure does not possess. Thus, as was mentioned above, the expression in the Bohr theory referring to electron transitions is not equivalent in meaning to the expression referring to spectral lines. In such cases, accordingly, since the defining and the defined expressions in explicit definitions are equivalent in meaning, it is most unlikely that rules of correspondence can provide such definitions.[5]

---

[5] Failure to observe that the language of theoretical physics is not equivalent in meaning to the language in which experimental procedures are formulated is the source of much confused puzzlement. It was Eddington's failure to note this point that made it possible for him to raise the question which of the two tables that allegedly confronted him as he sat down to write his book was the "real" one—the substantial and commonplace table of everyday experience, or the scientific table that is mostly emptiness and consists of sparsely scattered electric charges moving about with great speeds. (A. S. Eddington, *The Nature of the Physical World*, New York, 1928, pp. ix ff.) In point of fact, however, Eddington was not at all confronted with two tables. For the word "table" signifies an experimental idea that does not occur in the language of electron theory; and the word "electron" signifies a theoretical notion that is not defined in the language employed in formulating observations and experiments. Though the two languages may be coordinated at certain junctures, they are not intertranslatable. Since there is thus only one *table*, there is no issue as to which is the "real" one, whatever may be understood by this honorific

Another reason of perhaps even greater weight is that theoretical notions are frequently coordinated by rules of correspondence with more than one experimental concept. As has already been argued, theoretical notions are only implicitly defined by the postulates of a theory (even when the theory is presented by way of a model). There are therefore an unlimited number of experimental concepts to which, as a matter of logical possibility, a theoretical notion may be made to correspond. For example, the theoretical notion of electron transition in the Bohr theory corresponds to the experimental notion of a spectral line; but that theoretical notion can also be coordinated (via Planck's radiation law,[6] which is deducible from the Bohr theory) with experimentally determinable temperature changes in black-body radiation. Accordingly, in those cases in which a given theoretical notion is made to correspond to two or more experimental ideas (though presumably on different occasions and in the context of different problems), it would be absurd to maintain that the theoretical concept is explicitly defined by each of the two experimental ones in turn.

This lack of unique correspondence between theoretical and experimental notions deserves further comment and illustration. It is a familiar fact that theories in the sciences (especially though not exclusively in mathematical physics) are generally formulated with painstaking care and that the relations of theoretical notions to each other (whether the notions are primitives in the system or are defined in terms of the primitives) are stated with great precision. Such care and precision are essential if the deductive consequences of theoretical assumptions are to be rigorously explored. On the other hand, rules of correspondence for connecting theoretical with experimental ideas generally receive no explicit formulation; and in actual practice the coordinations are comparatively loose and imprecise.

Some examples will make clear the intent of these general remarks. In modern axiomatizations of geometry (as in the one by the German mathematician David Hilbert) a number of primitive terms (e.g., 'point,' 'line,' 'plane,' 'congruence') are implicitly defined by the postulates of

label. For an extended and vigorous critique of Eddington's philosophy of science, see L. Susan Stebbing, *Philosophy and the Physicists*, London, 1937.

[6] The radiation law, formulated in the theoretical terms of mathematical physics, states that

$$E_\lambda = \frac{hc^2}{\lambda^5} \left(\frac{1}{e^{hc/kT\lambda} - 1} \right)$$

where E_λ is the energy of radiation with wave length λ, h is Planck's constant, c the velocity of light, T the absolute temperature, and k the Boltzmann constant (a constant of proportionality in the equation of the kinetic theory of gases which relates the absolute temperature of a gas and the average kinetic energy of its molecules).

the system; and additional terms (e.g., 'circle,' 'cube') are explicitly defined with the help of the primitive ones. *Within* axiomatic geometry there are therefore precisely stated relations between the theoretical notions of the system. However, when the geometrical calculus comes to be used in some area of empirical inquiry, the coordination of these notions with experimental ideas is usually far from exact. For example, the word 'plane' as it is used in contexts of empirical inquiry is not a precisely defined term. Which surfaces are to count as planes are sometimes specified by rules for grinding bodies so that their surfaces will eventually fit snugly when placed adjacent to one another; sometimes by rules that simply involve perceptual judgments based on using the naked eye; sometimes by rules requiring the use of complicated optical instruments. The correspondence between the theoretical notion of plane and the experimental one is thus neither unique nor precise. Similarly, though the theoretical distance between two points is always a unique number (which may in fact be a so-called "irrational" number), the measured distance between two actual bodies is almost always a range of magnitudes that fall into a certain interval.

Let us look again from this perspective, but more closely, at the correspondence between the notion of wave length in the electromagnetic theory of light and the experimental notion of a spectral line. Even a cursory examination shows that the correspondence is not unique. For spectral lines are all of finite breadth, and the resolving power of optical instruments is limited. Accordingly, what is experimentally identified as a spectral line corresponds, not to a unique wave length, but to a vaguely bounded range of wave lengths. And conversely, a theoretically monochromatic beam of light (i.e., a beam of radiation composed of rays all with the same wave length) is coordinated in practice with experimentally determinable spectral lines that have a discernible width and that are therefore produced, from the standpoint of the theory, by polychromatic radiation.

The general point that emerges from these examples is that, though theoretical concepts may be articulated with a high degree of precision, rules of correspondence coordinate them with experimental ideas that are far less definite. The haziness that surrounds such correspondence rules is inevitable, since experimental ideas do not have the sharp contours that theoretical notions possess. This is the primary reason why it is not possible to formalize with much precision the rules (or habits) for establishing a correspondence between theoretical and experimental ideas.

If we ask, therefore, what formal pattern is exhibited by correspondence rules it is difficult to give a straightforward answer. In some cases, the rules appear to state the necessary and sufficient conditions for describing an experimental situation in theoretical language. Thus, if 'T' is a theoretical predicate and 'E' an experimental one, the rules may have the form '*x* is *T* if and only if *y* is *E*.' This seems to be a plausible way of rendering the rule that coordinates the theoretical notion of an electron jump with the occurrence of a spectral line. In other cases, the rule may state only a sufficient condition for using a theoretical notion. The rule then has the schematic form 'if *y* is *E* then *x* is *T*.' This seems to be the form of the rule implicit in applying the theoretical notion of 'plane' to an actual surface that conforms to an experimental specification of what it is to be a plane. In still other cases, the rule may supply only a necessary condition for the use of a theoretical term: 'if *x* is *T* then *y* is *E*.' For example, under the experimental conditions obtaining in a Wilson cloud chamber, the condensation of water vapor in fine lines appears to be a necessary condition for describing this effect in terms of the theoretical notion of the passage of alpha particles.

Correspondence rules may have still other forms. They may be given a meta-linguistic formulation, explicitly coordinating *expressions* with one another, rather than (as in the above discussion) what is *designated* by expressions; and they may have forms more complex than those mentioned. For example, a rule may state that from a statement of the form '*x* is *T*' one may deduce a statement of the form '*y* is *E*,' and conversely; or a rule may coordinate not just one but several theoretical notions simultaneously with a set of experimental ideas—a type of rule that seems to be involved in stating the way in which the geometrical terms 'point,' 'line,' 'plane,' etc., are to be employed in concrete experimental contexts.

It would not serve our purposes to pursue this matter any further. Enough has been said, however, to support the claim that rules of correspondence do not provide explicit definitions of theoretical notions in terms of experimental ideas, as well as to suggest that such rules are protean in form. But if this claim is indeed soundly based, it helps to fortify the distinction between experimental laws and theories, and at the same time it raises problems concerning the cognitive status of theories. Some of these problems will be explored in the following chapter.

2. A further point must now be made about the way rules of correspondence serve as links between theoretical and experimental ideas. The sketch given above of the Bohr theory of the atom will again serve to introduce the discussion. According to that account, although there are rules of correspondence for some of the notions employed in the theory, not all the theoretical notions are linked with experimental ideas. For example, there is a rule of correspondence for the theoretical notion of electrons in transition from one permissible orbit to another; but there is no such rule for the notion of electrons moving with accelerated

velocities on an orbit. Similarly, in the kinetic theory of gases, there is no correspondence rule for the theoretical notion of the instantaneous velocity of single molecules, although there is such a rule for the theoretically defined notion of the average kinetic energy of the molecules. Moreover, there is at present a correspondence rule for the notion of the number of molecules in a standard volume of gas under standard conditions of temperature and pressure (Avogadro's number); but Avogadro's number was not determined by experimental means until relatively late in the history of the kinetic theory, and until then there was no rule of correspondence for that theoretical notion.

The feature just noted in these examples of theories can be stated more generally but schematically as follows. Suppose the postulates of some theory T employ n primitive nonlogical terms 'P_1,' 'P_2,' . . . , 'P_n,' with the help of which a number of further theoretical terms 'Q_1,' 'Q_2,' . . . , 'Q_r,' can be explicitly defined. (Thus, to illustrate this general account, suppose that 'length,' 'mass,' and 'time' are the primitive terms of the theory, and that 'velocity' and 'kinetic energy' can be explicitly defined on the basis of these primitives.) However, though rules of correspondence must be added to the postulates if T is to have a scientific use, no such rules are introduced for all the 'P's or for all the 'Q's. It is even possible that there are correspondence rules for only some of the 'Q's, but for none of the 'P's. Accordingly, the theoretical notions of T are not all tacked down once for all to experimental concepts.

Most if not all theories in the natural sciences have this characteristic. At any rate, a theory having it possesses a flexibility that permits its extension to fresh areas of inquiry sometimes markedly unlike the subject matter for which the theory was originally devised. As has already been noted, the systematic explanation of a wide variety of experimental laws about qualitatively diverse materials is a distinctive achievement of theories. One of the ways a theory can be made to do this is by introducing new correspondence rules for notions for which none have been instituted previously, when advances in experimental research and technique make this possible. By contrast with alterations in the postulates of a theory, which in effect constitute a modification of the implicit definitions of the theoretical notions, the introduction of new correspondence rules does not change either the formal structure or the intended meaning of the theory, though new rules may enlarge the theory's range of application. Thus, the experimental determination of Avogadro's number (as a consequence of which this theoretical notion came to be linked with an experimental concept) did not involve any modification in the postulates of the kinetic theory of gases; but it did result in bringing the experimental investigation of crystal structure by means of x-rays into relation with that theory.

It is important to remember, moreover, that a theory is a human arti-
fact. Like other artifacts, a theory is likely to contain some elements that
are simply expressions of the special objectives and idiosyncrasies of their
human inventors, rather than symbols with a primary referential or representative function. This point was stressed by Heinrich Hertz in his
account of the requirements that physical theories ought to satisfy. Hertz
maintained that the sole task of physical science is to construct "images
or symbols of external objects" in such a way that the logical consequences of the symbols (i.e., of our conceptions of things) are always the
"images" of "the necessary consequents in nature of the things pictured."
Hertz thus made central the role of theory as an instrument for enabling
us to infer observable events from other observable events. However, he
clearly recognized that this instrumental requirement does not uniquely
determine the symbolism (or theory) that will achieve this objective.
He noted, in particular, that a theory will unavoidably contain what he
called "superfluous or empty relations"—symbols that are not representative of anything in the subject matter for which the theory is devised.
According to Hertz, these "empty relations" enter into our theories simply because the theories are complex symbols, "images produced by our
mind and necessarily affected by the characteristics of its mode of portrayal." [7]

Considerations of this general kind thus lead us to expect that not
every constituent notion in a theory will be linked with some experimental idea by a correspondence rule. In any event, the primary role of
many symbols occurring in theories is to facilitate the formulation of a
theory with great generality, to make possible logical and mathematical
transformations in a relatively simple manner, or to serve as heuristic
aids for the extended application of the theory. Illustrations of such symbols are the continuous variables and differential quotients of mathematical physics; these are extensively used, despite the fact that theoretical notions such as mathematically continuous density functions or
instantaneous velocities, when they are strictly construed, do not correspond to any experimental concepts. An indefinite number of further examples of such symbols can be found in the locutions used when a theory
is embedded in some convenient model—for example, in the language of
point-masses of analytical mechanics, of the ether of nineteenth-century
electromagnetic theory, of valence bonds of analytic chemistry, or of
"wavicles" of current quantum theory.

Since theories are constructed with a view toward explaining a wide
variety of experimental laws, it is clear that such an end can in general
be achieved only if a theory is so formulated that no reference is made

[7] Heinrich Hertz, *The Principles of Mechanics,* London, 1899 (reprinted New
York, 1956), p. 2.

in it to any set of specialized experimental concepts. For otherwise the theory would be limited in its application to situations to which just those concepts are relevant. Indeed, the more comprehensive the range of possible application of a theory, the more meager is its explicitly formulated content with respect to specialized details of some subject matter. Such details are left to be supplied by supplementary assumptions and correspondence rules, introduced as occasion requires when the theory is employed in different experimental contexts.[8] This does not mean, however, that scientific theories tend in the limit to become empty of all content as their range of application becomes more inclusive. It does mean that a theory seeks to formulate a highly general structure of relations that is invariant in a wide variety of experimentally different situations but that can be specialized by augmenting the fundamental postulates of the theory with more restrictive assumptions, so as to yield systematically a series of diversified subordinate structures.

Two examples, though not fully typical of all scientific theories, will illustrate this point, and will thereby make clearer the architecture of at least some theories. The first example is taken from analytical geometry. It is there shown that the biquadratic $ax^2 + 2bxy + cy^2 + 2dx + 2ey + f = 0$ is the equation of a conic section, where the variables 'x' and 'y' are the coordinates (or shortest distances from two fixed and mutually perpendicular lines taken to determine a frame of reference) of any point of the conic, and the coefficients (or "arbitrary constants") have fixed values but are otherwise unspecified (except for the requirement that not all of them are zero). Whatever properties all conics have in common can be deduced from this equation—for example, that a straight line intersects a conic in at most two points, or that two conics have at most four points in common. However, the structure common to all conics can be differentiated into more specialized structures by imposing additional conditions upon the coefficients of the equation. Thus, on the assumption that a, b, and c are not all zero and by stipulating that $b^2 - ac < 0$, the equation will formulate the structural properties of the ellipse, and of the circle as a special case of the ellipse if $b = 0$ and $a = c$. Under the requirement that $b^2 - ac = 0$, the equation will represent a parabola. With the condition that $b^2 - ac > 0$, the equation will represent a hyperbola. Finally, if $(b^2 - ac)f + (ae^2 + cd^2 - 2bde) = 0$, the equation will represent the "degenerate conic" consisting of a pair of straight lines. Accordingly, by specializing the arbitrary constants, different special structures are obtained, and their distinctive characteristics can be explored.

---

[8] Cf. W. F. G. Swann, "The Significance of Scientific Theories," *Philosophy of Science*, Vol. 7 (1940), pp. 273-87, and "The Relation of Theory to Experiment in Physics," *Reviews of Modern Physics*, Vol. 13 (1941), pp. 190-96; also L. Silberstein, *The Theory of Relativity*, London, 1924, pp. 296ff.

The second example is taken from the Newtonian theory of mechanics. According to the theory, a change in the momentum of a body (when referred to a suitable spatial frame of reference) is equal to the force acting on the body. This can be written as $ma = F$, where 'm' is the mass of the body, 'a' its acceleration at an instant, and 'F' is the force. A number of very general consequences about the motions of bodies can be formally derived from this fundamental postulate, even though the nature of the force that may act on a body is not stated. However, nothing can be inferred from the equation about the actual motion of a body unless, among other things, further assumptions are introduced about the force that is supposed to be acting—assumptions that in some cases at any rate include a rule of correspondence between the theoretical notion of force and certain experimental ideas. The fundamental postulates of Newtonian theory place very few formal restrictions on the kind of mathematical functions that may be used to express the character of forces. In practice, however, the functions are of a relatively simple kind. For example, in the study of vibratory motions, the general form of the force-function is: $F = Ar + Br^2 + Cr^3 + Dv + Ef(t)$, where '$r$' is the distance of the body from some designated point, 'v' is the velocity of the body along this line, '$f(t)$' a function of the time t, and 'A,' 'B,' 'C,' 'D,' and 'E' are arbitrary constants for which different numerical values are assigned according to the problem under consideration. Thus, if A is negative and the remaining constants are zero, the body undergoes simple harmonic motion without frictional resistance; if A and D are both negative and the remaining constants are zero, the body is undergoing damped harmonic motion; if A and D are both negative, E is not zero, B and C both zero, and $f(t)$ a periodic function of the time, the body is undergoing a forced vibration; and so on. In general, by specializing F in various ways, different experimental laws can be deduced from the fundamental equations of Newtonian mechanics.

Although these examples are not paradigmatic of all theories—since not all theories contain parameters that are specialized in the manner just indicated—the examples do illustrate one important way in which theories differ from experimental laws, as well as one technique by which some theories achieve comprehensive generality. For unlike terms occurring in experimental laws, theoretical notions employed in the basic assumptions of a theory may either not be associated with any experimental ideas whatever, or may be associated with experimental ideas that vary from context to context. The possibility of extending a theory to cover fresh subject matter depends in considerable measure upon this feature of theories. These examples also help to enforce the point that a theory remains otiose for scientific inquiry until it is linked by some correspondence rules to experimentally identifiable properties of a subject matter.

The Cognitive Status
of Theories

6

The preceding chapter recognized that the distinction between experimental laws and theories is not a sharp one, and that no precisely formulated criterion is available for identifying the statements to be classified under these rubrics. It was nevertheless argued that systems of explanation are achieved with the aid of those types of assumptions designated as theories, which are unmistakably more comprehensive than explanations obtained through the use of other assumptions called experimental laws; and it was urged that for this reason theories merit special attention.

Two features of theories have accordingly been discussed at some length. It was noted, in the first place, that theoretical notions are in general defined only implicitly by fundamental premises of a theory, whether the premises are formulated as abstract postulates or in terms of some model. In the second place, considerable stress was placed in consequence upon the necessity for rules of correspondence to link theoretical ideas with experimental concepts. On the other hand, some care was taken to make clear that the three components mentioned as usually present in a theory (an abstract set of postulates which implicitly define the basic terms of the theory, a model or interpretation for the postulates, and rules of correspondence for terms in the postulates or in the theorems derived from them) are not to be construed as separate items, introduced in succession at various stages in the actual construction of theories, but simply as features that can be isolated for purposes of analysis. It is in fact often quite difficult to state fully and with precision the abstract postulates, freed of all interpretations, that are embedded in a theory, or to formulate in detail the tacitly used rules of correspondence. Most theories, at any rate, are generated within the matrix of some model and are codified, with at best only casual mention of any rules of correspondence, in terms of an interpretation for their fundamental premises.

The account of theories presented thus far is nevertheless incomplete in at least two important respects. Enough has perhaps been said to make clear what is to be understood by a model (or interpretation) for a theory. However, there has been only negligible discussion of the rationale for having models, or of the role played by models in the construction of theories and in the expansion of their range of application. Furthermore, the point has been emphasized that rules of correspondence do not, in general, associate every theoretical concept employed in a theoretical explanation with some experimental notion. Nothing has been said, however, about the import of this fact for the much-debated issue of the cognitive status of theories, and in particular for the commonly held view that, since theories familiarly occur as premises in explanations, theories are assumptions into whose truth or falsity it is appropriate to inquire. It is to the discussion of these two groups of questions that the present chapter is devoted.

I. *The Role of Analogy*

The claim that a genuinely satisfactory scientific explanation must "reduce" the unfamiliar to the already familiar was judged in Chapter 3 to be dubious when taken at its face value. It was also acknowledged, however, that if the claim is properly interpreted it is not without merit but asserts a generally sound point. The suggestion was briefly made that explanations can be regarded as attempts at understanding the unfamiliar in terms of the familiar, insofar as the construction and development of explanatory systems are controlled, as they frequently are, by a desire to find and exploit structural analogies between the subject matter under inquiry and already familiar materials. This suggestion must now be amplified, and some types of analogies which may influence the construction as well as the subsequent use of theories must be examined.

Common speech is full of expressions that initially were employed in a more or less conscious metaphorical sense, though many of them have well-nigh lost their original meanings and are currently used in what is in effect a literal manner. It rarely occurs to us today when we employ the phrase "to foot the bill," for example, that at one time it expressed a felt similarity between the sum of a column of figures and the lower extremities of the human body. The widespread use of metaphors, whether they are dead or alive, testifies to a pervasive human talent for finding resemblances between new experiences and familiar facts, so that what is novel is in consequence mastered by subsuming it under established distinctions. In any event, men do tend to employ familiar systems of relations as models in terms of which initially strange domains of experience are intellectually assimilated. This is not always a consciously deliberate process in most contexts of experience. Similarities between the new and the old are often only vaguely apprehended without being carefully articulated. Moreover, little if any attention is generally paid to the limits within which such felt resemblances are valid. Accordingly, when familiar notions are extended to novel subject matters on the basis of unanalyzed similarities, serious error can easily be committed. Animistic explanations of physical events are well-known illustrations of such unwarranted extensions of conceptions from a domain in which their use is legitimate to domains in which it is not. Even in modern natural science words like 'force,' 'law,' and 'cause' are occasionally still used with decided anthropomorphic overtones that are echoes of their origins. Nonetheless, apprehensions of even vague similarities between the old and the new are often starting points for important advances in knowledge. When reflection becomes critically self-conscious, such apprehensions may come to be developed into carefully formulated analogies and hypotheses that can serve as fruitful instruments of systematic research.

In any event, the history of theoretical science supplies plentiful examples of the influence of analogies upon the formation of theoretical ideas; and a number of outstanding scientists have been quite explicit about the important role models play in the construction of new theories. For example, Huygens developed his wave theory of light with the help of suggestions borrowed from the already familiar view of sound as a wave phenomenon; Black's experimental discoveries concerning heat were suggested by his conception of heat as a fluid, and Fourier's theory of heat conduction was constructed on the analogy of the known laws of the flow of liquids; the kinetic theory of gases was modeled on the behavior of an immense number of elastic particles, whose motions conform to the established laws of mechanics; the conception of a potential function, first developed in the mechanics of point-masses, was extended by analogy into theories of hydrodynamics, thermodynamics, and electromagnetism; and nineteenth-century theories of electricity and magnetism were built in analogy to the mechanics of stresses and strains in an elastic solid. In each of these examples, as in many others that could be mentioned, the model served both as a guide for setting up the fundamental assumptions of a theory, as well as a source of suggestions for extending the range of their application.

Perhaps no scientist of first rank has been as clearly aware as was Maxwell of the place of analogies in the conduct of physical research and in the formulation of theories. In the opening remarks to his paper in which he first proposed a mathematical formulation of Faraday's ideas on lines of force, Maxwell gave an instructive account of the way in which analogies can be exploited in science. He described a "physical analogy" as "that partial similarity between the laws of one science and those of another which makes each of them illustrate the other." He noted, for example, that the change in the direction of light when it passes from one medium to another is identical with the altered direction of a particle when it passes through a narrow opening in which strong forces are acting. Although the analogy holds only for the direction but not for the speed of motion, he nevertheless regarded the analogy to be useful "as an artificial method" for the solution of a certain class of problems.[1] Maxwell also cited the analogy to which William Thomson (later Lord Kelvin) first called attention, between the theory of gravitation and the theory of heat conduction. Maxwell explained that

> the laws of conduction of heat in uniform media appear at first sight among the most different in their physical relations from those relating to attractions. The quantities which enter into them are *temperature, flow of heat, conductivity*. The word *force* is foreign to the subject. Yet we find that the mathematical laws of the uniform motion of heat in homogeneous media are identical in form with those of attractions varying inversely as the square of the distance. We have only to substitute *source of heat* for *centre of attraction, flow of heat* for *accelerating effect of attraction* at any point, and *temperature* for *potential,* and the solution of a problem in attractions is transformed into that of a problem in heat.

He therefore observed that

> the conduction of heat is supposed to proceed by an action between contiguous parts of a medium, while the force of attraction is a relation between distant bodies, and yet, if we know nothing more than is expressed in the mathematical formulae, there would be nothing to distinguish between the one set of phenomena and the other.·

The two subjects do indeed assume quite different aspects if additional facts are introduced. Nevertheless, Maxwell believed that the resemblance in mathematical form between some of the laws for these distinct subjects is useful "in exciting appropriate mathematical ideas." [2] He then

---

[1] James Clerk Maxwell, "On Faraday's Lines of Force," in *The Scientific Papers of James Clerk Maxwell*, Vol. 1, p. 156.

[2] *Ibid.*, p. 157.

went on to say that it was through the use of analogies of this kind that he developed his mathematical representation of the phenomena of electricity, employing as the model for this purpose the mathematical analysis of the motions of incompressible fluids.

The preceding illustrations and Maxwell's discussion suggest a classification of analogies into two broad types, which we may call "substantive" and "formal" analogies. In analogies of the first kind, a system of elements possessing certain already familiar properties, assumed to be related in known ways as stated in a set of laws for the system, is taken as a model for the construction of a theory for some second system. This second system may differ from the initial one only in containing a more inclusive set of elements, all of which have properties entirely similar to those in the model; or the second system may differ from the initial one in a more radical manner, in that the elements constituting it have properties not found in the model (or at any rate not mentioned in the stated laws for the model).

The various atomistic theories of matter illustrate the exploitation of this type of analogy. The fundamental assumptions of the kinetic theory of gases, for example, are patterned on the known laws of the motions of macroscopic elastic spheres, such as billiard balls. Similarly, part of electron theory is constructed in analogy to established laws of the behavior of electrically charged bodies. In this type of analogy, the system employed as a model is frequently a set of visualizable macroscopic objects. Indeed, when physicists speak of a model for a theory, they almost always have in mind a system of things differing chiefly in size from things that are at least approximately realizable in familiar experience, so that in consequence a model in this sense can be represented pictorially or in imagination.

In the second or formal type of analogy, the system that serves as the model for constructing a theory is some familiar structure of abstract relations, rather than, as in substantive analogies, a more or less visualizable set of elements which stand to each other in familiar relations. Mathematicians frequently employ such formal models in developing some new branch of their subject. A simple example is provided by the way rules are stated for manipulating fractional and negative exponents in algebra. These rules are so specified that the laws for operating with such exponents are formally the same as the laws for positive integral exponents. Thus, since $a^3 \times a^2 = a^{3+2}$, and $(a^3)^2 = a^{2\times 3}$, we also have $a^{-5} \times a^{\frac{2}{3}} = a^{-5+\frac{2}{3}}$ and $(a^{-5})^{\frac{2}{3}} = a^{\frac{2}{3} \times -5}$; and in general, $a^m \times a^n = a^{m+n}$, and $(a^m)^n = a^{n \times m}$, no matter whether m and n are positive, negative, integral, or fractional. Indeed, formally identical laws are obtained for irrational and complex exponents as well. The example cited is perhaps trivial. It nevertheless illustrates an important procedure that has been widely used in creating new areas of mathematical study—in the construction of geometries for n-dimensional "spaces," of many branches of higher algebra, of parts of modern function theory, and much else.

Formal models play a no less important role in mathematical physics. Maxwell's example of the identity in structure of the mathematics of gravitation theory and the equations of heat conduction is a case in point. More recent examples are supplied by the articulation of relativity theory and quantum mechanics, in both of which patterns of relations have been introduced in close analogy to important equations of classical mechanics. According to Newtonian mechanics, for example, the linear momentum of an isolated system remains constant, where the momentum is by definition the sum of the product of the mass and velocity for each body in the system, and the mass of a body is assumed to be independent of its velocity. But experiments in the early twentieth century showed that the mass of a particle moving with great speed varies with the velocity, so that the principle of the conservation of momentum does not appear to hold for such particles, and the notion of "mass" was appropriately redefined in relativity theory. In consequence, a principle formally like the classical one can be asserted even for bodies with high velocities. More specifically, the notion of "relativistic mass" was introduced in such a way that the relativistic mass of a body is a function jointly of the velocity of the body, of its "rest mass" (its mass at zero velocity), and of the velocity of light.[3] Nevertheless, though the relativistic mass of a body is not independent of its velocity, the relativistic mass (like the Newtonian mass) is equal to the ratio of the force acting on the body and its acceleration. Moreover, when the conservation principle is reformulated in terms of relativistic mass, it is in agreement with experimental finding. In short, a new notion of mass and a correspondingly new conservation principle of momentum were introduced into relativity theory under the guidance of a formal analogy. The example illustrates how the mathematical formalism of one theory can serve as a model for the construction of another theory with a more inclusive scope of application than the original one. In consequence, the old theory turns out to be a special case of the new one, while the new theory exhibits features that are "continuous" (because formally identical) with certain fundamental assumptions of the old.[4]

[3] The relativistic mass m of a body is given by the formula $m = m_0 / \sqrt{1 - v^2/c^2}$, where m_0 is the rest mass, v the velocity of the body, and c the velocity of light.

[4] The Schrödinger equation in quantum mechanics is another striking illustration of the use of formal analogies. According to the Hamiltonian form of the equations of motion in classical mechanics, the total energy W of a system is equal to the sum of

Thus far, attention has been exclusively directed to the role of models in articulating a new theory. It would be a mistake to conclude, however, that once the new theory has been formulated the model has played its part and has no further function in the use made of the theory. In the first place, the task of the theoretical scientist is not completed when he has simply formulated the main assumptions of a theory. Those assumptions must be explored for consequences that may lead to the systematic explanation of diverse experimental laws, for suggestions concerning directions to be followed in fresh areas of experimental inquiry, and for clues as to how the formulations of experimental laws need to be modified so as to enlarge the scope of their valid application. As long as experimental knowledge is incomplete and a theory continues to be fruitful as a guide to further research, these are tasks that are

the kinetic energy T and the potential energy V, so that

$$H(p,q) = T(p) + V(q) = W$$

where p is the momentum and q the position of a particle. For a single particle, this yields:

$$H(p,q) = p^2/2m + V(q) = W$$

Schrödinger's equation is obtained by replacing the momentum p by the differential operator $\dfrac{h}{2\pi i}\dfrac{\partial}{\partial q}$ and W by $-\dfrac{h}{2\pi i}\dfrac{\partial}{\partial t}$, and introducing the function $\psi(q,t)$ on which these operators are to be applied. We then obtain:

$$H\left[\left(\frac{h}{2\pi i}\right)\left(\frac{\partial}{\partial q}\right)q\right]\psi(q,t) = -\left[\left(\frac{h^2}{8\pi^2 m}\right)\left(\frac{\partial^2\psi}{\partial q^2}\right)\right] + V\psi = -\left(\frac{h}{2\pi i}\right)\left(\frac{\partial\psi}{\partial t}\right)$$

The following comment on this equation is of interest in the present context: "It must be recognized that this correlation of the wave equation and the classical energy equation . . . has only formal significance. It provides a convenient way of describing the system for which we are setting up a wave equation by making use of the terminology developed over a long period of years by the workers in classical dynamics. Thus our store of direct knowledge regarding the nature of the system known as the hydrogen atom consists in the results of a large number of experiments—spectroscopic, chemical, etc. It is found that all the known facts about this system can be correlated and systematized (and, we say, explained) by associating with this system a certain wave equation. Our confidence in the significance of this association increases when predictions regarding previously uninvestigated properties of the hydrogen atom are subsequently verified by experiment. We might then describe the hydrogen atom by giving its wave equation; this description would be complete. It is unsatisfactory, however, because it is unwieldy. On observing that there is a formal relation between this wave equation and the classical energy equation for a system of two particles of different masses and electric charges, we seize on this as providing a simple, easy, and familiar way of describing the system, and we say that the hydrogen atom consists of two particles, the electron and proton, which attract each other according to Coulomb's inverse-square law. Actually we do not know that the electron and proton attract each other in the same way that two macroscopic electrically charged bodies do, inasmuch as the force between the two particles in the hydrogen atom has never been directly measured. All that we know is that the wave equation for the hydrogen atom bears a certain formal relation to the classical dynamical equations for a system of two particles attracting each other in this way."—Linus C. Pauling and E. Bright Wilson, *Introduction to Quantum Mechanics*, New York, 1935, pp. 55-56, quoted by kind permission of the publishers. McGraw-Hill Book Company, Inc.

never finally done; and in all these tasks models continue to play important roles. In the historical development of the kinetic theory of gases, for example, the model for the theory suggested questions about the ratios of molecular diameters to the distances between the molecules, about various kinds of forces between the molecules, about the elastic properties of molecules, about the distribution of the velocities of the molecules, and so on. Such questions would perhaps have never been raised had the theory been formulated as an uninterpreted set of postulates. But in any case, these questions led to the deduction of a variety of consequences from the theory, some of which served as hints for reformulating experimental gas laws and for recognizing new ones. The model provided by the abstract pattern of relations associated with Newtonian mechanics served a similar function in the development of the nineteenth-century theories concerning the transmission of light through a hypothetical ether.[5] More generally, a model may be heuristically valuable because it suggests ways of expanding the theory embedded in it.

But in the second place, models for a theory also have a use in suggesting at what points rules may be introduced for establishing correspondences between theoretical and experimental notions. Were a theory stated as a set of uninterpreted postulates, exhibiting not even a formal analogy to some already familiar system of abstract relations, the formulation would provide no clues as to how the theory could be applied to concrete physical problems. The example of such an abstract calculus stated in the preceding chapter makes evident the difficulties that almost anyone would encounter in making fruitful uses of that calculus if he were totally unfamiliar with any model for the postulates. But although, as already noted, a model does not by itself establish correspondence rules for the terms of a calculus, it can often suggest which theoretical terms may be associated with experimental ideas. For example, the usual interpretation of the postulates for the kinetic theory of gases leads quite naturally to the association of the theoretical expression 'change in the total momentum of the molecules striking a unit surface' with the experimental notion of pressure; similarly, the model suggests that the theoretical expression 'product of the mass of each molecule and the total number of molecules' can be made to correspond to the experimental notion of the mass of a gas; and so on. Again, the interpretation of optical theory in terms of waves propagated in a medium invites the association of theoretical expressions referring to the amplitude of waves in the model with the intensity of the illumination; the wave interpretation also suggests the linking of theoretical expressions referring to the interference of waves with the dark lines (or absence of illumination) observed in certain experimentally generated patterns of light and shadow. Finally, the Bohr model of the atom suggests that those expressions in the mathematical formalism of the theory which are interpreted as representing electron jumps should be made to correspond to experimentally determinable spectral lines. Examples of this function of models can be supplied almost without limit, but the few illustrations cited suffice to show that, even after the various ideas of a theory have been formulated with the help of a model, the model continues to be of important service in both the extension and the application of the theory.

Emphasis has thus far been placed on the heuristic value of models in the construction and use of theories. The fact should not be overlooked, however, that models also contribute to the achievement of inclusive systems of explanation. A theory that is articulated in the light of a familiar model resembles in important ways the laws or theories which are assumed to hold for the model itself; and in consequence the new theory is not only assimilated to what is already familiar but can often be viewed as an extension and generalization of an older theory which had a more limited initial scope. From this perspective an analogy between an old and a new theory is not simply an aid in exploiting the latter but is a desideratum many scientists tacitly seek to achieve in the construction of explanatory systems. Indeed, some scientists have made the existence of such an analogy an explicit and indispensable requirement for a satisfactory theoretical explanation of experimental laws.[6] And conversely, even when a new theory does organize systematically a vast array of experimental fact, the lack of marked analogies between the theory and some familiar model is sometimes given as the reason why the new theory is said not to offer a "really satisfactory" explanation of those facts. Lord Kelvin's inordinate fondness for mechanical models is a notorious example of such an attitude; he never felt entirely at ease with Maxwell's electromagnetic theory of light because he was unable to design a satisfactory mechanical model for it. More recently, a distinguished physicist has argued that a theory for which no visualizable models can be given is just as good as one for which such models are available, provided that both theories enable us to handle experimental problems equally well; and he has made clear that in this latter respect the mathematical formalism of current quantum theory, for which no satisfactory model of this kind is known, is unusually successful. Nevertheless, he has also registered the uncomfortable sense of loss, shared by many physicists, because quantum theory offers no "explanation" of the experimental facts—a feeling he attributes to the circumstance that we can construct for the theory no physical model in which the "inter-

[6] Cf. Norman R. Campbell, *Physics, the Elements,* Cambridge, England, 1920, pp. 129-30.

play of elements [is] already so familiar to us that we accept them as not needing explanation." [7] It is a matter of historical record that there are fashions in the preferences scientists exhibit for various kinds of models, whether substantive or purely formal ones. Theories based on unfamiliar models frequently encounter strong resistance until the novel ideas have lost their strangeness, so that a new generation will often accept as a matter of course a type of model which to a preceding generation was unsatisfactory because it was unfamiliar. What is nevertheless beyond doubt is that models of some sort, whether substantive or formal, have played and continue to play a capital role in the development of scientific theory.

The formulation of a theory in terms of some model is nevertheless not free from dangers, and a model may be a potential intellectual trap as well as an invaluable intellectual tool. The chief dangers are twofold: some inessential feature of a model (especially a substantive model) may be mistakenly assumed to constitute an indispensable feature of the theory embedded in it; and the model may be confused with the theory itself. In consequence, the exploitation of the theory may be routed into unprofitable directions, and the pursuit of pseudo-problems may distract attention from the operative significance of the theory. Thus, the emission theory of light was built on the image of projectiles moving along a uniformly homogeneous straight line; and there is some reason to think that this image delayed the discovery of the periodicity of light. On the other hand, the wave theory of light was initially based on the model of sound waves; and the conception that light, like sound, is a longitudinal wave motion apparently stood in the way of further extensions of the undulatory theory of light for about a century, until with the adoption of a different model light waves were assumed to be transversal. Again, the sensation of strain in muscular effort was the original of the notion of force; and this model became the source of so many misconceptions that much effort has been required to rid the notion of its anthropomorphic associations. Similarly, some of the difficulties encountered in understanding current quantum theory are in part a consequence of the use of a particle model in stating the theory. The particles envisaged in the model are "classical" particles, each with a determinate position and velocity at any given instant. But, according to the theory, determinate "positions" and "velocities" cannot be simultaneously assigned to the subatomic "particles" postulated by the theory. These theoretical "particles" are therefore not classical particles, so that the model fails at this point to be of aid and is, on the contrary, a frequent source of misconceptions concerning the import of quantum theory.

It must nevertheless be acknowledged that there is no way of telling

[7] P. W. Bridgman, *The Nature of Physical Theory*, Princeton, 1936, p. 63.

in advance whether a given model will prove to be an obstacle to the fruitful development of theory, since it is usually only after a model has been tried that one can tell which of its features suggest inquiries leading into blind alleys and which are heuristically valuable. The only point that can be affirmed with confidence is that a model for a theory is not the theory itself. In consequence, the adequacy of a theory as an instrument for systematic explanation and prediction cannot be taken without further scrutiny to establish the physical reality of every aspect of the substantive model in terms of which the theory may be interpreted. This is obvious when several models are known for the same theory, but it is equally true when only a single model is available.[8] For ex-

---

[8] Henri Poincaré gave a famous proof that, if one mechanical explanation for a phenomenon can be given, an infinity of others can also be constructed. The proof consists in noting that the number of equations relating the coordinates of position and momentum of the masses in the hypothetical model with the experimentally determinable parameters of the phenomenon is greater than the number of such parameters. It then follows that the coordinates of the model can be chosen at will, subject only to the requirement that they satisfy some assumed law for them which is consistent with the equations. In detail, the argument is as follows: Let the parameters which can be determined experimentally and which specify the phenomenon under investigation be $q_1, q_2, \ldots, q_n$. These parameters are related to one another and to the time t by laws which we may suppose can be expressed as differential equations. Now suppose that there is a model consisting of a very large number p of molecules, whose masses are m_i and coordinates of position are x_i, y_i, z_i ($i = 1, 2, \ldots, p$). We assume that the principle of conservation of energy holds for the model, so that there is a potential function V of the $3p$ coordinates x_i, y_i, z_i; the $3p$ equations of motion for the molecules will then be

$$m_i \frac{d^2 x_i}{dt^2} = - \frac{\partial V}{\partial t}$$

with similar equations for y and z, while the kinetic energy of the system will be

$$T = \tfrac{1}{2} \Sigma m_i (\dot{x}^2 + \dot{y}^2 + \dot{z}^2)$$

so that

$$T + V = \text{constant}$$

The phenomenon will then have a mechanical explanation if we can determine the potential function V and can express the $3p$ coordinates x, y, z as functions of the parameters q.

But if we assume there are such functions, so that

$$x_i = \phi_i(q_1, \ldots, q_n)$$
$$y_i = \psi_i(q_1, \ldots, q_n)$$
$$z_i = \theta_i(q_1, \ldots, q_n)$$

the potential function V can be expressed as a function of the q_i alone, the kinetic energy T will be a homogeneous quadratic function of the q_i and their first derivatives $\dot{q}_i$, and the laws of motion of the molecules can be expressed by the Lagrangian equations:

$$\frac{d}{dt}\left(\frac{\partial T}{\partial \dot{q}_k}\right) - \frac{\partial T}{\partial \dot{q}_k} + \frac{\partial V}{\partial q_k} = 0 \quad (k = 1, 2, \ldots, n)$$

Accordingly, the necessary and sufficient condition for a mechanical explanation of the

ample, the interpretation of electromagnetic theory, proposed during the nineteenth century, in terms of mechanical stresses and vortices in a luminiferous ether, was in general not equated even by physicists of the day with the actual content of that theory. On the contrary, despite the acknowledged success of the theory in explaining a variety of experimental laws and in predicting accurately an extensive class of phenomena, in the judgment of leading physicists the "physical reality" of the ether was not thereby established.

This last example clearly illustrates the point that, though the evidence for a theory may be overwhelming, the evidence may nevertheless not be assessed as sufficient for asserting the physical existence of various elements in the substantive model in terms of which the theory is formulated. But the example also invites consideration of the question whether theories can be rightly regarded as asserting anything whatsoever, and if so what, and whether it is appropriate to characterize theories as either true or false statements. This is the question to which we now turn.

II. *The Descriptive View of Theories*

The cognitive status of universal statements in general, and of scientific theories in particular, has been the subject of a long and inconclusive debate. The issues raised in the controversy are complex and involve not only highly technical problems of logic and scientific fact but also large philosophical considerations about the nature of meaning and knowledge. An exhaustive treatment of the subject will therefore not be attempted here. Discussion of issues will be centered around three major positions that have been taken on the cognitive status of theories in physical science—on the question whether, and if so in what sense, theories may be viewed as either true or false statements.

According to the first and historically oldest account, a theory is lit-

---

phenomenon is that there are two functions $V(q_1, \ldots, q_n)$ and $T(\dot{q}_1, \ldots, \dot{q}_n, q_1, \ldots, q_n)$ satisfying these requirements, with the obvious proviso that the laws of the phenomenon can be transformed so as to take the indicated Lagrangian form. Such functions can be specified, if and only if

$$T(\dot{q},q) = \tfrac{1}{2}\Sigma m_i(\dot{x}_i^2 + \dot{y}_i^2 + \dot{z}_i^2) = \tfrac{1}{2}\Sigma m_i(\phi_i^2 + \psi_i^2 + \theta_i^2)$$

where $$\phi_i = \dot{q}_1\left(\frac{\partial\phi_i}{\partial q_1}\right) + \dot{q}_2\left(\frac{\partial\phi_i}{\partial q_2}\right) + \cdots + \dot{q}_n\left(\frac{\partial\phi_i}{\partial q_n}\right)$$

and similarly for ψ_i and θ_i.

But since the number p can be taken as large as one pleases, this condition can always be satisfied, and indeed in an infinite number of different ways.—Paraphrased from H. Poincaré, *Electricité et Optique*, Paris, 1890, pp. ix-xiv.

erally either true or false; and, although a theory can at best be established only as "probable," it is as significant to ask whether a theory is true or false as it is to ask a similar question about a statement concerning some individual matter of fact, such as the statement 'Krakatoa was destroyed by a volcanic eruption in 1883.' A corollary often drawn from this view is that when a theory is well supported by empirical evidence, the objects ostensibly postulated by the theory (e.g., atoms, in the case of an atomic theory) must be regarded as possessing a physical reality at least on par with the physical reality commonly ascribed to familiar objects such as sticks and stones.

A second (and historically the youngest) position on the cognitive status of theories maintains that theories are primarily logical instruments for organizing our experience and for ordering experimental laws. Although some theories are more effective than others for attaining these ends, theories are not statements, and belong to a different category of linguistic expressions than do statements. For theories function as rules or principles in accordance with which empirical materials are analyzed or inferences drawn, rather than as premises from which factual conclusions are deduced; and they cannot therefore be usefully characterized as either true or false, or even as probably true or probably false. However, those who adopt this position do not always agree in their answers to the question whether physical reality is to be assigned to such theoretical entities as atoms.

Finally, the third stand on the cognitive status of theories is a sort of halfway position between the other two. According to it, a theory is a compendious but elliptic formulation of relations of dependence between observable events and properties. Although the assertions of a theory cannot be properly characterized as either true or false when they are taken at face value, a theory can nevertheless be so characterized insofar as it is translatable into statements about matters of observation. Proponents of this position usually maintain, therefore, that in the sense that a theory (such as an atomic theory) can be said to be true, theoretical terms like "atom" are simply a shorthand notation for a complex of observable events and traits, and do not signify some observationally inaccessible physical reality.

This third view, which we shall consider first, is associated with the historically influential conception that the sciences never "explain" anything, but merely "describe" in a "simple" or "economical" fashion the succession and concomitance of events. Something has already been said about this conception, but it deserves further discussion. The conception was vigorously espoused by many nineteenth-century scientists in reaction to the development of atomistic theories in physics and chemistry, since these theories appeared to them not only to be unneces-

sary for systematizing the experimental facts but also to assign an unwarranted absolute priority to Newtonian mechanics.[9] Moreover, the
descriptive account of science was espoused by many thinkers who rejected the assumptions of classical rationalism and sought to emancipate
science from any dependence on unverifiable "metaphysical" commitments. In its inception, at any rate, the descriptive thesis was regarded
both as an accurate analysis of the nature of physical science and as
a weapon in the struggle against philosophies that were felt to hinder
the development of science.

As has already been noted, much of the debate on the adequacy of
the descriptive view of science has been verbal, because of the ambiguity of the word "description." The word has a wide spectrum of
meanings, none of which is privileged; and some critics of the descriptive view have apparently never taken to heart Humpty Dumpty's observation to Alice that a word means just what those using it choose to
mean by it. Nevertheless, some of the meanings of the word are easily
confounded, and proponents of the descriptive thesis have not always
distinguished them.[10] However, our present interest is not in these matters, but only in the descriptive view of science insofar as it is construed
as a thesis concerning the translatability of theoretical statements into
statements about observable things.

[9] These were the issues that played a central role in the development of what is
called the "science of energetics."

[10] It will suffice to illustrate two of them that are sometimes not distinguished.
Consider the experimental law that the period of a simple pendulum oscillating
through a small arc is proportional to the square root of its length. If a physicist
were to test the law by performing some experiments, the report of his findings
would probably include at least the following items: an account of the chronometer
employed, of the relevant features of the pendulum used, and of the manner in
which the periods of the pendulum were observed, plus a finite set of numbers,
perhaps represented by points on an accompanying graph, each of which is an
actual measure of a period for a given length of the pendulum. Although the language of the report may be technical and highly condensed, these items in the report
are descriptions in a customary sense of the word.

On the other hand, though the law of the simple pendulum may also be called
a description, it is a description in a somewhat different sense. Thus, the law asserts
a universal association of period and length, not only for periods and lengths of
pendulums actually examined, but for any pendulum whatever. Indeed, even if
pendulums with lengths of 100 feet and 400 feet respectively should never be constructed, the law asserts that the period of the first would be half the period of
the second. Moreover, the law is asserted on the assumption that the weight of the
string supporting the oscillating bob is negligible, and that such matters as the resistance of the air or the friction between the string and the point of suspension are
small enough to be neglected. However, these assumptions may not be realized in
actual experiments on pendulums, so that the law involves a deliberate "idealization"
or schematization of what actually takes place. Accordingly, if the law is said to
be a description, it is a description in a sense different from that in which the report
of an actual experiment is a description. For unlike the report, the law "describes"
something that may never take place.

The most radical form of the descriptive thesis is simply the consistent extension of the phenomenalist theory of knowledge to the materials of the sciences. According to this theory, the psychologically primitive and indubitable objects of knowledge are the immediate "impressions" or "sense contents" of introspective and sensory experience. Moreover, if the postulation of inherently unknowable (because observationally inaccessible) things is to be avoided, all expressions ostensibly referring to such hypothetical objects (which include the physical objects of common sense) must be *defined* in terms of these immediate data. In consequence, every empirical statement containing expressions other than those designating such data (or complexes of such data) must in principle be translatable without loss of verifiable meaning into statements about the succession or coexistence of the allegedly immediate objects of experience. Just as a statement about a nation (e.g., 'Germany invaded France in 1870') can be translated into a set of statements about the behavior of individual human beings, so a statement about the sun (e.g., 'The surface temperature of the sun is 3000° C') is translatable, according to this version of phenomenalism, into a class of statements concerning sense contents.[11]

An allied but in some ways less radical form of the descriptive view of science dissociates itself from the atomistic psychology which so often accompanies phenomenalism, as well as from the assumption that elementary sensory qualities are the ultimate simples into which everything else is to be analyzed. This version of the doctrine accepts the common-

11 This form of phenomenalism has its historical roots in the writings of Berkeley, Hume, and J. S. Mill. Ernst Mach also belongs to this company, at least with respect to one pronounced emphasis in his writings, as do Karl Pearson, Bertrand Russell (in one phase of his development), P. W. Bridgman, and Herbert Dingle. A fairly representative and compact statement of Mach's views is the following: "The world consists of colors, sounds, temperatures, pressures, spaces, times, and so forth, which now we shall not call sensations, nor phenomena, because in either term an arbitrary, one-sided theory is embodied, but simply *elements*. The fixing of the flux of these elements, whether mediately or immediately, is the real object of physical research." —Ernst Mach, *Popular Scientific Lectures*, Chicago, 1898, p. 209. The most complete formulation of Mach's phenomenalistic epistemology is contained in his *Analysis of Sensations*.

Pearson was considerably less subtle than Mach in stating this view, and did not hesitate to accept a straightforward "subjectivistic" formulation of it, which Mach explicitly declined. "There is no better exercise for the mind than the endeavor to reduce the perception we have of 'external things' to the simple sense-impressions by which we know them. . . . Beyond the sense-impressions, beyond the terminals of the sensory nerves, we cannot get. Of what is beyond them, of 'things-in-themselves' . . . we can know but one characteristic . . . [the] capacity for producing sense-impressions. There is no necessity, nay, there is want of logic, in the statement that behind sense-impressions there are 'things-in-themselves' *producing* sense-impressions."—Karl Pearson, *Grammar and Science*, Everyman ed., London, 1937, pp. 60-62. For a more recent formulation and defense of phenomenalism, cf. A. J. Ayer, *Language, Truth and Logic*, 2nd ed., New York, 1950, Chaps. 7 and 8.

sense notion that normally we directly observe sticks and stones and animals, the motions of bodies and the actions of men, and the like. It therefore takes ordinary "gross experience" as the starting point for its analyses, even though it recognizes that judgments based on such experience are frequently erroneous and must be corrected in the light of further reflection. The thesis which this version of the doctrine maintains is that all theoretical statements are in principle translatable, again without loss of meaningful content, into statements of the so-called "physicalistic thing language"—that is, into statements about the observable events, things, properties, and relations of common-sense and gross experience. Accordingly, on this conception of the doctrine also, the claim that theories are simply conveniently compendious descriptions is once more a thesis concerning the translatability of theoretical statements, though this time into the familiar language that formulates the materials of publicly verifiable experience.[12]

However, both versions of the descriptive view as here interpreted encounter serious problems.

1. The first version is beset by the standing difficulty of phenomenalism: that, although it is a thesis about the translatability of theoretical statements into the "language" of sense data, an autonomous language of bare sense contents actually does not exist, nor is the prospect bright for constructing one. As a matter of psychological fact, elementary sense data are not the primitive materials of experience out of which all our ideas are built like houses out of initially isolated bricks. On the contrary, sense experience normally is a response to complex though unanalyzed patterns of qualities and relations; and the response usually involves the exercise of habits of interpretation and recognition based on tacit beliefs

---

[12] This version of the descriptive view is sometimes suggested by Mach. Thus: "The communication of scientific knowledge always involves description, that is, a mimetic reproduction of facts in thought, the object of which is to replace and save the trouble of new experience. Again, to save the labor of instruction and of acquisition, concise, abridged description is sought. This is really all that natural laws are. Knowing the value of the acceleration of gravity, and Galileo's laws of descent, we possess simple and compendious directions for reproducing in thought all possible motions of falling bodies. A formula of this kind is a complete substitute for a full table of motions of descent, because by means of the formula the data of such a table can be easily constructed at a moment's notice without the least burdening of the memory."—*Popular Scientific Lectures,* Chicago, 1898, pp. 192-93. But a more explicit statement of this conception may be found among contemporary writers who subscribe to the doctrine known as "physicalism." Cf. Otto Neurath, "Universal Jargon and Terminology," *Proceedings of the Aristotelian Society,* Vol. 41, pp. 127-48; "Protokollsaetze," *Erkenntnis,* Vol. 3, pp. 204-14; "Radikaler Physikalismus und Wirkliche Welt," *Erkenntnis,* Vol. 4, pp. 346-62. See also Rudolf Carnap, "Testability and Meaning," *Philosophy of Science,* Vol. 3 (1936) and Vol. 4 (1937); however, Carnap has changed his views on a number of points since the publication of this article.

and inferences, which cannot be warranted by any single momentary experience. Accordingly, the language we normally use to describe even our immediate experiences is the common language of social communication, embodying distinctions and assumptions grounded in a large and collective experience, and not a language whose meaning is supposedly fixed by reference to conceptually uninterpreted atoms of sensation.

It is indeed sometimes possible under carefully controlled conditions to identify simple qualities that are directly apprehended through the sense organs. But the identification is the terminus of a deliberate and often difficult process of isolation and abstraction, undertaken for analytical purposes; and there is no good evidence to show that sensory qualities are apprehended as atomic simples except as the outcome of such a process. Moreover, though we may baptize these products by calling them sense data and may assign different labels to different classes of them, the use and meanings of those names cannot be established except by way of directions for instituting processes involving overt bodily activities. Accordingly, the meanings of sense-data terms can be understood only if the distinctions and assumptions of our commerce with the gross objects of experience are taken for granted. In effect, therefore, those terms can be used and applied only as part of the vocabulary of the language of common sense. In short, the "language" of sense data is not an autonomous language, and no one has yet succeeded in constructing such a language. However, if there is indeed no such language, the thesis that all theoretical statements are in principle translatable into the language of pure sense contents is questionable from the outset.

2.   But however this may be, further difficulties emerge in connection with the notion of translatability. In the familiar sense of the word "translatable," a statement in one language is translatable into another language only if there is a statement (or a finite conjunction of statements) in the latter equivalent in meaning (or logically) to the given statement. In this sense translations from one natural language into another are plentiful, despite occasional disagreements on the adequacy of proposed translations. For example, no one who understands French and English will seriously question that the English statement "At constant temperature, the volume of a given mass of a gas is inversely proportional to its pressure" is a translation of the French statement "*A une même température, les volumes occupés par une même masse de gaz sont en raison inverse des pressions qu'elle supporte.*"

Is there any evidence that every statement in science, and in particular every theoretical statement, is translatable in this sense either into a phenomenalistic language or into the language of gross experience? The evidence would be conclusive, if each subject-matter term employed in the sciences were actually introduced by way of an *explicit* definition (or by way of some other variant of substitutive definitions) whose subject-matter expressions all belong to the language of observation. For in that case, all terms in the sciences not occurring in this language would be eliminable in favor of those occurring in it. But in point of fact, as has already been noted, theoretical notions are not introduced in this way, so that neither version of the descriptive view of science is immediately warranted by considerations of actual scientific practice. The question remains whether, despite the facts of actual procedure, theoretical terms cannot in principle be eliminated in consonance with the descriptive thesis.

Proponents of this thesis have attempted to show that the answer is affirmative, and that with the help of various modern logical techniques the eliminations can be effected. These techniques include, among others, the devices associated with Bertrand Russell's definitions-in-use and his notion of incomplete symbols, most of which have found fruitful employment in formal logic and the foundations of pure mathematics. It is nevertheless very doubtful whether the use of these techniques in the analysis of statements in empirical science has thus far yielded any results that substantiate either version of the descriptive thesis. Examples of translations that can be effected with the help of these techniques are rarely taken from the concrete materials of the natural sciences; and even when they are, the translations are not carried through in more than outline. It is difficult to escape the conclusion that the descriptive thesis is not a claim about past achievements and that at best it is only a doubtfully realizable program for future analysis.[18]

3. There is indeed a general consensus that the outlook for establishing the thesis is dim when the word "translatable" is understood in its customary sense. In current discussions, at any rate, that thesis has been considerably weakened. The thesis is asserted not in the form men-

---

[18] Perhaps the most ambitious attempt to establish this thesis within the framework of a phenomenalistic theory of knowledge is to be found in Rudolf Carnap, *Der Logische Aufbau der Welt*, Berlin, 1928. But even here the required definitions of expressions occurring in the natural sciences are only sketched. Carnap has since abandoned not only his earlier phenomenalism but also the thesis that theoretical statements are translatable into a physicalistic language. Cf. his "Methodological Character of Theoretical Concepts," *Minnesota Studies in the Philosophy of Science* (ed. by Herbert Feigl and Michael Scriven), Minneapolis, 1956, Vol. 1. For a more recent effort to carry through Carnap's program, see Nelson Goodman, *The Structure of Appearance*, Cambridge, Mass., 1951. A detailed critique of Russell's attempt to establish the thesis is contained in Ernest Nagel, *Sovereign Reason*, Glencoe, Ill., 1954, Chap. 10.

tioned above, but in the sense that for every theoretical statement there is a *class* of observation statements which is logically equivalent to the given statement, thus leaving it open whether the class is finite or not. The point of this emendation, and the import of its consequences, will be evident from an example. Let us assume that the expression 'electric current' is a theoretical term, for which appropriate rules of correspondence have been established. It would then be generally acknowledged that the statement 'There is now an electric current in this wire' (asserted at a given time for a given wire) is not equivalent in content to, say, the conditional observation statement 'If the galvanometer on that shelf were introduced into this circuit, the pointer of the instrument would be deflected from its present position.' The equivalence does not obtain for at least two reasons. On the assumption that the theoretical statement implies anything at all about the behavior of any galvanometer, it implies not only a single statement about a particular galvanometer but an indefinitely large class of similar statements about all other such instruments. Accordingly, if the original statement about a wire is at all equivalent to statements about the behavior of galvanometers, the statement must be equivalent to an indefinitely large (perhaps infinitely large) class of them.

In the second place, the presence of an electric current in the wire is associated with observable phenomena other than the behavior of galvanometers. As is well known, optical, thermal, chemical, and other magnetic phenomena could also be used as evidence upon which to decide whether or not the wire is carrying a current. In consequence, the class of statements which is supposedly equivalent to the theoretical statement must also include statements about these additional phenomena as well. On the other hand, it is difficult to fix the membership of that supposed class, and it is certainly not possible to specify that membership once and for all and in detail. For we cannot foresee the experimental discoveries that may be made in the future, some of which may provide still further ways (at present unsuspected) of detecting the presence of a current in a wire. In consequence, statements about these still unknown but hypothetically relevant phenomena must also be included in the class equivalent to the theoretical statement, so that the variety and number of such member statements may be greater than those we can specify at any given time. Accordingly, the indicated emendation of the translatability thesis is consonant with the possibility that this supposed class is not only infinitely numerous but is also incapable of being definitely specified.[14]

---

[14] The emendation to permit the translation of a theoretical statement into an infinite class of other statements has been partly inspired by an analogous procedure in mathematics. It is therefore instructive to see how this procedure operates in

Whether a possibly endless procedure of specifying a supposedly equivalent but otherwise indefinite class of observation statements is to be called a process of "translation" for a theoretical statement is only a verbal issue. Such a procedure is in any event different from what is ordinarily understood by "translation" and from the sense of the word with which this discussion began. For though the class of observation statements into which a scientific theory is thus "translatable" is by postulation logically equivalent to the latter, it is a class whose members can never be completely ascertained, either in respect to their variety or in respect to their number.

4. A distinction, relevant to the present discussion, is sometimes drawn between two types of theories. The distinction was apparently first formulated explicitly in 1855 by W. J. M. Rankine, one of the founders of a school of physics which sought to develop thermodynamics as the basis for a unified system of natural science (called the "science of energetics"). Rankine declared that there are two methods of framing a physical theory. Theories formed by what he called the "abstractive" method allegedly formulate relations between properties common to classes of objects or phenomena "perceived by the senses" and do not postulate anything "hypothetical" or conjectural. Examples of such theories (variously designated as "abstractive," "phenomenological," or "macroscopic") are Newtonian mechanics and gravitation theory, Fourier's theory of heat conduction, and classical thermodynamics. Theories formed by the second or "hypothetical" method assert relations between hypothetical entities that are "not apparent to the senses"; and their empirical validity can be judged only indirectly, in terms of the agreement of their consequences with the results of observation and experiment. Familiar examples of such theories (for which "hypothetical,"

that domain. It can be shown in detail that statements about real numbers are translatable into statements about infinite classes of rational numbers. For example, the real number $\sqrt{2}$ can be defined as the set of rational numbers x such that $x^2 < 2$, the real number $\sqrt{3}$ as the set of rationals y such that $y^2 < 3$, and the real number $\sqrt{6}$ as the set of rationals z such that $z^2 < 6$. Furthermore, the product $\sqrt{2} \times \sqrt{3}$ is defined as the set of all rational numbers w such that $w = xy$, where x is a rational number with $x^2 < 2$ and y a rational number with $y^2 < 3$. The statement that $\sqrt{2} \times \sqrt{3} = \sqrt{6}$, which is about real numbers, can then be translated into a statement about infinite classes of rational numbers: "The set of all rational numbers each of which is the product of some rational whose square is less than 2 and of some rational whose square is less than 3 is identical with the set of all rational numbers whose squares are less than 6." It is clear, however, that in this case the infinite classes are completely determined, so that in this respect there is a marked difference between the mathematical example and the proposed translation of theoretical statements into a class of observational statements. The mathematical model is not an adequate guide to the analysis of theoretical statements in empirical science.

"transcendent," and "microscopic" are frequently used labels) are the molecular theory of gases, the wave theory of light, and the various atomic theories of chemical interaction. Newton's famous dictum *"Hypotheses non fingo"* is often understood to mean that he declined to entertain theories of this type. Rankine recognized the heuristic value of hypothetical theories, but he regarded their use as constituting only a preliminary stage in the development of abstractive ones. For he believed that abstractive theories possess distinctive advantages over hypothetical ones, in their freedom from assumption about "occult" components of physical phenomena, in their capacity to attain "that degree of certainty which belongs to observed facts," and in their greater promise of bringing "all branches of physics into one system." [15]

The subsequent history of physics has not confirmed Rankine's claims concerning the superior merits of abstractive theories. Indeed, the impressive successes of atomistic theories of matter in predicting new phenomena and in unifying systematically large parts of physics and chemistry have persuaded many distinguished scientists that one must turn from abstractive to microscopic theories for a "deeper" understanding of physical phenomena and for more adequate conceptions of "how things really work." [16] Nevertheless, proponents of the descriptive view of science generally regard abstractive theories as the ideal form of scientific theory, on the assumption that the translatability thesis holds for theories of this type even if not for microscopic theories. [17] It is therefore

[15] W. J. M. Rankine, "Outlines of the Science of Energetics," *Miscellaneous Scientific Papers*, London, 1881, pp. 209-28, first published in *Proceedings of the Philosophical Society of Glasgow*, Vol. 3, No. 6.

[16] Cf. Georg Joos, *Theoretical Physics*, New York, 1934, p. 457, and Enrico Fermi, *Thermodynamics*, New York, 1937, p. x. For a very informative presentation of much interesting material supporting this claim, see Emile Meyerson, *Identity and Reality*, New York, 1930.

[17] For example, Ernst Mach stated this position very explicitly in his *History and Root of the Principle of the Conservation of Energy* (Chicago, 1911, first published in German in 1872): "In the investigation of nature, we have to deal only with knowledge of the connexion of appearances with one another. What we represent to ourselves behind the appearances exists *only* in our understanding, and has for us only the value of a *memoria technica* or formula, whose form, because it is arbitrary and irrelevant, varies very easily with the standpoint of our culture" (p. 49). See further pp. 54-55, and also the statement: "The ultimate unintelligibilities [i.e., "the simplest facts to which we reduce the more complicated ones"] on which science is founded must be facts, or, if they are hypotheses, must be capable of becoming facts. If the hypotheses are so chosen that their subject (Gegenstand) can never appeal to the senses and therefore also can never be tested, as in the case of the mechanical molecular theory, the investigator has done more than science, whose aim is facts, required of him—and this work of supererogation is an evil. . . . In a complete theory, to all the details of the phenomena details of the hypothesis must correspond, and all rules for these hypothetical things must also be directly transferable to the phenomena. But then molecules are merely a valueless image" (p. 57).

desirable to examine briefly how the two types differ, and to evaluate the claim that the translatability thesis is sound for at least one of them.

That there is a *prima facie* difference between the two types is undeniable. For example, an abstractive theory such as Newtonian mechanics and gravitational theory (asserted for macroscopic objects) seems on the face of it to postulate no "hidden" conjectural mechanisms such as the molecular theory of heat so obviously does, and to be "closer" to the facts of observation and experiment than is the molecular theory. It would be an error to conclude, however, that Newtonian theory is not really a "theory" in the sense discussed in the previous chapter, and that it is really a set of experimental laws. For the fundamental notions of Newtonian mechanics are not experimental ideas, even though they are suggested by, and correspond to, experimental ideas; and they are only implicitly defined by the postulates of the theory. This is patent in the case of the notions of absolute space and absolute time, which are basic to Newton's formulation of the theory and which he distinguished sharply from the experimental ideas of relative space and relative time. But the point is also valid for other terms employed in Newton's theory, such as 'point-mass,' 'instantaneous velocity,' 'instantaneous acceleration,' and 'force.' Thus, when the expression 'instantaneous velocity of a point-mass' is strictly construed, it refers to the limit of an *infinite series* of ratios, so that the instantaneous velocity of a point-mass cannot be determined by overt experimental means.[18] The point is equally confirmed when other standard examples of abstractive theories are analyzed, such as Fourier's theory of heat conduction or classical thermodynamics. Accordingly, abstractive theories share with hypothetical theories those features that distinguish theories from experimental laws.

The difference between abstractive and hypothetical theories appears to lie elsewhere.[19] Whether or not a hypothetical theory is interpreted by way of some visualizable model, not all of its fundamental terms are associated by correspondence rules with experimental notions. On the other hand, each postulationally defined term in an abstractive theory seems to be coordinated by such rules with some experimental idea. Thus, Fourier's theory of heat conduction is formulated by a partial differential equation which contains, in mathematical notation, the fol-

[18] A more detailed account of the fundamental notions of mechanics will be found in the succeeding chapter.

[19] This difference is suggested by the penultimate sentence in the quotation from Mach in footnote 17. It has been elaborated independently by Norman R. Campbell in his *Physics, the Elements,* Chap. 6, as well as in his *What Is Science?* London, 1921, Chaps. 5 and 7. For a somewhat similar analysis, but one developed against the background of a different philosophy, cf. Henry Margenau, *The Nature of Physical Reality,* New York, 1950, Chap. 5.

lowing expressions: the 'coordinates of an arbitrary point in an infinitely long slab,' 'time,' 'temperature at a point,' 'density at a point,' 'thermal conductivity,' and 'specific heat.'[20] Each of these theoretical terms corresponds to an experimental notion. Similarly, Newtonian gravitational theory employs the ideas of mass, distance, time, and instantaneous acceleration, each of which is associated with some experimentally determinable magnitude.

It is this circumstance which gives to abstractive theories the appearance of being simply experimental laws, and which makes it relatively easy to provide a visualizable model for them. Moreover, in the past abstractive theories have in general been framed in close analogy to experimental laws previously established in limited fields of inquiry. For example, experimental studies on the conduction of heat preceded Fourier's analytical theory of heat; and the *experimental* ideas and laws which were first developed eventually suggested the *theoretical* notions and the mathematical form of the theory. A similar historical connection holds between other abstractive theories (such as Newtonian mechanics or Maxwell's electromagnetic field theory) and the findings of prior experimental investigations. Nevertheless, despite such strong analogies between abstractive theories and experimental laws, the analogies do not support the claim that those theories are simply experimental laws, for reasons already stated.[21]

Accordingly, abstractive and hypothetical theories are in the same boat, as far as their translatability into the language of observation is concerned. In any event, no one has yet succeeded in showing how either type of theory can be so translated, even in principle; and the translatability thesis remains for both of them, not a description of the established nature of any actual theory but a highly debatable program for analyzing theoretical statements. It follows that, on the view concerning the cognitive status of theories which we have been considering, truth and falsity cannot properly be predicated of any current physical theory—at least not until its alleged translatability into observational language is established. In effect, therefore, the view under discussion coincides

[20] The equation is:

$$\rho c \left(\frac{\partial \theta}{\partial t} \right) = \lambda \left(\frac{\partial^2 \theta}{\partial x^2} + \frac{\partial^2 \theta}{\partial y^2} + \frac{\partial^2 \theta}{\partial z^2} \right)$$

where ρ is the density, c the specific heat, θ the temperature, t the time, λ the thermal conductivity, and x, y, and z the coordinates of a point.

[21] A vigorous critique of the claim that abstractive theories introduce no "hypothetical" or conjectural assumptions, and a defense of the view that abstractive and hypothetical theories do not differ essentially as *theories*, is contained in Ludwig Boltzmann's essays "Ein Wort der Mathematik an die Energetik" and "Über die Unentbehrlichkeit der Atomistik in der Naturwissenschaft," both contained in his *Populäre Schriften*, Leipzig, 1905.

with the second position mentioned earlier, according to which theories are best regarded as instruments for the conduct of inquiry, rather than as statements about which questions of truth and falsity can be usefully raised.

III. *The Instrumentalist View of Theories*

The position which we shall call, for the sake of brevity, the "instrumentalist" view of the status of scientific theory has received a variety of formulations.[22] But, although there are important differences between some of them, it would not be relevant to the purposes of the present discussion to consider the formulations individually. In any event, the merits of the position do not belong to any one particular formulation exclusively. Its strength derives from its concern with the actual function of a theory in scientific inquiry, and from its ability, in consequence of this concern, to outflank a number of difficulties that embarrass alternative positions.

The central claim of the instrumentalist view is that a theory is neither a summary description nor a generalized statement of relations between observable data. On the contrary, a theory is held to be a rule or a principle for analyzing and symbolically representing certain materials of gross experience, and at the same time an instrument in a technique for inferring observation statements from other such statements. For example, the theory that a gas is a system of rapidly moving molecules is not a description of anything that has been or can be observed. The theory is rather a rule which prescribes a way of symbolically representing, for certain purposes, such matters as the observable pressure and temperature of a gas; and the theory shows among other things how, when certain empirical data about a gas are supplied and incorporated into that representation, we can calculate the quantity of heat required for raising the temperature of the gas by some designated number of degrees (i.e., we can calculate the specific heat of a gas). The molecular theory of gases is thus neither logically implied by nor (according to some proponents of the instrumentalist view) does it logically imply any statements about matters of observation. The *raison d'être* of the theory is to serve as a rule or guide for making logical transitions from one set of experimental data to another set. More generally, a theory

[22] Cf. C. S. Peirce, *Collected Papers*, Cambridge, Mass., 1932, Vol. 2, p. 354; 1933, Vol. 3, pp. 104-06; 1934, Vol. 5, pp. 226-28; Frank P. Ramsey, *The Foundations of Mathematics*, New York, 1931, pp. 194ff., 237-55; Moritz Schlick, *Gesammelte Aufsätze*, Vienna, 1938, pp. 67-68; John Dewey, *The Quest for Certainty*, New York, 1929, Chap. 8; W. H. Watson, *On Understanding Physics*, London, 1938, Chap. 3; Gilbert Ryle, *The Concept of Mind*, New York, 1949, pp. 120-25; Stephen Toulmin, *The Philosophy of Science*, London, 1953, Chaps. 3 and 4.

functions as a "leading principle" or "inference ticket" *in accordance with which* conclusions about observable facts may be drawn from given factual premises, not as a premise *from which* such conclusions are obtained.

Several consequences follow directly from this account.

1. The view that a theory is a "convenient shorthand" for a class of observation statements (whether finite or infinite in number), and the correlative claim that a theory must be translatable into the language of observation are both irrelevant and misleading approaches to understanding the role of theories. The value of a theory for the conduct of inquiry would not be enhanced if perchance it could be shown to be logically equivalent to some class of observation statements; and failure to establish such an equivalence for any of the theories in physics does not diminish their importance as instruments for analyzing the materials of experience with a view to solving concrete experimental problems and systematically relating experimental laws. From the perspective of the instrumentalist standpoint, moreover, it is no less gratuitous to ask whether a theory has a "surplus meaning" and what its "factual reference" is, over and above its meaning and reference as revealed by its organizing role in inquiry. For such questions in effect tacitly assume a modified version of the translatability thesis, according to which a theory, though not equivalent in meaning to a class of observation statements, must nevertheless be construed as equivalent to some other class of factual statements distinct from the theory itself. The questions thus invite a misguided search for answers, not within the actual context of inquiry in which a theory performs its functions, but in terms of arbitrary preconceptions as to how the import of theories is to be ascertained.

A simple example will perhaps make clearer the instrumentalist position on this point. A hammer is a deliberately contrived tool, with the help of which a variety of "raw materials" can be brought into definite relations, so as to yield such things as packing boxes, furniture, and buildings. The uses to which a hammer may be put cannot be specified once for all, so that the products of its use may increase both in number and in kind. In any event, we would think it nonsense were anyone to suggest that a hammer is in any familiar sense "equivalent" to the things produced or producible by its means; and we would also regard as curious the questions whether a hammer adequately "represents" the products already made with its help or whether, in addition to these products, the hammer designates a "surplus" set of further things that it could help to produce. According to the instrumentalist view of theories, however, theories are in important respects like hammers and other physical tools, even though this analogy obviously fails at many points.

Theories are intellectual tools, not physical ones. They are nevertheless conceptual frameworks deliberately devised for effectively directing experimental inquiry, and for exhibiting connections between matters of observation that would otherwise be regarded as unrelated.

It is therefore pointless even to attempt the translation of a theory into some determinate class of observation statements. For the function of a theory, like that of a physical tool, is to help organize "raw data" rather than to summarize or to duplicate such data. On this view, theories, like other instruments, do indeed have a "factual reference"—namely, to the subject matter for whose exploration they have been constructed and in which they have an effective role. Moreover, if a theory has a "surplus meaning" over and above the meanings associated with it because of the special uses to which it has already been put, it has such a meaning in one of two senses: either in the sense that it is interpreted in terms of some familiar model; or in the more pregnant sense that, as in the case of other instruments, its further uses, even if only vaguely entertained in imagination, may be more inclusive than those actually assigned to it at any given time. Current quantum theory, for example, brings into systematic order a wide range of physical and chemical phenomena. But physicists do not appear to believe that the use of the theory in connection with these phenomena exhausts its capacity to serve as a leading principle for analyzing and organizing still unexplored materials. On the contrary, physicists continue to enlarge the uses of the theory, on the basis of more or less vague suggestions provided by the theory; and, quite apart from the various models employed for interpreting the formalism of quantum mechanics, these suggestions constitute the operative "surplus meanings" of the theory.

2. It is common if not normal for a theory to be formulated in terms of ideal concepts such as the geometrical ones of straight line and circle, or the more specifically physical ones of instantaneous velocity, perfect vacuum, infinitely slow expansion, perfect elasticity, and the like. Although such "ideal" or "limiting" notions may be suggested by empirical subject matter, for the most part they are not descriptive of anything experimentally observable. Indeed, in the case of some of them it seems quite impossible that when they are understood in a literal sense they could be used to characterize any existing thing. For example, we can attribute a velocity to a physical body only if the body moves through a finite, nonvanishing distance during a finite, nonvanishing interval of time. But instantaneous velocity is defined as the *limit* of the ratios of the distance and time as the time interval diminishes toward zero. In consequence, it is difficult to see how the numerical value of this limit could possibly be the measure of any actual velocity.

There is nevertheless a rationale for using such limiting concepts in constructing a theory. With their help a theory may lend itself to a relatively simple formulation—simple enough, at any rate, to render it amenable to treatment by available methods of mathematical analysis. To be sure, standards of simplicity are vague, they are controlled in part by intellectual fashions and the general climate of opinion, and they vary with improvements in mathematical techniques. But in any event, considerations of simplicity undoubtedly enter into the formulation of theories. Despite the fact that a theory may employ simplifying concepts, it will in general be preferred to another theory using more "realistic" notions if the former answers to the purposes of a given inquiry and can be handled more conveniently than the latter.

On the other hand, the use of such limiting concepts in the formulation of a theory presents difficulties to the view that factual truth or falsity can be significantly predicated of the theory. For a factual statement is normally said to be true if it formulates some indicated relation either between existing things and events (in the omnitemporal sense of "exist") or between properties of existing things and events. However, if a theory formulates relations between properties that ostensibly do not (or cannot) characterize existing things, it is not clear in what sense the theory can be said to be factually true or false.

Analogous difficulties for this view are raised by the circumstance that in general a theory contains terms for which no rules of correspondence are given, whether or not an interpretation is provided for the theory on the basis of some model. In consequence, no experimental notions are associated with such terms, so that in effect those terms have the status of *variables*. However, though such terms enter into expressions having the *grammatical form* of statements, many of these expressions are in fact not statements at all but only *statement-forms*. Consider, for example, the expression 'For any x, if x is an animal and x is P, then x is a vertebrate.' This has the grammatical form of a statement; but, since it contains the otherwise unspecified predicate variable 'P,' it is a statement-form, not a statement, and cannot be characterized as either true or false. The statement-form yields a statement if, for example, the definite predicate 'mammalian' is substituted for (or associated with) the predicate variable.[28] The point can be illustrated by examples from actual physical theories. We have already noted that in the molecular theory of gases there is no correspondence rule for the expression 'the velocity of an individual molecule,' though there is such a rule for the expression 'the average value of the velocities of all the molecules.' Similarly, the expression $\psi(x, t)$ is employed in the Schrödinger equation in

---

[28] Another way of obtaining a statement from the statement-form is to "quantify" over the predicate variable, thus obtaining, for example, the statement "There is a property P, such that for every x if x is an animal and x is P, then x is a vertebrate."

quantum mechanics for characterizing the state of an electron. There is in effect a correspondence rule for the expression $\psi(x,t)\psi^*(x,t)$ (where ψ^* is the complex conjugate of ψ), but no such rule for $\psi(x,t)$ itself. On the face of it, therefore, theories containing such terms are statement-forms, and cannot be said to be true or false.

These and similar difficulties do not arise for the instrumentalist view of theories, since on this view the pertinent question about theories is not whether they are true or false but whether they are effective techniques for representing and inferring experimental phenomena. The fact that theories contain expressions which describe or designate nothing in actual existence, or which are not associated with experimental notions is indeed taken as confirmation for the claim that theories must be construed in terms of their intermediary, instrumental function in inquiry, rather than in terms of their adequacy as objective accounts of some subject matter. From the perspective of this standpoint, it is not a flaw in the molecular theory of gases, for example, that it employs limiting concepts such as the notions of point-particle, instantaneous velocity, or perfect elasticity. For the task of the theory is not to give a faithful portrayal of what transpires within a gas but to provide a method for analyzing and symbolizing certain properties of the gas, so that when information is available about some of these properties in concrete experimental situations the theory makes it possible to infer information having a required degree of precision about other properties.

Similarly, it is not a source of embarrassment to the instrumentalist position that in inquiries into the thermal properties of a gas we use a theory which analyzes a gas as an aggregation of discrete particles, although when we study acoustic phenomena in connection with gases we employ a theory that represents the gas as a continuous medium. Construed as statements that are either true or false, the two theories are on the face of it mutually incompatible. But construed as techniques or leading principles of inference, the theories are simply different though complementary instruments, each of which is an effective intellectual tool for dealing with a special range of questions. In any event, physicists show no noticeable compunction in using one theory for dealing with one class of problems and an apparently discordant theory for handling another class. They employ the inclusive wave theory of light, according to which optical phenomena are represented in terms of periodic wave motion, when dealing with questions of diffraction and polarization; but they continue to use the relatively simpler theory of geometrical optics, according to which light is analyzed as a rectilinear propagation, when handling problems in reflection and refraction. They introduce considerations based on the theory of relativity in applying quantum mechanics to the analysis of the fine structure of spectral lines; they ignore such considerations when quantum theory is exploited for analyzing the nature of chemical bonds. Such examples can be multiplied; and if they prove nothing else, they show at least that the literal truth of theories is not the object of primary concern when theories are used in experimental inquiry.

It does not follow, however, that on the instrumentalist view theories are "fictions," except in the quite innocent sense that theories are human creations. For in the pejorative sense of the word, to say that a theory is a fiction is to claim that the theory is not true to the facts; and this is not a claim which is consistent with the instrumentalist position that truth and falsity are inappropriate characterizations for theories. It is indeed possible to maintain, consistent with that position, that many of the models in terms of which theories have been interpreted are fictions (in some cases even explicitly introduced as fictions, as were some of Lord Kelvin's mechanical models of the ether). In maintaining this much, one is merely asserting either that there simply is no empirical evidence satisfying some assumed criterion for the physical reality of those models, or that in terms of this criterion the available evidence is negative. On the other hand, it is also consistent with the instrumentalist view to recognize that some theories are superior to others—either because one theory serves as an effective leading principle for a more inclusive range of inquiries than does another, or because one theory supplies a method of analysis and representation that makes possible more precise and more detailed inferences than does the other. However, a theory is an effective tool in inquiry only if things and events are actually so related that the conclusions the theory enables us to infer from given experimental data are generally in good agreement with further matters of observed fact. As in the case of other instruments, the effectiveness of a theory as an instrument, or its superiority to some other theory, is thus contingent on objective features of a subject matter and depends on something other than personal whim or preference.

3. The instrumentalist view of theories is illuminated by, and receives a measure of support from, an interesting theorem in formal logic, known as Craig's theorem.[24] We shall explain this theorem, but will omit technical complications and fine points. Let L be some "language," such as the language of physics, which contains not only locutions customarily included in the vocabulary of formal logic (e.g., 'if-then,' 'not,' and 'for every x'), but also a class O of expressions designating things and properties regarded as "observable" in some assumed sense of the word (e.g.,

[24] William Craig, "On Axiomatization within a System," *Journal of Symbolic Logic*, Vol. 18 (1953), pp. 30-32; and in a less technical form "Replacement of Auxiliary Expressions," *Philosophical Review*, Vol. 65 (1956), pp. 38-55.

'copper wire,' 'green,' and 'longer than'), as well as a class T of expressions that are counted as "theoretical" ones (e.g., 'electron' or 'light wave'). Every nonlogical expression of L is stipulated to belong to just one of the two classes O and T. Furthermore, L is supposed to be a "formal system," thus satisfying a number of conditions that in fact are not met by the actual language of physics. In the first place, the vocabulary of L is fully specified, and explicit rules are laid down for constructing statements out of the vocabulary. A statement all of whose component nonlogical expressions belong to O will be called an "observation statement"; a statement containing at least one expression belonging to T will be said to be a "theoretical" one. In the second place, the permitted inferences in L are codified in a fixed set R of rules of logical inference. Thirdly, L is axiomatized, in a manner familiar from geometry.

But something further needs to be said about this axiomatization. Let W be the class of all statements in L that are in fact true, whether because they are logically necessary or because they correctly formulate what is only contingently the case; and let W_O be the set of observation statements in W that are not certifiable as logically true, while W_T is the analogous set of theoretical statements in W. The axioms A of L will in general be a proper subclass of W, so that there are statements in W which are not logically equivalent to some of the axioms in A. Moreover, it is by now obvious that insofar as L is a tolerably faithful though idealized representation of the actual language of physics, the axioms will contain both theoretical and observation statements. Some observation statements will be included in the axioms, because not all true observation statements are derivable from theoretical statements alone. On the other hand, theoretical statements must also be included, because many observation statements cannot be asserted as true on direct experimental grounds (e.g., observation statements about past events), nor can such statements be inferred logically except with the help of some theory from other observations known to be true. In any event, the axioms A, together with all statements derivable from them in accordance with the rules of inference R, will constitute the class W of true statements of L. However, since by hypothesis only the statements of W_O formulate observable matters, we shall stipulate that the "empirical content" of L is codified by the class of statements W_O—a class which may be finite or infinite. Accordingly, other things being equal, no factual data concerning the primary subject matter of, say, physics, can provide grounds for choosing between two languages having an identical empirical content.

It is natural to ask, therefore, whether it may not be possible, after all, to construct a language having the same empirical content as L but containing no theoretical statements. We have already considered this question in connection with the thesis that theories are translatable into observation statements, and have concluded that the thesis has not been established. But the possibility is not thereby precluded that some other way can be devised for dispensing with theories without thereby diminishing the empirical content of a language.

It is to this point that Craig's theorem is relevant. Craig's approach is different from the one pursued by proponents of the translatability thesis. He does not propose to *translate* theories into observation statements, but in effect to *replace* a formal linguistic system containing theoretical expressions by another formal system having no theoretical terms and yet having the same empirical content as the initial system. More specifically, Craig shows how to construct a formal language L^* in the following manner: the nonlogical expressions of L^* are the observation terms O of L; the rules of inference R^* of L^* are the same as R (except for inessential modifications); and the only nonlogically true statements included in the axioms A^* of L^* are observation statements, which are specified by an effective procedure (too complicated to be described here) upon the true observation statements W_0 of L. It can then be proved that an observation statement S is a theorem of L if and only if S is a theorem in L^*, so that the empirical content of L^* is the same as that of L. Accordingly, whatever systematization of observation statements is achieved in L with the help of theories can apparently be achieved in L^* without theories. It therefore seems that from the standpoint of formal logic theories are not essential instruments for the organization of physics.

However, such a conclusion is not warranted by Craig's finding, as he himself notes. For quite apart from the difficulty that the language of physics is not a formal system, and is unlikely to become one because of its unpredictably changing character, two features of Craig's method for constructing the language L^* seriously diminish the significance of his theorem for scientific inquiry. In the first place, though the method shows how the axioms A^* of L^* can be effectively specified, it does not guarantee that the axioms will be finite in number (unless the class W_0 of true observation statements in L is finite). Nor does the method guarantee that, if A^* is either infinitely numerous or contains a finite but very large number of axioms, the axioms will be specified in a way making it psychologically possible for anyone to use them efficiently for deductive purposes. It is pertinent to recall that the usual axiomatizations of various subjects contain not simply a finite number of axioms but a relatively small number of axioms. If the number of axioms of the ordinary sort were even moderately large (for example, if a million axioms were needed for plane geometry), it would be humanly impossible to keep them all in mind, and it is doubtful whether significant theorems could be established.[25] Accordingly, the axioms for L^* specified by Craig's method may be so cumbersome that no effective logical use can be made of them.

In the second place, Craig's method proceeds in such a way that for each statement S in W_O there is an axiom in L^* logically equivalent to S. For example, if S is a true observation statement in L, then the conjunction S and S and . . . and S (in which S is repeated a certain calculable number of times) is an axiom in L^*. In short, all the true statements of L^* will in effect be axioms of L^*. This feature of the method suffices to make it quite valueless for purposes of scientific inquiry. Such a set of axioms for L^* provides no simplified formulation of the empirical content of L^*, and indeed only reformulates it, so that the axioms offer no advantage over a mere listing of all true observation statements. Moreover, in order to specify the axioms for L^* we would have to know, *in advance* of any deductions made from them, *all* the true statements of L^*—in other words, Craig's method shows us how to construct the language L^* only *after* every possible inquiry into the subject matter of L^* has been completed. The bearing of this point on the instrumentalist view of theories is patent. For the discussion calls attention to the fact that theories in science are important, not primarily because they may be true, but because they serve as guides to the investigation, formulation, and organization of matters of observable fact even *before* all observation statements are established as true (or probably true). One moral that can be extracted from Craig's theorem is that, whether or not truth and falsity are properly predicable of theories, this is at any rate not the exclusively relevant question in assessing the place of theories in science.

4. But it is time for noting some limitations in the instrumentalist standpoint. Proponents of this view often seem to believe that, if the instrumental role of theories is once established, theories are thereby shown to be improper subjects for the characterizations "true" and "false." There is, however, no necessary incompatibility between saying that a theory is true and maintaining that the theory performs important functions in inquiry. Few will deny that statements such as "The distance between New York and Washington, D. C., is approximately 225 miles" may be true, and yet play valuable roles in the plans of men. Indeed,

---

[25] This point is not blunted by the fact that various formal systems have been constructed on the basis of infinitely many axioms. For these systems employ what are called "axiom-schemata," each describing a distinctive *form* of an axiom that can be embodied in an infinite number of specific statements. However, though the number of axioms in such systems is infinite, the number of *axiom-schemata* is finite and quite small.

most statements that by common consent can be significantly affirmed as true or false can also be studied for the use that is made of them. In brief, it does not follow that theories cannot be regarded as "genuine statements" and cannot therefore be investigated for their truth or falsity, merely because theories have indispensable functions in inquiry.

Moreover, those who characterize theories as leading principles, as rules in accordance with which inferences are drawn rather than as premises from which conclusions are derived, often overlook the contextual nature of this distinction. It is today common knowledge that an inference such as the familiar one which proceeds from the premises 'All men are mortal' and 'The Duke of Wellington is a man' to the conclusion 'The Duke of Wellington is mortal,' makes tacit use of the purely logical rule of inference (or leading principle) known as the principle of the syllogism (a statement of the form 'x is P' is derivable from two statements of the form 'All S is P' and 'x is S'). The leading principle is not a premise in the inference, and the conclusion is drawn not from it but in accordance with it. The principle is, moreover, a formal one, since it refers only to the form of statements, irrespective of what subject-matter terms they may contain.

However, it is now also generally recognized that an argument sanctioned by a formal rule of inference can be reconstructed, so that the same conclusion can be obtained from a subset of the original premises, in accordance with a *material* leading principle that compensates for the premises which have been dropped. Thus, it is correct to infer 'The Duke of Wellington is mortal' from the single premise 'The Duke of Wellington is a man,' provided we adopt the material rule of inference "Any statement of the form 'x is mortal' is derivable from a statement of the form 'x is a man.'" The leading principle in this case is said to be a material one, because it mentions specific subject-matter terms that must occur in the class of inferences that the principle sanctions.

On the other hand, this procedure can in general be used in reverse, so that a material leading principle for an argument can be dropped and replaced by a corresponding premise. For example, the conclusion 'This piece of copper wire will expand' can be inferred from the premise 'This piece of copper wire will be heated' in accordance with the material leading principle "A statement of the form 'x will expand,' is derivable from a statement of the form 'x is copper and will be heated.'" But the same conclusion can be obtained without this leading principle, if we add the statement 'All copper expands on heating' to the original premise. It is clearly a matter of convenience in which of these alternate ways an argument is constructed. Accordingly, though the distinction between premises and rules of inference is both sound and important, a given statement may function as a premise in one context but may in effect be used as a leading principle in another context, and vice versa.

The point illustrated by these simple examples obviously holds for the more complex arguments in which theories play a fundamental role. There is little doubt, for example, that in many cases the wave theory of light is used, or can be construed, as a leading principle or technique for inferring statements about experimentally identifiable data from other such data. Nor is it disputable that this way of viewing the theory brings out a role it plays in inquiry which might otherwise be overlooked, or that this perspective on theories is a salutary antidote to dogmatic affirmations that some particular theory is the final truth about the "ultimate nature" of things. It nevertheless does not follow that theories do not or cannot also serve as premises in scientific explanations and predictions, as bona fide statements concerning which it therefore seems quite proper to raise questions of truth and falsity.

In point of fact, theories are usually presented and used as premises, rather than as leading principles, in scientific treatises as well as in papers reporting the outcome of theoretical or experimental research. Some of the most eminent scientists, both living and dead, certainly have viewed theories as statements about the constitution and structure of a given subject matter; and they have conducted their investigations on the assumption that a theory is a *projected map* of some domain of nature, rather than a set of *principles of mapping*. Much experimental research is undoubtedly inspired by a desire to ascertain whether or not various hypothetical entities and processes postulated by a theory (e.g., neutrons, mesons, and neutrinos of current atomic physics) do indeed occur in circumstances and relations stated by the theory. But research which is ostensibly directed to testing a theory proceeds on the *prima facie* assumption that the theory is affirming some things and denying others. In short, neither logic nor the facts of scientific practice nor the frequently explicit testimony of practicing scientists supports the dictum that there is no valid alternative to construing theories simply as techniques of inference.

Moreover, as has already been suggested, questions can be raised about a theory when it is regarded as a leading principle that are substantially the same as those which arise when the theory is used as a premise. For whether or not a material leading principle happens to be a theory, the principle is a dependable one only if the conclusions inferred from true premises in accordance with the principle are in agreement with facts of observation to some stipulated degree. In consequence, there is on the whole only a verbal difference between asking whether a theory is satisfactory (as a technique of inference) and asking whether a theory is true (as a premise).

The claim made by some proponents of the instrumentalist view, that no theory logically implies any statements about facts of observation, must similarly be qualified. The claim is obviously sound if a theory is construed to be a leading principle, since a rule of inference is not a premise in factual inquiries and is not the sort of thing of which one can say that it entails factual conclusions. The claim is also sound if it is understood to assert that, even if a theory is used as a premise, no instantial conclusions follow from the theory alone, but only when appropriate rules of correspondence are supplied for the theory and when suitable statements of initial conditions are added as premises. On the other hand, the claim is plainly mistaken if it maintains that no statements about matters of observable fact are implied by a theory even when this proviso is satisfied. For such a contention is contradicted every time a theory is used in the indicated manner—for example, when the wave theory of light is employed to account for the chromatic aberration of optical lenses.

One final comment on the instrumentalist view must be made. It has already been briefly noted that proponents of this view supply no uniform account of the various "scientific objects" (such as electrons or light waves) which are ostensibly postulated by microscopic theories. But the further point can also be made that it is far from clear how, on this view, such "scientific objects" can be said to be physically existing things. For if a theory is just a leading principle—a technique for drawing inferences based upon a method of representing phenomena—terms like 'electron' and 'light wave' presumably function only as conceptual links in rules of representation and inference. On the face of it, therefore, the meaning of such terms is exhausted by the roles they play in guiding inquiries and ordering the materials of observation; and in this perspective the supposition that such terms might refer to physically existing things and processes that are not phenomena in the strict sense seems to be excluded. Proponents of the instrumentalist view have indeed sometimes flatly contradicted themselves on this issue. Thus, while maintaining that the atomic theory of matter is simply a technique of inference, some writers have nevertheless seriously discussed the question whether atoms exist and have argued that the evidence is sufficient to show that atoms really do exist. Others have explicitly asserted that atoms and other "scientific objects" are generalized statements of relations between sets of changes, and cannot be individual existing things; but they have also declared that atoms are in motion, and possess a mass. Such inconsistencies suggest that those who are guilty of them are not really prepared to exclude, as improper, questions of truth and falsity concerning a theory. In any event, it is clearly not inconsistent to admit the logical propriety of such questions, and also to recognize the important instrumental function of theories.

IV. *The Realist View of Theories*

Are theories then "really" statements, of which truth and falsity are meaningfully predicable, despite the difficulties that have been noted in this view? Enough has already been said to suggest that, whether the question is answered affirmatively or negatively, the answer given may not be the exclusively reasonable one. Indeed, those who differ in their answers to it frequently disagree neither on matters falling into the province of experimental inquiry nor on points of formal logic nor on the facts of scientific procedure. What often divides them are, in part, loyalties to different intellectual traditions, in part inarbitrable preferences concerning the appropriate way of accommodating our language to the generally admitted facts. It is a matter of historical record that, while many distinguished figures in both science and philosophy have adopted as uniquely adequate the characterization of theories as true or false statements, a no less distinguished group of other scientists and philosophers has made a similar claim for the description of theories as instruments of inquiry. However, a defender of either view cannot only cite eminent authority to support his position; with a little dialectical ingenuity he can usually remove the sting from apparently grave objections to his position. In consequence, the already long controversy as to which of the two is the proper way of construing theories can be prolonged indefinitely. The obvious moral to be drawn from such a debate is that once both positions are so stated that each can meet the *prima facie* difficulties it faces, the question as to which of them is the "correct position" has only terminological interest.

1.   Let us consider the chief obstacles to each of the two views under discussion, beginning with those facing the conception of theories as true or false statements.

a.   There is in the first place the purely formal difficulty that a theory is not a statement, but only a statement-form. For if, as often happens, some terms of a theory are not associated with any correspondence rules, those terms are in effect variables, so that on the face of it the theory does not satisfy the grammatical requirements for statements. This difficulty can be met by a formal device, first explicitly proposed by Ramsey.[26] The device consists simply in introducing what are called "existential quantifiers" as prefixes for statement-forms, so that the resulting expression will formally be a statement. For example, the expression 'If a human being has the trait P, then such a person has blue eyes' is a statement-form;

[26] Frank P. Ramsey, *op. cit.*, pp. 212-36.

but by adding the prefix 'There is a trait P,' we obtain from it the statement 'There is a trait P such that if a human being has P, then the person has blue eyes.' Similarly, suppose that, although the terms 'mass' and 'acceleration' are associated with correspondence rules, the term 'force' is not. The expression 'If a body undergoes changes in motion, then the product of the mass and acceleration of the body is equal to the force F acting on it' is then in effect a statement-form, from which we can obtain the statement 'If a body undergoes changes in motion, then there is a (measurable) property F such that the product of the mass and acceleration of the body is equal to F.' More generally, let '$T(M,N,P,Q)$' be a theory whose theoretical terms 'M' and 'N' are associated with correspondence rules while its theoretical terms 'P' and 'Q' are not, so that '$T(M,N,P,Q)$' is by hypothesis a statement-form. Then 'There is a P and there is a Q, such that $T(M,N,P,Q)$' is a statement. Accordingly, since by using the Ramsey device the observational consequences that can be derived from a theory are not altered, that device suffices for outflanking the formal difficulty under discussion.

b. In the second place, there is the objection previously mentioned that theories are commonly formulated in terms of limiting concepts which characterize nothing actually in existence, so that at any rate nonvacuous factual truth cannot be claimed for such theories. This objection can be turned in a number of ways. A familiar gambit is to challenge the contention that limiting concepts do not apply to existing things. To be sure, we cannot, for example, ascertain by overt measurement the value of an instantaneous velocity of the magnitude of some length whose theoretical value is stipulated to be equal to the square root of 2. But unless accessibility to overt measurement (or more generally to observation) is made the criterion of physical existence, so it is sometimes said, this does not show that bodies cannot have instantaneous velocities or lengths with real number magnitudes. On the contrary, if a theory postulating such values is supported by competent evidence, then according to the rejoinder under discussion there is good reason to maintain that these limiting concepts do designate certain phases of things and processes. Since in testing a theory we test the totality of assumptions it makes, so the rejoinder continues, if a theory is regarded as well established on the available evidence, all its component assumptions must also be regarded. Accordingly, unless we introduce quite arbitrary distinctions, we cannot pick and choose between the component assumptions, counting some as descriptions of what exists and others as not.

There is another way in which the objection under discussion is sometimes countered. The rejoinder then consists in admitting that limiting concepts are simplifying devices, and that a theory employing them does not in general assert anything for which literal truth can reasonably be claimed. Nevertheless, existing things possess traits that often are either indistinguishable from the "ideal" traits mentioned in a theory or differ from such "ideal" traits by a negligible factor. In consequence, on this rejoinder to the objection, a theory is said to be true in the sense that the discrepancy between what a theory asserts and what even ultra-refined observation can discover is small enough to be counted as arising from experimental error.

c. A third type of difficulty for the conception of theories as true or false statements is created by the fact, to which attention has already been directed, that apparently incompatible theories are sometimes employed for the same subject matter. Thus, a liquid cannot be both a system of discrete particles and also a continuous medium, though theories dealing with the properties of liquids adopt one assumption in some cases and the opposing assumption in others.

The usual reply to this objection consists of two parts. One of them is essentially a repetition of the rejoinder mentioned in the preceding paragraph. A theory may be employed in a given area of inquiry, even though it is apparently incompatible with some other theory that is also used, because the former is simpler than the latter and because for the problems under discussion the more complex theory does not yield conclusions in better agreement with the facts than are the conclusions of the simpler theory. Accordingly, the simpler theory can be regarded as in a sense a special case of the more complex one, rather than as a contrary of the latter.

The second part of the reply is that, though incompatible theories may be used for a time, their use is but a temporary makeshift, to be abandoned as soon as an internally consistent theory is developed, more comprehensive than either of the previous ones. Thus, although there were serious discrepancies between the atomic theories employed at the turn of the present century to account for many facts of both physics and chemistry, these conflicting theories have been replaced by a single theory of atomic structure currently used in both these sciences. Indeed, inconsistencies between theories, each of which is nevertheless useful in some limited domain of inquiry, are often a powerful incentive for the construction of a more inclusive but consistent theoretical structure. Accordingly, a proponent of the view that theories are true or false statements can escape any embarrassment for his position from the circumstance that incompatible theories are sometimes employed in the sciences; he can insist on the corrigible character of every theory and refuse to claim final truth for any theory. He can freely admit that even a false theory may be quite useful for handling many problems; and he can join this admission with the claim that the succession of theories in some branch of science is a series of progressively better approximations to the unattainable but valid ideal of a finally true theory.

d. And finally, there is the objection currently raised against the position under discussion because of the difficulties encountered in interpreting quantum mechanics in terms of some familiar model. For example, theoretical as well as experimental considerations have led physicists to ascribe to electrons (and to other entities postulated by quantum theory) apparently incompatible and in any case puzzling characteristics. Thus, electrons are construed to have features which make it appropriate to think of them as a system of waves; on the other hand, electrons also have traits which lead us to think of them as particles, each having a spatial location and a velocity, though no determinate position and velocity can in principle be assigned simultaneously to any of them. Many physicists have therefore concluded that quantum theory cannot be viewed as a statement about an "objectively existing" domain of things and processes, as a map that outlines even approximately the microscopic constitution of matter. On the contrary, the theory must be regarded simply as a conceptual schema or a policy for guiding and coordinating experiments.

The rejoinder to this objection follows a familiar pattern. The fact that a visualizable model embodying the laws of classical physics cannot be given for quantum theory, so runs the reply, is not an adequate ground for denying that the quantum theory does formulate the structural properties of subatomic processes. It is doubtless desirable to have a satisfactory model for the theory. But the type of model that is regarded as satisfactory at any given time is a function of the prevailing intellectual climate. Even though current models for quantum theory may strike us as strange and even "unintelligible," there are no compelling reasons for assuming that the strangeness will not wear away with increased familiarity, or that a more satisfactory interpretation for the theory will not be eventually found. Moreover, the alleged unintelligibility of the present model stems in large measure from a failure to note that words like 'wave' and 'particle' used in describing it are being employed in an analogical manner. It is only in a Pickwickian sense that an electron is a particle (in the customary meaning of the word), just as it is in a stretched sense that $\sqrt{-1}$ is a number (in the sense in which the cardinal integer 3 is a number). An electron is said to be a particle (or alternatively, a wave) because some of the properties ascribed to electrons are analogous to certain properties associated with classical particles, or alternatively, with familiar water waves, even if the analogy fails for other properties. When the language of "particles" and "waves" is understood in terms of the way these words are actually used in the context of quantum mechanics, so it has been contended, not even the appearance of contradiction arises in the quantum-theoretical characterizations of electrons. But in any event, the basic issue is not whether a particular substantive model of subatomic processes is satisfactory. The basic issue is whether the relations between elementary constituents of physical objects and processes are more adequately stated by the mathematical formalism of quantum mechanics than by any other formal model available at present. On this issue, there is no disagreement among competent students that the answer is affirmative.

This sample of objections to the view that theories are true or false statements suffices to show that the view has dialectical resources for maintaining itself in the face of severe criticism. Undoubtedly the rejoinders to these criticisms can be met with counterrejoinders, though none to which defenders of the view under attack cannot offer at least a *prima facie* suitable reply. It would therefore not be profitable to continue this phase of the discussion any further. Let us turn instead to some of the criticisms of the instrumentalist position.

2. Two main difficulties have been noted in the instrumentalist position as usually formulated. The first of them is that much experimental research is directed to finding evidence for or against a theory—an undertaking which is apparently pointless if a theory is not a genuine statement but simply a formulation of policy or rule of procedure. However, this objection can be readily made innocuous. For it is sufficient to reply that a theory can indeed be "tested" by searching for evidence which will either "confirm" or "refute" it, but only in the sense that confirmatory or disconfirmatory evidence is sought for observational conclusions drawn from observational premises in accordance with the theory. As we have seen, the sole issue raised by this way of putting the matter concerns the relative convenience of employing *material* rather than *purely formal* leading principles in reconstructing deductive inferences.

The second and more serious difficulty is that a consistently held instrumentalist view apparently precludes its adherents from admitting the "physical reality" (or "physical existence") of any "scientific objects" ostensibly postulated by a theory. For if a theory employing such terms as 'atom' or 'electron' is merely a leading principle, it is incongruous to ask whether there "really are" atoms; and it is then acutely puzzling to say, as some physicists do, that because of the experimental evidence now "pointing" to the atom, "we are as convinced of its physical existence as of our hands and feet."

However, the force of this objection is unclear because of the notorious ambiguity if not obscurity of the expression 'physical reality' or 'physical existence.' In any event, writers using these phrases do not in general understand them in the same sense. It will therefore be useful to consider some of the different criteria that are commonly employed, whether explicitly or tacitly, when physical reality is either affirmed or denied of scientific objects such as electrons, atoms, electric fields, and the like.

a. For anything to be regarded as physically real, perhaps the most familiar requirement is that the thing or event be publicly perceived when suitable conditions for its observation are realized. In terms of this criterion, sticks, stones, flashes of lightning, the smells of cooking, and the like can be said to exist physically, but not the pains a man feels when he turns an ankle, nor the pink elephants a drunkard may experience in his delirium. However, most scientific objects are not physically real in this sense. Thus, although illuminated surfaces are physically real on this criterion, light waves are not; and, although the condensations of water vapor to form visible tracks in a Wilson cloud chamber are real, the alpha particles which (according to current physical theory) produce those tracks are not. Certainly it is not on this interpretation of "physically real" that we are as convinced of the physical reality of atoms as we are of our hands and feet. On the other hand, even if some hypothetical scientific objects were physically real in this sense— for example, if the genes postulated by current biological theory of heredity could be made visible—the role of the theoretical notions in science, in terms of which such objects are specified, would not be altered. It is of course quite possible that if we could perceive molecules, many questions still outstanding about them would be answered, so that molecular theory would receive an improved formulation. Nevertheless, molecular theory would still continue to formulate the traits of molecules in *relational* terms—in terms of relations of molecules to other molecules and to other things—not in terms of any of their qualities that might be directly apprehended through our organs of sense. For the *raison d'être* of molecular theory is not to supply information about the sensory qualities of molecules but to enable us to understand (and predict) the occurrence of events and the relations of their interdependence in terms of pervasive structural patterns into which they enter. Accordingly, in this sense of the phrase the physical reality of theoretical entities is of little import for science.

b. A second, widely accepted criterion of physical reality is close to being the polar opposite of the first, and has already been mentioned in passing. According to it, every nonlogical term of an assumed law (whether experimental or theoretical) designates something that is physically real, provided that the law is well supported by empirical evidence and is generally accepted by the scientific community as likely to be true. On this criterion, therefore, physical reality is ascribed not only to such experimentally identifiable items as the kinetic energy of a bullet, the strain in a body subjected to stresses, the viscosity of a liquid, or the electrical resistance of a wire but also to theoretical objects like light waves, atoms, neutrinos, and waves of probability. Anyone who employs this criterion will accordingly hold that many objects postulated by some accepted theory are physically existing things, even before any empirical evidence confirming detailed specific assumptions about those objects is available. This seems to have been the criterion adopted by many contemporary physicists who believed in the physical existence of antiprotons as postulated by quantum theory, although definite experimental evidence for them was lacking until recently. On the other hand, those who employ this criterion will deny physical reality to a scientific object once so characterized (such as the phlogiston substance postulated by the phlogiston theory of combustion) when the theory postulating that object is abandoned as unsatisfactory—unless, indeed, a different but acceptable theory postulates a closely analogous object.

c. A third criterion of physical reality sometimes employed is that a term designating anything physically real must enter into more than one experimental law, with the proviso that the laws are logically independent of each other and that none of them is logically equivalent to a set of two or more laws. This requirement can obviously be strengthened by demanding that there be a considerable number of such experimental laws. The rationale for this requirement is to characterize as physically real only things that can be identified in ways other than, and independently of, the procedures used to define those things. For example, the magnitude of the gravitational force of the earth on a body appears as the constant 'g' in Galileo's law for freely falling bodies. If this were the only law in which 'g' occurred, then on this criterion the term 'gravitational force' would not designate a physical reality. However, 'g' enters into a number of other experimental laws, such as the law for the period of the simple pendulum. Accordingly, physical reality can be ascribed to the gravitational force of the earth. On the other hand, the situation appears to be different in the case of the notion of electric field. We can determine the strength of an electric field in a region by introducing into that region a test body with known mass and electric charge, and measuring the force upon that body. The field strength is then defined as the ratio of the force to the charge on the body; and it is an experimental law that under specified conditions this ratio has the same constant value for any test body of relatively small dimensions. Though in this way the term 'electric field' enters into an experimental law, this seems to be the only experimental law in which the term does occur. If so, then according to the present criterion physical reality cannot be ascribed to electric fields.

The application of this criterion to scientific objects postulated by microscopic theories involves some complications, since theoretical terms do not occur in statements of experimental laws. It would take us too far afield to unravel these complications in any detail. In any event, it will suffice for present purposes to construe the criterion for the physical reality of theoretical entities as saying that the theoretical term ostensibly referring to such entities must be *associated* by correspondence rules with experimental concepts, and furthermore that these experimental concepts must enter into at least two logically independent experimental laws which can be derived from the theory. For example, in the kinetic theory of gases theoretical expressions such as the 'mass of a molecule,' the 'mean kinetic energy of molecules,' the 'number of molecules,' and the like are associated with experimental concepts such as the 'mass of a gas,' the 'temperature of a gas,' and the 'ratio of the product of the pressure and volume of a gas to its temperature.' These latter terms occur in several experimental laws, such as the Boyle-Charles' law, Dalton's law of partial pressures, or the law that at given temperature and pressure the difference of the two specific heats per unit volume is the same for all gases—all of which are laws derivable from the theory.

It is worth noting that on this criterion of physical existence, not every entity postulated by a theory will in general be said to exist, even if the theory as a whole is well confirmed by experiment and is accepted as likely to be true. Thus, some physicists were once doubtful about the physical existence of neutrinos, initially postulated to preserve the conservation of energy principle in quantum theory; and it is possible that this doubt was based on the fact that the term 'neutrino' did not conform to the requirement set by this criterion. Similarly, when Planck first introduced the theoretical notion of discrete quanta of energy in order to account successfully for the distribution of energy in the spectrum of black-body radiation, physicists (including Planck himself) were doubtful of the existence of such quanta. The situation was altered when the notion of energy quanta was associated with the constant 'h' that appeared not only in Planck's radiation law but also in other experimental laws concerning the photoelectric effect, the line spectra of the elements, the specific heats of solids, and so on, all of which were derived from theories containing the quantum hypothesis as one component assumption.

d. A fourth, and in some ways more restrictive, criterion of physical reality is often adopted. On this criterion, a term signifies something physically real, if the term occurs in a well-established "causal law" (whether theoretical or experimental), in some indicated sense of "causal." In a more specific version of this criterion, the term must describe what is technically called the "state of a physical system," so that if 'A_t' is the state-description of a system at time t, the causal law asserts that the given state is invariably followed (or preceded) by the state $A_{t'}$ at time t' later (or earlier) than t.[27]

For example, in mechanics the state of a system of particles is described by the set of numbers that specify the positions and velocities of the particles. The causal laws of mechanics enable us, given the positions and velocities of a set of particles at any initial time, to determine their positions and velocities at any other time. Accordingly, the mechanical state of a system is physically real. Similarly, the state of a system in quantum theory is described by a certain function (known as the Psi-function) of positions and energies of elementary particles, where the function is a solution of the fundamental wave equation of the theory. This equation in effect asserts that the Psi-state of a system at some given time is invariably succeeded by the calculable Psi-state of the system at any specified future time. Accordingly, on the present criterion, the Psi-state is physically real. On the other hand, since in quantum mechanics the coordinates of position and velocity of an individual elementary particle, such as an electron, do not constitute the state-description of the particle, they do not describe what is physically real. In the view of at least some physicists, physical reality cannot therefore be ascribed to individual electrons and other such subatomic entities.[28]

e. One final criterion of physical reality is worth noting, according to which the real is that which is invariant under some stipulated set of transformations, changes, projections, or perspectives. An elementary geometrical example will illustrate the general idea underlying this criterion. Imagine a circle painted on a sheet of glass on a horizontal plane and a small source of light at some distance perpendicularly above the center of the circle. The circle will then be projected as a shadow cast on a screen parallel to the glass, and this shadow will also be a circle. Suppose, however, that the glass is rotated through an axis passing through the glass and parallel to the screen, with the source of light

[27] The notion of "state" will be discussed more fully in the next chapter.
[28] Cf. the discussion of this point in the debate between two leading contemporary physicists, Erwin Schrödinger, "Are There Quantum Jumps?" *British Journal for the Philosophy of Science*, Vol. 3 (1952), pp. 109-23, 233-42; and Max Born, "The Interpretation of Quantum Mechanics," *British Journal for the Philosophy of Science*, Vol. 4 (1953), pp. 95-106.

and the screen remaining in their initial positions. The shadows on the screen will then no longer be circles; they will first assume the form of ellipses, and eventually take the form of parabolas. Under this projection, neither the shape nor the perimeter nor the area of the circle on the glass will be preserved in the circle's shadow: these are not invariant properties of the circle under projection. Nevertheless, some properties of the circle are invariant under such projection. For example, if a straight line is painted on the glass to intersect the circle, the shadow of the line will always intersect the shadow of the circle in two points. If the present criterion were applied to this example, we would have to say that neither the shape nor the perimeter nor the area of the figure on the glass is a physical reality, but that only properties of the figure invariant under projection (such as the one mentioned) are physically real.

It will be evident that on this criterion different sorts of things can be characterized as physical realities, according to which set of transformations is specified for this purpose. Thus, some writers have denied physical reality to immediate sensory qualities, since these vary with physical, physiological, and even psychological conditions. The title to such reality has been reserved by these thinkers for the so-called "primary qualities" of things, whose interrelations are independent of physiological and psychological changes and are formulated by the laws of physics. Similarly, the numerical value of the velocity of a body is not invariant when the motion of the body is referred to different frames of reference, so that on this criterion relative velocity is not a physical reality. Many writers on the theory of relativity have in fact maintained that spatial distances and temporal durations as conceived in pre-relativity physics are not physically real, since they are not invariant for all systems moving with respect to each other with constant relative velocities. Physical reality, according to these writers, must be ascribed only to those features of things that are formulated by the invariant laws of relativity physics (such as the relativistic kinetic energy of a body or its relativistic momentum). In an analogous manner, physical reality has been attributed to theoretical entities like atoms, electrons, mesons, probability waves, and the like because they satisfy some indicated condition of invariance.

To prevent possible misunderstanding, it is perhaps worth stressing that the criteria mentioned in the preceding discussion are intended to be explicative of what is supposedly meant in a number of contexts when something is said to be physically real. The ascription of physical reality in any of the senses distinguished must therefore not be misconstrued. It must not be understood as implying that a thing so characterized has a place in the scheme of things to be contrasted with certain other things having the invidious label of "mere appearance," or that in addition to satisfying the requirements specified by the corresponding criterion the thing is in some way more valuable or more fundamental than everything not so characterized. Many scientists as well as philosophers have indeed often used the term "real" in an honorific way to express a value judgment and to attribute a "superior" status to the things asserted to be real. There is perhaps an aura of such honorific connotation whenever the word is employed, despite explicit avowals to the contrary and certainly to the detriment of clarity. For this reason it would be desirable to ban the use of the word altogether. As things stand, however, linguistic habits are too deeply ingrained and too widespread to make such a ban possible. Accordingly, these cautionary remarks have been added in order to make clear that any invidious contrasts which may be suggested by the word "real" are irrelevant to the present discussion.

In any event, this brief list of criteria does not exhaust the senses of 'real' or 'exist' that can be distinguished in discussions about the reality of scientific objects. It is long enough, however, to indicate that a proponent of the instrumentalist view of theories cannot give an unambiguous answer to the ambiguous question whether it is congruous with his position to accept the physical reality of such things as atoms and electrons. But the list is also long enough to suggest that there are at least some senses of the expressions 'physically real' and 'physically exist' in which an ironically minded instrumentalist can acknowledge the physical reality or existence of many theoretical entities.

More specifically, if the third of the above criteria is adopted for specifying the sense of 'physically real,' it is quite patent that the instrumentalist view is entirely compatible with the claim that atoms, say, are indeed physically real. In point of fact, many instrumentalists themselves urge such a claim. To make the claim is to assert that there are a number of well-established experimental laws related in a certain manner to one another and to other laws by way of a given atomic theory. In short, to assert that in this sense atoms exist is to claim that available empirical evidence is sufficient to establish the adequacy of the theory as a leading principle for an extensive domain of inquiry. But as has already been noted, this is in effect only verbally different from saying that the theory is so well confirmed by the evidence that the theory can be tentatively accepted as true.

Proponents of the instrumentalist position may, of course, reserve judgment about whether other theoretical entities postulated by the theory really do exist, since the requirements for their physical reality as set by the criterion adopted may not be clearly satisfied. But on such particular issues proponents of the view that theories are true or false statements may have similar hesitations. It is therefore difficult to escape the conclusion that when the two apparently opposing views on the cognitive status of theories are each stated with some circumspection, each can assimilate into its formulations not only the facts concerning the primary subject matter explored by experimental inquiry but also all the relevant facts concerning the logic and procedure of science. In brief, the opposition between these views is a conflict over preferred modes of speech.

Mechanical Explanations and the Science of Mechanics

7 The preceding chapters have been concerned almost exclusively with a number of over-all questions focal in current as well as older analyses of the character of explanations conforming to the deductive pattern. Additional issues arise, however, when the structure of explanations in various special areas of science is examined, even when attention is restricted to explanations of the deductive type. We must therefore consider a number of these more particular issues, but we shall have to discuss them in the context of the special systems of explanation in which they are generated. One such system is classical theoretical mechanics. Classical mechanics continues to be a basic part of modern physics as well as to illustrate an important type of physical explanation, despite the great changes that have taken place in physical science since the turn of the century. Accordingly, the present chapter discusses what is meant by a mechanical explanation and examines critical methodological issues raised by the axioms of the theory of mechanics.

I. *What Is a Mechanical Explanation?*

Mechanics is the first of the natural sciences to have achieved a unified system of explanation for the phenomena it claims for its province. Long before recorded history men had learned to use simple machines, such as levers and wheels, for lightening labor and accomplishing otherwise impossible architectural and industrial feats. In any event, by shrewd observation and trial-and-error methods, much information was accumulated concerning the mechanical properties of physical things. Nevertheless, the explicit formulation of mechanical laws, based on systematic analyses of pervasive mechanical relations, apparently did not begin until classical antiquity. One branch of mechanics, namely statics, reached an advanced stage of development by the time of Archimedes in the third century B.C. However, attempts to extend such analyses to cover the motions of bodies not in equilibrium were not entirely successful until the signal achievements of Galileo and Newton. A long line of subsequent workers—D'Alembert, Lagrange, Laplace, Gauss, and Hamilton, to mention only a few of the most illustrious names—finally recast and elaborated the fundamental principles of the science, and applied them to a surprisingly large number of diverse domains.

By the middle of the nineteenth century mechanics was widely acknowledged as the most perfect physical science, embodying the ideal toward which all other branches of inquiry ought to aspire. Indeed, it was the common assumption of outstanding thinkers, physicists as well as philosophers, that mechanics is the basic and ultimate science, in terms of whose fundamental notions the phenomena studied by all other natural science could and should be explained. "In true philosophy," Huygens declared in the seventeenth century, "one conceives the cause of all natural effects in terms of mechanical motions. This, in my opinion, we must necessarily do, or else renounce all hopes of our comprehending anything in Physics." Huygens' conviction was repeatedly echoed by outstanding scientists during the next 250 years; thus Hertz declared that "All physicists are unanimous in holding that the task of physics is to reduce the phenomena of nature to the simple laws of mechanics." [1] As recently as 1909 Painlevé, an eminent French mathematician, maintained that "Mechanics is the necessary foundation for the other sciences, to the extent at any rate that these latter wish to be precise." [2] Huygens was giving voice to the belief, shared by a substan-

[1] Christian Huygens, *Treatise on Light,* Chicago, n.d., p. 3; Heinrich Hertz, *Die Prinzipien der Mechanik,* Leipzig, 1910, p. xxix.
[2] Paul Painlevé, *Les axiomes de la mécanique,* Paris, 1922, p. 3. This essay first appeared in 1909. For an illustration of philosophic opinion, one may cite Wundt's claim that "Mechanics is the beginning and foundation for all explanatory natural science. It is the most general natural science, in so far as one attempts to reduce, on the strength of the postulate of the permanence of material substance, all natural phenomena given to external sense to the phenomena which mechanics studies, that is to motions of bodies and of their parts."—Wilhelm Wundt, *Logik,* 3rd ed., Vol. 2, p. 274. Note also the views of Kirchhoff and Helmholtz. Kirchhoff declared that "The highest object at which the natural sciences are constrained to aim, but which they will never reach, is . . . in one word, the reduction of all the phenomena of nature to mechanics."—Quoted by J. B. Stallo, *Concepts of*

tial portion of modern scientists, that explanations in terms of mechanics are the only alternative to obscurantist philosophy and the verbal physics of a decadent scholasticism.

The historical importance of the science of mechanics alone would make it worth careful study, but there are additional reasons for devoting special attention to it. In the first place, it exhibits in a relatively simple fashion the kind of logical integration which other branches of science aim to achieve; and it therefore illustrates distinctions of logic and method that are exemplified in other theories of science only in ways overlaid with greater technical complications. In the second place, its one-time pre-eminence as the most universal and perfect science, and its subsequent decline from this position, have provoked vigorous controversy concerning the adequacy of scientific method as traditionally conceived and practiced; and these discussions cannot be understood without clear notions concerning the nature and limits of mechanical explanations. Thus, it is an often repeated claim that many of the assumptions and modes of analysis associated with classical mechanics— for example, assumptions concerning the "strictly causal" or "rigorously deterministic" character of natural processes, or concerning the possibility of developing adequate theories which construe complex processes in terms of more elementary ones—are no longer supported by recent developments in the natural sciences and must be abandoned in favor of different conceptions of scientific method. In the third place, though mechanics has declined from the position of eminence it once occupied, new claimants for the position of a universal science of nature have appeared, to which all other sciences are to be "reduced." But these various claims can be understood only if the distinctive features of explanations in "mechanical terms" are at least relatively clear, and only if the circumstances under which a theory may serve as a universal system of explanation are made plain. An examination of the character of mechanical explanations thus promises substantial rewards, and it is to this task that we now turn.

1. What, then, is a mechanical explanation? Although the words "mechanical" and "mechanics" were employed by Huygens and his successors in a fairly precise sense, they are used ambiguously in much popular and even technical discussions and are at best only loosely defined terms. It will be well to note briefly at the outset the variety of contexts in which they frequently occur, and to specify subsequently the

Modern Physics, New York, 1884, p. 18. Helmholtz maintained that "the object of the natural sciences is to find the motions upon which all other changes are based, and their corresponding motive forces—to resolve themselves, therefore, into mechanics."—*Ibid.*

sense of "mechanics" and of "mechanical" which is germane to the science of mechanics. The words often occur in discussions of levers, pulleys, and pendulum clocks, but they are no less common in accounts of the modern automobile, electric clocks, and the photographic camera. Again, innumerable books take for their explicit subject matter the mechanics of such diverse processes as hearing, breathing, the transmission of hereditary traits, or the operation of political organizations; and researches which proceed on the assumption that biological organisms are physicochemical compounds are frequently characterized as illustrations of "mechanical materialism." Moreover, perfunctory responses of human beings to various social situations in which they may be involved are also sometimes described as "mechanical"; and certain musical and poetic compositions, as well as certain *theories* of music and poetry, are often described in the same way.

It is safe to say that there is no core of precise meaning common to these various usages of the words "mechanics" and "mechanical." It is indeed obvious that the sense of "mechanical" when used in judgments evaluating human performances is quite alien to the sense of the word in contexts of theoretical analysis in the natural sciences. Moreover, even in these latter contexts it does not always carry the sense associated with it in the *science* of mechanics. As the above examples make plain, it is commonly employed not only in analyses of problems studied specifically by that science, but also of thermal, electromagnetic, optical, chemical, physiological, and social processes which are usually not explained in terms of the characteristic notions of that discipline. In one broad usage of "mechanical," any answer to questions such as "How does it work?" or "How is it done?" is apparently a mechanical explanation, whatever may be the determining factors of the processes under discussion to which the answer calls attention. Accordingly, in this broad sense of the term all the sciences of nature provide mechanical explanations, to the extent that all the special sciences seek to discover the conditions under which things and events occur and to formulate the laws that express such relations of dependence. However, when the word is used in this over-all fashion, the above quoted expression of Huygens' conviction asserts little more than a truism. For even in the contexts of inquiries into human behavior, the only dissenters from that interpretation are likely to be those who believe that one is "killing meanings by explanations" whenever one seeks the various conditions upon which the "inner life of man" depends. Accordingly, to appreciate what Huygens meant and what historians of ideas have in mind when they characterize certain periods of scientific development as dominated by the ideal of mechanical explanation, we must examine the sense of "mechanical" which is specific to the classical science of mechanics.

However, though the standard definitions of the science of mechanics provide important clues to what this sense is, they are not very revealing without much further analysis. The customary definitions are variants of Maxwell's definition of mechanics as the science of matter and motion,[8] and these certainly delimit in a general way the scope of the science; for example, chemical reactions are *prima facie* excluded from its domain. However, there are few if any branches of physics which could not be viewed as inquiries into the motions of matter. For example, iron filings in the presence of a bar magnet assume definite positions, as does a magnetized needle in the presence of a wire carrying an electric current. But although these examples illustrate matter in motion, so that on the basis of Maxwell's definition they ought to fall within the province of mechanics, in point of fact they are often excluded from it. The proposed definition does not therefore make entirely clear what are the actual limits of the science of mechanics—indeed, the word "matter" is far too imprecise for defining anything clearly—and we must look deeper for an adequate account of the character of mechanical explanations.[4]

2. The most direct and satisfactory method for ascertaining the scope of a science and the distinctive character of its explanations is to turn to the comprehensive laws and theories—whenever such theories are available—that constitute at a given stage of the development the ultimate premises of its explanations. Fortunately, it is possible to do just this in the case of classical mechanics, for the content of the science is confined

[8] *Matter and Motion* is the title of J. C. Maxwell's book, first published in 1877. The following are some typical definitions by other writers: "Die reine Mechanik . . . [ist die] Lehre von denjenigen Erscheinungen, bei welchen ausschliesslich Bewegungen ins Auge zu fassen sind, als sie sich mit der Bewegung materielle Punkte, starre, flüssiger und elastische feste Körper beschäftigen."—Gustave Kirchhoff, *Vorlesungen über Mathematische Physik, Mechanik,* Dritte Auflage, Leipzig, 1883, p. iii. "Die Mechanik ist die Lehre von der Bewegungen der Naturkörper, d.h., der Ortsveränderungen . . . derselben, welche mit keinerlei Änderung ihrer übrigen Eigenschaften verbunden ist."—Ludwig Boltzmann, *Vorlesungen über die Prinzipien der Mechanik,* Leipzig, 1897, Vol. 1, p. 1. "Die Mechanik ist die Lehre von der Bewegung."—A. Voss, "Grundlegung der Mechanik," in *Encyklopädie der mathem. Wissenschaften,* Leipzig, 1901, Band 4, Part I, p. 12. "Die Mechanik ist die Lehre von den Bewegungsgesetzen materieller Körper."—Max Planck, *Einführung in der Allgemeinen Mechanik,* Leipzig, 1921, p. 1. "Mechanics . . . is defined in the specific sense as the study of the laws of motion of material bodies, i.e., the relative changes of position of such bodies with time."—Nathaniel H. Frank, *Introduction to Mechanics and Heat,* New York, 1939, p. 3.

[4] A useful hint concerning the actual subject matter of mechanics is supplied by the etymology of the word "mechanics." The word is derived from the Greek expression for a contrivance, where the contrivances are devices for lifting and moving weights, such as levers, inclined planes, wedges, and the wheel and axle. The study of such machines, with a view to discovering the various advantages they possess, is still regarded as a distinctive task of the science of mechanics.

fairly well within the framework of ideas provided by the Newtonian fundamental "axioms" or "laws" of motion. It will be sufficient, therefore, to examine these axioms, and to elicit from them the essential features of mechanical explanations.[5]

Newton's three axioms or laws of motion were stated by him as follows:

LAW I. Every body perseveres in its state of rest, or of uniform motion in a right line, unless it is compelled to change that state by forces impressed thereon.

LAW II. The alteration of motion is ever proportional to the motive force impressed; and is made in the direction of the right line in which that force is impressed.

LAW III. To every action there is always opposed an equal reaction: or the

[5] Some comments are in order concerning the choice of the Newtonian formulations as the basis for the present discussion. There are, of course, other formulations than Newton's of the theory of mechanics, for example those of Lagrange and Hamilton. These alternative formulations make it possible to analyze many complex problems with greater ease and flexibility than can be done in terms of the Newtonian one. Nevertheless, these alternatives are mathematically equivalent to the Newtonian schema, and nothing would be gained by using one of these generally less familiar alternatives as our point of departure. Again, some of these alternative systematizations of mechanics adopt as primitives theoretical notions which are different from those used by Newton. For example, in the Newtonian system the fundamental notions are those of space, time, force, and mass; in the system proposed by exponents of the science of energetics, the basic ideas are those of space, time, energy, and mass; and in the Hertzian representation of mechanics, the fundamental notions are those of space, time, and mass. There thus appears to be a lack of unanimity among physicists themselves as to which are the cardinal ideas of mechanics; and to the extent of this disagreement there may also be divergences on what constitutes the essential character of a mechanical explanation. In point of fact, however, this absence of unanimity is not serious for our purposes, and arises from circumstances that are analogous to the differences between alternative formulations of Euclidean geometry which employ different primitive ideas in building up the system. For, although a system of mechanics may reject the notion of force as a primitive theoretical idea, and may even dispense entirely with the use of the word 'force,' it is always possible to introduce the term into the system by way of a nominal definition. Moreover, as will soon be evident, differences between formulations of mechanics of the kind just indicated do not prejudice in any way the main outcome of the analysis of mechanical explanation. Finally, there are formulations of the theory of mechanics which adhere fairly closely to the Newtonian schema but which (like Boltzmann's formulation) make more explicit than did Newton the various assumptions upon which the system is developed. It seems therefore that one of these more careful formulations should be made the basis for the present discussion. However, while such more recent statements of the theory of mechanics are invaluable for the discussion of certain issues raised by the science, the refinements they introduce are not obviously pertinent for our present problems; and, when need for them arises, recourse to them will be made.

There are several recent attempts to introduce modern standards of rigor into the axiomatizations of Newtonian mechanics. Cf. J. C. C. McKinsey, A. C. Sugar, and Patrick Suppes, "Axiomatic Foundations of Classical Mechanics," *Journal of Rational Mechanics and Analysis*, Vol. 2 (1953), pp. 253-72; Herbert A. Simon, "The Axioms of Newtonian Mechanics," *Philosophical Magazine*, Vol. 33 (1947), pp. 888-905.

mutual actions of two bodies upon each other are always equal, and directed to contrary parts.

When these axioms are translated into current terminology and current notation of mathematical analysis, the Newtonian theory of mechanics asserts that:

a. If the external forces F acting on a body (whose momentum along a straight line is mv) are zero, then the time-rate of change in mv (which may happen to be zero in the limiting case, so that the body is at rest relative to that line) is also zero. That is, if $F = 0$, then $\frac{d(mv)}{dt} = 0$, where v is a vector or directed magnitude and m is the mass. In classical mechanics it is assumed that the mass of a body, which Newton called its "quantity of matter," is an invariable property of a body and is not affected by its motion. Accordingly, the formula for the first axiom can be written as: If $F = 0$, then $m\frac{dv}{dt} = 0$, or finally as: If $F = 0$, then $\frac{dv}{dt} = 0$.

b. If the external force acting on a body with mass m is F, then the time-rate of change in the momentum mv of the body is proportional to the magnitude of F, and has a direction that falls on the straight line along which F itself falls. That is, $\frac{d(mv)}{dt} = kF$, where 'k' is a constant of proportionality and F and $\frac{d(mv)}{dt}$ are vectors having the same direction. (The formula for the second axiom can be transformed into: $m\frac{dv}{dt} = kF$. By a suitable choice of units k can be set equal to unity; and if the time-rate of change in velocity is called the acceleration of the body and is represented as a vector a, the second law can be stated in the familiar form: $ma = F$.) [6]

c. If F_{AB} is the force a body B exerts on a body A, then there is a force F_{BA} which A exerts on B, such that F_{BA} is equal in magnitude but opposite in direction to F_{AB}. That is, $F_{AB} = -F_{BA}$, where the F's are vectors or directed magnitudes.

Two important sets of questions are immediately raised by these axioms: (1) What is to be understood by the various key terms in the

---

[6] To be quite explicit about the matter, on Newton's view the motion of a body undergoes alteration if there is a change either in its speed along a straight line or in the direction in which it moves. Accordingly, a body is undergoing acceleration during any period of time if its speed is increased or diminished or if the direction of its motion is altered. (If its speed is diminished it is undergoing negative acceleration.)

formulation? When a body is said to be at rest or in motion along a straight line, what are the "straight lines" with respect to which the body is assumed to be at rest or in motion, and how is the "time" of motion specified? (2) What is the status of these axioms? Are they "generalizations" from experience, are they propositions whose truth can be established a priori, or are they "definitions" of one kind or another? We shall pursue neither of these problems for the moment, though they will receive considerable attention presently. They are mentioned here only to indicate the difficult ground that must be traversed if a reasonably complete account of the structure of mechanics is to be achieved.

However, two related observations need to be made at this point. Newton asserts in the second law that the direction of acceleration of a body under the action of a force is along *the* straight line upon which the force itself acts. However, if a body has any appreciable spatial dimensions and if the force acts upon the whole of it, there is no *unique* straight line that determines the direction of the acceleration—for the different parts of the body are then accelerated along distinct straight lines. Accordingly the axioms of motion must be supposed to be formulated for so-called "point-masses"—for bodies whose masses are in theory concentrated at a "point." The application of the axioms to the motions of actual physical bodies, which are clearly not point-masses, thus assumes an extension of the fundamental theory to cover the motions of systems of point-masses that are subject to more or less rigid mutual constraints. Such an extension involves some difficult mathematics, though it does not require the introduction of any new theoretical ideas: the theoretical mechanics of solid bodies, fluids, and gases can be developed on the foundation supplied by the mechanics of point-masses, with the proviso that bodies with appreciable volumes are construed as systems of an indefinitely large number of point-masses. But the facts just noted make it evident that the axioms of motion are *theoretical* statements, in the sense of "theory" previously discussed; they are not statements about relations between experimentally specified properties, but are postulates implicitly defining a number of fundamental notions that are otherwise left unspecified by the postulates of the theory.

The second observation corroborates the point just made. Although the Newtonian axioms do not mention it explicitly, they do tacitly assume that spatial dimensions and temporal periods can be subdivided indefinitely, so that the magnitudes associated with them may be vanishingly small. The axioms also assume that the velocities and accelerations ascribed to point-masses are those which point-masses possess in the limiting case when the temporal periods involved approach zero—in brief, the axioms assume *instantaneous* velocities and accelerations for point-masses. Let us first be clear why such assumptions appear to be needed.

Suppose we wish to determine the velocity of an automobile moving along a straight and level road; and suppose we measure the distance it covers in one hour, find it to be 30 miles, and conclude that the velocity is 30 miles per hour. But it is evident that the automobile may be moving with a variable speed during that hour, and that the indicated velocity may not actually be the speed of the car during any part of the 30-mile journey. The velocity of 30 miles per hour may thus represent what is only the *average* speed of the vehicle. If we wish to obtain a more detailed account of the car's behavior, we would have to measure the speed during shorter periods of time, say during periods of one minute each; and we might find that during a certain minute the velocity is one mile per minute, while during another minute the velocity is one-quarter mile per minute. However, the point made about the car's possibly variable speed during one hour can obviously be repeated for one-minute temporal intervals; and still more smaller intervals could be taken—say of one second each—during which the speeds are successively determined.

Now as a matter of empirical fact this procedure of taking shorter and shorter periods for measuring speeds cannot be continued indefinitely, for there is a lower limit to the experimental discriminations we can make both of spatial and temporal intervals. However, the theory of mechanics seeks to provide a completely general analysis of the motions of bodies which is independent of the actual state of experimental technology; and the theory aims to formulate the structure of relations that characterizes bodies at all points of their motions. Newton therefore ignored the practical lower limit to the subdivision of distances and periods, and formulated the theory on the assumption that point-masses have limiting (or instantaneous) velocities and accelerations as the time intervals are diminished without limit. Indeed, Newton invented his "method of fluxions"—now called the differential and integral calculus—for the sake of handling such "instantaneous" aspects of the motions of bodies; and his axioms of motion, when stated in the language of mathematical analysis, take the form of differential equations of the second order.[7] But these facts simply confirm the remarks of the preceding para-

---

[7] A few further explanations must be supplied for the reader unfamiliar with the ideas of the differential and integral calculus. The fundamental notion of the calculus is that of the limit of an infinite series, whether of numbers or of functions. The notion of a limit of an infinite series of numbers can be illustrated as follows. Consider the infinite series: 30, 22½, 20, 18¾, . . . , whose terms are obtained from the formula $15(1 + 1/n)$ by assigning to n the values 1, 2, 3, 4, . . . , in succession. For the sake of visualization, we may suppose each term in the series to be the average velocity of a car, when the time intervals during which its velocity is being measured are successively 1 hour, 30 minutes, 15 minutes, 7½ minutes, and so on. Whatever the value of n is, the corresponding term in the series will differ from 15 by no more than $15/n$. Thus, if n has the value 10, the corresponding term 16½ differs from 15 by $15/10$; if n has the value 1000, the corresponding term

graphs that the axioms of motion when asserted with strict universality are not experimental laws, but that on the contrary they are theoretical postulates for which rules of correspondence must be supplied before they can be said to have any definite empirical content.

3. Let us consider finally what light the axioms of motion throw on the problem under discussion: the distinctive features of mechanical explanations. The axioms can be examined either with respect to their mathematical *form* or with respect to the kinds of *terms* they relate, and we shall adapt the discussion to this distinction.

What is meant by the *form* of a mathematically expressed statement is more easily illustrated than formulated, and a few examples will make the intent clear. The law of linear thermal expansion of solids is commonly expressed as $l + l_0[1 + k(T - T_0)]$, where l_0 is the length of the solid at some initial absolute temperature T_0, l its length at some arbitrary temperature T, and 'k' the coefficient of linear expansion, which

$15\frac{9}{200}$ differs from 15 by 15/1000, and so on. Accordingly, if a sufficiently large value is taken for n, all the terms in the series after a certain one has been reached will differ from 15 by less than any small positive quantity we may antecedently specify. Thus, if we wish to find a term in the series such that all terms following it will differ from 15 by less than 1/1,000,000, we must let n be equal to or greater than 15,000,000. In this example, 15 is the limit of the infinite series: it is the number such that the differences between it and successive terms of the series will gradually be less than any preassigned small positive number. The general definition of the limit of a series of numbers takes the following form: Let $x_1, x_2, \ldots, x_n, \ldots$ be an infinite series of numbers, and let ϵ be any small positive number. Then l is said to be the limit of the series if, for any specified ϵ, there is a term x_N in the series such that all terms following it (i.e., for all terms x_n, where $n > N$) will differ from l by less than ϵ.

Now let s be the distance a car travels, and to fix our ideas suppose the distance is related to the time t by the function $s = t^2$; that is to say, suppose that after t seconds the car has traveled $s = t^2$ feet. Now let the time be increased by an interval Δt (read "delta t"), so that the distance the car travels is increased by Δs. Accordingly, the car will travel a total distance of $s + \Delta s$ feet in $t + \Delta t$ seconds. It is evident that we must have $s + \Delta s = (t + \Delta t)^2 = t^2 + 2t\,\Delta t + \Delta t^2$. It will also be clear that the additional distance Δs the car covers as a consequence of traveling for the additional time Δt is given by the equation: $\Delta s = 2t\,\Delta t + (\Delta t)^2$. To obtain the velocity of the car during this additional part of its journey we need only divide Δs by Δt, so that $\Delta s / \Delta t = 2t + \Delta t$. This relation will hold no matter how large or small the interval of time Δt is taken, and by assigning different numerical values to Δt we will obtain an infinite series of velocities $\Delta s / \Delta t$. If, however, Δt is made progressively smaller so that it will approach zero as a limit, the ratio $\Delta s / \Delta t$ will also approach a limit—in this case $2t$. The limiting value of $\Delta s / \Delta t$ is represented by 'ds/dt,' and is called the first differential coefficient (or the first derivative) of s with respect to t. It is the instantaneous velocity of the body. It must be carefully noted that the differential coefficient 'ds/dt' is not an ordinary fraction with numerator 'ds' and denominator 'dt'; the expression must be considered as if it contained just one symbol which represents the limit of an infinite series of ratios.

Just as the instantaneous velocity of a body is the limit of an infinite series of velocities (and is represented by the first differential coefficient of the distance with respect to the time), so the instantaneous acceleration of a body is the limit of an

is constant for all bodies of the same substance but varies with different substances. The equation can be rewritten as $l - l_0 kT - (l_0 - l_0 kT_0) = 0$, and is a linear equation in the two variables '*T*' and '*T*.' Galileo's law for freely falling bodies, relating the velocity v a body possesses after falling t seconds from a position with its initial velocity v_0, is $v - v_0 = gt$, where '*g*' is a constant. This equation can also be rewritten as $v - gt - v_0 = 0$, and is also a linear equation in the two variables '*v*' and '*t*.' These two laws, each expressed as a relation between two variables, have the same mathematical form; they are obviously both special cases of the linear "matrix" in two variables: $ax + by + c = 0$, where '*x*' and '*y*' are the two variables, and '*a*,' '*b*,' and '*c*' are so-called "arbitrary constants." It should be noted that all these equations, in addition to containing variables, the numeral '0,' and the "arbitrary constants," also contain certain constant expressions which represent specific numerical relations and operations—namely, the relational sign '=,' the sign of algebraic addition '+,' and the (suppressed) sign '×' for mul-

infinite series of accelerations. But the acceleration of a body per unit of time is the rate of change of velocity per unit time; and accordingly, if we are considering the instantaneous velocities of a body at different instants, the instantaneous acceleration will be the limit of the rates of change of the instantaneous velocities as the intervals between the instants at which these velocities are considered become progressively smaller. Thus, the instantaneous acceleration of a body will be represented by the first differential coefficient of the instantaneous velocity with respect to the time; and in consequence the instantaneous acceleration will be represented by the *second* differential coefficient of the *distance* with respect to the time. Thus the instantaneous acceleration is written as $d^2 s/dt^2$.

It is the aim of the differential calculus to develop rules for obtaining the differential coefficients of any function. Thus, we have already seen that the first derivative with respect to the time of $s = t^2$ is $ds/dt = 2t$; and it can easily be shown that the second derivative with respect to the time of $s = t^2$ is $d^2 s/dt^2 = 2$. Each of these equations, $ds/dt = 2t$, and $d^2 s/dt^2 = 2$, is called a differential equation, simply because it contains a differential coefficient. The first of these equations is said to be a differential equation of the first order, and the other equation a differential equation of the second order, while the differential equation $d^2 s/dt^2 = 2ds/dt$ is said to be of the third order. Thus, the order of a differential equation is the order of the highest differential coefficient contained in it. The fundamental equations of the science of mechanics are differential equations of the second order.

As already indicated, the task of the differential calculus is to find the differential coefficient of any function with respect to some indicated variable. But there is a converse problem: given a differential equation, to find the functional relation between the variables contained in it, such that the expression of the function no longer contains any differential coefficients. This converse problem involves the process of integration, and in general it raises more difficult mathematical questions than the original problem of differentiating a function. We cannot even enter into a schematic discussion of it, and we shall leave the matter with some examples. Given the differential equation: $ds/dt = 2t$, the relation between the variables s and t which satisfies the equation is given by the function $s = t^2 + a$, where a is any constant. The solution of the differential equation $d^2 s/dt^2 = 2$ is given by the equation $s = t^2 + at + b$, where a and b are any constants. Thus the solution of a differential equation of the first order contains one arbitrary constant, and the solution of a second-order differential equation contains two arbitrary constants.

tiplication. Accordingly, two statements may be said to exhibit the same mathematical form with respect to a specified set of variables if both can be obtained from a common matrix by substituting the corresponding variables and special arbitrary constants for those occurring in the matrix.

Consider next Boyle's law relating the volume V and pressure p of an ideal gas at constant temperature: $pV = k$, where 'k' is a constant. This is a quadratic equation in the two variables 'p' and 'V,' and has a different form from the equations of the preceding paragraph. But consider also the simplest specialization of the general economic law of demand, according to which the demand for a commodity increases with a fall in price and diminishes with a rise in price, the specialization consisting in the assumption that the demand D and price P vary inversely.[8] This special case can be written as: $DP = c$, where c is a constant, obviously with the same form as Boyle's law. For both laws are substitution instances of the general matrix $xy = a$, where 'x' and 'y' are variables and 'a' is an arbitrary constant.

Now these various examples of laws possessing the same form suffice to make one point quite clear—that two laws may have the same form—without thereby implying that either one can serve as an explanatory premise for the other. The fact that the law of thermal expansion has the same form as the law for freely falling bodies supplies not the slightest reason for supposing that the former can be explained with the help of the latter. It is of course abstractly possible that one law of a given form may explain a second law of the same form. But if this should happen to be the case, it is not a consequence *merely* of their formal similarity.

This observation suggests the further conclusion that the distinctive feature of mechanical explanations is not to be found in the mathematical form of the axioms of motion. But we must examine this suggestion on its own merit. Unlike the above examples of experimental laws, the axioms of motion must be formulated as differential equations, as has already been noted. It will suffice for our purposes if we confine our attention to the second axiom. Assume, therefore, that a single point-mass is acted on by a force F, that the spatial coordinates 'x,' 'y,' and 'z' specify its position with respect to three mutually perpendicular axes of reference, and that the components of the force along these axes are F_x, F_y, and F_z. The second axiom may then be written as $m\,\dfrac{d^2x}{dt^2} = F_x$, with similar equations for the other components of the force. The axiom can thus be stated as a set of linear differential equations of the second order. Is it this fact that distinguishes as mechanical any explanation in which the axiom is a premise?

---

[8] Cf. Alfred Marshall, *Principles of Economics*, 8th ed., London, 1930, p. 99.

Since the axiom says nothing about the specific character of the force that may act on point-masses, let us assume for the moment that the force-function F may be specified in any number of ways, depending on the nature of the problem under discussion, and may therefore involve reference to a variety of magnitudes. Now there are theories in physics whose mathematical form is identical with that of the second axiom of mechanics, but which are nonetheless sometimes distinguished from mechanics. For example, the theory of electrostatics has the form of linear differential equations of the second order, and yet explanations in terms of this theory are not always counted as mechanical explanations. Again, Maxwell succeeded in transforming the fundamental equations of electromagnetic theory so that they assume the form of the Lagrangian equations of mechanics, an alternative formulation of the Newtonian axioms. But it does not follow from the fact that such a transformation can be made, and no physicist supposes that it does, that the laws of electricity and magnetism have thereby been explained by the theory of mechanics.

Indeed, the more general point can be made that there are some differential equations which play a fundamental role in several branches of physics, although these different areas of inquiry are in general not thereby regarded as falling into the province of a common theory. For example, the partial differential equation, known as Fourier's equation,

$$\frac{\partial u}{\partial t} = a \left(\frac{\partial^2 u}{\partial x^2} + \frac{\partial^2 u}{\partial y^2} + \frac{\partial^2 u}{\partial z^2} \right)$$

can be used to formulate the fundamental theory in hydrodynamics, heat conduction, static and current electricity, and magnetism. But this indicates no more than that diverse subject matters exhibit structures of relations that are abstractly or formally indistinguishable. It does not signify that what is distinctive of the corresponding theories for each of these domains is exhaustively rendered by the *formal* structure of the theory. The formal identity of distinct theories is of course a highly important fact about them. Such an identity makes it possible to employ mathematical techniques developed for one area of inquiry in many other areas as well; and the formal analogies between different theories, as well as the imagery that may be associated with the formalism, can be of immense heuristic value in the conduct of research.[9]

One final observation is in order. Although the axioms of motion do have the form of second-order linear differential equations, it is of interest to conjecture whether physicists would regard a mere change in

[9] Cf. Ernst Mach, "On the Principles of Comparison in Physics," in *Popular Scientific Lectures*, pp. 236-58; also his "Die Ähnlichkeit und die Analogie als Leitmotiv der Forschung," in *Erkenntnis und Irrtum*, Leipzig, 1920, pp. 220-31.

this form as sufficient ground for declaring such an altered theory to be no longer a theory of mechanics. Thus, suppose it were found that Newton had been mistaken in his assumption that the motion of bodies can be analyzed in terms of the time-rate of change of momenta, and that a more satisfactory theory can be developed in terms of the time-rate of change of *accelerations*. The fundamental equations of motion would then be *third-order* differential equations. However, it seems unlikely that, were this the essential difference between the new theory and the old, the former would cease to be regarded as a theory of mechanics. In point of fact, the alteration in the form of the equations of motion required by the general theory of relativity is much more radical than the one suggested by this hypothetical example. Nevertheless, explanations in terms of the amended theory continue to be regarded by most physicists as mechanical explanations.

4. In the light of all this, it seems reasonable to conclude that it is not in consequence of their mathematical form that the axioms of motion are to be viewed as the premises of a distinctive science. We must therefore turn to the second alternative mentioned earlier, and examine the kinds of *terms* the axioms relate, in order to ascertain the characteristic features of mechanical explanations.

But a serious difficulty now confronts us. It arises from the circumstance that, though the axioms (or the text that usually accompanies them) state explicitly what is the general character of some of the terms they relate, they do not do this for all the terms. Thus the second axiom asserts that the time-rate of change of the momentum of a body is proportional to the impressed force; and it is clear from this formulation that a certain relation is asserted to hold between the mass and the acceleration of a body on the one hand, and the impressed force on the other hand. But unless something further is said about the force, the second axiom cannot contribute anything to the analysis of actual motions. Special assumptions, such as the one made in Newton's gravitational theory, must be introduced concerning the force-function if the analysis is to make headway. Now, the difficulty is that neither the axioms nor the explanatory text Newton supplied for them indicate even in a general way what limitations, if any, are to be imposed on the force-function; and this information, which is crucial for fixing the distinctive features of mechanics, can be obtained only from a survey of the major types of problems to which the axioms have been traditionally applied. It is for this reason that perhaps no straightforward answer can be given to the question, "What is a mechanical explanation?"

We have already noted that two of the terms mentioned in the axioms of motion are the mass and the instantaneous acceleration of a body.

According to classical mechanics, mass is simply an "additive" property of bodies that is not altered with changes in the motions of bodies, and is manifested as the resistance a body offers to alterations in its velocity. Let us assume that the notion of mass is sufficiently clear, and that appropriate methods are available for assigning numerical values to masses. Consider next the notion of an instantaneous acceleration. This is defined as the limit of a series, each of whose terms is the ratio of the difference of two instantaneous velocities and a time interval; and an instantaneous velocity is defined as the limit of a series, each of whose terms is the ratio of a distance along a straight line and a time. Let us waive for the present all questions concerning how straight lines, distances, and times are to be specified, and assume that these notions are also sufficiently clear; but in any event, instantaneous accelerations and velocities presuppose nothing more than certain mathematical operations upon the measures of spatial and temporal relations. Accordingly, the first fruit of our analysis is that the axioms of motion involve reference to at least three kinds of magnitudes, namely, measures of space (including distances, angles, areas, and volumes), of time, and of mass.

There remains the more difficult problem of making clear what sort of characteristics are involved in the notion of force. Newton himself mentioned three different "origins" for impressed forces—percussions, pressures, and centripetal (central) forces; and this brief list suggests the kind of force-functions that are typical of the science of mechanics. However, we can obtain a somewhat better survey of the types of force-functions employed in classical mechanics by examining some of the larger modern treatises on the subject.[10] These are usually divided into four parts: (a) the mechanics of point-masses, which is the foundation for all the rest; (b) the mechanics of rigid bodies; (c) the mechanics of elastic or deformable bodies; and (d) the mechanics of liquids and gases.

a. Two main types of force-functions are employed in the mechanics of point-masses: *positional* forces, which depend only on the relative positions and masses of the point-masses in the system under consideration, but which often also depend on certain coefficients that characterize the elements of the system; and *motional* forces, which are functions not only of relative positions, masses, and such coefficients but also of the relative velocities of the point-masses and of certain temporal periods. Let us consider each of these.

Positional forces can be divided into two subgroups: *central* forces, where the accelerations between any pair of bodies are always directed toward a fixed point; and *constraining* forces, where the point-masses

[10] E.g., A. G. Webster, *The Dynamics of Particles, and of Rigid, Elastic and Fluid Bodies,* New York, 1922; and Georg Joos, *Theoretical Physics,* New York, 1934.

are constrained to move on some specified surfaces or curves. Perhaps the best-known example of a central force is Newton's gravitational force. Familiar examples of constraining forces are the forces acting on a simple pendulum; in this case the force-function is specified in terms of a variable distance and a coefficient whose value can be calculated from the gravitational constant and certain purely geometrical magnitudes.

Various types of force-functions are used in the case of motional forces. The study of damped vibrations (which is illustrated by a pendulum moving through a resisting medium such as air) requires a force-function which depends on the variable distance and variable velocity of a body, and upon two constant coefficients. One of these constants can be calculated from the gravitational constant and the geometry of the physical system; the other is the coefficient of damping, whose value depends on the particular medium in which the vibration is taking place. The force-function employed in the study of forced vibrations (or resonance) is specified in terms of factors already mentioned in connection with damped vibration, to which must be added a variable time and certain other constants that are functions of the geometry and periodicity of the physical system.

b. We come next to the mechanics of rigid bodies, which treats of such matters as the rotation of solids of various shapes around fixed points and axes, the vibration of compound pendulums, and the sliding and rolling of bodies on surfaces. Since for purposes of theoretical analysis a rigid body is viewed as an infinite aggregate of point-masses whose mutual distances are constant, the mechanics of rigid bodies can be developed from the mechanics of point-masses; and the force-functions of rigid bodies can be regarded as compounded out of those employed in the mechanics of point-masses with the help of various mathematical operations. However, in addition to the variables and constants of the kind already mentioned, the force-functions of rigid mechanics usually also contain a coefficient of friction. This coefficient is a constant for a given pair of bodies, but its value varies with the kinds of surfaces that the bodies may come into contact with during motions.

c. Still other kinds of coefficients are required in the mechanics of deformable bodies, that is, bodies whose constituent point-masses are capable of relative displacement. This part of mechanics analyzes among other things the impacts of bodies, their compression under pressures, and their elongation under strains. The most familiar of these additional coefficients is the linear coefficient of elasticity (or Young's modulus), whose value varies with different materials. However, several other coefficients of an analogous sort are required in more complex problems concerning deformations.

d. Finally, two coefficients that play important roles in the mechanics of fluids and gases are the coefficients of viscosity and surface tension, both of which vary with the kinds of substances under discussion.

Although we have now sketched the topics commonly covered in treatises devoted specifically to mechanics, we have by no means exhausted the uses of the Newtonian analysis of motion. For example, a body carrying an electric charge moves under the influence of another electrically charged body in a manner that can be formulated by the Newtonian equations. Despite the fact that in this case the force-function will contain terms referring to the magnitudes of electric charges (in accordance with the laws of electricity), explanations of such motions are also frequently called mechanical explanations. A similar observation holds for magnetizable bodies moving under the influence of magnets. Accordingly, in addition to the magnitudes previously cited, the force-functions in what are often regarded as mechanical explanations may also mention as determining factors electric charge, magnetic strength, and several other items. In short, a survey of actual physical practice shows that a large variety of problems can be successfully handled in terms of the Newtonian equations of motion, and that a considerable number of distinct factors may enter into the specification of the force-function. In consequence, the labels "mechanics" and "mechanical explanation" have a wide but by no means precise scope. Nevertheless, as we shall see in a moment, they can be given a narrower or a more inclusive connotation, according to various restrictions that may be imposed on the composition of force-functions, if the latter are to count as "mechanical" force-functions.

5. Let us, however, first summarize the outcome of this survey. The force-functions employed in mechanics are specified in terms of some or all of a set of "parameters," which are either *variables* or *constant coefficients*. The variables are in all cases spatiotemporal magnitudes: distances, angles, time intervals, velocities, and the like. The constant coefficients are of three main types: universal constants, such as the gravitational constant, which has the same value whatever the materials investigated; constants having different values in different problems, but which (like the constants required in the analysis of motions under constraint) can in principle be calculated from the universal constants and the geometry of the physical system under consideration; and co-

efficients like those of mass, elasticity, viscosity, electric charge, and magnetic strength, having different values for different bodies or materials, but whose magnitudes cannot in general be evaluated from such geometrical considerations and must be ascertained in some independent manner.

There appears to be but one universal constant employed in mechanics. In classical mechanics the mass of a body (a constant of the third type) is the Newtonian mass. It is an "intrinsic" property of a body and does not depend on the velocity of the body. Moreover, if m_1 and m_2 are the masses of two bodies, respectively, the mass of the system consisting of those two bodies is $m_1 + m_2$. On the other hand, in relativity theory the mass of a body is no longer constant, but is a function of its relative velocity and is no longer "additive" in the sense just indicated. For the sake of simplicity we shall assume that it is the Newtonian mass of a body to which reference is made in what immediately follows, but the discussion would not be substantially altered if by "mass" we understood the relativistic mass of a body. It is difficult to supply an exhaustive enumeration of constants of the third type. However, on the assumption that a list of such constants can be constructed,[11] it is possible to state what is characteristic of a mechanical explanation, in the sense of classical mechanics.

In the most inclusive sense of the phrase, a mechanical explanation is one that satisfies the following three conditions, which we designate as M: (a) Its ultimate premises assert that the time-rate of change in the momentum of a physical system is a function of the magnitude and direction of the forces acting on it. (b) The direction of the change in momentum of a body is along the direction of the impressed force; and the direction of such a change associated with several forces is along the direction of the vectorial sum of the component forces. (c) The forces are specified exclusively in terms of the spatiotemporal magnitudes and relations of bodies, a universal constant, and a number of constant coefficients (assumed to be listed exhaustively) whose values depend on the individual properties of a given system of bodies.[12]

[11] The *Handbook of Chemistry and Physics,* by Charles D. Hodgman and Norbert A. Lange, various editions, supplies tables of values for eight coefficients that clearly fall into this category.

[12] This account of the nature of mechanical explanation does not differ in substance from the traditional definition of mechanics as the science of matter and motion, or from the frequent characterization of mechanics as the science that deals with those properties of things that are "definable" in terms of mass, length, and time. However, the present account does attempt to make explicit what these more customary and compact formulations assert, and at the same time to correct some of the obvious defects in these formulations. It is not sufficient to say, for example, that mechanics occupies itself exclusively with properties definable in terms of mass, length, and time, for there is no *prima facie* reason for maintaining that, say, the

In point of fact, however, more restrictive conditions have sometimes been proposed upon an explanation if it is to qualify as a mechanical one. Let us consider some of the historically more important ones. Although it was Newton who propounded the theory of gravitation, he did not regard it as ultimately satisfactory because it involved the notion of "action at a distance"—a notion he regarded as "so great an absurdity that I believe no man, who has in philosophical matters a competent faculty of thinking, can ever fall into it." For he maintained that "it is inconceivable that inanimate brute matter should, without the mediation of something else which is not material, operate upon and affect other matter without mutual contact." [18] What Newton apparently desired, and what was made quite explicit by Descartes and his followers, was a theory of mechanics which employs only force-functions that correspond to contact action. Accordingly, if Newton's dissatisfaction with action at a distance is taken seriously, a special restriction will in effect be imposed on explanations which are to count as "genuinely" mechanical.

Nevertheless, the gravitational force-function is specified entirely in terms of the universal gravitational constant, distances, and coefficients of mass. It is therefore more economical in its use of distinct types of parameters than are the force-functions for contact action. For these latter usually involve not only spatial variables and coefficients of mass but also coefficients of elasticity, of friction, and of viscosity. On the other hand, those who have sought to restrict "genuine" mechanical explanations to explanations in terms of contact action have sometimes maintained that the various specific differences between substances (which are represented in the force-functions of contact action by these various special coefficients) must themselves be ultimately explained exclusively in terms of spatiotemporal differences (or at most in conjunction with differences in the distribution of masses) in the microscopic structures of those substances. Indeed, the Cartesian physics is the expression of just such an extreme ideal. "The notion of body," Descartes maintained, "consists not in weight, nor in hardness, nor color and so on, but in extension alone. . . . There is therefore but one matter in the whole universe, and we know this by the simple fact of its being extended. All the variations in matter, or diversity in its forms, depends

linear coefficient of elasticity is so definable while the coefficient of linear thermal expansion and the coefficient of electric resistance are not. It may well be the case that in terms of overt laboratory procedure the coefficients of all so-called properties of matter are defined on the basis of relations between lengths, masses, and times. It is nonetheless the case that not all branches of physics thereby become parts of mechanics.

[18] *Isaac Newton's Papers and Letters on Natural Philosophy* (ed. by I. Bernard Cohen), Cambridge, Mass., 1958, pp. 302-03.

on motion. . . . Motion is the transference of one part of matter or one body from the vicinity of those bodies that are in immediate contact with it, and which we regard as in repose, into the vicinity of others." [14] In point of fact, an important part of the history of modern theoretical physics consists in attempts to show that specific material constants, such as the coefficients of viscosity, can be explained in terms of a theory of mechanics of this more stringent type.

Accordingly, we can distinguish in the historical literature of physics at least three senses of "mechanical explanation," although additional distinctions could easily be made. We state these in order of increasing stringency. [15]

a. The least demanding requirements for counting an explanation as mechanical are the three conditions M stated above (page 170). These conditions neither require nor exclude the postulation of submicroscopic particles or processes (such as atoms or vortices in some hypothetical medium) in order to explain the motions of macroscopic phenomena. Since in general the parameters occurring in the theoretical premises of explanations satisfying these conditions are of different kinds (e.g., the constant coefficients may be coefficients of mass, electric charge, elasticity, friction, etc.), let us call such theories *unrestricted mechanical theories.* For the most part, mechanical explanations in physics are of this unrestricted type.

b. A more demanding requirement for a mechanical explanation is that it satisfy the first two of the conditions M but that the force-function be specified exclusively in terms of spatiotemporal variables, the universal constant of gravitation, and coefficients of mass. The Newtonian theory of gravitation provides explanations of this type. It is evident, however, that if a theory of mechanics satisfying this requirement is to be adequate for the usual range of problems treated in classical mechanics, it must postulate submicroscopic particles and processes. These postulated entities must then be analyzable in terms of the axioms of motion; and the spatiotemporal organizations of their masses must account for the specific differences between the properties of macroscopic bodies. On the other hand, a theory satisfying the present requirement does not necessarily have to employ force-functions having an identical form in every problem. It may adopt a force-function like the

---

[14] René Descartes, "The Principles of Philosophy," Part 2, Principles 4, 23, and 25, in *The Philosophical Works of Descartes,* translated by E. S. Haldane and G. R. T. Ross, Cambridge, England, 1931, Vol. 1, pp. 255, 265, 266.

[15] Cf. C. D. Broad, "Mechanical Explanation and Its Alternatives," *Proceedings of the Aristotelian Society,* London, 1919, Vol. 19, pp. 85-124. The discussion throughout this first section of the chapter leans heavily on Broad's paper.

one in Newtonian gravitational theory for dealing with one area of problems, and a force-function having a different form for another area. Since the parameters occurring in theories conforming to the present requirement are narrowly limited to what are often regarded as the distinctively mechanical ones, we shall call such theories *pure mechanical theories.* The familiar definition of mechanics as the science whose fundamental magnitudes are space, time, and mass can therefore be regarded as an elliptic formulation of the defining traits of such theories. The traditional conception of mechanics as the universal science of nature seems to have adopted pure mechanical theories as the ideal which the sciences should seek to realize.

c. Finally, an even more stringent condition is imposed if, in addition to satisfying the requirements for pure mechanical theories, a theory of mechanics is expected to employ only force-functions having a single prescribed form. For example, the force-function may be limited to the form associated with central forces (such as the inverse-square form of Newtonian gravitational theory), or it may be restricted to having the form for contact forces between perfectly elastic bodies. We shall call such theories *unitary mechanical theories.* Cartesian physics envisaged a mechanics something like this type, although, as has already been mentioned, the Cartesian ideal imposed even more stringent requirements, since according to it spatiotemporal parameters were the only permissible ones in an ultimately satisfactory theory. During the nineteenth century both Helmholtz and Kelvin sought to develop unitary mechanical theories in considerable detail, though only with limited success. What historians of ideas call the "mechanical conception of nature" appears to be the once widely entertained view that all phenomena of physical if not of animate nature can eventually be explained by a unitary mechanical theory.

This lengthy discussion thus shows that several answers can be given to the question, "What is a mechanical explanation?" Some of the answers are less precise than others, since, as we have seen, it is not possible to define sharply the class of unrestricted mechanical theories without first supplying an exhaustive list of the various constants that may occur in such theories, that is, without an exhaustive survey of all branches of science in which the Newtonian axioms (or their equivalents) play an explanatory role. The discussion suggests that similar questions about other branches of science—for example, "What is a biological explanation?" or "What is a sociological explanation?"—are not likely to receive answers with fewer qualifications or greater precision than the question we have been considering. Nevertheless, the discussion makes clear that there is a core of common meaning in all the senses of "mechanical explanation" that have been distinguished. Moreover, it illustrates a mode of approach for characterizing what is distinctive of various explanatory systems in different branches of science, and thereby for examining important methodological problems concerning relations of dependence between different explanatory systems. However, before considering such problems, we must first deal with a number of crucial issues raised by the axioms of mechanics.

II. *The Logical Status of the Science of Mechanics*

The Newtonian theory of mechanics has had a long and successful career, certainly a longer one than other modern physical theories of comparable scope. It is now a commonplace that the range of its explanatory powers is less extensive than was once supposed, and that the analyses it offers are in point of fact incorrect when applied to bodies with relative velocities that are sizable when compared with the velocity of light. Nevertheless, Newtonian mechanics will doubtless continue to be accepted during the foreseeable future as at least a remarkably good first approximation to an accurate theory for a large class of phenomena, and as a basis for many important practical arts.

Despite the unusually distinguished and successful role Newtonian mechanics has played in the history of modern science, its foundations have been under vigorous debate since Newton first formulated his axioms of motion. Moreover, although the axioms have received more than two centuries of critical attention from outstanding physicists and philosophers, there still is wide disagreement about what the axioms assert and what their logical status is. The axioms (or their logical equivalents) have been claimed to be a priori truths, which can be asserted with apodictic certainty; to be necessary presuppositions of experimental science, though incapable either of demonstration by logic or refutation by observation; to be empirical generalizations, "collected by induction from phenomena"; to be general hypotheses suggested by facts of observation, but no more than probable conjectures relative to the experimental evidence confirming them; to be concealed definitions or conventions, without any empirical content; or to be guiding principles for the acquisition and organization of empirical knowledge, but not themselves genuine instances of such knowledge.

The number of these alternative interpretations of the status of the axioms of motion is both impressive and disconcerting. For even if some of these alternatives are dismissed at once as obviously no longer tenable, enough remain to suggest that issues are at stake which are relevant to the logic of science in general, and not only to the science of mechanics.

It is the aim of the present section to examine some of the alternative views concerning the axioms of motion, partly for the sake of a firmer grasp of the logical problems arising in the science of mechanics, but in large measure in order to clarify further the general structure of theoretical explanations.

1. *The first axiom of motion.* It requires little to see that the first axiom of motion, taken in isolation from some needed textual explanations, is seriously incomplete as a statement intended to have an empirical content. To say that a body will persevere in its state of rest or uniform motion in a straight line, unless compelled to change its state by forces impressed on it, is to say nothing definite, if nothing further is stated as to (a) what is the spatial frame of reference to which the motion of a body is referred, (b) what is the system of chronometry used in measuring velocities, and (c) what are the marks by which the presence or absence of impressed forces is determined. For example, a body moving with uniform velocity relative to one system of spatial reference frames and clocks will possess an accelerated motion relative to a suitably chosen different system of reference. Accordingly, each of these questions has been the subject of much dispute, and each has been answered in different ways.

a.   Let us assume for the present that the spatial frame of reference to which the motion of a body is to be referred—whether this frame is Newton's "absolute space" or the "fixed stars" or something else—has been made sufficiently explicit; this question will occupy us at some length in Chapter 9. By way of introducing some of the difficulties involved in the remaining questions, let us consider one of the historically famous and influential arguments, whose object was to establish the first axiom of motion by a priori reasoning.[16] In his important *Traité de dynamique,* first published in the eighteenth century, D'Alembert declared:

> A body at rest will remain so, as long as an external cause does not move it. For a body cannot be brought into motion by itself, since there is no reason why it should move in one direction rather than another. From this it follows that once a body is put into motion by any cause, it cannot of itself either accelerate or retard this motion.
> A body put into motion by some cause must continue to move uniformly

---

[16] There have been many such arguments, proposed among others by Euler, Kant, Laplace, and Maxwell. The one discussed in the text has been selected because it is more explicitly formulated than are the more familiar ones. For a recent survey of discussion of the first law, see G. J. Whitrow, "On the Foundations of Dynamics," *British Journal for the Philosophy of Science,* Vol. 1 (1951), pp. 92-107.

and in a straight line, provided that no new cause, different from the one which put it into motion, acts on it. That is to say, as long as no cause different from the one initiating the motion acts on the body, it will continue to move forever in a straight line and traverse equal spaces in equal times.

For either the indivisible and instantaneous action of the force at the beginning of the motion is sufficient to make the body cover a certain distance, or the body requires for its motion the continued action of the moving force.

In the first case, it is clear that the space traversed can only be a straight line described uniformly by the body. For once the first instant is passed, the action of the moving force no longer exists (by hypothesis) and yet the motion continues. It will therefore necessarily be a uniform motion, since a body cannot by itself accelerate or retard its own motion. Moreover, there is no reason why the body should deviate to the right rather than to the left. Accordingly, in the first case (where we assume that the body is incapable of moving by itself for a certain time, independently of the moving force) it will move by itself during this time uniformly and in a straight line. However, a body which can thus move for a certain time, must continue to move forever in the same manner, if nothing prevents it from doing so. For suppose a body starts to move at point A and is capable by itself of traversing the line AB.

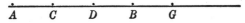

Take any two points C and D on this line which lie between A and B. Now the body at D is in precisely the same state as when it was at C, except that it is at a different place. Hence the same thing must happen to the body at D as happened to it at C. But at C it is (by hypothesis) capable of moving uniformly to B. Therefore at D it will be able to move itself uniformly to G, where DG = CB, and so on. Accordingly, if the initial and instantaneous action of the moving cause is able to move the body, the latter will be moved uniformly in a straight line, as long as no new cause prevents it.

In the second case, since it is assumed that no new cause different from the moving cause acts on the body, nothing will therefore determine the latter to increase or diminish. It follows from this that the continued action of the moving cause will be uniform and constant, so that during the time it is acting the body will move uniformly in a straight line. But from the same reason which makes the moving cause act uniformly and constantly during a certain time, continuing always so long as nothing hinders its action, it is clear that this action must ever remain the same and produce constantly the same effect.[17]

However, the argument founders at a crucial point, even if we overlook the difficulties in D'Alembert's tacit supposition that the notions of absolute rest and absolute velocity are physically significant. D'Alembert

[17] Jean D'Alembert, *Traité de dynamique*, Paris, 1921, Vol. 1, pp. 3-6.

simply assumes that a force is required to account for changes in the *uniform velocity* of a body (where the state of rest is a special case of uniform velocity), but that none is required to account for changes in the mere *position* of a body. But this is to beg the entire question at issue. Why should uniform velocity be selected as the state of a body which needs no explanation in terms of the operation of forces, rather than uniform rest or uniform acceleration (such as motion along a circular orbit with constant speed) or, for that matter, some still different state of motion of the body (for example, the constancy of the time-rate of change of the acceleration)? On purely a priori grounds these alternatives have equal merit, and none of them is logically self-contradictory; indeed, the Aristotelian mechanics of sublunary motion was based on the first of these alternatives, while the theory of celestial motions was based on the second.

Or consider D'Alembert's use of the so-called "principle of sufficient reason" (or principle of symmetry) to establish the conclusion that a body cannot put itself into motion, nor can it by itself accelerate or retard any motion it may possess, since if it could there would be no "reason" for the asymmetries that would thus be produced. But an analogous argument from symmetry could be used to show that, when a body which has been moving under the action of forces is freed from their influence, it will continue to move in an accelerated manner. Suppose, for example, that a body is moving with a constant speed along a circular orbit, so that it is undergoing acceleration. On the Newtonian theory the body must then be subject to a force directed to the center of the circle. Suppose now that this central force is eliminated. On the Newtonian analysis the body must then continue to move with the same speed along the *tangent* to the circle. However, one could argue for a different conclusion on the basis of considerations of symmetry: What "reason" is there for the body to change the character of its motion? Why should it move along the tangent rather than, say, along the radius to the tangent? For if it moves along the tangent, it will move to the left (or right) of the positions it would occupy were it to remain on the circle, and similarly for any other path than that of the circle. Accordingly, the body must continue to circulate on its original orbit. Such an argument is of course worthless. It is worthless simply because one can always exhibit in a given state of motion of a body a variety of distinct symmetries and asymmetries; and purely logical considerations do not suffice to determine which of these symmetries constitutes the actual determinants of the body's motion.

b. But although D'Alembert's argument does not establish what he believed it did, the argument can be (and has been) construed as establishing something else. What is the criterion, it may be asked, for asserting that a body is under the action of no forces? Suppose the answer is: the perseverance of the body in its state of rest or uniform motion in a straight line. If this is the criterion adopted for the absence of forces, then indeed D'Alembert's argument does establish the first axiom of motion by a priori reasoning. However, the axiom is then a concealed definition, a convention which specifies the conditions under which one will *say* that there are no impressed forces acting on a body. D'Alembert's argument is then an overlong and perhaps misleading proof of the truism that "Every body continues in its state of rest or uniform motion in a straight line, except in so far as it doesn't." [18]

Another ground has sometimes been advanced for maintaining that the first axiom is simply a definition. For, as already noted, in addition to assuming a definite spatial frame of reference, in its usual formulation the axiom takes for granted a definite system of chronometry as well as a criterion for the absence of forces. If, then, some method not involving the explicit or tacit use of the axiom were available for identifying the absence of forces, the axiom could be construed as an implicit definition of "uniform motion" or "equal time intervals." This is precisely the position adopted by some physicists, for example, by Kelvin and Tait when they reformulated the law to read "The times during which any particular body, not compelled by force to alter the speeds of its motion, passes through equal spaces are equal." [19]

But do these considerations establish the conclusion that the first axiom is "really nothing but" a concealed definition? Such a view was certainly not shared by Newton, D'Alembert, and other outstanding contributors to the science of mechanics. We must therefore consider what can be said in support of the interpretation of the first axiom as a statement having in some sense an empirical content.

Such an interpretation is possible only if the claim can be made good that absence of forces and equality of times can be identified without recourse to the first axiom. This claim rests partly on historical considerations, partly on considerations of actual scientific practice. Thus those who support the claim rightly note that, long before the first axiom of motion was formulated, men employed the notion of force and of the equality of times, and even developed methods for measuring them, however ill-defined and crude these notions and methods may have been. It is at least plausible that the idea of force had its origin in the felt

---

[18] Cf. A. S. Eddington, *The Nature of the Physical World*, New York, 1928, p. 124. Although Newton did not suppose that his first axiom is definitional in character, his exposition sometimes appears to commit him to just such a view. On this point see p. 187 below.

[19] William Thomson (Lord Kelvin) and P. G. Tait, *Treatise on Natural Philosophy*, Cambridge, England, 1883.

strains in muscles during physical exertions, and subsequently became associated with the behavior of beams, liquids, ropes, and springs exposed to various loads and pressures; [20] and the history of chronometry supplies many examples of mechanisms used to define and measure equality of times—for example, water clocks, hourglasses, standard candles—which were not constructed or evaluated on the basis of the axioms of motion. There is therefore overwhelming evidence *against* the thesis that the absence of forces or the equality of times can be determined *only* on the basis of the first axiom.

Further discussion of the definition and measurement of forces must be delayed until we turn to the second axiom of motion. But one general feature exhibited in the history and practice of time measurements now requires attention. It will perhaps appear as self-evident that if the first axiom is not to collapse into a definition, there must be a way for measuring time that is independent of the use of the law. But in any event, it is clear that if some periodic process is selected as a clock, in terms of which the equality of intervals of time is to be defined, the problem emerges as to which process is to be chosen for this purpose. For different periodical mechanisms do not appear to be equally good as clocks, since some beat our periods more "regularly" or "uniformly" than do others. The question thus naturally arises whether there is some way of identifying clocks which are "absolutely regular" or whether in the end a "true equality" of times must not perhaps be defined in terms of the first axiom (or some other theoretical postulate) so that the latter becomes a definition after all. It was a difficulty of this sort which led Newton to distinguish between "absolute" and "relative" time, but his definition of the former is useless as a practical basis for chronometry, quite irrespective of the question whether his definition is even "meaningful." [21]

c. However, since physics is an obviously flourishing science, it is clear that this difficulty can be resolved in some fashion, and we must now indicate schematically how this is done. To fix our ideas, suppose that a water clock of definite construction is specified at the outset as the measure of time. With its help we may then seek to establish laws connecting various processes with the time during which they run their

[20] Cf. Max Jammer, *Concepts of Force,* Cambridge, Mass., 1957.

[21] Newton's definition is "Absolute, true and mathematical time, of itself, and from its own nature flows equably without regard to anything external, and by another name is called duration: relative, apparent, and common time, is some sensible and external (whether accurate or unequable) measure of duration by the means of motion, which is commonly used instead of true time; such as an hour, a day, a month, a year."—*Mathematical Principles of Natural Philosophy,* Florian Cajori ed., Berkeley, Calif., 1947, p. 6.

course, the time being defined in terms of the standard water clock; and we may find that there is a rough regularity in the way in which these processes develop. For example, we may discover that a pendulum requires approximately the same number of (water-clock) temporal units to complete one oscillation, and that the distance a ball rolls down an inclined plane is always approximately proportional to the square of the (water-clock) time. But we may also find that, though there is this rough regularity, the pendulum on some occasions requires a longer (water-clock) time to complete a swing than on others, no matter how carefully we conduct our investigation and even after various disturbing factors that can affect the motions of pendulums (such as friction at the point of suspension, air resistance, and the like) have been identified and minimized; and similarly in the case of the ball rolling down the inclined plane. Now we could leave the matter there and conclude that these physical processes exhibit only an approximately uniform behavior, so that the laws we formulate for them are only approximately true. But there is another alternative open to us, namely, to declare the water clock to be "inaccurate" and adopt a different clock as our standard.

Now what do we mean by saying that the water clock is "inaccurate" if, as we are assuming, there is no instrument for measuring "absolute time"? And in what direction shall we look for a new clock if we decide to abandon the water clock as our standard? The general answer to the first question is that the water clock is inaccurate in the sense that, on the one hand, if it is taken as the standard, a large class of processes must be judged to exhibit irregularities in the time required by them to complete their cycles—irregularities which are apparently inexplicable in terms of identifiable disturbing factors—and on the other hand, if some other clock is adopted as the standard these irregularities disappear or are appreciably diminished. The answer to the second question is that we will seek as clocks periodic mechanisms which make it possible to compare and differentiate with respect to their respective periods an increasingly wider range of processes, and which enable us to establish with increasing precision general laws concerning the duration and development of these processes.

To see this more clearly, let us suppose that the water clock is abandoned as the standard measure of time, and a pendulum of specified construction is adopted for this purpose. And suppose further that many processes (e.g., the rolling of balls down an inclined plane, the transmission of sounds, the rotation of the earth, various chemical transformations) which appeared to be irregular when the water clock was employed as a standard now exhibit a greater if not perfect regularity. A definite advantage has thus been obtained from the adoption of a new timekeeper. For in consequence of this change, dependencies are discovered between the periods of various processes which might either have escaped our attention entirely had we retained the old clock or have required formulations so complex as to make them practically worthless. But it is obvious that there is no necessary limit to this process of abandoning one standard measure of time in favor of another, and that further gains may be won if the pendulum is replaced by, say, the rotating earth as the standard clock. The procedure here outlined illustrates what has been called the process of "successive definition," a process repeatedly encountered in the history of modern science.[22]

However, there is a final point that now demands attention, a point that bears directly on the logical status of the first axiom of motion. For suppose the rotating earth is adopted as the standard clock and that the first axiom is repeatedly confirmed with greater precision by suitable experiments and observations in many areas of investigation—all this on the assumption, of course, that the absence of forces can be independently identified. We might find, nevertheless, that there are some experimental deviations from what the axiom leads us to expect which cannot be put down to random errors of observation and which cannot be attributed to any identifiable factors that disturb those processes. Again a choice is open to us. We could conclude that the first axiom and the consequences entailed by it are only approximately true, and we could perhaps modify the axiom in some suitable manner. Such a modification might in turn require a considerable overhauling of other parts of the theory and involve complicating the formulations of many laws. On the other hand, we could accept the first axiom as fully accurate but attribute the experimental discrepancies from it to slight "inaccuracies" in the rotating earth as a timekeeper. But instead of specifying some other periodic mechanism to serve as the standard clock, we might now adopt the first axiom as the *criterion* for the equality of temporal periods—two times being *defined* as equal if during them a body moving under the action of no forces covers equal distances along a straight line. On this alternative procedure, therefore, though the first axiom was initially accepted on experimental grounds, it seems finally to acquire the status of a principle for *interpreting* experimental data, or of a *convention* which implicitly defines the equality of times.

Suppose this second alternative is adopted. Does this signify that the first axiom ceases to have any empirical content and is simply an arbitrary stipulation for measuring time? No straight and simple *general* answer to the question can be given, for the question as stated is essentially incomplete. The issue it raises can be settled only when some definite formulation of the theory of mechanics is assumed, a formulation which must include not only the axioms of motion but also a careful specification of the coordinating definitions adopted for its basic terms. For, as has been repeatedly noted, until an appropriate number of basic terms of the theory are associated with experimentally specifiable concepts, *all* the theoretical assumptions are abstract postulational definitions which possess an empirical relevance only proleptically. Now it is certainly possible to develop the theory of mechanics postulationally, so that in that formulation the first axiom is an arbitrary implicit definition of the equality of times; but it is equally possible to formulate the theory so that the first axiom does have an empirical content. It is, however, not always easy to know relative to which formulation of mechanics the status of the axiom is at issue. There is no one official formulation of the theory, and in different contexts different modes of articulating it may be assumed. Indeed, even in a single treatise, alternative foundations for the theory may be tacitly employed for different problems. Such shifts in modes of approach are not necessarily signs of confusion. They may illustrate only the flexibility with which definitions and empirical statements can sometimes be interchanged within a highly systematized and close-knit body of knowledge.

In point of fact, however, even in those formulations of the theory of mechanics in which the first axiom appears to have a purely definitional status, there are important (though sometimes neglected) empirical assumptions that control its adoption for this role. After all, even when the equality of times is defined in terms of the earth's rotation rather than in terms of the axiom, the discrepancies between what the axiom leads us to expect and what we actually find are not overwhelming. In a large number of instances falling into the range of application of mechanics, the outcome of observations actually made is the same, whether the earth is taken for the timekeeper or whether the period of its observed rotation is "corrected" in the light of the definition of equality of times supplied by the axiom.

Moreover, and this is the crucial point, although the axiom may indeed function as a convention for defining equality of times in terms of the behavior of a *given* physical system (upon which, by hypothesis, no external forces are acting), it is not by *convention* that other such systems exhibit similar regularities during time intervals defined as equal by the motion of the first system. Suppose, for example, that we adopt as our standard clock a given body A which is under the action of no impressed forces, and that by definition two temporal periods are said to be equal when A covers equal distances along a straight line during each of those periods. Thus far we are using the first axiom simply as a convention, so that the statement that A moves with uniform velocity is "true by definition." But suppose further that another body B is also moving along a straight line in the absence of any impressed forces. It is then clearly not a matter to be settled by a convention whether B requires equal times (as defined by A's motion) to traverse equal distances, for whether this is actually the case can be decided ultimately only by observation of B's motion. Accordingly, the first axiom can be said to be a convention only in the limited sense that it may be used to define the equality of times in the context of the motion of a *particular* physical system. The axiom cannot be correctly said to be merely a convention if, when this definition is adopted, an indefinitely large class of systems exhibits periodicities of motion essentially like the periodicities of the system taken as the standard one, so that in consequence any one of these systems is as suitable for the role of a standard clock as is the initial system. In short, when the equality of times is once defined by the motion of a given body, the axiom is not "true by convention" if in fact a great many bodies are found to move in conformity with the axiom.

Some conventions there undoubtedly must be in theoretical science, for terms do not get defined by themselves. Moreover, the exact locus in the articulation of a theory may not be fixed and may vary with the particular formulation a theory receives. In consequence, a *sentence* that is employed in one formulation of a theory or in one context of usage as a convention or form of definition may function in some other formulation or context as a statement of matters of empirical fact. It is nevertheless a patent error to conclude that such a sentence (an illustration of which may be the sentence formulating the first axiom of motion) is nothing but a convention in *all* contexts, or that because part of the empirical meaning of a statement is fixed by a convention the statement is itself simply a convention.

d. However, there still remains the question whether, even if the first axiom is not deliberately employed as a convention, it can rightly be regarded as a statement possessing an empirical content. That the axiom has no such content has been often maintained, quite apart from the fact that it is formulated in terms of the notion of point-masses and instantaneous velocities. This contention is frequently based on the claim that no bodies are exempt from the action of all forces, and that in point of fact no bodies have ever been observed which move with unchanged velocities over indefinitely long distances.[23] The claim is undoubtedly

[23] Cf. Henri Poincaré, *Foundations of Science*, New York, 1921, p. 94. "Many elementary text-book writers content themselves with observing that when a hockey puck slides on ice, the smoother the ice the farther the puck travels with a given blow before coming to rest. They then ask to imagine that the ice becomes in the

just, and is fatal to the view that the first axiom is an inductive generalization from observed instances—in the manner in which, say, "All crows are black" is a generalization based on the observation of a number of black crows. But, though the axiom is not an inductive generalization in this sense, may it not have an empirical content nevertheless, and may it not rest on experimental evidence of a more indirect sort?

Two lines of argument are sometimes used to support an affirmative answer to these questions. The first runs briefly as follows. It may be true that bodies are always under the action of some forces, and that none has ever been observed to retain indefinitely a constant velocity. However, some bodies can be found which are subject to fewer forces, or to forces of smaller magnitudes, than are other bodies; and some bodies can be progressively if not completely isolated from the influence of forces. If such bodies are viewed as occupying positions in a series according to the degree of isolation they exhibit, then the motions of bodies occupying positions further along in the series deviate less from a state of uniform velocity than do the motions of bodies at earlier positions. The first axiom formulates this complex set of facts in terms of a postulated *limiting* motion, were the series ideally prolonged without limit. However, the axiom must not be read with a sort of myopic literalness; it should not be construed as asserting that there are in fact bodies under the action of no forces, or as requiring for its validity the existence of such bodies. For the language of limits must be handled with care. In physics, as in mathematics, the assertion that a series of terms has a limit is often best construed as simply a way of stating a relational property characterizing the unquestionably existing members of the series, rather than as a statement which affirms the (possibly doubtful) occurrence of some term not initially assumed to be a member of the series. Accordingly, the first axiom does have an empirical content, for the axiom formulates certain identifiable relational characteristics of the actual motions of bodies, all of which are under the action of forces, when these bodies are serially ordered.

limit perfectly smooth—an ideal surface which has no effect on the puck. The assertion is thus made that the puck would continue indefinitely in a straight line with constant velocity. As a suggestive illustration one can hardly criticize this, though it is well to point out that the surface must be idealized beyond the limit suggested, i.e., it must be made infinite in extent and, more important still, must be flat, i.e., cannot be on the surface of the earth. A perfectly smooth sheet of ice level with the earth will not answer, for in this case the path will not be a straight line at all. It would not even be a great circle, owing to the effects of gravity coupled with the rotation of the earth. In other words, the illustration which sounds at first not bad proves very unfortunate on closer inspection. Probably much the same thing would be true for any attempted large-scale phenomenal illustration of the first law of Newton."—R. B. Lindsay and H. Margenau, *Foundations of Physics*, New York, 1936, p. 89.

The second line of argument is in part an implicit criticism of the first. It begins by noting that it is in general impossible to specify the content of one part of a theory independently of the theory as a whole. In particular, it maintains that we cannot experimentally verify the first axiom in isolation from the entire theory of mechanics, if for no other reason than that any such proposed verification involves assumptions concerning the forces that may be acting on bodies, and therefore involves the use of other parts of the theory of mechanics. The correct way of putting the question, accordingly, is whether the *theory* of mechanics has an empirical content, where the "theory of mechanics" must be understood to include not only the three axioms of motion together with the coordinating definitions for its various terms, but also the special assumptions that are customarily made concerning the force-function. When the question is stated in this way, however, the answer is a clear affirmative, since no one seriously doubts that the theory does have much to say about the constitution of actual motions. In consequence, since the first axiom is essentially involved in the analyses of motions made by the theory, it too has an empirical content. For example, the theory analyzes the motions of a planet attributed to the gravitational force of the sun by resolving the force into two components, one along the tangent to its orbit and the other along a line directed to a fixed point, the center of mass of sun and planet. In conformity with the theory, however, the motion along the tangent is then assumed to be in accordance with the first axiom, so that accelerations in the motion of the planet at any point of its theoretical orbit must be directed toward the center of mass of the system. Since such an analysis is highly successful, in the sense that the theoretical orbit deduced in conformity with these assumptions is in good agreement with observed positions of the planet, the evidence which thus confirms the theory as a whole confirms the first axiom as well.

Accordingly, the claim that the first axiom is a general hypothesis requiring confirmation from experiment, and that it therefore does have an empirical content, is not without at least a *prima facie* foundation. But we shall not evaluate this claim, nor relate it to the various counterclaims that have already been discussed, until we have examined the remaining axioms of motion.

2. *The second and third axioms of motion.* It is convenient to consider the remaining two axioms together. As with the first axiom, we shall assume that the spatial frame of reference to which they refer the motions of bodies has been specified in some satisfactory manner. Accordingly, since the measurement of time in classical mechanics has already

been sufficiently discussed, only two further notions remain to be examined, that of force and of mass.

a.   The notion of force has been the source of much difficulty in the foundations of mechanics. As has already been noted (page 179), it doubtless originated in familiar experiences of human effort; and much of the language in ordinary treatises of physics suggests that, when bodies are said to "attract" or "repel" one another or to "exert forces" upon one another, something like the strains we feel in our own organisms is being imputed to the dynamic transactions of inorganic nature. Indeed, the suggestiveness of ordinary physical language goes even further, and expressions like the "action of forces" seem to envisage forces as substantial "entities" with a "being" or mode of existence in their own right independent of the bodies upon which they may act. Much of the critical work on the foundations of science, especially during the nineteenth century, has been directed to the elimination of such anthropomorphic notions from physics; and probably no physicist today, even when he uses anthropomorphic language, intends it to be taken seriously or as more than a convenient mode of speech.

This critical housecleaning made clear that forces conceived in analogy to felt exertions or substantial agents have no identifiable role in the theory of mechanics and could therefore, by Occam's celebrated razor, be banished from that science as so much useless lumber. An essential requirement for a concept in a quantitative discipline is that it be associated with ways of recognizing and measuring the property it formulates; and an anthropomorphic notion of force satisfies the first part of this condition only within severely limited areas, and the second part not at all. But, although there is complete unanimity among physicists on the need and efficacy of the housecleaning, there is far less agreement as to what, if anything, should replace the banished idea in mechanics.

Newton's own discussion of the notion of force presents a curious puzzle. His explicit definition of "impressed force" is that it "is an action exerted upon a body, in order to change its state, either of rest, or of moving uniformly in a right line." [24] This formulation does not, in so many words, *equate* "impressed forces" with "change in the state of motion of a body"; on the contrary, it associates forces with *actions* (or *causes*) which change the momenta of bodies, so that these latter changes appear to be simply the *effects* of force. However, Newton offers no general measure for forces except in terms of changes in momenta; and however else forces may be identified, they are to be meas-

[24] Isaac Newton, *op. cit.*, p. 2.

ured in terms of the accelerations to which they give rise. On the other hand, the second axiom asserts that a change in momentum is proportional to the impressed force. But clearly, if impressed forces are measured in terms of changes in momenta, then what the axiom appears to assert is simply that the change of momentum of a body is proportional to the change in momentum. Far from being an axiom of motion, the second axiom seems on this analysis to collapse into a blatant logical truism.

There can be little doubt that Newton did not intend anything like this by his axiom. But whatever he did mean by it, the view that the second axiom is simply a nominal definition of the term "force" has been widely adopted, especially by those physicists who believe such a definition of "force" to be the only alternative to an anthropomorphic and "metaphysical" account of the notion.[25] This view is fostered by the common practice of stating the second axiom in the equational form $F = ma$, which suggests that what is being asserted is an *identity*, and therefore that the formula expresses an analytical truth. It is of course obvious that those who define "force" in this way must supply an independent definition of "mass" not involving the use of the second axiom; for the definition of "mass" which is sometimes proposed as the ratio of force to acceleration would make the explication of "force" as mass times acceleration a circular one. It is also clear that, if the second axiom is taken to be a definition, the first axiom must also be regarded as a convention, since then there is no way of recognizing the absence of impressed forces except in terms of the uniform motions of bodies.

What is to be said for the view that the second axiom is simply a definition? Some of the points noted in our discussion of the status of the first axiom are relevant to the present question. There can be no doubt that a consistent formulation for the theory of mechanics can be given such that, if 'mass' and 'acceleration' are either taken to be primitive terms in the system or defined without reference to forces, the term 'force' is defined as 'mass times acceleration.'[26] In such a formulation

---

[25] Ernst Mach appears to have been the first explicit proponent of this view. See his article "Über die Definition der Masse," which appeared in 1868 and is included in his *History and Root of the Principle of the Conservation of Energy*, Chicago, 1911, pp. 180-85. Kirchhoff took a similar view, as did Boltzmann.

[26] For example, in Mach's reformulation of mechanics (*Science of Mechanics*, La Salle, Illinois, 1942, p. 304), he replaces Newton's definitions and axioms by the following:

a) *Experimental proposition.* Bodies set opposite each other induce in each other, under certain circumstances to be specified by experimental physics, contrary *accelerations* in the direction of their line of junction.

b) *Definition.* The mass ratio of any two bodies is the negative inverse ratio of the mutually induced acceleration of those bodies.

c) *Experimental proposition.* The mass ratios of bodies are independent of the

there is indeed no need for retaining the word 'force' except as a convenient abbreviation for a longer expression, since whenever the word does occur it can be replaced without loss of meaning by its defined equivalent. On this formulation of the theory, Newton's second axiom can be omitted without loss, since the axiom then states an analytical truth. Accordingly, if the claim that the second axiom is nothing but a definition maintains no more than that the theory of mechanics *may* be formulated in the indicated manner, it is based on solid foundations.

However, those who interpret the second axiom as a definition frequently wish to maintain more than this. They often assume that there is no alternative to this interpretation, on pain of relapse into a "metaphysical" conception of force. It is this more radical claim that must now be examined, and an attempt will be made to show that it is a mistaken one.

Those who adopt the view that the second axiom as it occurs in the Newtonian formulation of mechanics does have an empirical content are faced with two questions: (i) Is it possible to supply a *general* measure of force which is independent of the use of the second axiom? (ii) And in any case, is it possible to construe the axiom so that it does not reduce to a definition, without introducing any anthropomorphic or other suspected meanings for the word 'force'? An affirmative answer to the first question entails an affirmative answer to the second. However, as will be argued, a negative answer to the first question (if, for example, it turns out that it is not *always* possible to measure forces without relying on the second axiom) does not necessarily require a negative answer to the second one. But let us see what can be said on each of these issues.

i. It has already been noted that even before Newton, men recognized the occurrence of *statical* forces—that is, of forces associated with bodies in equilibrium—and developed methods for measuring them. For example, the primitive notion of weight is that of such a force, and weights can be measured with the help of levers and springs without requiring the assumption of the second axiom of motion. Now in some cases it is possible to employ the notion of statical force in situations in which bodies are not in equilibrium and to investigate experimentally the rela-

character of the physical states (of the bodies) that condition the mutual accelerations produced, be those states electrical, magnetic, or what not; and they remain, moreover, the same, whether they are mediately or immediately arrived at.

d) *Experimental proposition.* The accelerations which any number of bodies *A, B, C, . . .* induce in a body *K* are independent of each other. (The principle of the parallelogram of forces follows immediately from this.)

e) *Definition.* Moving force is the product of the mass value of a body with the acceleration induced in that body.

tions between the magnitudes of such a force and the accelerations they produce. A familiar example is that of a small weight moved by a long spring coil to which it is attached; the tension (or static force) in the spring can be measured by its extension, and the variations in the momentum of the weight at different portions of its path can also be measured. Hence it is possible in principle to determine experimentally whether or not the acceleration of the weight at different portions of its path is proportional to the corresponding extension of (and therefore to the static force exerted by) the spring.[27] Accordingly, there are circumstances in which it is possible to identify and measure forces independently of the second axiom, and in consequence to seek experimental evidence for the latter.

However, it is not in general possible to do this, either because technical means may not be available for measuring the static forces assumed to be present in a given situation or because the notion of static forces cannot be significantly extended to many cases involving the motion of bodies. The first alternative raises no fundamental issue, and need not keep us; but the second does. If we impute the accelerated motion of a planet to a force acting on it, there seems to be no way of identifying such a force with a central force which is measurable, even in principle, by experimental means that do not assume the second axiom. It is fanciful imagination, not physics, which would measure the imputed force acting on a planet by means of an extended spring connecting the planet and the sun. In such cases—and they constitute the large majority of cases analyzed by the theory of mechanics—the magnitude and direction of the hypothetical forces acting on bodies are calculated from the accelerations they induce in these bodies. Accordingly, the answer to the first question is in the negative: physics has not found it possible, thus far at any rate, to supply a general measure of force which is independent of the second axiom of motion.

ii. But does it follow from this that the axiom must be regarded as simply a definition of "force," even in those cases in which no independent measure of force is available? The supposition that it must be so regarded arises partly from the circumstance that in the *explicit* formulation of the axiom it is not customary to say anything further about the force-function *F*, although in point of fact the function is *tacitly* assumed to be of restricted type and to satisfy certain implicit conditions. As was noted in the first part of the present chapter, when the axiom is actually used for the analysis of a problem, a specific force-function

[27] This example is given by Norman R. Campbell, *Physics, the Elements,* Cambridge, England, pp. 559-60. Cf. also Otto Hölder, *Die Mathematische Methode,* Berlin, 1924, p. 410, for a similar example.

must be adopted which has a definite form and contains explicitly stated variables and constants. The axiom does not explicitly prescribe the specific character of the function, which may vary from one class of problems to another class; the physicist at work on a problem must rely on his own ingenuity and good fortune to find a function appropriate for the case at hand.

Nevertheless, the physicist's choice is tacitly confined to a fairly limited class of functions, however vague the boundaries of this class may be. The force-function will in general depend exclusively on the relative distances of the physical system under investigation from other systems, on certain material constants (that may be either universal constants or specific for a given system), and possibly on the relative velocities of the systems or on the magnitudes of certain temporal intervals. Moreover, the function will normally have such a form that its numerical value tends to decrease as the relative distances mentioned in it increase. And finally, the function will in general be required to have a relatively "simple" form, even if the "simplicity" tacitly demanded cannot be articulated precisely, may be almost entirely a psychological matter, and is likely to change as mathematical techniques for solving differential equations improve. Indeed, unless some condition of simplicity however vaguely conceived is imposed on the force-function, the axiom runs the risk of being trivially true. For it is easily demonstrable that if no restrictions are placed on the complexity of a mathematical function, a force-function can always be constructed whose numerical values are equal to the changes in the momentum of a body. In brief, what the second axiom can be taken to assert is that there are determinants for changes in momenta which can be formulated in a relatively simple manner and can be specified in terms of the spatial configurations and certain physical properties of bodies. Accordingly, if the class of functions to which F is restricted is designated by 'K,' then instead of stating the axiom in the form that gives it the appearance of a definitional equivalence (namely, "The force F is equal to the product of the mass and the acceleration"), it is more clear and less misleading to formulate it on the present interpretation of the axiom as "For every change in the momentum of a body, there is a force F such that F is a member of K, and $F = ma$."

However, two additional points bearing on this interpretation should be noted. In the first place, there is an obvious sense in which the notion of force plays only an auxiliary role in mechanics. On the present interpretation, as well as on the view that the second axiom is simply a definition, the term 'force' is only a convenient means for making general statements. For even in the above emended formulation of the second axiom, the axiom as it stands does not suffice for solving mechanical problems; the solution can be found only after a definite force-function has been adopted. In consequence, the differential equations upon which the actual solution of problems depends simply connect changes in momenta on the one hand with a number of variable magnitudes and constants related in a certain manner on the other hand; and these differential equations dispense entirely with the word 'force.' Thus, if the Cartesian coordinates of a planet with mass m are x_1, y_1, z_1, and of the sun with mass M are x_2, y_2, z_2, and if the variable distance between the two bodies is r, the differential equations of motion take the form:

$$m\frac{d^2x_1}{dt^2} = G\frac{mM(x_1 - x_2)}{r^3}$$

with similar equations for the remaining coordinates. They assert that the time-rate of change in the momentum of each body is proportional to the product of their masses and inversely proportional to the square of their distances; and they neither mention the word 'force' nor presuppose its use. From this perspective, therefore, there is no fundamental difference in ultimate outcome between the present interpretation of the second axiom and the view that the latter is only a nominal definition of the word 'force.' However, it is undoubtedly convenient to retain the word in the exposition of the general theory of mechanics. For it is useful to have an expression which covers the various force-functions that may be employed in different problems, especially since the class of such functions is only vaguely delimited and cannot be exhaustively enumerated. With its help, moreover, it is possible to establish many general theorems that are valid for large classes of physical systems to which the theory of mechanics is applicable, quite irrespective of the particular character of the assumed forces—for example, the theorem that if no forces are acting on a system of bodies, the sum of their momenta is conserved throughout their motions.

In the second place, although on the present interpretation the second law does have an empirical content, it is nevertheless incapable of ever being decisively refuted by any conceivable experiment. For the axiom does not specify a definite force which will account for some particular acceleration; it merely asserts that *there is* a force which satisfies certain tacitly assumed conditions and which it is the business of the physicist to specify in detail. But a statement of the form 'There is a force F such that . . .' can be shown to be false only if it is possible to establish its contradictory, namely, a statement of the form 'For all forces F, it is not the case that . . .'; and in general one could establish the latter only if one could examine exhaustively all possible force-functions satisfying the stipulated conditions. It is clear, however, that such an examination could never be completed, for the number of abstractly possible force-functions is not fixed and may exceed any finite limit. Accordingly, although the axiom may be confirmed through the discovery of appropriate force-functions which successfully account for the acceleration of bodies, it can never be shown to be false.

Now it is for this reason that the second axiom is frequently regarded not primarily as an assertion about the conditions under which accelerations occur but rather as a compact formulation of a specialized *guide* for research, as a *methodological rule* which directs the physicist what to look for when he is analyzing the motions of bodies. For in respect to the feature of being incapable of conclusive refutation, the second axiom is very much like such a rule. Any number of failures on the part of a physicist to find what the axiom directs him to obtain, do not necessitate abandoning the search and scuttling the rule. For the rule may nonetheless be a good one because research conducted in accordance with it may have been generally rewarded with success and because even a rule that is only sometimes useful may be better than no rule at all. In point of fact, the second axiom viewed as a regulative principle has been eminently fruitful in guiding the construction of a systematic body of warranted knowledge; and if it continues to be accepted as a rule of procedure, it is clearly not because it is an arbitrary and unfounded rule for investigating the motions of bodies. On the other hand, though the axiom is literally not refutable by such investigations, consistent failure in certain domains to discover what the axiom directs us to find may make it advisable to abandon it as a methodological rule, whether permanently or only temporarily, and to replace it by a more useful directive. This has indeed been the fate of the second axiom.

b. Let us turn finally to the notion of mass and the third axiom of motion. Newton was especially careful to indicate what he believed was the experimental basis of this axiom. He cited a number of experiments by others as well as by himself which confirmed the thesis that, when one body is acted upon by another, the change in the momentum in the first is equal in magnitude but oppositely directed to the change in the momentum of the second. However, the experimental determination of these magnitudes obviously presupposes the measurement of mass, and Newton's account of this notion is notoriously unsatisfactory. He defined the 'mass' of a body (or its 'quantity of matter') as the product of its density and volume; but since he nowhere indicates how density is to be measured, and since the density of a body is commonly defined and measured in terms of the body's mass and volume, his account of mass is quite useless.[28] What, then, is to be understood by 'mass' (which must be clearly distinguished from 'weight'), and how are masses to be measured?

It is sometimes said that by the 'masses of bodies' we are to understand nothing more than the set of numerical coefficients which satisfy the equational form of the third axiom, so that on this view the axiom is simply another convention—this time for defining the relative masses of bodies. Thus, if two bodies A and B induce in each other the relative accelerations a_{AB} and a_{BA} (where a_{AB} is the acceleration of A induced by B, and analogously for a_{BA}), then on the view under consideration the masses of A and B are two numbers m_A and m_B so selected that $m_A a_{AB} = -m_B a_{BA}$. If this view is correct, then Newton was clearly engaged in a useless search when he attempted to supply an experimental foundation for his third axiom.

However, we are by now perhaps sufficiently familiar with the limitations of the conventionalistic interpretation of the first two axioms to be chary of a similar reading of the third one. And indeed, though in point of fact there is a definitional component in this axiom, it is not this component that is central to it. It is certainly possible to proceed in the manner just indicated, namely, introduce two numbers m_A and m_B so that for a given set of mutual accelerations of the two bodies the equation $m_A a_{AB} = -m_B a_{BA}$ is satisfied, and call these numbers the 'masses' of the two bodies. But how can we be sure that these numbers will always be *positive;* or that their ratio is a constant whatever the relative positions and velocities of the bodies; or that the mass coefficients so defined are independent of all special properties of the bodies (such as their chemical, thermal, or magnetic traits); or that the masses are additive; or that the masses assigned in this way to two bodies A and B are consistent with the masses so assigned to the pair of bodies A and C as well as to the pair B and C? The immediately obvious answer is that we cannot be sure of any of these things if 'mass' is defined in the proposed manner. Accordingly, the proposed definition of 'mass' does not assign a sense to the word such as is actually associated with it in mechanics; and the third axiom is not simply a convention for defining it.

In order to see more clearly the empirical assumptions that are involved in the use of the term 'mass' in mechanics, let us present in brief outline the definition of 'mass' that has by now become standard and that, if misconstrued, appears to establish the thoroughgoing conventionalistic character of the third axiom.[29] Assume once more any two bodies

[28] But, although the definition is useless as a way of measuring mass, there is a point to it. Newton wished to distinguish mass from weight, for mass is a property which, unlike weight, is invariant under the motions of bodies. His proposed definition can be regarded as an attempt to formulate the invariance of the property.

[29] Mach was the first to propose this definition (*op. cit.* in footnote 25 above), and it has been widely adopted, though Mach's paper was at first denied publication.

A and B which are isolated from the influence of all others (possibly by being moved to distances "sufficiently great" from all others) and which induce in each other the accelerations a_{AB} and a_{BA}. But this time let us take it to be an *experimental* fact (and not a definition) that the ratio of these accelerations is negative, that it is constant for the given pair of bodies whatever be their positions and velocities, and that it does not depend on the special material properties of the bodies. Suppose that the value of this constant is $-k_{BA}$, so that $a_{BA} = -k_{BA}a_{AB}$; and suppose further that, when the acceleration of a third body C is compared with that of B under analogous experimental conditions, the constant ratio of these accelerations is $-k_{CB}$, so that $a_{CB} = -k_{CB}a_{BC}$. The question now arises whether it is possible to deduce from these experimental data the constant $-k_{CA}$, which is the ratio of the accelerations of the two bodies A and C—whether, that is, $a_{CA} = -k_{CA}a_{AC}$ *follows* from the other equations. The answer is a decided negative, since any value of k_{CA} is *logically* compatible with the values of the other two constants. Let us, however, assume as an additional experimental fact that the constants obtained in the indicated way for *any* set of three bodies are always so related that $k_{CA} = k_{CB}k_{BA}$—that is, the ratio of the accelerations of the bodies C and A is always equal to the product of the ratio of accelerations of C and B and the ratio of accelerations of B and A. But suppose next that B and C are combined to form a single system (B^*C). What relation is there between the constant ratio of the mutually induced accelerations of this system and body A (that is, between the constant in the equation $a_{(B^*C)A} = -k_{(B^*C)A}a_{A(B^*C)}$ and the other constants mentioned? Again only experiment, and not formal logic, can decide; but let us assume as a third and final experimental fact that in general $k_{(B^*C)A} = k_{BA} + k_{CA}$.

We are now ready to define 'mass.' Let us *call* the constants 'k_{BA},' 'k_{CA},' . . . , the 'relative masses' of the bodies B and A, C and A, and so on; this is entirely a matter of definition. But in virtue of the first set of experimental facts assumed in the preceding paragraph, the relative masses of bodies are invariable under their motions, and are always positive. Next let some arbitrarily selected body, say A, be designated as the standard body with the mass m_A, where m_A is a positive number; m_A could, for example, be assigned the unit value 1. This again is entirely a matter of convention. However, as a consequence of the first set of experimental facts, all other bodies B, C, . . . will then be associated with a unique set of positive numbers m_B, m_C, . . . , which will be called the 'coefficients of mass' or simply the 'mass' of these bodies. *Calling* these numbers the masses of bodies is of course a matter of definition; but the fact that, relative to the initial choice of standard body and the assignment of a numerical value for its mass, these numbers do not vary with the motion of the bodies, is not a matter of definition. Moreover, in virtue of the second set of experimental facts mentioned in the previous paragraph, the only difference that would arise if some body other than A were taken as the standard mass would be a change in scale. For example, suppose that A is taken to have a unit mass and that in consequence B has a mass of 3 and C of 6; if B were to replace A as the unit mass, A would have a mass of $\frac{1}{3}$, and C of 2. Finally, as a consequence of the third set of experimental assumptions, masses combine additively, that is, the mass of the system $(B^\circ C)$ consisting of the bodies B and C is equal to $m_B + m_C$.[30]

If now we replace the constants in the equation expressing the first set of experimental assumptions by the ratios of masses as thus defined, we obtain for every pair of mutually induced accelerations equations of the form $m_{AB} = -m_B a_{BA}$. But this is simply the expression for the third axiom. What, then, is the status of this axiom? Is it simply the consequence of the definition of mass? Is it a convention for such a definition? The answer is now plain. The *special mathematical form* of the axiom is indeed a consequence of the definition. For instead of defining the coefficients of mass in the manner indicated, it would be possible to define them as some function of the numbers m—for example, by assigning $m_A{}^2$ or $1/m_A$ for the mass coefficient of A. And for each of these alternative ways of assigning numerical values as the measures of mass, the axiom would receive a somewhat different mathematical formulation; for example, for the alternatives mentioned, the corresponding equations would be $m_A{}^2 a_{AB} = -m_B{}^2 a_{BA}$, and $m_B a_{AB} = -m_A a_{BA}$, respectively, instead of $m_A a_{AB} = -m_B a_{BA}$. Nevertheless, the axiom is not a consequence simply of the *definition* of 'mass,' for it is a consequence of the definition conjoined with the factual assumption that for any pair of bodies the ratio of their mutually induced accelerations is a negative constant which is independent of the positions, velocities, and special properties of the bodies. Similarly, the axiom in its traditional formulation undoubtedly

[30] This account of the definition of 'mass' is based on the assumption that masses can be assigned to bodies in pairs, so that all bodies except the two under consideration at a given time have supposedly been removed to great distances. This is clearly an unrealistic assumption; for example, the bodies constituting the solar system cannot be moved around to suit our needs, although their masses can in fact be determined. Mach's procedure for assigning coefficients of mass outlined in the text runs into difficulties and must be modified if it is used to assign such coefficients to an arbitrary number of bodies which cannot be treated in pairs. Nevertheless, the simplified account in the text suffices for our purposes, and the points that emerge are not affected essentially when a method adequate for more complex cases is used as the basis for discussion. For a discussion of some of the limitations of Mach's procedure, cf. C. G. Pendsen, "A Note on the Definition and Determination of Mass in Newtonian Mechanics," *Philosophical Magazine*, Ser. 7, Vol. 24 (1937); "A Further Note . . . ," *op. cit.*, Vol. 27 (1939); "On Mass and Force in Newtonian Mechanics," *op. cit.*, Vol. 29 (1940); H. A. Simon, *loc. cit.*, and "Discussion: The Axiomatization of Mechanics," *Philosophy of Science*, Vol. 21 (1954).

has served as a guide for constructing a satisfactory definition of 'mass,' for the definition is so constructed that it makes possible a formulation of the axiom freed from unclarities. Nevertheless, the axiom is not literally the definition of 'mass'; the definition of 'mass' outlined above actually equates the 'relative masses' of bodies with the 'negative inverse ratio of their mutually induced accelerations.' But, as has been repeatedly emphasized, the constancy of this ratio is not a matter of definition, and it is in the affirmation of this constancy that the main burden of the third axiom consists.

3. *Concluding remarks.* We must now bring together the main points that have emerged in this discussion of the logical status of the axioms of motion, and present some conclusions.

a. We examined one argument directed toward establishing the first axiom by a priori reasoning, and found it to be seriously in error. There are other arguments to the same end, with respect to the other axioms of motions as well as the first; but examination would show that none of them is more cogent than the one explicitly discussed.[31] Indeed, in the light of the important emendations which the Newtonian axioms have received in general relativity theory, it is safe to conclude that no such arguments can be successful. This conclusion can be extended to the fundamental assumptions of other theories, in other branches of physics and other departments of science. The history of the sciences, especially in recent years, supplies overwhelming evidence for the quite general thesis that no theory in the positive sciences has the status of an a priori truth.

We must, however, consider briefly some arguments which aim to show, not that any special set of axioms of motion are necessary truths, but that mechanics—conceived somewhat loosely and generally as the theory of the motions of bodies—is the inescapable presupposition of all other sciences. The view that all change whatever "can be nothing else but motions of the parts of that body which is changed" was advanced as far back as Hobbes. This thesis was also urged by Leibniz, made axiomatic by the founders of mechanics, and continued to dominate the minds of physicists and philosophers even after Newtonian mechanics lost its prestige as the universal science of nature.[32] The thesis has been

---

[31] Other alleged proofs of the a priori necessity of the first axiom have been given by L. Euler, *Letters to a German Princess;* I. Kant, *The Metaphysical Foundations of Natural Science;* and J. C. Maxwell, *Matter and Motion.* An alleged proof for the a priori character of the second axiom may be found in Paul Natorp, *Die logischen Grundlagen der exacten Wissenschaften,* Leipzig, 1923, pp. 367-72, and of the third axiom in the work of Kant previously cited.

[32] Thomas Hobbes, "Elements of Philosophy Concerning Body," in *The Metaphysical System of Hobbes,* sel. by Mary W. Calkins, Chicago, 1910, p. 75; G. W.

defended on a priori grounds as well as on the basis of general empirical considerations.

One form of the a priori argument was advanced by the philosopher and psychologist Wilhelm Wundt. The substance of his reasoning is as follows: Suppose that we see an object undergoing a qualitative change, for example, altering its color or temperature. However, although we perceive the change, we nevertheless assume that in some sense the object remains the *same* object. As far as our actual intuition of what has happened is concerned, Wundt went on to maintain, the change is manifested simply as the disappearance of one object characterized by one set of qualities, and the appearance of another object possessing a different set of qualities. Our conviction that the two objects are identical must therefore rest on our relating the two sets of qualities in some conceptual manner. Thus, our *intuition* of change yields *two* objects, while our *conception* of change postulates only *one*. How then is our intuition to be reconciled with our conception? The attempted reconciliation by way of postulating an underlying unchanging substance is unsatisfactory; for such a substance is then something unknowable and transcendent to experience. A solution to the difficulty must therefore be sought within experience itself, by finding some phenomenal characteristics of objects which can be intuited as changing and which nevertheless leave the objects unaltered. But according to Wundt, the only respect in which an object can be *perceived* to change and yet also be perceived as self-identical is in its *motion*. "Changes in position are the only intuitable changes in things, in spite of which the things remain self-identical." In consequence, every change must be reduced to motion. Once this point is established, it is child's play to make out a *prima facie* plausible case for the priority of mechanics to every other branch of natural science.[38]

Wundt's argument is a curious one. Although it is apparently based on a supposed incompatibility between our perceptual intuition and our conception of change, in point of fact it derives entirely from confusing different *conceptions* (or tacit definitions) of the "self-identity" of objects. Does it make any sense to place in opposition our intuition and our conceptions of change, if by hypothesis the intuitions involve no conceptualizations whatever of what is immediately experienced? Can we significantly assert that our intuitions of qualitative change reveal simply the substitution of one "object" for another "object" unless we intuit or perceive the objects in terms of some conceptual scheme? For

Leibniz, *Hauptschrifte zur Grundlegung der Philosophie* (Buchenau-Cassirer ed.), Leipzig, Vol. 1, p. 326.

[38] Wilhelm Wundt, *Die Prinzipien der Mechanischen Naturlehre*, 2nd Ed., Stuttgart, 1910, pp. 177-80; and also his *Logik*, Vol. 2, p. 274.

example, when we see an "object" change its color from blue to red, what is the "object" that we see? Is it a piece of *litmus paper*? But if this is the way the object is characterized, it is the *same* object that we perceive before and after the change in color, and not *two* objects; for in its customary sense the notion of a piece of litmus paper does not necessitate the invariance of the color. However, if the object that is assumed to be seen is characterized as a piece of *blue* litmus paper, it is a different object that is perceived after the change. Accordingly, the answer to the question whether the object has changed depends on the implicit categorial scheme employed in characterizing the perceived situation. On the other hand, if the claim is made that no conceptual scheme of any sort is being used, then it is improper to describe the perception of change in terms of changes in *objects*. Moreover, it is simply a begging of the question to maintain that we perceive an object to retain its self-identity when we perceive only a change in its position. A bit of wire which at one moment is seen as straight and at another moment as circular, or a surface which at one moment is seen against a white background and at another moment against a blue one may not in fact be perceived as self-identical throughout the motion. Accordingly, if Wundt's argument can be taken as a fair sample of attempts to prove the priority of mechanics on a priori grounds, such attempts must be judged as unsuccessful.

However, the priority of mechanics has also been maintained on the basis of more empirical considerations. Perhaps the strongest and most interesting argument of this sort to be advanced is based on the claim that in the end the experimental evidence for all theories is obtained through the use of instruments whose construction and operation can be understood only in terms of mechanics. Instruments like beam balances and pendulum clocks clearly illustrate this claim. But even instruments like voltmeters and thermometers, which may be used to test laws that do not fall within the science of mechanics, involve mechanical principles in their construction: the mechanics of rigid bodies in designing voltmeters or in obtaining glass tubes of uniform diameters, or the elementary mechanics assumed in the physical geometry required for obtaining equidistant intervals on the instrumental scales. Now it can be readily admitted that perhaps in all the apparatus employed in the natural sciences mechanical laws are tacitly assumed. But are mechanical laws the only ones that are thus involved? Does not the operation of a voltmeter also involve specific electromagnetic laws? And even in the case of instruments that appear to be exclusively mechanical contrivances (such as beam balances), is it not often essential to analyze their operation in terms of the influence of temperature or magnetic variations, that is, in terms of laws which are *prima facie* not laws of mechanics? In the history of physics, mechanics is the branch of the science that was the first to have been developed and to reach maturity; and the instruments employed in the earlier history of physical inquiry were analyzed exclusively in terms of mechanics. Nevertheless, it was eventually discovered that mechanical laws do not provide a sufficient basis for understanding and controlling the behavior of such instruments. The historical priority of mechanics does not establish for this discipline a logical priority.

In consequence, we must conclude that neither the axioms of motion nor the inherent priority of mechanics can be demonstrated by a priori reasoning.

b. The claim that one or the other of the axioms is nothing but a definition, or a "truth" that is certifiable simply by recourse to definitions, has been repeatedly made. Indeed, the conventionalistic thesis has sometimes been radically extended, so that all theories and even ostensibly experimental laws are construed simply as "disguised definitions," and at best as rules of action rather than as statements to be judged as true or false in the light of empirical evidence. For example, the statement that lead melts at 327° C is commonly regarded as an experimental law, and there is little doubt that its acceptance is based on a large number of carefully constructed experiments. Suppose, however, that a chemist were to find a substance whose properties are indistinguishable from those of lead, except for the fact that it has a different melting point. This discovery would then presumably disprove the law. But according to radical conventionalism, the law could nevertheless be retained despite its apparent incompatibility with observed fact. For the chemist could refuse to classify the substance as lead, give it a new name, and so safeguard the law. Were the chemist to do this, it would be clear that the "law" is nothing more than a definition (or part of the definition) for the term "lead." Moreover, so the argument continues, even if the chemist should not in fact proceed in this manner, the mere possibility that he *could* do so suffices to show that it is entirely a matter of stipulation or convention whether or not the statement about the melting point of lead is to be included in the class of "true laws." Nor is the example completely grotesque and tailored to fit the requirements of a thesis. When substances were discovered possessing all the chemical properties of lead but having different densities, physicists did not abandon the law that lead has a uniformly constant density under standard conditions. On the contrary, these various "lead-like" substances were classified as "isotopes" of lead, each possessing a definite and constant density, where in general a chemical element is said to have two or more isotopes if its atomic nuclei may differ in the number of their neutrons. The law was thus retained, essentially through the device of redefining the term "lead."

We shall postpone a general consideration of the conventionalistic thesis until we have examined the issues raised by the adoption of a system of geometry, so that the thesis may be discussed within the special context in which it was first developed. For the present we shall evaluate the thesis in connection with the axioms of motion.

i. Insofar as the axioms are regarded simply as a set of formal postulates, whose nonlogical terms are neither interpreted nor associated with experimental notions by way of correspondence rules, the axioms cannot properly be said to be either true or false. The axioms are then just parts of an abstract calculus, to be manipulated in accordance with rules that take into account only the purely syntactical features of the given system of signs. Moreover, even if an interpretation is supplied for the axioms, the interpretation may be in terms of notions that are in turn defined by way of ideal, limiting processes; and in that eventuality the interpreted axioms are not assertions about experimentally ascertainable relations between physical bodies. In either case, the axioms are thus far only a framework into which experimental concepts may be fitted. If nothing further is done with the axioms, the view that they are "conventions" is well taken.

ii. But even when suitable rules of correspondence are supplied for the theoretical terms of mechanics, by adopting the axioms we adopt a certain mode of analyzing the motions of bodies and ignore other logically possible approaches to the study of motions. For example, the axioms require us to find determinants for the accelerations of bodies, and not for their velocities. However, the observed motions of bodies can be analyzed in a variety of ways, direct observation of motions does not prescribe any particular way of doing so, and some schema of conceptualization must be adopted in order to formulate experimental laws of motion. The Newtonian axioms constitute one such schema, though other schemata are abstractly possible, as the history of science indicates. Indeed, the actually observed motions of bodies do not conform with perfect precision to the experimental laws of classical mechanics; and other general assumptions, logically distinct from the Newtonian ones, can be formulated that agree with the observed facts within the same limits of accuracy that characterize the accepted laws. The axioms are therefore clearly not formulations of what is actually observed; and they undoubtedly function in inquiry as general principles for interpreting what is observed. Accordingly, the conventionalistic thesis is on firm ground in denying that the axioms are inductive generalizations from observed facts and in regarding them as one schema among others for analyzing what often appear as complex and irregular motions—with a view to achieving a relatively simple system of laws about the motions of bodies.

iii. Not only are the axioms not inductive generalizations; they cannot be refuted with demonstrative certainty by experimental findings. For by introducing special, if *ad hoc,* assumptions, it is in principle always possible to reaffirm the axioms as valid, despite seemingly contrary evidence. In this respect also the axioms are like guiding principles. They may indeed be abandoned when the guidance they supply repeatedly fails to solve some class of problems. But they also may be retained in the face of such failures, on the logically impeccable ground that past failures do not imply continued failure in the future.

iv. On the other hand, although there are ways of formulating the theory of mechanics so that one or more of the Newtonian axioms in effect become definitions, it is also possible to state the theory so that the axioms possess an empirical content. In point of fact, we have construed the axioms in this latter manner, without rejecting as illegitimate alternate ways of reading them. We have argued in connection with the first axiom that, while the axiom may function as a convention, in connection with a specific body, to define the equality of temporal periods, it is an empirical fact and not a convention if the motions of *other* bodies conform to the axiom. We have noted in connection with the second axiom that, although it is in general not possible to measure forces directly, so that the magnitude of forces can be calculated in many problems only by means of the axiom, the axiom asserts that there are determinants (or forces) of a certain kind for every change in the momenta of bodies. Despite the fact that this assertion cannot be conclusively refuted by observation, the axiom on this reading of it is not a definition. Finally, we have maintained in connection with the third axiom that, although it can be used to define the coefficients of mass of bodies, the coefficients so defined are related to one another in a manner that reflects certain empirical characteristics of the motions of bodies formulated by the axiom.

Accordingly, the thesis that the axioms are simply conventions cannot be sustained without serious qualifications. To be sure, conventions and definitions there must be in the articulation of scientific theories. However, there are several ways of articulating the theory of mechanics in such a way that the different formulations are logically equivalent to one another. Each mode of formulation may require the introduction of conventions at points that are distinctive of the particular mode. It may therefore happen that a *sentence* used in one formulation of the theory to state contingent matters of fact may be employed in some other formulation as a defining convention. But the shift in status of a sentence, from the statement of a law in one context of usage to the codification of a convention in another context of usage, can usually be effected only if some other sentence, initially having the role of expressing a definition, is given the altered function of stating a law. In any event, it is not possible to ascertain what empirical content, if any, any one of the axioms of mechanics has, without reference to the other axioms and to the way the theory to which they belong as component parts is codified. It is the system of theoretical assumptions taken as a whole that fixes the meanings of the terms occurring in them, and that determines whether a given sentence in the theory has the status of a convention or of a statement about matters of fact. In short, if any axiom possesses an empirical content, the axiom possesses it not in isolation but only by virtue of being a component part in the total theory and only in the sense that, when suitable rules of correspondence are supplied for a sufficient number of theoretical notions mentioned in the postulates or theorems of the system, various generalized statements entailed by the theory can be subjected to experimental control.

It thus becomes evident that no brief and simple answer can be given to the question: What is the logical status of the Newtonian axioms of motion? It is quite certain that the axioms are not a priori truths to which there are no logical alternatives; and it is equally clear that none of them is an inductive generalization, in the sense of a generalization that has been obtained by extrapolating to all bodies interrelations of traits found to hold in observed cases. But beyond these negative characterizations of the axioms, a reasonably satisfactory answer to the question requires reference to the place the axioms occupy in some particular codification of the theory of mechanics, and to the uses to which the axioms are put in various special contexts. What can perhaps be asserted quite generally is that, on the one hand, the Newtonian axioms can often play the role of schemata for analyzing the motions of bodies or of stipulations for defining certain experimental notions, and, on the other hand, when the axioms are coupled with additional assumptions (among others, with assumptions concerning force-functions) they can be correctly construed as statements possessing a definite empirical content.

Space and Geometry

8 Even a casual examination of the Newtonian axioms of motion makes clear that some frame of spatial reference must first be stipulated before the axioms can be employed to analyze the motions of bodies. The first axiom asserts that a body continues to move with constant speed along a straight line unless some force is impressed on the body. The second axiom declares that the acceleration of a body (that is, its change of speed along a straight line or its departure from rectilinear motion) is proportional to the impressed force. What is to be understood, however, by "straight line" in these statements, and with respect to what frame of reference is a motion to be judged as rectilinear? These questions, raised but postponed in the previous chapter, must now be discussed. They have been under critical consideration since the time of Newton, and the difficulties in the Newtonian answers to them have finally led in the present century to the development of a non-Newtonian mechanics. But the logical issues they involve are relevant to the study of the structure of explanation in general, and not only in mechanics. Although we shall take the axioms of mechanics as the starting point of our discussion, we shall eventually be concerned with these more general considerations.

I. The Newtonian Solution

Neither Newton nor his contemporaries had any reason for supposing that a doubt could arise as to what is to be understood by "straight line" in his formulations of the axioms of motion, for the only theory of geometry known at that time was the system of Euclid. It was therefore taken for granted that a line is straight if it conforms to the conditions specified in Euclidean geometry. Let us assume for the present that Euclidean geometry offers no difficulties. We shall return to the nest of problems in this assumption both in the present and the next chapter.

However, no such unanimity existed concerning the spatial frame to which the motions of bodies are to be referred. Even in Newton's day vigorous debates took place over this question. It might seem at first glance that *any* frame of reference could be chosen, and that only convenience in handling special problems ought to dictate the selection. But a more careful examination of Newtonian theory reveals such a conception to be mistaken. It is of course true that in actual practice a variety of different frames of reference are used and that considerations of convenience do control the choice of reference frame. Thus in some problems it is convenient to take the earth for this purpose, in other problems the sun, and still others the fixed stars; and in each case, within the limits of accuracy demanded by the corresponding problem, the analysis of motions effected by using Newton's axioms may be in good agreement with experimental findings. Nevertheless, from the point of view of Newtonian theory, these various practical reference frames are not equally satisfactory, and none of them is wholly suitable. We must understand clearly why this is so.

To fix our ideas, suppose we examine the motion of a body released from an initial position of rest relative to the earth and falling freely in the earth's gravitational field somewhere north of the equator. If we assume the earth to be a reference frame permitted by Newtonian theory, then according to the theory the body should fall with accelerated speed along a line directed toward the earth's center of mass. On the other hand, if the sun is taken as a theoretically allowable reference frame for describing the body's motion, the theoretical trajectory will no longer be a straight line but a more complex curve. For now the body must be regarded as sharing the earth's diurnal rotation as well as annual revolution around the sun, and, instead of falling along the line just described, the body will move along a curve that is generally to the east of this line. Furthermore, if one of the fixed stars is next adopted as a permissible frame of reference, the theoretical trajectory of the body will be still different and more complex. For not only is the body part of a physical system (i.e., the earth) which both rotates around an axis and revolves around the sun but it is also part of the solar system which is accelerated with respect to some of the stars. However, the stars themselves are "fixed" only by courtesy, so that the theoretical trajectory of the body will in general vary with the star (or system of stars) employed as the frame of reference. To be sure, the differences between these various trajectories are often slight, and, since they can be neglected in many practical problems, it does not much matter in those cases which of the alternative reference frames is selected. The point nevertheless remains that in theory, and sometimes in practice, it is not a matter of indifference which frame of reference is adopted for the study of motions. For the magnitude of the acceleration that a physical system undergoes, and therefore the forces that must be assumed (in accordance with the second axiom) to be acting on the system, depend essentially upon the reference frame with respect to which the acceleration is specified.

Let us be more explicit. If the earth is taken as the fixed frame of reference, the assumed force which is to account for the motion of a freely falling body must be proportional to the acceleration of that body relative to the earth. If the force is assumed to be simply the gravitational force of the earth, the trajectory of the body should be a straight line directed to the earth's center of mass. But in point of fact the body is deflected from that path; and as long as the earth is regarded as "fixed," there appears to be no ready way of explaining this circumstance, short of introducing *ad hoc* "deflecting forces" to account for it. The situation is altered, however, if the sun is taken as the reference frame. For now the indicated deflection is immediately explained in terms of the rotational acceleration of the earth. The general conclusion to be drawn from this example is therefore as follows: When a certain spatial reference frame is adopted, the Newtonian axioms suffice to analyze many types of motions of bodies, if forces of a relatively simple form are assumed as determinants of accelerations. On the other hand, if an arbitrary reference frame is adopted the forces that must be assumed are in general enormously complex, vary in no easily specifiable way from case to case, and have the earmarks of *ad hoc* hypotheses. Accordingly, if forces are not to be introduced in an arbitrary manner, if the determinants of accelerations are to be specified in a uniform way for extensive classes of motions rather than postulated in different ways for different special problems, there must be a privileged or "absolute" frame of reference to which the motions of bodies must be referred. At any rate, so Newton believed, and the remarkable success of his system of mechanics persuaded several generations of physicists that he was right.

The point just made can be formulated in a more technical manner. Since this technical formulation uses a notion that plays a fundamental role in the construction of physical theories, it is desirable to present that formulation in outline. Suppose that the motion of bodies is referred to a spatial reference frame S, so that the distances of an arbitrary point-mass from three mutually perpendicular axes determined by S are x, y, and z. Then the differential equations of motion of a point-mass with mass m are $m \dfrac{dx^2}{dt^2} = F_x$, with similar equations for the other coordinates, where F_x is one component of a definite force-function. For example, if the point-mass m is in the gravitational field of a body with mass M and spatial coordinates x_1, y_1, z_1, then $F_x = \dfrac{GmM(x - x_1)}{r^3}$, where r^2 (the square of the distance between the two bodies) $= (x - x_1)^2 + (y - y_1)^2 + (z - z_1)^2$. Now let S′ be any other reference frame that is moving with respect to S in any arbitrary manner; for example, it may be rotating with respect to S or it may be moving with accelerated speed. Let x', y', z', etc., be the coordinates of the bodies referred to S′. The coordinates of S will then be related to those of S′ by equations of transformation which will in general involve the time. To fix our ideas, suppose that S′ is moving with respect to S with a constantly accelerated speed, so that the coordinates of the two systems are related by the equation:

$$x' = x + v_x t + \frac{a_x t^2}{2}$$

(with similar equations for the other two coordinates), where v_x is the x-component of the velocity of S′ with respect to S at time $t = 0$, and a_x is the x-component of the constant acceleration of S′. A simple calculation shows that the differential equations of motion of the body referred to S′ have the form:

$$m \frac{d^2 x'}{dt^2} = \frac{GmM(x' - x_1')}{r'^3} = m \frac{d^2 x}{dt^2} + a_x m$$

It is thus clear that in S′ the force acting on the point-mass m differs from the force in S by a quantity proportional to the constant acceleration of S′ relative to S. In brief, the equations of motion are in general *not invariant* under a transformation of coordinates from one frame of reference to another; and in particular, they are not invariant for two reference systems which are relatively accelerated. Accordingly, if S is a reference system in which, for example, the first axiom is satisfied by a certain body, that body will not satisfy the axiom if its motion is referred to S′. Thus suppose that a body, say the star Arcturus, is far removed from the influence of other bodies, so that when its motion is referred to a certain frame of reference, say the frame defined by the constellation Orion, its motion is along a straight line with constant velocity. But if Arcturus is referred to coordinate axes fixed in the earth, its motion is no longer rectilinear and uniform, but is accelerated; and by hypothesis, there is no identifiable force which accounts for its motion when it is so referred.

It was considerations of this kind, including the general *noninvariance* of the equations of motion under transformations to arbitrary reference frames, which persuaded Newton that motions must be referred to a privileged frame of reference that he called "absolute space." "Absolute space," according to him, "in its own nature and without regard to anything external, always remains similar and immovable." Absolute space is thus nonsensible and is not a material object or relation between such objects. It is an amorphous receptacle within which all physical processes occur and to which physical motions must be referred if they are to be understood in terms of the axioms of mechanics. On the other hand, Newton declared that

> Relative space is some movable dimension or measure of the absolute spaces; which our senses determine by its position to bodies; and which is vulgarly taken for immovable space. . . . Absolute motion is the translation of a body from one absolute place to another; and relative motion, the translation from one relative place into another. . . . But because the parts of space cannot be seen, or distinguished from one another by our senses, . . . instead of absolute places and motions we use relative ones; and that without any inconvenience in common affairs; but in philosophical disquisitions, we ought to abstract from our senses, and consider things themselves, distinct from what are only sensible measures of them. For it may be that there is no body really at rest, to which the places and motions of others may be referred.[1]

In effect, Newton was prepared to admit that *kinematically* all motion is relative, but he maintained that, when considered *dynamically* and in terms of the forces determining them, motions must be referred to absolute space as the frame of reference.

Newton supported his assumption of an absolute space by theological and general philosophical arguments, but he also adduced what he believed was incontrovertible experimental evidence in its favor. He recognized quite explicitly that it is impossible to ascertain by any mechanical experiments whether a body is really at rest or moving with uniform *velocity* with respect to absolute space. For the differential equations of motion *are invariant* (i.e., their form is preserved) in all reference frames that have a uniform velocity (with rest as a limiting case) relative to absolute space. In consequence, it is not possible to distinguish experi-

---

[1] Isaac Newton, *Mathematical Principles of Natural Philosophy*, Florian Cajori ed., Berkeley, Calif., 1947, Book 1, Scholium.

mentally between absolute and relative uniform *velocity*.[2] On the other hand, Newton maintained that it *is* possible to distinguish by means of mechanical experiments between absolute and relative *acceleration*, and therefore to decide experimentally whether or not a body has an accelerated motion with respect to absolute space. The evidence he offered for this conclusion included the now-famous bucket experiment. Since Newton's interpretation of this experiment has been the focus of much subsequent criticism, we shall describe it.

A bucket filled with water is suspended from a rope, so that the rope when twisted becomes the axis of rotation for the bucket. At the outset, the water and the sides of the bucket are relatively at rest, and the surface of the water is (approximately) a plane. The bucket is then rotated. The water does not begin to rotate immediately, so that for a time the bucket has an accelerated motion with respect to the water. Nevertheless, the surface of the water during this interval remains a plane. Eventually, however, the water also acquires a rotary motion, so that finally it is at rest relative to the sides of the bucket. But now the surface of the water is concave in shape, and no longer flat. The bucket is next made to stop its rotation abruptly. However, the water does not cease to rotate immediately, and for a time has an accelerated motion relative to the sides of the bucket. Nonetheless, during this period the surface of the water continues to remain concave in shape. Finally, when the water also ceases to rotate and comes to rest relative to the bucket, its surface once more becomes a plane.

Accordingly, as Newton construed the experiment, the surface of the water can be a plane, whether it is at rest or in accelerated motion with respect to the sides of the bucket. Similarly, the surface of the water may be paraboloidal in shape, whether it is at rest or in accelerated motion relative to the bucket. He therefore concluded that the shape of the surface is independent of its state of motion relative to the bucket. On the other hand, he regarded the paraboloidal surface as a *deformation* of its normal shape, and therefore a consequence of *forces* acting on the water. According to the second axiom, however, such forces must be accompanied by *accelerated* motions. Since the state of motion of the water relative to the bucket had already been eliminated as irrelevant,

---

[2] This follows directly from what has already been said. If in the above discussion S is the frame of reference supplied by absolute space, and S' is any reference frame moving with uniform velocity relative to S, then the equations of transformation from S to S' are: $x' = x + v_x t + x_0$, where x_0 is the component along the x-axis of the distance between the origins of the two systems at time $t = 0$, with similar equations for the other coordinates. But under these transformations the differential equations of motion *are* invariant, so that it is impossible to determine whether a body is at rest or in uniform motion with respect to S. The fact that the equations of motion are invariant in all frames of reference moving with uniform velocity relative to one another is commonly called the "Newtonian principle of relativity."

Newton concluded that an acceleration relative to *absolute space* must be taken as the manifestation of the deforming forces acting on the water. In essentials Newton's argument is therefore as follows: Deformations of surfaces are evidence of impressed forces; impressed forces give rise to accelerated motions; but the deformations of surfaces are independent of the *relative* accelerations of bodies; hence the accelerations in question must be *absolute* accelerations. Since it is possible to establish by mechanical experiments whether bodies undergo deformations, it is possible to distinguish experimentally between absolute and relative accelerations, and so to identify experimentally motions that are accelerated with respect to absolute space.

Now there is something extremely puzzling about an assumption according to which it is in principle impossible to discover by mechanical means whether a body is at rest or in uniform velocity with respect to a reference frame, although it is allegedly possible to ascertain whether the body has an accelerated motion relative to that reference frame. For if a body has an acceleration with respect to a given coordinate system, it follows that the body must also have a relative velocity. If it is possible to identify the former experimentally, it seems quite mysterious why it is impossible to identify the latter. An assumption about the world, yielding a consequence that is inherently incapable of verification by experiment, appears to many minds as eminently unsatisfactory and paradoxical. Some writers have therefore concluded that the notion of absolute space is physically "meaningless." In any event, the Newtonian solution of the problem of reference frames for motions was generally regarded as something of an Achilles' heel in his system of mechanics. Although the system was accepted for more than two centuries, it was accepted primarily because a more satisfactory solution was not available.

But let us examine Newton's interpretation of the bucket experiment. Newton's argument was severely criticized by Ernst Mach, who showed that it involved a serious *non sequitur*. Newton noted quite correctly that the variations in the shape of the surface of the water are not connected with the rotation of the water relative to the sides of the *bucket*. But he concluded that the deformations of the surface must therefore be attributed to a rotation relative to *absolute space*. However, this conclusion does not follow from the experimental data and Newton's other assumptions, for there are in fact two alternative ways of interpreting those data: the change in the shape of the water's surface is a consequence either of a rotation relative to absolute space or of a rotation relative to *some system of bodies different from the bucket*. Newton adopted the first alternative, on the general assumption that inertia (i.e., the tendency of a body to continue moving uniformly along a "straight line") is an inherent property of bodies which they would continue to possess even if the remaining physical universe were annihilated.

Mach called attention to the second alternative. He argued in substance that inertial properties are contingent upon the actual distribution of bodies in the universe, so that nothing can be significantly predicated of a body's motion if the rest of the universe is assumed to vanish. He therefore maintained that it is entirely gratuitous to invoke a rotation relative to absolute space in order to account for the deformation of the water's surface, but that on the contrary it is sufficient to take a coordinate system defined by the fixed stars as the frame of reference for the rotation. Accordingly, if Mach's general approach is adopted, and if an adequate theory of mechanics can be constructed in conformity with it, it is not necessary to assume the puzzling asymmetry between absolute velocity and absolute acceleration that is so central to Newtonian theory. In terms of Mach's approach, there still may be fundamental differences between various frames of reference. Thus, the Newtonian axioms may be valid when the motions of bodies are referred to some of these reference frames, but not valid for other frames of reference. There may thus be a class of "privileged" reference frames even on Mach's view, so that motions relative to them may be called "absolute" while others are only "relative." However, absolute velocity in this sense is in principle as verifiable as is absolute acceleration.[8]

There is another way of analyzing the bucket experiment that helps make clearer just what is at stake and that throws additional light on the logical status of theories. Suppose we adopt a reference frame S, rotating relative to the earth in such a fashion that its axis of rotation is parallel to the axis of rotation of the bucket, and its constant angular velocity equal to the maximum angular velocity of the bucket. The following are then the observed data in the experiment: At the outset, the water has an accelerated rotation relative to S, with its surface a plane. Eventually, however, the water ceases to have this acceleration, its surface then being paraboloidal. Moreover, after the bucket has been abruptly stopped from rotating with respect to the earth, so that the water is finally at rest relative to the bucket, the water is once more accelerated relative to S and once more has a plane surface. Accordingly, the surface is paraboloidal only when it is at rest with respect to S, and it is a plane only when it is accelerated relative to S. The character of the water's surface is thus independent of its state of motion relative

---

[8] Cf. Ernst Mach, *Science of Mechanics*, La Salle, Ill., 1942, Chap. 2, Sec. 4, pp. 271-98. Frames of reference belonging to the privileged class are commonly called "inertial" or "Galilean" frames or reference. As is well known, Mach's critique of Newton profoundly influenced Einstein and prepared the way for the latter's general theory of relativity.

to the bucket, but not independent of its state of motion relative to S. On this analysis, therefore, a plane surface is associated with accelerated motion (relative to S), while a concave surface is connected with a state of rest (relative to S).[4]

Why not assume, in the light of this, that the "normal" surface of the water is paraboloidal and that it is the "abnormal" plane surface which is the "deformed" one? The answer is that, were this assumption adopted, we would also have to complicate in a serious way the Newtonian equations of motion. If S were generally selected as the reference frame for all motions, the angular velocity of S relative to any given system under investigation would then enter into the law about the latter. Since different systems generally possess different angular velocities relative to S, no simple formula would embrace these various special laws. The field of invariance of the differential equations of motion would in fact be extremely limited. On the Newtonian proposal of a frame of reference, or on Mach's alternative to it, the equations of motion are invariant for all so-called "Galilean frames." That is, if the equations are satisfied when motions are referred to some particular reference frame, they are satisfied in all reference frames having a constant velocity with respect to the first. On the other hand, if the equations are satisfied when motions are referred to S, they will be satisfied only in those reference frames at rest with respect to S. In short, with S as the frame of reference for all motions, the specific force-functions that would have to be supplied in order to analyze motions in terms of the Newtonian axioms would be different for nearly every special problem and would have to be invented *ad hoc* for each case.

Is not the supposition absurd, it may nevertheless be asked, that the water is in a deformed state when its surface is a plane? Do not deformations take place only when forces are acting? Is it not therefore an experimental fact that the paraboloidal surface is a consequence of such forces, and hence of the rotation of the water with respect to some frame of reference, rather than of its state of rest relative to S? Similarly, does not the rotation of the plane of Foucault's pendulum and of the axis of a gyroscope, or the flattening of the earth at its poles, or the deflection of a freely falling body from a rectilinear path to the earth's center provide *experimental* evidence that the earth must be rotating? Accordingly, is it not quite untenable to hold, as the previous paragraph suggested one might hold, that the water in the bucket and the earth itself are assumed to be "absolutely accelerated" merely because the

---

[4] For this way of analyzing the experiment, see Peter G. Bergmann, *Introduction to the Theory of Relativity*, New York, 1942, p. xiv. For a similar analysis, but one employed as an argument for the earth's absolute motion, cf. J. C. Maxwell, *Matter and Motion*, Art. 105, pp. 84-86.

equations of motion receive a simple, invariant form when those assumptions are made?

These queries bring us to the crux of the present discussion. The fundamental point must constantly be kept in mind that, even if the water in the bucket is declared to have an "absolute acceleration" when its surface is concave, it is not at all necessary to assume, as Newton did, that this rotation (or the rotation of the earth) takes place with respect to absolute space. Mach's critique of Newton is conclusive on this matter. The frame of reference relative to which the acceleration is said to take place can be taken to be defined by the system of the fixed stars, or by some other system of physical bodies, as it actually is in practice. The rotation of the plane of Foucault's pendulum, for example, does not establish the earth's rotation with respect to absolute space, but only with respect to the fixed stars. If the stars were concealed from us by clouds permanently surrounding the earth's surface, so that their existence would be unsuspected by us, Foucault's experiment would show only that the earth is rotating relative to the plane of the pendulum.

It is nevertheless conceivable (indeed, it happens to be the case) that, when the motions of bodies are referred to coordinate frames provided by physical bodies, the motions do not conform with complete precision to the axioms of motion. To put it differently, it is conceivable that no physical coordinate frame is a Galilean or "inertial" frame. If we decide to retain the Newtonian axioms in unmodified form, we can then introduce an "ideal frame," with respect to which the motions of bodies are in strict agreement with the axioms but to which *physical* frames of reference will at best be only good approximations. The rationale for this procedure is that, unless we adopt inertial frames for analyzing the motions of bodies in terms of the Newtonian axioms, the experimental laws of motion would undoubtedly be more complex and less manageable than if inertial frames were employed. Accordingly, the primary objective of using inertial frames, whether they are actually realized in physical systems or are only ideal constructions, is to effect a simplification in the formulation of laws. It is a happy circumstance that there are in fact physical systems which are at least approximate realizations of inertial frames. Were this not the case, the science of mechanics might perhaps have never been developed.

However, none of this can be validly construed to mean that the laws established for motions referred to inertial frames are "more real" or "more objective" than would be the less simple and noninvariant laws that might be established without introducing such frames. On the contrary, it can be shown that, if a set of relations can be affirmed to hold of a system of bodies when their motions are referred to an inertial frame, there must be definite relations between those bodies when the motions are referred to noninertial frames, even though the formulation of the latter relations may be more complex and more difficult to achieve than the formulation of the former.

For example, it is often convenient in analytical geometry to represent curves by so-called "parametric equations," in which the coordinates of points on a curve are expressed as functions of some auxiliary variable. Such parametric equations frequently make it possible to analyze the properties of a curve with much less trouble than if the curve is represented by an equation which relates the coordinates to one another directly. Nevertheless, it would be absurd to maintain that the parametric equations are "more correct" or "truer" than the equation relating the coordinates directly, or that the latter formulates the curve in a manner more "objective" (or less "objective," as the case might be) than do the parametric equations. Thus, a plane curve whose parametric equations in terms of the auxiliary variable 't' is $x = t^2 - 2t$, $y = t^4 + t^2 - 2t$ can also be represented by an equation which connects its coordinates directly, namely, $(y - x^2 - 9x - 8)^2 = (x + 1)(4x + 8)^2$. In many problems the former equations are much more manageable than is the latter, although the two modes of representation have the same geometric content. Analogously, the differential equations for the motion of a planet in the sun's gravitational field, when the motion is referred to the fixed stars as the coordinate frame, assume the familiar form involving the inverse square of the distance between the sun and planet. However, it is a mathematical consequence of this fact that the motion can be referred to, say, the earth as the frame of reference, so that differential equations can in principle be stated for the planet's motion when the latter is studied in this manner. These differential equations will in general be forbiddingly complex, but they will nonetheless formulate the planet's motion as objectively and completely as do the initial equations.

The introduction of inertial frames as the basis for analyzing the motions of bodies required great creative imagination, for the motions of bodies as directly observed do not exhibit patterns of changes that obviously require the use of such frames. The notion of inertia is thus not the product of "abstracting" from manifest traits of sensory experience, in the way in which the idea of a circle is commonly supposed to be such a product. On the other hand, the notion of inertia has become so completely a part of our intellectual heritage and equipment that, unless we make considerable effort to do so, it is difficult to conceive of an alternative way for interpreting the "observed facts" of motion. Moreover, the idea of inertial frames is indissolubly associated in Newtonian mechanics with the invariance of the equations of motion under transformation from one inertial frame to another. However, what is invariant is often tacitly identified with what is "objectively real," with what is permanent and not subject to spatiotemporal limitations, with what is universal.[5] Accordingly, the invariance of the equations of motion, when the motions are referred to inertial frames, gives to inertial frames a quality of importance over and above the importance they possess in making possible the analysis of mechanical phenomena in terms of a relatively simple set of force-functions. It is at least plausible that the intellectual discomfort sometimes produced by the suggestion that the water in the bucket experiment is "deformed" when the surface is flat stems in part from a disinclination to adopt frames of reference which would enormously restrict the range of invariance—and hence the "objectivity"—of the equations of motion.

It is worth recalling, finally, that the forces postulated by Newton's second axiom as determinants of accelerations cannot, in general, be measured independently of the accelerations. As was noted in the preceding chapter, the force-functions employed in Newtonian mechanics are in the main assumed hypothetically; they are explicitly characterized only by the general requirement that their magnitudes be proportional to the changes in the moment of bodies, and that they have the same direction as do these changes. Accordingly, the stimulus that usually leads to the search for forces and to the construction of force-functions is the fact that some physical system is undergoing accelerated motion. It is therefore putting the cart before the horse if we claim that we can always decide whether a body is accelerated or deformed, by ascertaining through independent experimental means what forces are impressed on it. The contrary is certainly very frequently the case. However, if we must first agree whether or not a body is accelerated or deformed before we can have grounds for believing that a force is acting on it, then at least in such cases we must first adopt a frame of reference for motions as well as a system of geometry for measuring them, before we can inquire whether a body is accelerated or deformed. Newton's procedure in assigning logical priority to the selection of a frame of reference, with respect to which motions are to be analyzed in terms of his axioms, was thus entirely cogent, however faulty may have been his arguments for absolute space.

We have now indicated at sufficient length why the question of adopting a spatial frame of reference is an important one in Newtonian mechanics, and we have also discussed the rationale for Newton's solution of the problem. We must next turn to no less important issues that arise in considering the use of geometry as a system of spatial measurement.

II. *Pure and Applied Geometry*

If we wish to determine the length of a room or the height of a moderately sized house, the usual procedure is to lay off some measuring rod (for example, a yardstick or steel tape) against the object to be measured, and so ascertain the number of times the unit length is contained in the distance under consideration. This normal method obviously assumes that the measuring rod has already been calibrated in accordance with certain rules, that the edge of the rod is straight, and that the rod is unaffected in any relevant way while it is being repeatedly moved in the process of measurement. These assumptions raise difficult questions, which we shall ignore for the present. But it is clear that this method of measuring distance is not always feasible. We usually cannot estimate the width of broad rivers in this way, nor the distances between places separated by tall mountains. We certainly cannot employ this method for measuring the distances between the stars, or the dimensions of atoms and other submicroscopic objects.

In many practical problems, and in most scientific ones, the measurement of spatial magnitudes cannot therefore be effected by such a "direct" procedure. In general, spatial measurements are made only indirectly, and require among other things the use of geometric theory. For example, if we want to determine the length of wire needed to run a line from the ridgepoles of two buildings 80 feet apart, one of which is 30 feet tall and the other 50 feet, we would most likely *calculate* the required length with the help of the Pythagorean theorem. For the length of wire needed is the hypotenuse of a right triangle whose remaining sides are 80 feet and 20 feet, respectively, so that the length in feet is equal to the square root of $80^2 + 20^2$, or $20\sqrt{17}$—approximately 83 feet.

But what justifies us in using the Pythagorean theorem in this example? The obvious answer is that the theorem is a logical consequence of the axioms of Euclidean geometry, so that if these axioms are accepted the theorem is fully warranted. However, the question is not fully resolved by this answer. For an exactly similar question must be put concerning the axioms. The axiomatic formulation and deductive development of Euclidean geometry have the great advantage that if the question can be satisfactorily answered for the axioms it need not be faced again for any of the theorems. Nevertheless, the question must be seriously faced. What then are the grounds for accepting the axioms? In discussing such grounds we shall be compelled to examine issues that bear directly upon the logical status of theories in general, and not only upon the status of geometry.

1. Let us briefly review some of the opinions that have been held on this question. It is well known that geometry originated in the practical arts of land measurement among the ancient Egyptians. A number of useful formulas were discovered by them, which enabled their surveyors, the *harpedonaptai*, to fix definite boundaries between fields and to calculate their areas. Their formulas were simply a collection of independent rules of thumb, and the discovery that they were connected by relations of logical implication was apparently the achievement of the ancient Greeks. The Egyptian formulas were analyzed, some geometric figures were defined in terms of other figures, and additional relations between the bounding surfaces and edges of bodies were established. Moreover, after several centuries of such effort, it was shown that, if a small number of propositions about magnitudes in general, and geometric figures in particular, are accepted without proof, an indefinite number of other propositions—including those previously established—can be deduced from them. Euclid's *Elements* was thus a theoretical codification of the art of mensuration which had its roots in practices with a long prior history, and for centuries Euclid was accepted as a model of logical rigor and as the ideal form of a theoretical science.[6]

Geometry came to be employed, even before the complete emergence of modern science, as a basis not only for surveying, but also for astronomy, architecture, instrument making, and some of the engineering as well as fine arts. Newton was therefore able to regard geometry as simply a branch of a universal mechanics. As he himself put it,

> To describe right lines and circles are problems, but not geometric problems. The solution of these problems is required for mechanics; and by geometry the use of them, when so solved, is shewn; and it is the glory of geometry that from those few principles, fetched·from without, it is able to produce so many things. Therefore geometry is founded in mechanical practice, and is nothing but that part of universal mechanics which accurately proposes the art of measuring. But since the manual arts are chiefly conversant in the moving of bodies, it comes to pass that geometry is commonly referred to their magnitude, and mechanics to their motion.[7]

On this view, accordingly, the axioms of geometry are true statements about certain features of physical bodies, features assumed to be specifiable in terms of definite physical procedures. Geometry is thus a hypothetico-deductive discipline which asserts that, *if* certain configurations

---

[6] It is now well known that Euclid's *Elements* do not conform to modern standards of logical rigor, for many of his theorems cannot in fact be deduced from his axioms, and additional ones must be supplied.

[7] Newton, *op. cit.* In a passage preceding this quotation, Newton asserted that "Geometry does not teach us to draw [right lines and circles, upon which mechanics is founded], but requires them to be drawn; for it requires that the learner should first be taught to describe these accurately, before he enters upon geometry."

are right lines, circles, and so on, then they must possess the properties enunciated in the various theorems.

However, two related questions are now forced on our attention, concerning which Newton had nothing explicit to say. Just what are the procedures which serve to specify, and if necessary to construct, straight lines, planes, circles, and the other figures that constitute the alleged subject matter of geometry? And in any case, on what grounds can we claim that the axioms and theorems of geometry are true of the figures that are thus identified? Newton simply referred the first question to "practical mechanics" and did not consider at all the second one. But neither question has any easy answer, and each runs into what seem like insuperable difficulties.

Straight lines can be readily constructed if we once possess a straightedge, and circles can be easily drawn if we use a pair of compasses whose points remain at a constant distance from one another. But how do we establish the "straightness" of an alleged straightedge or the constancy of the distance between compass points? On what evidence do we claim to know that the assumptions about straight lines and circles contained in the Euclidean axioms actually do hold of the figures obtained in this way? It will not do simply to say "Make measurements on these figures, and see if they conform to the Euclidean requirements." For to make measurements we must possess instruments that have straight edges and possess constant distances between their parts. We thus appear to be caught in a hopeless infinite regress. Neither does it seem satisfactory to fall back upon a direct inspection of an edge to determine whether it is straight, even if one adopts the somewhat sophisticated procedure of "sighting" along it in the manner of carpenters when they plane a piece of wood. Such direct inspection can be used only when relatively small segments of lines and surfaces are under consideration; the conclusions obtained by this procedure are not uniform for different observers or for the same observer at different times; and the procedure may even involve the same type of regress already noted. For when an edge is judged to be straight by direct inspection, what standard is employed in making the judgment? If it is some *image* of straightness, the original problem seems to present itself once more with respect to this image. On the other hand, if an edge is said to be straight on the basis of sighting along it, does not the judgment rest upon the tacit postulate that optical rays are rectilinear? An infinite regress thus appears to be unavoidable. Indeed, the regress cannot be circumvented until it is recognized, as we shall soon see, that the questions generating it are ambiguous, and that they confound issues concerning matters of *empirical fact* with issues concerning matters of *definition*.

However this may be, Newton's conception of geometry as simply a branch of an empirical science of mechanics is by no means the only view that has been taken of the subject. In classical antiquity most of the axioms were regarded as self-evident necessary truths, and the lack of "obviousness" of the parallel postulate was the chief stimulus to centuries of effort at demonstrating it from self-evident premises. Leibniz, a contemporary of Newton, explicitly maintained the Platonic doctrine that the "truths of geometry," like those of arithmetic, are certifiable as necessary without the need for an appeal to sensory experience. According to him, geometric truths are "innate, and are in us virtually, so that we can find them there if we consider attentively and set in order what we already have in the mind, without making use of any truth learned through experience or through the traditions of another." [8] Nevertheless, with some doubtful exceptions the ancients regarded geometry as dealing with the spatial properties of material bodies, even if Plato and his followers maintained that those properties are only imperfect actualizations of the eternal objects of geometrical inquiry. Just when the view was first advanced that geometry is the science of the structure of *space* (or "pure extension"), rather than of the spatial properties of material bodies, is an unsettled historical question. But by the time of Newton this view was already influential. This conception received a forthright statement from Euler, in the eighteenth century, who declared:

> Extension is the proper object of geometry, which considers bodies only insofar as they are extended, abstractedly from impenetrability and inertia; the object of geometry, therefore, is a notion much more general than that of body, as it comprehends, not only bodies, but all things simply extended, without impenetrability, if any such there be. Hence it follows that all the properties deduced in geometry from the notion of extension must likewise take place in bodies, inasmuch as they are extended.[9]

The conception of geometry as an a priori science of the structure of space was given a different turn by Kant, in his attempt to find a *via media* between the aprioristic rationalism of Leibniz and the sensationalistic empiricism of Hume. Although there is room for some doubt concerning the interpretation of many details in Kant's doctrine, its general import is that Euclidean geometry formulates the structure of the form of our external intuition. Accordingly, the axioms of Euclid and their consequences are apodictic truths concerning the spatial form of all possible experience. Kant's views on the nature of geometry have been highly influential, not only with professional philosophers but with mathe-

[8] G. W. Leibniz, *New Essays Concerning Human Understanding* (transl. by A. G. Langley), Chicago, 1916, p. 78.
[9] L. Euler, *Letters to a German Princess* (transl. by Brewster), Vol. 2, p. 31.

maticians and physicists as well. Although important currents of philosophic thought in the nineteenth century rejected the Kantian conception and argued for an empirical interpretation of the status of geometry, Kant's influence did not diminish until subsequent developments in logic, mathematics, and physics made his views progressively more untenable. For the view of geometry as a system of a priori knowledge concerning space had the incomparable advantage over its competitors that it appeared to explain, as alternatives to it did not, why Euclid was the only known system of geometry and why mechanics (at that time still the most perfectly developed branch of theoretical physics) was so inextricably dependent on that system.

2. However, before turning to these later developments and their consequences for a philosophy of geometry, we must make explicit a distinction that has already been briefly noted and is of paramount importance in what follows. In geometry, as in every deductive argument and every deductively formulated discipline, two questions must be sharply distinguished. The first is: Do the alleged theorems of the system follow logically from the axioms? To answer it and to discover new theorems implied by the axioms are among the prime concerns of mathematicians. To resolve it, no laboratory experiments or other empirical studies need be undertaken; the only equipment required is the technique of logical demonstration. The second question is: Are any of the axioms or theorems factually or materially true? This question does not fall within the jurisdiction of the mathematician *qua* mathematician; and answers to the first question can be sought irrespective of the answers that may be given to the second. Answers to the second question can in general be supplied only by the physicist or other empirical scientist, provided that the axioms and theorems refer to identifiable empirical subject matter. This proviso is crucial, and we must therefore discuss it at some length.

It has been a matter of common knowledge since Aristotle that the validity of a syllogistic demonstration does not depend on the special meanings of the terms occurring in its premises and conclusions. Accordingly, if a syllogistic argument is valid, it remains valid when the original terms are replaced by others. In evaluating the validity of a syllogism, it is therefore permissible to ignore completely the meanings of the specific subject-matter terms and to consider only the formal structure of the constituent statements. Formal structure can be considered most simply and effectively by replacing the specific subject-matter terms by variables. The resulting expressions will then contain only such words or symbols that signify logical relations or operations. Thus, when such replacements are made in the statement 'All men are mortal,' the resulting expression is 'All A's are B's,' in which the words 'all' and 'are' retain their customary meanings while no specific meanings are associated with the variables 'A' and 'B.' [10] However, the expression 'All A's are B's' is clearly no longer a statement concerning which it would be significant to ask whether it is true or false. The expression has only the *form* of a statement, one which becomes a statement when words having definite meanings are substituted for the variables. We shall call such expressions "statement-forms." A statement-form may be defined for present purposes as an expression containing one or more variables, such that if subject-matter terms are substituted for the variables the resulting expression is a statement, that is, an expression concerning which it is significant to raise questions of truth or falsity. Accordingly, to evaluate the validity of a syllogism it is sufficient to consider the statement-forms of which its premises and conclusion are instances. It is thus clear that, when we are concerned with the question whether the conclusion of a syllogism follows logically from the premises, it is irrelevant to ask whether these statements are true or false.

What has just been said about the syllogism obviously applies to any deductive argument. In particular, when Euclidean geometry is discussed as a demonstrative discipline, we can ignore the meanings of the specific geometrical terms in the axioms and theorems of the system, replace those terms with variables, and pursue the task of proving theorems by attending only to the logical relations between the resulting statement-forms. However, although this point is elementary, it appears not to have occurred to any of the ancient mathematicians and philosophers, despite the fact that they were quite familiar with it in connection with syllogistic arguments. But in any event, it is of greatest importance to distinguish between geometry as a discipline whose sole aim is to discover what is logically implied by the axioms or postulates, and geometry as a discipline which seeks to make materially true assertions about a specific empirical subject matter. In the former case, mathematicians explore logical relations between statements only insofar as the latter are instances of statement-forms, so that the meanings of specific subject-matter terms are in principle irrelevant. In the latter case, the nonlogical terms occurring in the axioms and theorems must be associated with definite elements in some subject matter, so that the truth or falsity of various statements belonging to the system may be suitably investigated. Geometry when studied in the first sense as simply a de-

[10] It is indeed possible to continue the process of abstracting from meanings still further, and to replace words like "all," "are," and other logical particles by signs which are governed by stated rules of operation. But it is not germane to the present discussion to pursue this possibility, although some of the outstanding achievements of recent logical studies are a consequence of developing this suggestion.

ductive system is often called "pure geometry"; when studied in the second sense as a system of factual truth, it is commonly called "applied" or "physical geometry."

Let us illustrate the main burden of this discussion by considering a formulation of Euclidean geometry that meets modern standards of logical rigor, for example, Oswald Veblen's axiomatization.[11] Veblen assumes a class of objects called "points," a triadic relation between points called the relation of "lying between" and a binary relation between pairs of points called "congruence." He then imposes upon these objects and relations a number of carefully formulated conditions, in the form of sixteen assumptions or axioms. He also defines in terms of the initial (or primitive) subject-matter expressions a number of other expressions, such as 'line,' 'plane,' 'angle,' and 'circle,' employing in this process ideas that belong to general logic (such as that of set or class). These defined expressions are introduced primarily for the sake of convenience and can be eliminated in favor of the primitive terms. The defined expressions may therefore be ignored in what follows. Let us now conjoin these sixteen axioms, so that they are components in a single but very complicated statement. The axioms can then be represented by the abbreviation A(point, between, congruent). On the other hand, let us represent any statement that can be formulated in terms of the primitive expressions of the system by T(point, between, congruent), although in general the primitive terms will not all occur in every such statement. The aim of demonstrative or pure geometry may then be said to be that of finding statements 'T' such that 'T(point, between, congruent)' is a logical consequence of 'A(point, between, congruent).'

However, the deducibility of 'T' from 'A' cannot depend on any special meanings associated with the expressions 'point,' 'between,' and 'congruent.' These terms may therefore be replaced by variables with which no meanings of any sort need be associated. Accordingly, the conjoined postulates of pure geometry in the Veblen axiomatization could in principle be stipulated as the statement-form '$A(R_1, R_3, R_2)$' where 'R_1' is a predicate variable (or unary relation variable), 'R_3' a triadic relation variable, and 'R_2' a binary relation variable. The task of the pure geometer is then to ascertain which statement-forms '$T(R_1, R_3, R_2)$' are logical consequences of the statement-form '$A(R_1, R_3, R_2)$.'

On the other hand, neither the pure geometer nor the physicist can investigate the truth or falsity of the statement-forms 'A' and 'T' for the patent reason that, since they are not statements, it is not even significant to ask whether they are true or false. Moreover—and this is the chief burden of the present discussion—it may be equally impossible to

[11] See his essay, "The Foundations of Geometry," in *Monographs on Topics of Modern Mathematics* (ed. by J. W. A. Young), New York, 1911.

inquire into the truth or falsity of the Veblenian axioms, even if the latter are formulated in terms of the familiar expressions 'point,' 'between,' and 'congruent' rather than variables, unless these familiar expressions are associated with definite, empirically identifiable physical objects or relations between such objects. Indeed, mathematicians often employ those familiar expressions, without thereby assigning to them any specific meanings involving such reference to empirical subject matter. Thus, although Veblen uses those expressions in his formulation of the geometric axioms, he is careful to note that the reader may associate with them "any meaning" or "any image" he pleases, as long as these meanings and images are consistent with the conditions imposed upon the use of the expressions by the axioms. The point of the proviso mentioned on page 219, in connection with answers to the question whether the axioms of geometry are factually true, is therefore this: The material truth or falsity of geometric axioms and theorems can be investigated only if, for the nonlogical terms occurring in the axioms and theorems, rules of correspondence or coordinating definitions are supplied which associate with those terms empirically identifiable elements in some subject matter.

3. In the light of this distinction between pure and applied geometry, let us now reconsider some of the views previously mentioned concerning the logical status of geometry.

a. The claim that the propositions of geometry are a priori, logically necessary truths is ambiguous, and can be construed in at least three senses. It can be understood to mean (1) that the statements of *pure geometry* are a priori and logically necessary, where a statement of pure geometry is of the form: If $A(R_1, R_3, R_2)$, then $T(R_1, R_3, R_2)$; or (2) that each of the postulates and theorems of *pure geometry* has this character; or finally (3) that the statements of *applied geometry*, whether axioms or theorems, are a priori and logically necessary.

On the first interpretation, the claim is obviously correct. But it is also trivial, since whenever a conclusion demonstrably follows from a premise, the conditional statement whose antecedent clause is the premise and whose consequent clause is that conclusion will always be a logically necessary truth. On the other hand, the claim is absurd when it is understood in the second sense. For if the postulates and theorems of pure geometry are taken to be statement-forms, they cannot be regarded as either true or false, and a fortiori neither as necessarily true nor as necessarily false.

There remains for consideration only the third way of construing the claim. The issue then resolves itself into the question whether the

Veblenian postulates are necessary truths for *every* interpretation of the primitive terms or only for *some,* and, if the latter, what the character of such interpretations is. The point of the question will be clearer if we first compare two different statement-forms: the statement-form 'If no S is P, then no P is S' with the statement-form 'No S is P.' It will be evident that, no matter what subject-matter terms are substituted for the variables 'S' and 'P' in the first, the resulting statement will invariably be a logically necessary truth, for example, the statement 'If no triangles are equilateral figures, then no equilateral figures are triangles,' even though the antecedent clause of this conditional statement happens to be false. On the other hand, the second statement-form will yield a necessary truth for some substitutions of subject-matter terms for the variables, but not for others, for example, the statement "No triangles are circles' is a necessary truth, while 'No triangles whose vertices are any three fixed stars are figures with an area less than two square miles' is not. Similarly, an inspection of Veblen's postulates (or of any other postulates for Euclidean geometry) shows that none of them formulates a necessary truth under every interpretation of the primitive terms. For example, Veblen's second axiom postulates that for any three points x, y, z, if y lies between x and z, then z does not lie between x and y. If we now replace the term 'point' with the term 'number,' and the relational expression 'y lies between x and z' with the relational expression 'y is greater than the difference of x and z,' we obtain the statement 'For any three numbers x, y, and z, if y is greater than the difference of x and z, then z is not greater than the difference of x and y,' which is clearly false (since, for example, although 4 is greater than the difference of 7 and 5, nevertheless 5 is greater than the difference of 7 and 4), and hence not a necessary truth. Accordingly, if the axioms are necessary truths, they are such only under certain interpretations of their primitives but not under others.

Let us, therefore, examine some proposed interpretations of the geometric axioms, and first the one contained in Euclid's *Elements.* Euclid prefaced the formal development of his system with a large number of "definitions." Some of these are definitions of terms like 'triangle' and 'circle' on the basis of what are obviously the primitive terms of the system, such as 'point' and 'line'; the other definitions are explications of these primitives. In effect, however, these explications are proposed *interpretations* of the primitives, and are presumably intended to instruct us as to the objects or relations which are designated by the primitives. For example, a point is said to be "that which has no part," a line is declared to be a "breadthless length," and a straight line is explained as "a line which lies evenly with the points on itself." These explanations undoubtedly suggest in a vague way the sorts of things to which the various terms are to be applied. Nevertheless, they are hardly explicit enough to permit us to identify without serious question what things are designated by the corresponding terms. What is it, for example, that has no parts? It can be no ordinary material object, though it may possibly be a corner of solids having sharp edges, or perhaps even an experienced pain of short duration. Moreover, even if we assume that we know just what things are to be counted as "breadthless lengths," when does such a thing lie evenly with the points on itself? It seems therefore profitless to ask whether on Euclid's own interpretation of his axioms they are true.

It might be objected, however, that this is all useless hairsplitting, since we know very well what is meant by 'point' and 'straight line.' Points and straight lines, so it might be said, are of course not material things; they are, however, *limits* of physical objects that can nevertheless be conceived and entertained in imagination. Moreover, we can perform experiments in imagination upon points, lines, and other geometrical objects; and when we do so, we find that we cannot form our images except in conformity with the Euclidean axioms. It has been maintained, for example, that the statement 'Two straight lines cannot intersect in more than one point' cannot be established by perceptual observation, but only by exercising our imagination. As one writer has put the argument,

> For in the first place it is only through imagery that we can represent a line starting from a certain point and extending indefinitely in a certain direction; and in the second place, we cannot represent in perception the infinite number of different inclinations or angles that a revolving straight line may make with a given straight line. We may, however, by a rapid act of ocular movement represent a line revolving through 360° from any one direction to which it returns. In this imaginative representation the entire range of variation, covering an infinite number of values, can be exhaustively visualized because of the continuity that characterizes the movement. It is only if such a process of imagery is possible that we can say that the axiom in its universality presents to us a self-evident truth.[12]

Two comments must be made on this general position. In the first place, if geometric objects are taken to be merely conceptual or imaginary ones, the fundamental problem under discussion has not even been broached. For that problem concerns the manner in which the conceptual structure of pure geometry may come to be used in physics and the various practical arts. Nothing is contributed to the solution of this problem either by repeating that points and lines are concepts or by identifying them with images. Just what is the relevance of lines entertained in imagination for astronomy or for the construction of precision instruments, both of which must make extensive use of geometry?

[12] W. E. Johnson, *Logic*, Vol. 2, London, 1922, p. 202.

In the second place, the argument from the alleged facts of mental experiments has no force whatsoever. When we perform experiments in imagination upon straight lines, in what manner are these lines envisaged? We cannot employ any arbitrary images of lines in the experiment. We must *construct* our images in a certain manner. However, if we examine the mode of their construction in those cases in which we allegedly intuit the imagined figures as Euclidean, we soon notice that the Euclidean assumptions are tacitly being used as the *rules of construction*. For example, we can certainly imagine two distinct lines with two points in common. But such lines do not count as straight lines, simply because they do not satisfy the Euclidean requirements of straightness, so that we seek to form our images so as to satisfy those requirements. Or to change the illustration, it is possible to "prove" that all triangles are isosceles—a result known to be incompatible with Euclidean postulates—with the help of suitably drawn diagrams. However, the alleged demonstration is a spurious one because (as we usually say) the diagrams have not been drawn "correctly"—where the standards of correctness are supplied by Euclidean geometry itself. Accordingly, if the Euclidean postulates serve as the rules for constructing our mental experiments, it is not at all surprising that the experiments invariably conform to the rules. In short, if Euclidean axioms are used as implicit definitions, they are indeed a priori and necessary because they then specify what sorts of things are to be counted as their own instances.

b.   The conception of geometry as a branch of experimental science seems highly plausible, if only because of the origins of geometry in the practical arts of measurement. This plausibility is not diminished by the difficulties we have now canvassed in the view that geometry is a body of a priori knowledge about the structure of space. For measurements can be conducted only with material instruments and not with parts of space. No account of applied geometry is therefore adequate which creates a puzzle out of the fact that geometry functions as a theory of mensuration. On the other hand, as has already been noted, the Newtonian view of geometry as the simplest branch of mechanics appears to have difficulties of its own; and we must now try to decide whether these difficulties are quite as insuperable as they seem.

It will be useful to distinguish between two general ways of employing geometry within experimental science. (i) The first and historically earlier approach consists in specifying *independently of Euclidean geometry* certain edges, surfaces, and other configurations of material bodies, and then showing that as a matter of observed fact the things so specified conform to the Euclidean axioms within the limits of experimental error. (ii) The second approach consists in using the Euclidean postulates as implicit definitions, so that no physical configurations (whether discovered or deliberately contrived) are *called* "points," "lines," and so on, unless they satisfy the postulates within certain limits of approximation. Both approaches face similar logical and empirical problems, but each approach introduces a distinctive emphasis into the discussion, and each assigns a different status to Euclidean geometry.

i. Let us consider the first approach more closely. Euclidean geometry and theoretical physics are certainly not more than 3000 years old. There surely was a time, therefore, when men engaged in various practical affairs did not have available the knowledge contained in these systems. Let us imagine ourselves placed into situations these men faced. Although we would have no inkling of geometry, we could nevertheless distinguish between different forms of surfaces—initially perhaps only by unaided sight or touch, but eventually perhaps by more reliable procedures. For example, some surfaces are noticeably rounded in one or more directions, others are less so, and still others appear to be quite flat. However, these discriminations are somewhat rough, and there may be no complete agreement among us as to which surfaces are the flattest. Moreover, as long as appropriate technologies are lacking, it is only by chance that we would run across such flat surfaces.

But suppose that mechanical skills become developed, and that we learn how to grind or cut bodies so that the surface of one body can be made to fit snugly upon the surface of another body. It may finally occur to us to take three bodies, and grind their surfaces until any two of them can be fitted smoothly on each other. This procedure provides what seems like a good objective criterion for surfaces having maximum flatness, and in any case we decide to call surfaces satisfying it 'plane surfaces.' It clearly would make no sense to ask whether such surfaces "really" are planes, for they are planes by *definition*, and by hypothesis there is no other standard for being a plane surface than the one stated. It is also worth noting that in judging whether two surfaces do fit snugly upon one another we may use some optical test, for example, the test that no light shines through when the surfaces are in close fit. Nevertheless, though we may employ such an optical test, we would *not* be assuming, tacitly or otherwise, that the propagation of light is "rectilinear," so that our procedure is not in fact circular. We would simply be employing a type of observable fact as a condition for *saying* that the surfaces fit snugly. It is essential to note, therefore, that thus far the *only* issue of fact at stake when a surface is declared to be a plane is whether the surface satisfies the indicated condition of fitting closely on other surfaces. In particular, it should be observed that no assumptions associated with Euclidean geometry are involved in assigning the label 'plane' to such surfaces.

We may now proceed in a similar manner to construct types of edges which we decide to call 'straightedges' or 'straight lines,' for example, by grinding two plane surfaces on a body so that they have a common edge. Furthermore, with the help of planes and straight lines we may construct other figures for which labels like 'point,' 'triangle,' 'quadrilateral,' and the like are introduced. Also, two straightedges may be defined as being of equal length if they can be made to coincide end to end, and a unit length may be specified by selecting some particular straightedge for this purpose.[18]

It is now possible to construct additive scales of length, angle, area, and volume. But except for one point, the details of the constructions will be omitted. In specifying a scale of length, as well as in making measurements on the basis of such a scale, it will in general be necessary to transport the unit length repeatedly. The question may therefore arise whether in the course of its motion it might not undergo a change in length. "How do we know," it might be asked, "that when a straightedge is moved from one place to another its length remains the same? How do we know that if two straightedges are equally long in one place, and one of the edges is carried to another place, the two edges continue to have the same length?"

These questions are worth considering here because they typify a frequent confusion between what are issues of fact and what are matters of definition. It *is* a question of empirical fact whether, if two straightedges are equally long at one place (i.e., if they can be made to coincide end to end) and are then transported along the same or different paths to some other place, they are equally long at the new place. Let us assume that this indeed in general is the case. On the other hand, it is not a question of empirical fact whether, if two straightedges are equally long at one place and one of them is moved to some other place, the two edges continue to be equally long. In terms of the procedure we have adopted this question can be answered only by making a decision and introducing a *definition*. In particular, it is not a question of *knowing* (that is, having observational evidence which would enable us to establish) whether or not the standard unit length alters its length when transported from place to place; this is a matter that, within the framework of assumptions we have adopted, can be settled only by a *stipulation*. It is therefore essential to distinguish the issue whether two edges that are equally long at one place continue to be equally long when both are transported along the same or different routes to another place, from the issue whether two straightedges equally long at one place continue to be equally long when only one of them is transported

[18] The above method of defining planes and straight lines is developed by W. K. Clifford, *The Common Sense of the Exact Sciences,* New York, 1946, Chap. 2; and also by N. R. Campbell, *Measurement and Calculation,* London, 1928, pp. 271-78.

to another place or whether the length of the standard unit is invariant under motion. The first issue can be decided by an appeal to observation, and thus involves questions of *knowledge;* the second issue cannot be settled in this way, and involves questions of *definition.*

"But is it not the case," so our imaginary critic might reply, "that we frequently do ascribe a change in the length of a body after it has been transported, and that we often take precautions against such changes? Indeed, we even ascribe such changes when they remain at the same place, and try to prevent alterations in length (as in the case of the standard meter or yardstick) by keeping the bodies in carefully controlled environments." The answer to this query is obviously affirmative. However, this answer is predicated on rejecting the simplifying factual assumption made in the preceding paragraph, according to which two straightedges equally long at one place (as judged by the coincidence of their respective end-points) will continue to be equally long at any other place, no matter by what paths they are transported from one locality to another. Let us therefore drop this assumption, and thereby complicate the discussion.

We must now suppose that we have learned to distinguish between various kinds of bodies, for example, between different sorts of wood, metals, and stones. We shall also assume that we know how to identify various physical sources of change in the shapes and relative sizes of bodies, such sources as compressions or variations in temperature. To fix our ideas, suppose that at time t_1 and place P_1 two straightedges a and b are equally long, where a is made of maple wood and b of copper. Assume further that at a later time t_2 the straightedge b is longer than a, but that in the interim there has been a rise in the temperature of both bodies. Let us also assume that after much experience we have come to recognize that when different substances are exposed to an identical temperate change, their relative lengths are altered, and in fact by unequal amounts for different pairs of substances. Accordingly, on the hypothesis that the sole identifiable source of change has been an increase in temperature at P_1, we atttribute the alteration in relative lengths of a and b to the rise in their temperature. We are not saying, it should be noted, that the length of a has remained constant and that the length of only b has increased; we are saying only that b has become longer *relative* to a.

Suppose next that, although a and b remain equally long when they are at P_1, they are unequally long when they are transported to P_2, whether by the same or different paths. This change in relative length may again be accounted for in terms of variations in temperature that one or both of the bodies have undergone. We have used temperature changes as the source of alterations in the relative lengths of straightedges, but what has been said about temperature can obviously be repeated for other sources of change that can be experimentally identified. In any event, however, the earlier assumption that two straightedges equally long at one place remain equally long when transported to another place must now be amended. In its amended form the factual assumption contains the proviso that, when straightedges are transported from one place to another, all the known sources of changes in relative lengths are kept constant, so that features of the environment, known experimentally to be relevant to alterations in relative lengths and shapes of bodies, are the same in the initial positions of the straightedges as in their final positions. Within the framework of this amended assumption, it then makes sense to say that, on being transported from P_1 to P_2 the length of a body changes (*relative* to some specially designated body) or that two bodies equally long at P_1 cease to be so when one of them but not the other is moved to P_2.

One further point in this amended account of problems of spatial measurement requires brief attention. For the objection might be made that the discussion is based on a circular and self-defeating procedure. We have been outlining a way of instituting a scale of lengths, allegedly without employing any assumptions of Euclidean geometry; and we have indicated the need for stipulating the conditions under which two straightedges are said to be of equal length. We have, however, assumed that it is possible to detect whether or not any changes occur in these conditions, for example, whether or not the temperatures of two straightedges are the same and remain constant. Must we not therefore have thermometers in our possession, and must we not in consequence have scales of length, *before* we can detect such changes or constancies? Does not the proposed construction of a scale of length assume that the end-product of the construction is available *antecedently* to the construction? And if so, is not the procedure patently circular?

Despite appearances to the contrary, no such circle need occur. For in point of fact, it is possible to determine whether there are changes in the temperature of bodies (and more generally, whether there are changes in any of the physical conditions upon which variations in relative lengths of bodies depend), without using instruments, such as the thermometer, which employ a previously established scale of lengths. For example, at a primitive level of investigation we might rely entirely on the sensitivity of our own bodies to changes in temperature within certain ranges. At a more advanced stage of knowledge we might use as a detector of temperature changes the unequal expansions or contractions of two straight rods made of different substances. It is essential to observe that in this case we would use, not a quantitative measure of linear expansion or contraction (for this would indeed involve us in a circular argument), but only the qualitative fact that two such rods initially of equal length become unequally long in variable temperature fields. At a still more sophisticated level of knowledge, we could recognize changes in temperature by using the fact that, when two different metals form a closed circuit, a magnetized needle near the circuit will be deflected when the temperature at the juncture of the metals is altered. The construction and use of such complicated detectors involves details into which we cannot enter. However, even the schematized account we have given of them suffices to indicate that an additive scale of lengths can be constructed without circularity and without employing some antecedent theory of geometry.

Once an additive scale of lengths has been instituted, and with these various difficulties out of the way, we can then construct certain figures that will be called 'circles,' and with the help of these figures a scale of angular measure. We have therefore outlined how, in principle, a class of figures and certain measures for them can be specified without using any assumptions of Euclidean geometry. The remaining problem is whether these figures (and others that can be constructed in an analogous manner) satisfy the axioms and theorems of Euclidean geometry; or, conversely, whether Euclidean geometry, when its terms 'point,' 'line,' and so on are interpreted as referring to the similarly named constructed figures, is true of them. This problem, however, is a straightforward empirical one, and there is no way of knowing the answer to it before making actual empirical inquiry. Moreover, it is clear that the evidence obtainable from such inquiry will at best show only an approximate agreement between Euclidean statements and the constructed figures. For in the first place, uncontrollable disturbing factors cannot always be eliminated in conducting measurements, so that random or experimental "errors" are likely to occur. In the second place, measuring instruments are capable only of limited discriminations. For example, at a given stage of technological development we are unable to distinguish between lengths falling below a certain minimum extension. On the other hand, Euclidean geometry postulates an unlimited discriminability of lengths when it asserts that certain lengths have relative magnitudes which can be expressed only by irrational numbers. Accordingly, no overt measurement can determine whether, as geometric theory requires, the magnitude of certain lengths is indeed irrational. And finally, Euclidean statements sometimes make assertions that cannot possibly be shown to hold of actual figures by direct measurement. For example, the statement that if the alternate interior angles formed by a transversal to two lines in a plane are equal the lines *never* intersect, is such a statement. For every plane that we may construct is of finite extent, and we cannot therefore determine by observation or overt measurement whether two lines do not intersect no matter how far they are produced. Nevertheless, within regions accessible to experiment, and subject to the qualifications mentioned, the agreement between figures constructed in the manner outlined and the statements of applied Euclidean geometry is in point of fact excellent. In consequence, and until fairly recently, the theory of mechanics as well as other branches of physics has been based squarely on the assumption that Euclidean geometry is true of a class of physical configurations constructed in a way more or less analogous to the one we have sketched. Moreover, although a different system of geometry is employed in Einsteinian relativity theory, the engineering arts as well as the manufacture of laboratory instruments will undoubtedly continue to make that assumption during the foreseeable future.

ii. We have now completed our discussion of the first approach to geometry mentioned on page 225. We must now examine the second alternative, according to which Euclidean postulates are used as implicit definitions for certain figures that constitute the domain of application of those postulates. Our examination will be relatively brief, since most of the relevant problems have already been discussed.

The essential difference between these alternative approaches is that, while on the first of them expressions like 'point,' 'line,' and so on are applied to physical configurations that are constructed or identified in accordance with rules specifiable *independently* of the Euclidean axioms, on the second approach those expressions are applied only to such configurations that satisfy the Euclidean requirements. On the former approach, accordingly, we are in principle committed to abandoning Euclidean geometry if actual observations and measurements on independently specified lines, angles, circles, and so on reveal a significant discrepancy between the properties of these figures and what Euclidean geometry leads us to expect. On the latter approach, on the other hand, we are in principle committed to retaining Euclidean geometry at all costs, and to alter our methods for constructing figures should those methods not yield configurations in conformity with Euclid. On the first alternative, Euclidean geometry is a system of contingent, a posteriori statements concerning spatial properties of bodies antecedently classified and named. On the second alternative, Euclidean geometry is a system of a priori rules for classifying and naming such properties.

Let us indicate in outline how Euclidean geometry can be used in this latter way. If we accept the Euclidean postulates as implicit definitions, we must find or construct figures that will satisfy the conditions stated by the postulates. Suppose, then, we begin by constructing surfaces, edges, and so on in the manner proposed in connection with the above discussion of the first approach. However, we are not yet entitled to *call* those configurations 'planes,' 'straight lines,' and the like, and must first make observations and measurements on them. We may find that the outcome of such an inquiry shows those figures to possess properties in good agreement with what Euclidean geometry requires of planes, straight lines, etc. In that eventuality, we are entitled to the hypothesis that those figures *are* planes, straight lines, and the rest. On the other hand, suppose that the outcome of the inquiry shows those figures to possess traits that deviate considerably from Euclidean requirements. For example, suppose that the angle sum of certain three-sided figures differs from two right angles (as defined by a stated scale of angular magnitude) by more than 10°, a difference far greater than a possible experimental error. In this eventuality, the constructed figures would not receive the familiar geometrical names, and in particular the three-sided figure would not be called a "triangle." On the contrary, we would modify our rules for constructing figures and for measuring their spatial magnitudes until we obtained configurations that were at least approximately Euclidean.

It may turn out, however, to be extremely difficult to construct Euclidean figures, and that no matter how much we alter the rules for manufacturing the desired kinds of surfaces and straightedges we rarely if ever succeed in obtaining anything even approximating Euclidean planes and straight lines. Such a situation would still not "disprove" Euclidean geometry, though it might make the retention of Euclidean geometry as a theory of mensuration highly inconvenient. We could, of course, put up with the inconvenience, and resign ourselves to the fact that calculations about spatial dimensions on the basis of that theory are rarely if ever in agreement with the results of overt measurement. Two other alternatives would nevertheless be open to us. We might succeed in developing *physical theories* based on Euclidean geometry, so that our persistent failure to construct (or find) Euclidean configurations would be systematically explained by those theories, while at the same time the spatial magnitudes of bodies as determined by actual measurement would be in good agreement with the numerical values calculated from those theories. Or alternately, we might abandon Euclidean geometry as the a priori system of rules for classifying and naming spatial configurations, and devise some other system of pure geometry with this end in view.

The discussion thus indicates that the apparently incompatible conceptions of Euclidean geometry as an empirical science on the one hand, and of Euclidean geometry as a system of a priori rules on the other hand, can both be accepted as legitimate. Geometry is a branch of empirical science when planes, straight lines, etc., are constructed or identified as features of physical bodies in accordance with rules that can be formulated and applied without reference to Euclidean geometry. Euclidean geometry is a system of a priori rules when the construction or identification of configurations that are to bear Euclidean labels is guided and controlled by the Euclidean postulates. Nevertheless, on each approach both empirical as well as a priori assumptions play a role. On the first alternative the rules for constructing the figures designated as 'planes,' 'straight lines,' etc., are a priori, and the Euclidean statements are empirical. On the second alternative the Euclidean postulates are a priori, and the assertions that certain figures (constructed or identified in accordance with specified rules) are planes, straight lines, etc., are empirical. In brief, the difference between the two alternatives is a difference in the locus at which conventions or definitions are introduced into a body of knowledge.

Geometry and Physics

9 The Newtonian conception of geometry as the simplest branch of mechanics was based on the tacit assumption that Euclidean geometry is the only theory of spatial relations that can provide a theory of mensuration. Since Newton's day, however, a large number of alternative pure geometries have been constructed. In consequence, the assumption that Euclidean geometry is the exclusively correct analysis of spatial relations is no longer tenable, and in fact some of these non-Euclidean geometries have come to be employed in developing non-Newtonian theories of mechanics. The logical status of geometry, discussed in the previous chapter on the Newtonian supposition that there are no alternatives to the Euclidean system, thus requires further study. It is to the further examination of the roles played by questions of empirical fact and questions of definitional stipulation in selecting a geometry as a theory of mensuration in physics, that the present chapter is devoted. We shall first give an account of the major alternatives to Euclid, and of their relations to one another as well as to the Euclidean system. This will require the presentation of some technical but unavoidable mathematical details. We shall then examine the considerations that enter into the choice of a geometry for purposes of developing a physical theory, and discuss the merits of the thesis that a system of geometry is at bottom only a set of conventions for making spatial measurements.

I. *Alternative Geometries and Their Interrelations*

The construction of non-Euclidean systems of pure geometry was the direct outgrowth of attempts to demonstrate Euclid's parallel postulate by showing it to be a consequence of his remaining assumptions. Euclid's form of the parallel postulate, which did not appear to be self-evident, as the others were alleged to be, asserts that if two straight lines in a plane are cut by a third, so that the sum of the interior angles on one side of the transversal is less than two right angles, the two lines will meet on that side when they are sufficiently prolonged. Its inclusion among the axioms used to be regarded as the great "scandal" of Euclidean geometry, but efforts to demonstrate it without assuming something equivalent were uniformly unsuccessful. However, the mere failure to deduce it from the remaining postulates of the system does not constitute a proof that it is impossible to deduce it from them. A revolution in mathematics occurred when just such a proof of impossibility was finally produced. The proof marked not only the end of more than two millenniums of futile effort, but also the beginning of non-Euclidean geometries—that is, of geometries that deny one or more of the Euclidean postulates—and eventually of non-Newtonian mechanics. In the present section we shall briefly describe two of the alternative pure geometries and examine their relations to the system of Euclid.[1]

1. a. There is an elementary technique for establishing the logical independence of a given statement from certain others. Let 'A_1,' 'A_2,' . . . , 'A_n,' be a set of axioms for Euclidean geometry, and suppose we wish to show that it is impossible to deduce 'A_1' from the others. Since, as we have already noted, the deducibility of a statement does not in general depend on the special meanings of their subject-matter terms but only on their formal structure, we may suppose that the axioms are a set of statement-forms. Now either 'A_1' is deducible from the remaining postulates or it is not. If it is, then if 'A_1' were replaced by a postulate 'A_1^*' formally incompatible with 'A_1' (for example, by the contradictory or a contrary of 'A_1'), then the new set of postulates would be an *incon-*

---

[1] Several lines of mathematical inquiry contributed to the development of non-Euclidean geometry, and each of them throws a special light on the internal structure and the interrelations of these alternative geometrical systems. One of the approaches is the axiomatic method; a second is the method of differential invariants, developed by Riemann as a generalization of certain basic ideas introduced by Gauss; a third is the method of projective definition of distance, associated with the names of Cayley and Klein. However, all but the axiomatic method assumes considerable mathematical training. We shall therefore concentrate on the axiomatic method, though something will nevertheless be said about the other two approaches to non-Euclidean geometry.

sistent set, that is, would yield incompatible consequences. On the other hand, if 'A_1' is logically independent of the other postulates, then the new set 'A_1^*,' 'A_2,' . . . , 'A_n' will generate a *consistent* system of consequences. However many theorems are established in the new system, none of them will be formally incompatible either with the new postulate or with any other theorem deduced from the entire set. Accordingly, the problem whether 'A_1' is logically independent of 'A_2,' 'A_3,' . . . , 'A_n' is reduced to the problem whether the set of postulates 'A_1^*,' 'A_2,' . . . , 'A_n' is a consistent set, where 'A_1^*' is the contradictory or a contrary of 'A_1.'

But how is the consistency of a set of postulates to be established? This is by no means an easy problem, and quite complicated logical and mathematical techniques may be required for solving it in particular cases. Two major lines of attack have been developed for handling it. The first and historically older method is to find an *interpretation* for the predicate variables occurring in the postulates, so that on this interpretation the statement-forms are converted into *true statements*. Thus, if the statement-forms 'A_1^*,' 'A_2,' . . . , 'A_n' can be converted into true statements by a suitable substitution of subject-matter terms for their predicate variables, the set is shown to be consistent. In consequence, 'A_1' is proved to be independent of the remaining postulates. The second method is a more formal one. It consists in showing that a given postulate set is *as consistent* as a certain other set whose consistency is taken for granted. This is shown by correlating the statement-forms in the first set with statement-forms in the second in such a way that if a contradiction could be deduced in the former, a contradiction would also have to occur in the latter.

We shall postpone further discussion of this second method until page 249, and for the present we shall illustrate and examine only the first. Consider, therefore, the following set of three statements:

S_1:   For any two distinct integers, either the first is greater than the second, or the second is greater than the first.

S_2:   For any two integers, if the first is greater than the second, then the second is not greater than the first.

S_3:   For any three integers, if the first is greater than the second, and the second is greater than the third, then the first is greater than the third.

These are all true arithmetical statements. We may nevertheless wish to know whether the first is deducible from the other two. With a view to answering this question, we therefore replace the specific subject-matter terms in them, and so obtain the following three statement-forms:

A_1:   For any two distinct elements x and y belonging to a class K, either x has the relation R to y, or y has the relation R to x.

A_2: For any two elements x, y in K, if x has the relation R to y, then y does not have R to x.

A_3: For any three elements x, y, z in K, if x has the relation R to y and y has R to z, then x has R to z.

Now construct a statement-form $A_1{}^*$ that is formally incompatible with A_1. For example, we may take for $A_1{}^*$ the following:

$A_1{}^*$: There are at least two distinct elements x, y in K, such that x does not have R to y and y does not have R to x.

Accordingly, A_1 is independent of A_2 and A_3 (and hence any instance of the first, such as S_1, is formally independent of the corresponding instances of the other two, such as S_2 and S_3) if the set of statement-forms $(A_1{}^*, A_2, A_3)$ is a consistent set. To establish its consistency we look for an interpretation for the predicate variables 'K' and 'R,' so that the statement-forms in the set are converted into statements which we have good reasons to believe are true. Thus, let us substitute the predicate 'human being' for the predicate variable 'K,' and the relation term 'being an ancestor of' for the predicate variable 'R.' On this substitution the statement obtained from $A_1{}^*$ is: There are at least two human beings such that neither is the ancestor of the other. This statement is patently true. The statements obtained from the remaining statement-forms are also clearly true. It thus follows that the set $(A_1{}^*, A_2, A_3)$ is consistent, so that A_1 cannot be deduced from the other two statement-forms—and hence S_1 cannot be deduced from S_2 and S_3.

b. It was through the use of the techniques just outlined that the first system of pure non-Euclidean geometry was constructed. In the third decade of the nineteenth century, Lobachewsky and Bolyai, two mathematicians working independently of each other, developed a system of geometry which was based on a *contrary* of Euclid's parallel postulate. Euclid's version of this postulate is equivalent to the more familiar Playfair's axiom, according to which there is just one straight line through a given point which is parallel to a given line. We shall therefore discuss Lobachewsky's innovation on the supposition that Playfair's axiom is used as the parallel postulate for Euclidean geometry instead of Euclid's original formulation.

Lobachewsky replaced the parallel postulate by the assumption that through a given point there are two parallels to a given line. From this new set of postulates he deduced a large number of mutually consistent theorems, many of which are *prima facie* incompatible with comparable theorems in Euclid. For example, in the Lobachewskian geometry, the angle sum of a triangle is not constant for all triangles (as it is in Euclid), is always less than two right angles, and diminishes as the areas of triangles increase.² Again, no triangles having unequal areas are ever similar. Also, the ratio of the circumference of a circle to its diameter is not constant for all circles, is always greater than π, and increases with their areas. Neither Lobachewsky nor Bolyai established the internal consistency of the new geometry, and the consistency of the system remained an open question for some time. Finally in 1869 Beltrami assigned meanings to the geometrical predicates of the system, so that the Lobachewskian postulates were interpreted as statements about lines and curves on certain saddle-back surfaces.³ However, these statements were demonstrably true within Euclidean geometry. Accordingly, the possibility of a non-Euclidean geometry as internally consistent as the system of Euclid was established beyond any reasonable doubt.

We shall say nothing further about Beltrami's interpretation because it does not lend itself to a simple exposition. It will nevertheless be useful to have before us in sufficient detail another interpretation of the Lobachewskian postulates, first proposed by Poincaré, for two-dimensional (or plane) Lobachewskian geometry. Consider the interior points (to be called 'L-points') of a fixed circle O with radius k in a Euclidean plane. All other points in the plane, whether on the circumference of O or external to it, are excluded from the class of L-points. Through any two L-points there is a unique circle orthogonal (i.e., at right angles) to O. The arcs of such circles falling *within* O will be called 'L-lines.' Through any L-point outside a given L-line two L-lines can be drawn

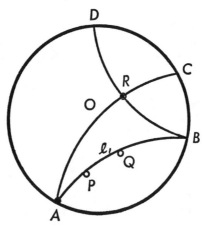

² In fact, the difference (commonly called the "defect") between two right angles and the angle sum is proportional to the area.

³ They are the surfaces of revolution obtained by rotating the plane curve called the "tractrix" around its asymptote as the axis of rotation. The tractrix is defined as the curve such that the segment of the tangent extending from the point of contact to the intersection of the tangent with a given line is of constant length.

which meet the given line on the circumference of O; and such lines will be called the 'L-parallels' to the given line. Thus in the drawing P and Q are any two L-points which determine a unique L-line l_1. From any L-point R not on l_1 two L-lines can be drawn which meet l_1 on the circumference of O at the non-L-points A and B; these are the L-parallels to l_1 through R. It is clear that any L-line through R which falls inside the angle ARB will intersect l_1, while any L-line through R that falls inside the angle BRC will not intersect l_1. We also define the 'L-distance' between two L-points as a certain function of those points and the points of intersection with O of the L-line determined by the given L-points.[4] (The 'L-angle' between two intersecting L-lines is defined as the angle between the tangents to the two L-lines at their intersections.) Moreover, a closed figure formed by three L-lines will be called an 'L-triangle'; and a closed figure all of whose L-points have a constant L-distance from a fixed L-point will be called an 'L-circle.' Other L-figures can be defined in an analogous manner.

It can be shown, however, that if in the Lobachewskian postulates we substitute for the word 'point' the expression 'L-point,' and for the term 'straight line' we substitute the expression 'L-line,' and so on, all the resulting statements are demonstrable in Euclidean geometry. For example, the Lobachewskian theorem concerning the angle sum of a triangle mentioned on page 237 then asserts the following: The angle sum of a Euclidean figure bounded by the arcs of circles orthogonal to a fixed circle is less than two right angles, the defect being proportional to the area of the figure. But this assertion is demonstrably true in Euclidean geometry. It follows that the Lobachewskian system is consistent —or, at any rate, as consistent as Euclidean geometry. For were the former inconsistent, a contradiction would also occur in that part of Euclidean geometry which develops the properties of circular arcs orthogonal to a fixed circle.

This rather skeletal interpretation of Lobachewskian plane geometry can be clothed with some fanciful flesh. Imagine the interior of the circle O to be inhabited by two-dimensional creatures, so that the circumference of O is the limit of their world. Suppose, also, that in this universe the absolute temperature is a maximum at the center of O but decreases

---

[4] This function is proportional to the logarithm of the anharmonic ratio of the four points mentioned. Thus, if P and Q are any two L-points, and A and B are the intersections with O of the L-line determined by P and Q, then the L-distance between P and Q is by definition equal to

$$\frac{k}{2} \times \log \left(\frac{PA}{PB} : \frac{QA}{QB} \right)$$

It is evident that the L-distance between an L-point and any point on the circumference of O is infinite.

with the distance r from the center, so that the absolute temperature T at any point is given by the formula $T = c(k^2 - r^2)$, where 'c' is some proportionality constant. Assume, furthermore, that all bodies in this universe have the same coefficient of thermal expansion, and that thermal equilibrium is established instantly between a body and its environment as the body moves from place to place. It then follows that the length of any measuring rod will be proportional to its absolute temperature. Accordingly, to a spectator who is not part of this curious world, a rod transported toward the circumference of O will appear to shrink progressively. An inhabitant of this world can therefore never reach its boundaries. For as judged by the spectator, an inhabitant's body and steps become smaller as he moves out toward the circumference, although he will not be aware of such a contraction. In effect, therefore, to an inhabitant all points on the circumference of O are at an "infinite distance" from any point within the universe. Moreover, as can be demonstrated, for inhabitants the "shortest distance" between any two points in their universe will not be the Euclidean straight line connecting the points. The shortest distance for them will be the arc of the circle passing through those points and orthogonal to the circle O. (Indeed, if we make the additional supposition that the velocity of light at any point in this universe is also proportional to the absolute temperature at that point, light will travel along such arcs.) And finally, through a point not on a given straight line in this universe, infinitely many straight lines can be drawn that do not intersect the given line. On the other hand, both the lines through that point intersecting the given line on the circumference of O will be parallel to the given line, since they meet the given line at "infinitely remote" points. In brief, in this universe the inhabitants will find that the geometry of bodies is a Lobachewskian one.

c. The Lobachewsky-Bolyai pure geometry is not the only alternative to Euclid. For the Euclidean parallel postulate can be replaced by a contrary that is different from the one adopted in the Lobachewskian system. A distinctive non-Euclidean geometry is in fact obtained if the Euclidean postulate is replaced by the assumption that through a point outside a given straight line there are *no* parallels to the latter. In this case, however, other Euclidean postulates must also be amended, for example, the postulates that a straight line can be prolonged indefinitely, and that two points always determine a unique line. The pure geometry obtained when these alterations are made is called "Riemannian," although Riemann arrived at it by developing Gauss's notions of curvature and geodesic rather than by using the axiomatic method. The following are some examples of theorems in Riemannian geometry: The angle sum of a triangle is always greater than two right angles, and the excess is proportional to the area of the triangle; all straight lines are of finite length, and two straight lines always enclose an area; the ratio of the circumference of a circle to its diameter is always less than π and increases as the area of the circle diminishes.

It is quite easy to supply a true interpretation for the postulates of Riemannian geometry, and thereby to establish the internal consistency of the system. To this end, consider the surface of a Euclidean sphere S, and call its points 'R-points.' The arcs of great circles on S will be called 'R-lines'; a closed figure on S bounded by three R-lines, an 'R-triangle'; and a closed figure on S such that the R-lines from every point on the circumference of the figure to a fixed R-point are all equal, an 'R-circle.' If we now substitute in the Riemannian postulates the expression 'R-point' for 'point,' the expression 'R-line' for 'straight line,' and so on, the resulting statements are demonstrable in the Euclidean geometry of the sphere. For example, the first of the above-mentioned theorems of pure Riemannian geometry is thus converted into the statement that the angle sum of a spherical triangle is greater than two right angles, by a quantity proportional to the area of triangle. This is a well-known theorem in Euclidean spherical geometry. It follows that Riemannian geometry is consistent, or at any rate as consistent as the Euclidean system.

d. The Lobachewskian and Riemannian geometries do not exhaust the alternatives to Euclid that can be constructed. They are the two most familiar types of non-Euclidean systems; and other types would need for their description more mathematical equipment than we have been assuming. However, at least a bowing acquaintance with other approaches to non-Euclidean geometry is desirable if some of the logical problems raised by modern physics are to be understood adequately. Accordingly, we turn to a highly simplified account of these other approaches.

As already mentioned, the method used by Riemann in constructing his type of non-Euclidean geometry (and after Riemann, but independently of his work, by Helmholtz) was based on certain ideas first developed by Gauss in the latter's study of various kinds of surfaces and their intrinsic properties. Gauss showed, in the first place, that given any surface, it is possible to express the equation of any figure on it in terms of a coordinate system embedded wholly in that surface. Gauss also showed how the notion of 'straight line,' regarded as equivalent to the 'path of shortest distance between two points,' can be generalized so as to apply to curves lying on arbitrary surfaces. Paths of shortest distance are called 'geodesics.' Accordingly, if the geodesics on a surface are defined, the rules are thereby given for measuring lengths on that surface.

On a Euclidean plane, for example, the geodesics are Euclidean straight lines, and lengths are measured with straightedges. On the surface of a sphere, geodesics are arcs of great circles, and lengths are measured by edges that are small arcs of great circles. On the surface of a right cylinder, on the other hand, the situation is more complicated, for the geodesics are of various types, and differ according to the *direction* in which one moves from a given point. Thus, from an arbitrary point in a direction parallel to the axis of the cylinder the geodesic is a Euclidean straight line; in a direction perpendicular to the axis the geodesic is a circle; and in intermediary direction it is a spiral. The situation is even more involved with more complex surfaces, such as that of an egg or a doughnut.

In general, the nature of a geodesic on a surface is different for different points on the surface and for different directions at a given point. It turns out, however, that the character of geodesics depends intimately on a certain "intrinsic" property of the surface. This property may vary from point to point on the surface but is unaltered (or is invariant) when the surface is deformed without stretching or tearing. Thus, in the case of a plane this property is unaltered when the plane is rolled up to form a cylinder or a cone. This property is said to be "intrinsic" to the surface, in the sense that it is definable exclusively in terms of coordinate systems lying entirely *in* that surface and requiring reference to nothing *outside* the surface. For reasons of analogy, Gauss called this property the 'curvature' of the surface at a point—a label that eventually proved to be misleading to nonmathematicians. The relation between geodesics and curvature is of such a kind that, given the form of the geodesics through a point on a surface, the curvature of the surface at that point can be deduced. Accordingly, if we know how to measure distances along the shortest paths through a point on a surface, we can calculate the curvature of the surface at that point. In consequence, if different rules for measuring lengths (and hence for specifying geodesics) are adopted, different values for the curvature of a surface are obtained.

Let us examine the analogy which led Gauss to introduce the notion of curvature in connection with surfaces. Consider first the curvature of curves. A circle with radius R is said to have a curvature of $1/R$, for this is an index of the amount the circumference "bends away" from the tangent at any point. It is clear that a circle has a constant curvature. For other curves, the curvature at a point is defined as the curvature of what is called the "osculating circle" at that point. The osculating circle at a point on a curve is the circle that passes through the given point and two "adjacent" points. A more precise definition of this circle is as follows: Let P be some given point on a curve, and M and N two other points on it; these three points determine a unique circle. Now keep P fixed, but let M and N move toward P. The circles determined by these points will in general be different. But when M and N finally coincide with P, a limiting circle will be obtained; this is the osculating circle at P.

It is useful to distinguish between a positive and negative direction in which the radius of the osculating circle is drawn to the point of contact with the curve; accordingly, the curvature of a curve at a point may be negative or positive or even zero. For example, an ellipse has a variable positive curvature, since the radii of the osculating circles at various points on the ellipse will not be constant in magnitude, but all are directed toward the interior of the ellipse. An equiangular spiral has a constant positive curvature. A straight line has a constant zero curvature (a straight line may be conceived as a circle with an infinite radius, so that the curvature or the reciprocal of this radius is zero). A cubical parabola has a variable curvature, which is sometimes positive, sometimes negative, and sometimes zero.

Consider next an arbitrary surface. Draw a line normal to the surface at any point on it, and imagine a plane containing the normal and intersecting the surface. With the normal as the axis, rotate the plane. In each of its positions the plane will in general intersect the surface in a curve. Indeed, the segments of the curves in the immediate neighborhood of the foot of the normal will be the geodesics of the surface at that point. Now it can be shown that in general the curvature C_1 of one of the geodesics is the *maximum* of all geodesics through that point, while the curvature of C_2 of a certain other geodesic is the *minimum*. Gauss called the product $K = C_1C_2$ the 'curvature of the surface at a point'; and it is easy to see that K may be positive, negative, or zero, and may have a constant value for all points on the surface or a different value at different points.

Thus, a sphere with radius R has a constant positive curvature of $1/R^2$. A plane has a constant zero curvature, but so has a right cylinder as well as a right cone. The saddle-back surface obtained by rotating a tractrix around its asymptote as axis has a constant negative curvature of $-1/R^2$, where R is the radius of the largest circular section of the surface. The surface of an egg has a variable positive curvature, the curvature being greater at points near the acutely shaped end of the egg than at points near the other end. There are also saddle-back surfaces which have a variable negative curvature.

A remarkable outcome of Gauss's analysis is the important theorem that two surfaces have the same geometry in regions not "too large" if and only if in those regions the surfaces have the same curvature. For example, if by "straight line" we understand "geodesic on a surface," then the geometry of the Euclidean plane is identical with the geometry of a limited portion of a right cylinder. Thus, on both of these surfaces the angle sum of a triangle is equal to two right angles, and the ratio of the circumference of a circle to its diameter is equal to π. On the other hand, the geometry on the surface of a sphere is different from that on a plane or on a saddle-back surface. Thus, the angle sum of a spherical triangle is greater than two right angles, whereas the angle sum for a plane triangle is always equal to two right angles.

The above explanation of the notion of curvature, whether for a curve or a surface, may easily lead to the assumption that a curve (or surface) has a curvature only because it is a figure in a "space" of higher dimensions than it is itself. For example, the curvature of a circle was defined as the reciprocal of its radius, so that we are apparently required to go outside the *one-dimensional* circumference of the circle into the *two-dimensional* plane. Similarly, the curvature of a two-dimensional surface was explained in terms of a plane passing through a normal to the surface, so that the notion of the curvature of such a surface seems to involve reference to a *third* dimension. There can indeed be little question that Gauss was led to his analysis of curvature by regarding curves and surfaces as embedded in spaces of higher dimensions; and this way of presenting some of the Gaussian ideas has undoubted heuristic and pedagogic advantages. Nevertheless, it would be a serious error to suppose that the only way of defining the curvature of a curve or of a surface is by way of reference to an embedding space of higher dimensionality. On the contrary, the curvature of a curve (and similarly of a surface) can be defined exclusively in terms of relations between magnitudes that pertain to the curve (and correspondingly to the surface) itself. Accordingly, the notion of curvature is entirely independent of even implicit reference to spaces of higher dimensionality.

However, the precise definition of curvature, stated entirely in terms of relations between magnitudes belonging to a figure and involving not even a tacit reference to anything outside the figure, cannot be given here, since such a definition requires the use of more advanced mathematical techniques than those with which most readers are familiar. We shall therefore merely assert that as a matter of fact such a definition can be given.[5] An analogy may nevertheless be helpful in this connec-

---

[5] The general structure of the definition is as follows: Let S be an arbitrary surface, and u and v the coordinates of any point on it with respect to a coordinate system lying entirely on the surface. Then the elementary distance ds between any two closely adjacent points on S is defined by $ds^2 = E du^2 + 2F du dv + G dv^2$, where E, F, and G are certain functions of coordinates that depend on the method of measuring lengths adopted on S. If L, N, and M are certain functions of E, F, and G, and hence of the coordinates, then the curvature K of S at a given point is defined as

$$K = \frac{(LN - M^2)}{(EG - F^2)}$$

The following approach may also be taken. The areas of the surface of a small

tion. An ellipsoid is often defined as the surface generated by an ellipse rotated around its major axis. It does not follow, however, that this is the only way an ellipsoid can be defined; it could be defined, for example, as a surface every point on which, when represented by Cartesian coordinates in a certain coordinate system, satisfies the equation $x^2/a^2 + y^2/b^2 + z^2/c^2 = 1$. Moreover, it would be a blunder to conclude that because an object (say a jelly bean) is an ellipsoid, it must have been produced by rotating an ellipse. Similarly, what is known as the "social contract theory" in political philosophy is usually stated in terms of a supposititious formation of political organizations at some historically remote time, as if before that time men existed without any social institutions. However, the intent of the theory is not to advance a historical thesis, but to analyze the structure of political obligations. The historical language in which the social contract theory is couched is thus an expository device, so that it is a mistake to evaluate the adequacy of the theory as if it were a statement about historical origins. The above exposition of the notion of curvature must be viewed in exactly the same way, as employing a form of statement that is heuristically valuable but is not to be construed literally. In any event, the fundamental point to be observed is that the curvature of a curve or of a surface can be defined without introducing considerations about dimensions of space higher than those of curves and surfaces, respectively.

Gauss's analysis of curvature did not go beyond that of the curvature of surfaces. It was the signal achievement of Riemann to have generalized the Gaussian ideas, so that the notions of geodesic and curvature could be employed in connection with spaces having any number of dimensions. In particular, Riemann's work has made it possible to define geodesics and curvatures of three-dimensional manifolds or continua, without assuming that such manifolds are embedded in a four-dimensional space. As in the case of two-dimensional surfaces, the Riemann curvature of three-dimensional continua may be positive, negative, or zero, and it may be constant for all points or may vary from point to point. Moreover, there is an intimate connection between the geometry of a space and its curvature. Thus the geometry conforms to the requirements of the Riemannian type of non-Euclidean geometry when the curvature of the space is constant and positive. The geometry is Lobachewskian when the curvature is constant and negative. The geom-

sphere, and its volume V, are given by the formulas

$$S = 4\pi r^2\left(1 - \frac{kr^2}{3} + \cdots\right); \quad V = \left(4\pi \frac{r^3}{3}\right)\left(1 - \frac{kr^2}{5} + \cdots\right)$$

where r is the radius of the sphere and k the curvature of the "space." Cf. H. P. Robertson, "Geometry as a Branch of Physics," in *Albert Einstein: Philosopher-Scientist* (ed. by P. A. Schilpp), Evanston, Ill., 1949.

etry is essentially Euclidean when the curvature is uniformly zero. Since the curvature of a manifold depends on what lines in it are defined to be the geodesics, the curvature is thus contingent on the rules adopted for measuring lengths. The point of cardinal interest that emerges from the Riemannian approach to the construction of non-Euclidean geometries is therefore this: the type of geometry required is a consequence of the rules adopted (or tacitly employed) for making spatial measurements. The importance of this point will become evident presently.

e. We have thus far indicated two approaches to the construction of non-Euclidean geometries: the axiomatic approach, and the Riemannian approach in terms of geodesics, curvature, and measurement. There is, however, a third method that must be briefly explained. This method emphasizes the differences in the *transformations* under which what is defined as the "distance" between two points is invariant in the alternative geometries.

This third approach, developed by Cayley and Klein in the latter third of the nineteenth century, views the alternative geometries we have been considering from the standpoint of projective geometry. These various geometries, including the Euclidean, are characterized as "metrical" because all of them make essential use of the notion of congruence, that is, of the equality of line segments, angles, areas, and volumes. Projective geometry, on the other hand, dispenses entirely with this notion, and studies only those properties of figures that are invariant under projection. For example, let a triangle be projected from one plane to another, that is, from some point outside either plane; then straight lines are drawn through the points on the triangle and continued until they meet the second plane, so that an image of the given triangle is formed on the second plane. In general, neither the lengths of the sides nor the magnitudes of the angles nor the area of the second triangle will be the same as those of the corresponding items in the first triangle. Nevertheless, some properties of the given figure do remain invariant under this transformation by projection. For example, any set of points that are collinear in the first figure will correspond to collinear points in the second; and any lines that are concurrent in the first figure will correspond to concurrent lines in the second. As another example, consider the projection of a circle from one plane to another. The figure on the second plane corresponding to the circle on the first will in general not be a circle. But the second figure will always be a conic section, and lines meeting on the circumference of the circle will be transformed into lines meeting on the circumference of the second figure.

It is possible to state the content of pure projective geometry more generally, and for our purposes it is essential to do so. Given any four points whose positions on a straight line are specified by four coordinates x_1, x_2, x_3, x_4 in that order, a certain ratio can be formed from the difference of these coordinates, called the "anharmonic ratio" of those points. This ratio is the double ratio:

$$\frac{x_1 - x_3}{x_2 - x_3} : \frac{x_1 - x_4}{x_2 - x_4}$$

The coordinates can be introduced in a purely projective manner, without making any use of the notion of congruence or distance. Accordingly, the difference between coordinates (for example, $x_1 - x_3$) must not be construed as a measure of distance between the corresponding points. In an analogous way it is possible to define the anharmonic ratio of four concurrent lines lying in the same plane, as well as the anharmonic ratio of four planes that have a line in common. It turns out, moreover, that anharmonic ratios are invariant under projective transformations, where such transformations can themselves be represented algebraically by homogeneous linear transformations of coordinates. Projective geometry can therefore be characterized as the theory of those transformations that leave anharmonic ratios invariant.

Projective geometry can itself be developed in an axiomatic manner. The postulates for projective geometry contain no assumptions either about congruence or about parallelism. Accordingly, projective geometry is neutral with respect to the three metric geometries we have been considering, and its axioms and theorems are compatible with any of the metric geometries. Indeed, projective geometry is more general than any one of the metric geometries, since it is concerned with structures of relations common to all three of the metric systems. The question naturally arises, therefore, whether by suitable specializations of the general transformations employed in projective geometry, the three metric geometries can be exhibited as three special cases of the general theory. The answer is in the affirmative; and, indeed, our interest in projective geometry is confined entirely to making evident in a general way the grounds for this answer.

A sufficient set of postulates for projective geometry can be laid down in alternative ways, where each way will employ certain terms as primitive or undefined. We shall not enter into any details in this connection. Let us suppose, however, that a certain set of postulates is adopted, which employ as primitive terms the expressions 'x is a point,' 'y is a line,' 'x lies on y,' and 'x is between w and z.' Other terms can then be defined with the help of these and the postulates, such as 'plane,' 'triangle,' and 'anharmonic ratio'; and, in particular, purely projective definitions can be given for certain structures of points, lines, and planes which are designated as 'conics' and 'quadrics' (i.e., surfaces like the ellipsoid in three-dimensional space). Furthermore, although within projective geometry it is not possible to distinguish between the familiar types of conic sections (i.e., between circles, ellipses, hyperbolas, and parabolas), it is possible to distinguish between "real" and "imaginary" conics. Real conics are those whose coordinates are real numbers; imaginary conics are those whose coordinates can only be complex numbers. It can also be shown that an arbitrary straight line will intersect a conic in two points, whether real or imaginary.

Let us now confine ourselves to plane projective geometry and stipulate that a given conic (in a plane), to be called the "absolute," is to remain invariant under all projective transformations (i.e., under all homogeneous linear transformations). That is, points on this conic must be transformed into points on the conic. Also, let x_1 and x_2 be the coordinates of any two points on a straight line that intersects the absolute in the points whose coordinates are a and b. The anharmonic ratio of these four points will then be invariant under projective transformations. Finally, let us define the "distance" between the two points x_1 and x_2 as the product of a certain constant k and the logarithm of this anharmonic ratio. It can be shown that distance so defined has the familiar additive properties of distance as ordinarily understood. For example, if A, B, and C are any three points on a straight line such that B is between the other two, then the projectively defined distance between A and C is equal to the sum of the distance between A and B and the distance between B and C. The magnitude of the angle formed by two straight lines can be defined in an analogous way.

We come, finally, to the major outcome of the projective approach. The measure of distance and angle as defined projectively satisfies the requirements of one or the other of the three metric geometries, depending on the special character of the conic that is taken to be the absolute. If the absolute is imaginary, the geometry is Riemannian; if the absolute is imaginary but degenerates into a pair of imaginary lines, the geometry is Euclidean; if the absolute is real, the geometry is Lobachewskian. Accordingly, the three metric geometries can indeed be viewed as specializations of an inclusive projective geometry, so that the differences between the metric geometries may be regarded as generated by different definitions of distance.[6]

[6] For a full account of the projective approach, cf. Felix Klein, *Vorlesungen über Nicht-Euklidische Geometrie,* Berlin, 1928. An important point is worth noting in connection with the projective approach to non-Euclidean geometry. The projective formulas defining distance and measure of angle have the same algebraic form in each of the three metric geometries. These formulas therefore enable us to establish a one-to-one correspondence between statements in one geometry and statements

2. We have thus far described some of the distinctive features of each of the three metric geometries but have said little about their relations to one another. It might seem at first sight that little need be said in this connection, since each of the three systems is incompatible with each of the others, and that this exhausts the matter. The situation is, however, somewhat more complex and requires fuller discussion.

a. We must first be clear in what sense the three metric geometries are mutually incompatible. Suppose that the primitive terms of pure Euclidean geometry E are $P_1{}^E, P_2{}^E, \ldots, P_5{}^E$ (e.g., 'point,' 'line,' 'plane,' 'lies on,' 'is between'), and that with their help an indefinite number of further terms are defined $D_1{}^E, D_2{}^E, \ldots$. Similarly, let the primitive and defined terms of pure Lobachewskian geometry L be $P_1{}^L, P_2{}^L, \ldots$, and $D_1{}^L, D_2{}^L, \ldots$; and analogously, let the terms of pure Riemannian geometry R be $P_1{}^R, P_2{}^R, \ldots$, and $D_1{}^R, D_2{}^R, \ldots$. We shall call the primitive terms of the three systems having the same *subscripts* the "corresponding" primitives. Suppose also that the defined terms in each of the three systems are defined in a precisely analogous way on the basis of the primitive terms of the corresponding system.[7]

We now assume that one of the axioms of E, namely $A_1{}^E$, is the statement-form: If x is $P_1{}^E$ and y is $P_2{}^E$, there is just one z which is a $P_2{}^E$ such that z lies on x and has the relation $D_1{}^E$ to y. On the other hand, the axiom $A_1{}^L$ is the statement-form: If x is a $P_1{}^L$ and y is a $P_2{}^L$, there are exactly two z's that are $P_2{}^L$ such that each z lies on x and has the relation $D_1{}^L$ to y. Moreover, the axiom $A_1{}^R$ of R is the statement-form: If x is a $P_1{}^R$ and y is a $P_2{}^R$, there is no z that is a $P_2{}^R$ such that z lies on x and has the relation $D_1{}^R$ to y. It will then be obvious from an inspection of the formal structures of these three postulates that if the *same* interpretation is assigned to *corresponding* terms in the three systems, it is impossible for an interpretation to satisfy more than one of the systems.

More generally, let S_1 and S_2 be any two postulational systems, and let the primitive and defined terms of one be made to correspond in some one-to-one fashion to the primitives and defined terms of the other. If

in each of the others in such a way that the deductive relations between statements in one system are the same as the deductive relations between the corresponding statements in each of the other systems. Accordingly, the consistency (or inconsistency) of one system (e.g., of Euclidean geometry) carries with it the consistency (or inconsistency) of each of the others. The projective approach thus illustrates the second method of establishing consistency, which was mentioned in the text on page 236.

[7] Thus, suppose that 'x has $D_1{}^E$ to y' is defined as 'x and y are both $P_2{}^E$; and there is a v which is a $P_3{}^E$ such that x and y both lie on v; and there is no w which is a $P_1{}^E$ such that w lies on both x and y.' Accordingly, '$D_1{}^E$' is defined in a manner analogous to the usual definition of 'parallel.'

now a postulate A or theorem T of S_1 is formally incompatible with a postulate A' or theorem T' of S_2, then no true interpretation can be given to both systems which interprets *corresponding* terms in the same manner.

b. It is essential to note, however, that the three pure metric geometries are incompatible only on the proviso that the same interpretation is given to corresponding terms. The three systems are by no means necessarily incompatible if *different* interpretations are supplied to corresponding terms, or if the same interpretation is given to terms that are *not* corresponding ones.

Before applying this observation to the three metric geometries, let us use it in connection with Euclidean geometry alone. As is well known, there are several alternative sets of postulates for Euclidean geometry, each set employing different terms as primitive. For example, the postulates given by Veblen (call them E_V) are formulated with the help of the primitive terms 'point,' 'between,' and 'congruent.' On the other hand, the postulates given by Huntington (call them E_H) are formulated with the help of the primitive terms 'sphere' and 'inclusion.' [8] However, despite the differences in the postulates and primitives of E_V and E_H, the systems developed on the basis of these alternative foundations are logically *equivalent*, so that they are foundations for the same system of geometry. Thus, it is possible to define in E_V certain terms 'sphere$_V$' and 'inclusion$_V$' that have the same formal properties in E_V as the terms 'sphere$_H$' and 'inclusion$_H$' have in E_H; and conversely and analogously for certain terms that can be defined in E_H. Accordingly, when a suitable correspondence is established between the terms of E_V and E_H, either set of postulates can be deduced from the other set. On the other hand, were the term 'point' in E_V made to correspond to, say, the term 'sphere' in E_H, the two systems would not only be not equivalent but would on the contrary be incompatible. Clearly, then, the question whether two systems of pure geometry are compatible is contingent on the manner in which a correspondence is established between their respective terms.

Let us return to the three pure metric geometries. It will suffice to consider just two of them, say the Euclidean and Riemannian systems. We have already shown that the Riemannian is as consistent as is the Euclidean by giving an interpretation to the primitive terms of Rieman-

---

[8] Cf. E. V. Huntington, "A Set of Postulates for Abstract Geometry," *Mathematische Annalen*, Vol. 73 (1913), pp. 522-59. Huntington's first postulate reads: If x, y, z are spheres, and x is included in y and y in z, then x is included in z. This makes it impossible to interpret the "inclusion" relation in the same way as Veblen's "x lies on y" relation, for the latter is not transitive.

nian geometry which converted its postulates into theorems of Euclidean spherical geometry. Let us ask, however, what we have done by giving this interpretation, in the light of the discussion in the preceding paragraph. In their customary formulations, both Euclidean and Riemannian pure geometries contain the term 'straight line' which, though it may be associated with certain images, functions in the two pure geometries as an uninterpreted term. Indeed, the formal properties of anything that is a straight line as stipulated by the Euclidean axioms are quite different from the formal properties of Riemannian straight lines. It follows that if 'straight line' (in Euclid) and 'straight line' (in Riemann) were taken to be *corresponding* terms to which the same interpretation is to be given, it would be logically impossible to give them an interpretation satisfying *both* systems. Evidently, therefore, the consistency of the Riemannian system is established, not by taking 'straight line$_E$' and 'straight line$_R$' as the corresponding terms of the two systems, but by finding some *other* term in Euclid (namely, 'arc of a great circle on a sphere') as the corresponding term for 'straight line$_R$.'

Once this point is grasped, it becomes clear that the essential step in proving the consistency of the Riemannian postulates is not adequately stated by saying that the proof consists in giving a geometrical interpretation yielding a valid Euclidean theorem. For the proof rests at bottom on defining a term with the help of Euclidean primitives, such that this term possesses within the Euclidean system the same formal properties that 'straight line' possesses within the Riemannian system. Accordingly, the argument for the consistency of Riemannian geometry can be stated in a purely formal way. What the argument establishes is that, given any Riemannian postulate having a certain logical structure into which the primitive terms of the system enter, a statement-form can be found within the Euclidean system having the same logical structure as the Riemannian postulate but into which either primitive or defined terms of Euclid enter. It follows immediately that, if a common interpretation is given to the terms of the two systems which correspond in this way, an interpretation converting the Euclidean postulates into true statements will automatically convert the Riemannian postulates into true statements.

c. It is clear, however, that this procedure can be reversed, and an interpretation can be supplied for the Euclidean postulates that will transform them into Riemannian theorems. In this reversed procedure the Euclidean term 'straight line' will of course not correspond to the Riemannian 'straight line,' for if it did, the Euclidean theorem 'The angle sum of a triangle is equal to two right angles' would be converted into the like-sounding Riemannian expression, which is incompatible with Riemannian postulates. Nevertheless, though there can be no Riemannian triangles whose angle sum, as defined by Riemannian rules for measuring angles, is equal to two right angles, there are *other* Riemannian figures, bounded by lines that are not Riemannian straight lines, which do have this angle sum.

It follows that the Euclidean and Riemannian pure geometries are not "inherently" incompatible. On the contrary, they are formally intertranslatable, in the following quite general sense. Let S_1 and S_2 be two deductive systems. The first employs the p primitive terms: $P_1^1, P_2^1, \ldots, P_p^1$, while the second employs the q primitive terms: $P_1^2, P_2^2, \ldots, P_q^2$, where p may not be equal to q. Moreover, the first is based on the m postulates: $A_1^1(P_1^1, P_2^1, \ldots, P_p^1), \ldots, A_m^1(P_1^1, \ldots, P_p^1)$ while the second is based on the n postulates: $A_1^2(P_1^2, \ldots, P_q^2), \ldots, A_n^2(P_1^2, \ldots, P_q^2)$, where again m may not be equal to n. Suppose, also, that it is possible to define in S_2 a set of terms $D_1^2, \ldots, D_p^2$ such that the statement-forms $A_1^1(D_1^2, \ldots, D_p^2), \ldots, A_m^1(D_1^2, \ldots, D_p^2)$ are deducible from the postulates of S_2; and suppose, finally, that it is possible to define in S_1 a set of terms $D_1^1, \ldots, D_q^1$ such that the statement-forms $A_1^2(D_1^1, \ldots, D_q^1), \ldots, A_n^2(D_1^1, \ldots, D_q^1)$ are deducible from the postulates of S_1. Under these assumptions the two systems S_1 and S_2 will be said to be "formally intertranslatable." [9] In this sense, therefore, not only are the Euclidean and Riemannian geometries formally intertranslatable, but also the Euclidean and Lobachewskian systems.

d. This conclusion is in fact illustrated by the alternative ways we have outlined for developing the non-Euclidean geometries. Thus, the approach via the notions of geodesic and curvature makes evident that the differences between the three geometries are metric ones, so that the *prima facie* mutual incompatibility of the systems is the consequence of adopting different rules or metrics for the measurement of spatial magnitudes. The approach via projective geometry enforces this point, and in addition provides us with what are in effect formulas of translation, so that with their help the correspondence between terms in the three systems can be established for translating each system into one of the others.

We may therefore conclude that the differences between the three systems of pure geometry we have been considering are differences in *notation*. They are three systems for codifying the same things in different ways, or different things in the same way. Thus, the term 'triangle' is employed in all three systems. However, the things correctly desig-

[9] It is of course assumed throughout this discussion that the logical principles employed in making deductions are the same in both S_1 and S_2.

nated as triangles on the basis of the requirements of one of the systems will be correctly designated by a *different* term in one of the other systems; on the other hand, the things correctly described as triangles within the framework of one system will *not* be things correctly described as triangles in another of the systems. It is thus certainly possible to regard the three pure geometries as alternative systems of rules for employing such terms as 'triangle,' 'circle,' 'distance,' and the like.

But if this is the outcome of our discussion, is not the result trivial, and does it not indicate that non-Euclidean pure geometries are without scientific importance? The answer to both parts of the query is negative. The construction of alternative "grammars" or systems of usages for familiar geometric locutions has in fact made it possible to analyze and organize spatial relations from different perspectives. Moreover, such possibilities not only have become the basis for advancing our knowledge of various spatial structures into which bodies enter but have also provided significant conceptual frameworks for developing more inclusive and unified theories of physics. We shall now examine, if only in outline, just how such more unified physical theories have in fact been based on non-Euclidean systems of measurement.

II. *The Choice of Geometry*

In the light of the preceding discussion, we can assume that at least three alternative systems of pure geometry are available to us. Each of these systems, when suitably interpreted in terms of certain features and behaviors of physical objects, can serve as a theory of spatial measurement.[10] How are we to choose between these alternatives, and what grounds if any are there for adopting one of them rather than another?

It should now be clear that the question covers two distinct issues. Since the three major types of pure geometry are intertranslatable, no interpretation converting the statement-forms of one system into true statements can fail to do likewise for the other two systems. The sole difference between the three systems of statements obtained in this way is that the *same facts* receive *different formulations.* Accordingly, if the question is taken to mean "Given a certain class of spatial properties and relations of bodies, which *language* shall we use to formulate them, and what reason is there for preferring one language to another?" the answer is obvious. "As far as the empirical facts to be codified and predicted are concerned," we are compelled to reply, "it will make not an iota of difference which language we adopt. However, we may find one

---

[10] These alternatives include not only the three metric geometries we have discussed but also metric geometries which assume *variable* curvatures. For the sake of simplicity we shall concern ourselves mainly with the former.

language *more convenient* than another, perhaps for several reasons. For example, we may find the Euclidean language psychologically simpler than the others, if only because we are more accustomed to it. Again, we may find that we need to refer to certain spatial traits of bodies more frequently than to other traits, and that the analytical formulations of the former in Euclid are more economical and involve less calculation than in the non-Euclidean systems. In any event, the grounds for adopting one geometry rather than another are not to be found in the spatial or physical structures of bodies, but in the greater practical advantages that one system of analysis and notation may have over the others."

The above question understood and answered in this manner is one phase of the philosophy of science known as "conventionalism," for which Henri Poincaré was a vigorous spokesman; we shall examine his views presently. However, the above question can be understood in a somewhat different sense, so that the answer to it will take a different turn. Suppose we are concerned with a class of edges, surfaces, volumes, and the like, which we decide to call "straight lines," "planes," "spheres," and so on. Suppose, further, that a one-to-one correspondence has been established between the subject-matter terms of the three pure geometries in a manner such that, when these already meaningful expressions "straight line," "plane," "sphere," and so on are substituted for the corresponding terms of the pure geometries, the resulting three systems of *statements* are mutually incompatible. The above question can now be taken to mean "Since the alternative *applied* geometries cannot all be true, is there any way of deciding between them, and are there considerations based on the empirical facts that may make the adoption of one system quite compelling?" So construed, the question does not admit of as ready an answer as on the earlier reading of the question. We must discuss some of the complex problems to which this altered sense of the question gives rise.

1. At first sight, the question as to which of the three systems of *applied* geometry is true appears to be decidable entirely on the basis of matters of empirical fact. Nevertheless, as has already been noted, the problem is complicated by the circumstance that the empirical truth of a geometry can be significantly discussed only if two procedures are antecedently instituted. In the first place, various objects called "straight lines," "planes," and so on must first be constructed or identified by means of rules of construction or identification specifiable *independently* of the alternative geometric systems. Unless this is done, then either there is no subject matter to be investigated by empirical methods or the subject matter consists of configurations that satisfy by *initial stipulation* one or the other of the three geometries. In the second place,

an overt, empirical procedure must be specified for making spatial measurements, so that in particular rules must be laid down for defining what bodies are to count as "rigid." Unless this is done, it is impossible to ascribe numerical values to spatial magnitudes, and hence to test experimentally any of the applied metric geometries. Something has already been said in the preceding chapter about both of these requirements, but the question of rigidity of bodies requires further attention.

a. The notion of rigidity is integrally involved in any theory of spatial measurement. When such measurements are made, bodies must be moved from place to place or reoriented in their positions. In consequence, the possibility must be noted that the relative spatial magnitudes (e.g., their relative lengths) may be altered because of the effects of various physical influences. The problem thus arises whether the straightedge adopted as the standard unit length suffers deformations in the process of measurement. For unless suitable precautions are taken against such deformations, the numerical values assigned to spatial magnitudes in the measuring procedure will in general depend on the particular time at which a measurement is made, as well as on the particular materials used in the construction of measuring instruments. For example, if measurements are taken in a region with nonuniform temperature, the geometrical properties that bodies are found to have will vary depending on whether the measuring rod is made of steel or of brass. To prevent such an uncomfortable multiplicity of incompatible results, and to assure sets of measured values independent of the special substances used in constructing measuring instruments, one of two things must be done. Either the measuring instruments as well as what they measure must be maintained in certain standard conditions throughout their entire histories, or correction factors must be introduced into the values obtained by actual measurement, so as to discount the effects of various deforming forces acting on bodies. Either precaution tacitly involves the notion of a *rigid* body which in theory is isolated from influences that may alter the relative lengths of physical objects, and which by *definition* therefore has an unchanging length.

A fundamental point must be noted in this connection. If the notion of rigidity is to be specified in experimental terms, but antecedent to the institution of a system of geometry, the influences to be counted as productive of deformations in bodies must perforce be influences detectable on the basis of their *differential* effects on different kinds of substances. In consequence, if there were a deforming "force" incapable of being screened or insulated and acting alike on all bodies whatever their composition, there would be no way of recognizing its presence by experimental means. For example, if two rods, one wooden and the other metal, were equally long in one environment, and were found to remain congruent to one another after being transported somewhere near the seat of such an alleged force, the rods could not be used to identify experimentally the presence of the force; and similarly for other pairs of objects of different make. Alleged "forces" of this kind have been labelled "universal forces," to distinguish them from the more familiar "differentiating forces" of common experience and laboratory practice. We shall soon consider whether there is any rationale for ever assuming universal forces. Meanwhile, however, the possibility of such forces can be ignored in setting up a definition of "rigidity." A body is usually said to be rigid if, and only if, it is isolated from all differentiating forces.

There is, of course, no inherent necessity for defining "rigidity" in just this way. It would be possible, for example, to call a body rigid when it is insulated only against the effects of temperature changes but not against those of humidity and mechanical stresses. Indeed, though we are familiar with many physical influences causing changes in the relative lengths of bodies, we cannot be entirely certain that we have discovered all such causes. If we adopt the definition of "rigidity" proposed at the end of the previous paragraph, we do so because we have certain aims in view: to obtain a system of mensuration independent of the special substances employed in constructing measuring instruments, and to formulate numerical laws in a manner more general than would otherwise be possible. Accordingly, as we discover new types of differentiating forces we revise our criteria of rigidity, primarily to achieve such generality of statement. In short, although the definition of rigidity is undoubtedly suggested by experimental facts, it is not necessitated by them; and its adoption rests on decisions we make with a view to attaining certain scientific objectives.

But in any event, when devising a schema of spatial measurement in physical science, it is customary to abstract from the great mass of physical and chemical properties that differentiate bodies. Moreover, the schema is set up in such a manner that the values to be assigned to spatial magnitudes are supposedly obtained through the use of ideally rigid measuring instruments. In consequence, the effects of variable differentiating forces upon both the instruments and the objects of measurement are systematically discounted. The rules we actually adopt for constructing scales of spatial measurement and for conducting procedures of such measurement are thus based upon numerous factual assumptions—assumptions about directly observable relations of congruence between surfaces and edges of bodies, as well as about a large variety of physical properties of bodies.[11] It follows that the numerical

[11] Helmholtz saw this point clearly. He declared: "The axioms of geometry are not simply principles which refer only to purely spatial relations. They refer to

values finally assigned to spatial magnitudes will in general not be the "raw numerical data" of overt measurement. These raw data require analysis, with a view to correcting them in the light of effects produced by assumed differentiating forces on the instruments and objects of measurement. In sum, therefore, the geometrical properties predicated of a figure on the basis of overt measurement (for example, that a triangular figure has an angle sum close to two right angles) are predicated on the assumption that all deformations produced by any differentiating forces have in principle been eliminated.

b. Let us return briefly to the problem of constructing or identifying straight lines, planes, and other figures that constitute the subject matter of geometry. The procedure discussed in the preceding chapter for constructing such figures has only a severely limited use, since it cannot be employed except to manufacture moderately sized physical configurations on the earth's surface. That procedure obviously does not suffice as a basis for a system of mensuration which enables us to determine the heights of mountains, the width of the oceans, or the magni-

magnitudes. But one may talk of magnitudes only if one specifies a definite procedure, according to which magnitudes are to be compared, analyzed into parts, and measured. All spatial measurements, and therefore all magnitudes based on space, thus presuppose the possibility of the motion of spatial configurations, whose forms and magnitudes are assumed to be invariant under the motion. Such spatial configurations are usually designated in geometry as geometrical bodies, surfaces, angles and lines, because one abstracts from the physical and chemical differences exhibited by material bodies. One assumes for them as the only physical property that of rigidity. But we have no criterion for the rigidity of bodies other than that the relations of congruence are invariant at all times under translations and rotations. . . .

"The axioms of geometry therefore express not only spatial relations, but at the same time state something about the mechanical relations of rigid bodies undergoing motions. One could regard the notion of a rigid spatial configuration as a transcendental concept, constructed independently of all experience, and which does not necessarily correspond to experience—just as our notions of material body do not correspond precisely to the concepts we obtain by induction. If we assume such an ideal rigidity, a follower of Kant could regard the axioms of geometry as given apriori in a transcendental intuition, which can be neither confirmed nor refuted by experience—since it would be in terms of them alone that we would decide whether a given body is to be judged as rigid or not. But in that case, the axioms of geometry would no longer be synthetic propositions in Kant's sense. For they would then only assert as much as is implied analytically by the concept of rigid body for measurement, since only such configurations could be judged as rigid which satisfy the axioms.

"On the other hand, if we supplement the axioms of geometry with the other propositions about the mechanical properties of bodies—if only with the law of inertia, or the proposition that under otherwise constant conditions the mechanical and physical properties of bodies are independent of their positions—the system of propositions thus obtained then has a genuine content, which is to be confirmed or refuted by experience."—"Über den Ursprung und die Bedeutung der Geometrischen Axiome," *Vorträge und Reden*, 3rd ed., Brunswick, 1884, Vol. 2, pp. 28-30.

tudes of astronomical distances and areas. Accordingly, that procedure must be supplemented by additional rules, which will specify in a more inclusive manner what figures are to count as straight lines, planes, and so on.

A rule of this kind that is generally adopted in physics in effect identifies straight lines as the paths of light rays in optically homogeneous media. Such a rule is implicit in the use of theodolites and telescopes for measuring distances and angles. However, the adoption of this rule seriously complicates the problem of exhibiting the grounds upon which choices may be made between alternative geometries. It is obvious, for example, that the expression 'path of light ray' codifies a *theoretical* notion, not an *experimental* one. We observe illuminated bodies, not light rays; and the concept of light ray is part of a theory introduced to explain the observable visual facts. The rule identifying straight lines with paths of light rays is thus part of the theory of optics. But it is in general not possible to test experimentally and in isolation from one another the individual special assumptions of the theory; and experimental evidence usually confirms or refutes the theory as a whole, rather than some particular component of it. Accordingly, the special assumption that light is propagated along, say, Euclidean straight lines cannot be controlled by performing some allegedly crucial experiment.

To be sure, that part of optical theory known as "geometrical optics" operates with a relatively small number of assumptions, among which the Euclidean character of optical paths in homogeneous media plays a commanding role. There is in fact a vast quantity of evidence, among others that obtained from the study of lenses, which makes this particular assumption practically inescapable—at least within the range of inquiry for which geometrical optics is relevant. Moreover, there is a certain amount of overlapping between the things said to be "straight lines" in accordance with the rules for constructing straightedges of rigid bodies, and the things said to be "straight lines" in accordance with the rules that identify them with certain optical paths. Thus, a line that is straight because it is the edge of a surface ground in a specified manner is also straight (within the limits of experimental error) in the sense that it corresponds to the line of sight. It is nevertheless patent that there are optical rays—for example, a light ray from a star to the earth—which cannot be directly compared with edges of solids.

In consequence, the numerical values obtained by actual measurement of most optical figures (such as the value obtained for the angle sum of a stellar triangle) are open to a variety of interpretations. It is not a simple matter to separate out those components of these numerical values representing features to be counted as "truly" geometrical properties from components representing the effects of some deforming physical influence. On the other hand, this difficulty is in principle no different than the problem of deciding on the basis of experimental evidence whether light is a vibratory or a corpuscular process. It is in fact possible to introduce corrections into the data to compensate for the effects of differentiating forces. Accordingly, for certain ranges of stellar magnitudes, and within the limits of assumptions that are supported by experimental evidence, it is feasible to determine whether or not a given class of optical figures are Euclidean. Until the second decade of the present century, the evidence for the Euclidean character of the path of light rays appeared to be overwhelmingly conclusive. Even today, relatively short optical paths, or optical paths far removed from gravitational fields, are generally accepted as excellent approximations to the Euclidean requirements. Except for some reservations to be noted shortly, the conception that applied geometry is a branch of natural science, and is to be judged as true or false on the basis of empirical evidence, seems well founded.

c. It will be instructive to consider at this point an objection sometimes raised against the very suggestion that measurement could conceivably establish a claim that optical or other figures are non-Euclidean. The objection takes its point of departure from the sound observation that all measuring instruments (e.g., meter rods, protractors, telescopes) are in fact so constructed that their relevant parts appear to be in excellent agreement with Euclid, and that these instruments are used within a framework of theoretical assumptions (e.g., geometric optics) which take Euclidean geometry for granted. But if this is so, the objection continues, it is self-refuting to suppose that numerical values obtained by means of such instruments can serve as possible evidence for the non-Euclidean character of any physical configuration. In particular, it is self-contradictory to maintain that measurements made with the help of such "Euclidean" instruments can establish that the relevant parts of those instruments possess a non-Euclidean structure.[12]

In point of fact, however, there is nothing incoherent in the supposition against which the objection is directed. On the face of it, there is nothing paradoxical in maintaining that an instrument whose geometry is by hypothesis Euclidean, when used to measure the spatial magnitudes of some *other* configuration, nevertheless yields numerical values such that this other figure is thereby shown to possess a non-Euclidean structure. Moreover, though the three metric geometries may be formally incompatible on a given interpretation of their primitive terms, the discrepancies between what they may then assert about configurations with

12 Cf. Hugo Dingler, *Das Experiment*, Munich, 1928, pp. 86ff.

"relatively small" dimensions may in fact fall far below the threshold of empirical detection. We have already noted that, for example, the angle sum of a physical triangle is less than, equal to, or greater than two right angles, according as the figure is a Lobachewskian, Euclidean, or Riemannian triangle, and also that the defect or excess of the angle sum is proportional to the area of the figure. However, if the triangle is not too large, the theoretical defect or excess may be so small that actual measurement may be unable to detect any significant deviation from zero. Accordingly, even when questions are waived concerning the possible action of deforming forces upon the triangle, experimental measurements on a sufficiently small triangle will not be able to determine whether it is a non-Euclidean rather than a Euclidean figure. In short, actual measurements can discriminate between the geometric types to which a figure belongs only if the figure is of vast astronomical dimensions.

It follows that though, on the basis of experimental data, a given instrument (e.g., a meter rod) may rightly be judged as possessing a Euclidean structure, the evidence is entirely compatible, because of the small size of the instrument, with the assumption that its structure is non-Euclidean. On the other hand, further investigations of configurations of very great size may yield data difficult to reconcile with the hypothesis that such large figures are Euclidean. In consequence, the initial assumption predicating a Euclidean structure to the measuring instrument may be revised, without, however, impugning the experimental data upon which that assumption was initially based. More generally, therefore, an instrument may be correctly regarded as an excellent first approximation to Euclidean standards, and may nevertheless be judged on the basis of more inclusive evidence and the requirements of theoretical consistency to possess a non-Euclidean structure. In sum, it is not self-contradictory to suppose that our measuring instruments, though built according to Euclidean specifications, might be discovered by measurement to be non-Euclidean.

2.   We must finally examine the view that a system of applied geometry is simply a set of "concealed definitions" or "conventions" for measuring spatial relations, and is not an empirical science. This view was vigorously argued especially by Henri Poincaré, who in fact advanced the more comprehensive thesis that most if not all the general "principles" of physics (such as the principle of inertia) are conventions.[18] Although we shall discuss Poincaré's views only insofar as they refer explicitly to geometry, the analysis and conclusions reached can be ex-

[18] H. Poincaré, *The Foundations of Science*, New York, 1921: *Science and Hypothesis*, Part II; *The Value of Science*, Chaps. 3 and 4.

tended without essential modifications to the more general form of the conventionalistic thesis.

a. Poincaré's argument for the definitional status of geometry is somewhat obscured by his not distinguishing clearly between pure and applied geometry. Moreover, he also assumed that the subject matter of geometry (presumably pure geometry) is an ideal "space," upon which, in the nature of the case, no experiments are possible; and it is uncertain whether he meant by this no more than that pure geometric statements are formulated in terms of "limiting" concepts such as breadthless lines and mathematically continuous curves. In any event, he maintained that, since the alternative pure metric geometries are intertranslatable, we are free to choose any one of them as a way of codifying spatial relations, so that our choice is in effect a choice between alternative conventions for naming such relations. Poincaré declared

> In space we know rectilinear triangles the sum of whose angles is equal to two right angles; but equally we know curvilinear triangles the sum of whose angles is less than two right angles. The existence of one sort is not more doubtful than that of the other. To give the name of straights to the sides of the first is to adopt Euclidean geometry; to give the name of straights to the sides of the latter is to adopt the non-Euclidean geometry. So that to ask what geometry it is proper to adopt is to ask, to what line is it proper to give the name straight?[14]

This part of Poincaré's argument thus claims only the formal intertranslatability of the three systems of statement-forms constituting the three systems of pure geometry. The thesis he establishes by this argument is simply the thesis that choice of notation in formulating a system of pure geometry is a convention. We have already recognized that, so understood, the conventionalistic thesis is undoubtedly correct.

However, Poincaré also argued for the definitional status of *applied* as well as of *pure* geometry. He maintained that, even when an interpretation is given to the primitive terms of a pure geometry so that the system is then converted into statements about certain physical configurations (for example, interpreting 'straight line' to signify the path of a light ray), no experiment on physical geometry can ever decide against one of the alternative systems of physical geometry and in favor of another. But the grounds he advanced for this claim are far from clear.

Poincaré sometimes wrote as if the grounds for this thesis about physical geometry were identical with those for the thesis about pure geometry. "Is the position tenable," he asked, "that certain phenomena, possible in Euclidean space, would be impossible in non-Euclidean space,

[14] *Ibid.*, p. 235.

so that experience, in establishing these phenomena, would directly contradict the non-Euclidean hypothesis?" According to him, however, this question was precisely equivalent to the query, "Are there lengths expressible in meters and centimeters, but which cannot be measured in fathoms, feet and inches, so that experience, in ascertaining the existence of these lengths, would directly contradict the hypothesis that there are fathoms divided into six feet?" [15] His answer was that the suppositions envisaged in both questions are obviously absurd, and that it is "impossible to imagine a concrete experiment which can be interpreted in the Euclidean system" and not in a non-Euclidean one.

On the other hand, Poincaré sometimes seemed to rest his claim about physical geometry on different considerations. On these occasions he called attention to the difficulty, if not the impossibility, of putting to a crucial experimental test a *single component* in a complex body of theory. He declared, for example, that if astronomers should find some stars to have negative parallaxes (a state of affairs *prima facie* incompatible with Euclid but in accordance with Riemann), two courses would be open to us: "We might either renounce Euclidean geometry, or else modify the laws of optics and suppose that light does not travel rigorously in a straight line." Poincaré believed everyone would regard the second alternative "as the more advantageous." According to him, therefore, a decision between alternative geometries cannot be made on the basis of evidence concerning their truth or falsity; a decision must rest on considerations concerning their relative convenience and simplicity. He therefore concluded that "Euclidean geometry is, and will remain, the most convenient," because of its greater "simplicity" and its generally good concordance with the properties of natural solids. [16]

b. How conclusive is Poincaré's argument? He was undoubtedly correct in maintaining that, if pure geometry is used as a system of implicit definitions, so that it provides the schematism and nomenclature for classifying spatial relations, the system can be retained in the face of all experimental findings. By their very nature implicit definitions cannot be characterized as true or false; and Poincaré was right in holding that they must be evaluated in ways other than by an appeal to experimental facts about the spatial properties of bodies. Nevertheless, this sound point is not the only issue raised and settled by Poincaré's analysis of the status of geometry. There is also the fundamental question whether, once an interpretation is given to the primitive terms of a pure geometry, so that it then is a system of *physical* geometry, the latter is nothing more than a "concealed definition." Poincaré did not uniformly distinguish this question from the one about the status of pure geometry; and in consequence his discussion of physical geometry leaves much to be desired.

Einstein remarked in commenting on Poincaré's conventionalistic philosophy of geometry that, although in his judgment Poincaré was right when viewed *sub specie aeternitatis,* in the perspective of actual history Poincaré's analysis must be qualified and that a physical geometry does indeed require evaluating in the light of empirical evidence.[17] We must now show in outline why such a qualification is needed, and why Poincaré had neither logic nor history on his side when he maintained that Euclidean geometry would never be abandoned.

Let us imagine a resolute defender of Euclidean geometry, and let us consider the price he might have to pay if he insists on retaining Euclid at all costs. Since the Euclid he wishes to defend is Euclid as an applied or physical geometry, he will construct or find physical configurations satisfying Euclidean requirements within the limits of experimental error. Suppose he has no trouble in doing this when dealing with bodies of moderate size; but suppose that in order to make measurements on configurations having astronomical dimensions, he adopts the hypothesis that the paths of light rays are Euclidean straight lines. Let us assume, however, that optical triangles of very large size consistently fail to satisfy Euclidean expectations, for example, because the angle sums of such triangles are invariably much greater than two right angles. Euclid's defender will of course not abandon Euclidean geometry for this reason, but he will doubtless try to explain the discrepancy. He can do this only by maintaining that the sides of the stellar triangles are not really Euclidean straight lines, and he will therefore adopt the hypothesis that the optical paths are deformed by some fields of force. He may in fact obtain evidence for the existence of differentiating forces, whose identifiable presence accounts for the deviation of light rays from rectilinear paths in agreement with the accepted physical theory of light. In that eventuality, everything is in order.

Suppose, however, that Euclid's protagonist does not succeed in locating such fields of differentiating forces. Since he is resolute in his commitments, he will still not abandon Euclid. But in this altered situation he will postulate forces that produce the same deformations in all bodies whatever their composition, and in all light rays whatever their wave lengths or amplitudes. In short, he will assume the existence of *universal* forces in order to account for the discrepancy between the measured angle sum of the stellar figure and the Euclidean angle sum. However, the sole ground he will have for believing in the existence of

[17] Albert Einstein, *Geometrie und Erfahrung,* Berlin, 1921, p. 8.

such forces is the fact that if they are postulated the indicated discrepancy can be explained. Accordingly, one possible consequence of the initial resolution to retain Euclidean geometry at all costs is that universal forces may have to be assumed in articulating adequate physical theories. On the other hand, if the introduction of such forces is excluded, perhaps on grounds of some methodological rule, the defender of Euclid will be compelled, under circumstances like the one we have imagined, to abandon Euclid in favor of one of its alternatives.[18]

We can state this result in another way. It is an experimental fact that we can find or construct rigid bodies whose spatial properties are good approximations to Euclidean requirements. However, such bodies are of moderate dimensions, and their rigidity is defined in terms of their being insulated against the effects of differentiating forces. Let us assume that no large-scale configurations occurring in nature conform to Euclid within the limits of experimental error, but let us suppose further that all attempts to explain this state of affairs in terms of the operations of differentiating forces are uniformly unsuccessful. It is then still possible to retain Euclidean geometry even for large-scale configurations, but only by postulating universal forces to account for the systematic "deformations" in such configurations that just prevent the latter from ever exhibiting Euclidean geometric properties. Nevertheless, universal forces have the curious feature that their presence can be recognized *only* on the basis of geometrical considerations. The assumption of such forces thus has the appearance of an *ad hoc* hypothesis, adopted solely for the sake of salvaging Euclid.[19] Indeed, the "deformations" in bodies which must be attributed to universal forces in order to save Euclid have a markedly geometrical rather than physical character. The deformations persist even if all differentiating forces are eliminated; and they are construed to be "alterations" in the "natural" shapes and spatial dimensions of bodies, only because the criterion of rigidity that is now tacitly employed is the possession by a body of just those geometrical properties prescribed by Euclid.

In any event, even if we admit universal forces in order to retain

---

[18] The distinction between "universal" and "differentiating" forces is employed with great clarifying effect by Hans Reichenbach in his *Philosophie der Raum-Zeit Lehre*, Berlin, 1928 (in English, under the title *The Philosophy of Space and Time*, New York, 1958). However, both the distinction and the terminology are of older date. It was used by F. A. Lindemann in his introduction to the English translation of Moritz Schlick's *Space and Time in Contemporary Physics*, New York, 1920, and is implicit in the writings of Helmholtz (e.g., in the essay quoted in footnote 11).

[19] "Universal force" is not to be counted as a "meaningless" phrase, for it is evident that a procedure is indicated for ascertaining whether such forces are present or not. Indeed, gravitation in the Newtonian theory of mechanics is just such a universal force; it acts alike on all bodies and cannot be screened.

Euclid, we would not achieve our scientific objectives were we simply to baptize the discrepancies between Euclidean requirements and the actual geometrical properties of bodies as "deformations produced by universal forces." For if we wish to predict and explain systematically the actual geometrical properties of bodies, we must incorporate the assumption of universal forces into the rest of our physical theory, rather than introduce such forces piecemeal subsequent to each observed "deformation" in bodies. It is by no means self-evident, however, that physical theories can in fact always be devised that have built-in provisions for such universal forces. Furthermore, even if this could be done, it does not follow that the resulting *total system* of physical theory, even though formulated within the framework of the "simple" Euclidean geometry, would necessarily be "simpler" and "more convenient" than would a *total system* of physics based on a "less simple" non-Euclidean geometry. Poincaré was therefore overlooking something very important when he supposed that the allegedly greater "simplicity" of Euclidean geometry is the exclusive consideration in choosing it rather than its rivals. In point of fact, the subsequent history of physics illustrates Poincaré's oversight. The general theory of relativity is formulated within the framework of a Riemannian type of geometry, but that theory has abandoned Euclid because by doing so it has achieved a more inclusive and "simpler" theory of mechanics than was possible when Euclid was used as the foundation for articulating classical mechanics.

3. It will be useful to summarize our discussion of the status of geometry in a series of brief conclusions.

a. When the notions of rigidity, plane surface, straightedge, and spatial congruence are suitably defined in terms of overt experimental procedures, rigid bodies of moderate size can be constructed whose spatial properties are in good practical agreement with Euclidean requirements. Accordingly, for an extensive class of bodies geometry is an experimental science, a branch of elementary mechanics. Although this domain of objects does not exhaust the field of actual application of geometry, it is nevertheless an important one. It includes a large part of the spatial measurements made in ordinary affairs of life as well as in the engineering arts; and it also embraces the construction of scientific instruments, whose calibrations require some kind of spatial measurement. Moreover, since the measurement of large as well as minute distances, areas, and volumes (many of which are very remote from the earth) is ultimately dependent on the use of such primary instruments, this domain—in which geometry is an experimental science—possesses an obvious priority over other fields of application of geometric theory.

In this domain, however, there is no effective room for choice between alternative metric geometries. For the discrepancies between the theoretical values of geometric magnitudes specified by these alternatives are far too minute in the case of moderately sized configurations to permit an experimental discrimination of them. If Euclidean geometry rather than one of its rivals is accepted as true in this domain of objects, it is accepted partly for the historical reason that the Euclidean system was the first to be developed, and partly because it appears to be psychologically simpler than the alternatives to it.

b.   There are nevertheless areas of application for geometry in which it is not in our power to construct physical configurations in conformity with prescribed experimental rules. In these areas the assumption that, for example, certain configurations are Euclidean straight lines is a hypothesis which cannot be tested directly or decisively. On the contrary, such geometrical hypotheses must be handled as components in a complex physical theory; they cannot be tested in isolation from other physical assumptions. Accordingly, the decision whether a given geometry is true of the objects in this area is normally controlled by the over-all adequacy of the various theories into which that geometry enters as a component.

Nevertheless, the decision is in practice not an arbitrary one, and rests in large measure on empirical considerations. To be sure, it is abstractly possible to retain a particular geometry in the face of apparently incompatible empirical evidence, by making suitable alterations in other parts of physical theory. However, the needed alterations may require the introduction of *ad hoc* assumptions, which in turn may not lend themselves to a systematic integration into the rest of physics. Accordingly, an unshakable commitment to a particular geometry may become an obstacle to the development of more inclusive and more integrated systems of physical theory.

c.   There is one undeniable sense in which a geometry can be rightly regarded as a set of conventions. A geometry is a set of conventions when it functions as a system of implicit definitions that fix the usage and the range of permissible application of such familiar terms as 'plane,' 'straight line,' and the like. Moreover, since the three alternative metric geometries are formally intertranslatable, whatever can be expressed in one of them is also expressible, though in different terminology, in each of the other two. Accordingly, no conceivable experiment can supply evidence to show that one such geometry is less able than another to serve as a vehicle for formulating a theory of spatial measurement. When a geometry is used in this manner, its "conventional" status is thus primarily a notational conventionalism.

On the other hand, issues of a different order emerge as soon as we inquire whether the mode of spatial analysis, implied by the adoption of such a notational convention, yields formulations of geometrical relations that can serve as the basis for adequately comprehensive and manageably simple physical theories. These issues cannot be settled by instituting conventions, but require the consideration of questions of empirical fact.

III. *Geometry and Relativity Theory*

In Newtonian mechanics the appropriate frame of reference for the motions of bodies is absolute space, and Euclidean geometry is employed as the theory of spatial mensuration. We have noted that the Newtonian notion of absolute space is beset by difficulties, and that the empirical evidence does not require its adoption as the frame of reference for the analysis of motions. Moreover, we are now also familiar with alternatives to Euclidean geometry, so that we are not compelled, as was Newton, to regard Euclid as the only basis for a theory of mechanics. However, it is a distinctive feature of the Einsteinian theory of general relativity that it dispenses both with absolute space as well as with Euclidean geometry in developing its analysis of the motions of bodies. Our discussion of Newtonian mechanics and of the logical status of geometry will be rounded out if we briefly examine in what manner relativity mechanics achieves its objectives without the assumptions that occupy so central a place in Newtonian theory.

The title "theory of relativity" for the system of mechanics Einstein developed is in some ways unfortunate, for it has undoubtedly misled many as to the actual import of the theory. In any event, Einstein succeeded in formulating a theory of mechanics in such a manner that its equations of motion are *invariant* over a more inclusive class of reference frames than are the Newtonian counterparts of those equations. It will be recalled that the classical equations of motion are valid for motions referred to inertial or Galilean frames of reference, and that they retain their form when motions are referred to any one of a set of reference frames moving with uniform velocity relative to one another. However, the Newtonian equations do not provide an altogether satisfactory analysis of motions when a noninertial reference frame is employed, that is— to use Newtonian language, when a reference frame is used that is accelerated with respect to absolute space. On Newtonian theory there is thus a class of privileged reference frames, over which the equations of motion are invariant. The singular achievement of the Einsteinian general theory of relativity, on the other hand, is that it assigns no such privileged status to any class of reference frames, so that the motions of bodies may be referred to an arbitrarily selected system of spatial coordinates. The fundamental equations of motion of this theory are invariant under the class of all continuous (and differentiable) transformations which establish correlations between the coordinates of different reference frames.

We shall not be concerned here with the technical and difficult details of Einstein's achievement, and we can indicate only in bare outline the main features of relativity theory. Einstein arrived at it in two steps. In the special theory of relativity he extended the Galilean-Newtonian principle of invariance so that it would be satisfied not only by the equations of mechanics but also by Maxwell's equations for electrodynamical fields. With this end in view, he made a careful analysis of the conditions under which spatial and temporal measurements are instituted in physics, and he showed that the magnitudes assigned to lengths and temporal durations depend in an essential way upon the state of relative motion of the bodies being measured. He concluded that, on the assumptions that light signals are employed in making spatial and temporal measurements and that the velocity of light is independent of the velocity of its source, if a body is moving with uniform velocity relative to a reference system S, then the lengths and durations on that body as measured in S are definite functions of this relative velocity. Einstein's analysis required a revision of the Newtonian assumption that the mass of a body is independent of its velocity relative to the system in which the mass is being measured, and in consequence important modifications had to be made in the Newtonian equations of motion. The net outcome of the special theory is that the emended equations of motion, as well as Maxwell's field equations, are invariant in all inertial frames of reference.

However, the special theory still assigns a privileged position to a special class of reference frames in formulating the equations of both mechanics and electrodynamics. This appeared anomalous to Einstein, since *kinematically* (i.e., when changes in the position of bodies are analyzed, without reference to any forces as determinants of such changes) all motions are relative. He therefore undertook to develop a theory of *dynamics* which would be free from this limitation and whose fundamental equations would retain their form whatever frame of reference was adopted for analyzing the motions of bodies.[20]

Einstein took his point of departure from the distinction in Newtonian mechanics between two kinds of mass: inertial mass, which is associated

[20] In technical language, the equations of motion are to be covariant for all reference frames.

with a body's resistance to changes in its velocity; and gravitational mass, which is associated with a body's behavior in gravitational fields and is commonly referred to as a body's weight. However, despite this theoretical difference, experiment shows that the numerical measures of the inertial and gravitational masses of a body are equal to one another. This equivalence is not explained by Newtonian theory. Einstein was not content to leave it as a brute contingent fact, and sought to account for it. He interpreted this equivalence to mean that a body does not possess two distinct kinds of mass, but that the property a body exhibits under certain conditions as inertia is manifested under other conditions as weight. With this interpretation as a fundamental postulate of his new theory, Einstein showed how a gravitational field (provided it is not too large in extent) can always be construed as an inertial field. Accordingly, where Newtonian theory explains the motion of a body on the assumption that the body is in the gravitational field of (or is attracted by) a second body, the new theory accounts for the motion by assuming a relative acceleration between two bodies and so dispensing with a special gravitational force. Moreover, Einstein succeeded in formulating the equations of motion in such a manner that they retained their form irrespective of the coordinate system selected as the frame of reference. In this formulation, bodies (and light rays in particular) moving without constraints follow paths that are always geodesics (i.e., paths of "shortest distance") relative to an arbitrary reference frame. But the geometry required for this formulation is a type of Riemannian metric of positive but variable curvature. In the limiting case, however, when gravitational fields are absent, the paths of light rays as well as of freely moving bodies are Euclidean straight lines.

General relativity theory thus involves a fusion of geometry and mechanics more intimate than the connections between them in Newtonian theory. Indeed, the word "geometry" as employed in relativity theory covers a far more inclusive set of relations than does the word in its Newtonian usage. For example, in relativity theory geometrical invariants refer to temporal as well as to strictly spatial traits of objects. In point of fact, the fundamental invariant of the theory is so constituted that, when certain parameters contained in it are assigned special values depending on the distribution of material objects in a given region, the trajectories of light rays and freely moving bodies can be deduced as the geodesics of that region. By contrast, the fundamental equations of motion in classical mechanics are not derivable from the geometrical framework of Newtonian theory. Nor are the force-functions employed by that theory in various problems determined by the geometrical properties of the objects under study; on the contrary, the introduction of a given force-function is the introduction of an additional and independent assumption. In general relativity theory, on the other hand, the distribution of bodies in a region determines the geometry of the region; and the equations of motion are derivable from the geometry so determined. It is thus evident that the comprehensive geometry of general relativity theory, containing the geometry of Newtonian mechanics as a limiting case, is a branch of physics. It follows that the adoption of this geometry rather than one of the alternatives to it cannot be a matter requiring only a decision between alternative conventions, but must rest on experimental evidence.

Some comments on three issues that have been raised in connection with the general theory of relativity may help to clarify several points in the above brief account of the theory.

1. The theory has been criticized on the ground that, unlike the special theory of relativity, its basic concepts—and in particular its free use of arbitrarily selected spatiotemporal coordinate systems—do not have experimental (or "operational") meaning. In a discussion of the relativity equations of motion for particles, P. W. Bridgman has argued that nowhere does the theory give "operational" definitions for assigning coordinates, whether in connection with these equations or other matters. According to him, moreover, no operational criteria are stated by the theory for deciding whether the physical happenings supposedly studied by different observers located in different reference frames are the "same" occurrences. The theory is in consequence accused of failing to analyze the "intuitively recognizable elementary happenings" in terms of which it assumes physical situations to be capable of exhaustive characterization. Bridgman therefore concludes that the general theory of relativity is not complete, and that it operates with nebulous assumptions which conceal a questionable philosophy.[21]

Bridgman is unquestionably right in noting that the sheer mathematical formalism of relativity theory does not supply the physical content of the theory. He is equally right in maintaining that, if the theory is to count as a branch of physics, appropriate coordinating definitions must be supplied relating theoretical terms with experimental concepts. Nevertheless, it is not reasonable to ask, as Bridgman appears to be doing, that every theoretical term be associated with an overt experimental procedure. It is even less reasonable to require that each component assumption of a theory be capable of independent experimental test. Few theories of classical physics satisfy the first of these proposed requirements, although they do not thereby fail to be adequate physical theo-

[21] P. W. Bridgman, *The Nature of Physical Theory*, Princeton, 1936, Chap. 7; see also his essay, "Einstein's Theories and the Operational Point of View," in *Albert Einstein: Philosopher-Scientist* (ed. by P. A. Schilpp), Evanston, Ill., 1949.

ries; and perhaps no theory of modern physics fulfills the second condition. As we have repeatedly noted, a sufficient condition for a theory to be testable and to perform its function in inquiry is that enough of its theoretical notions be associated with coordinating definitions so that a variety of the logical consequences of its postulates can be experimentally controlled. There can be no serious doubt that general relativity theory fully meets this requirement.

One other point needs to be recognized in this connection. The equations of motion of general relativity theory are invariant under a much more inclusive class of transformations than are the equations of motion either in classical mechanics or in the special theory of relativity. In the nature of the case, therefore, general relativity theory formulates a common structure of a wider variety of motions than are formulated by these other theories, so that it abstracts from many differences between physical systems explicitly recognized by the latter ones. Accordingly, the rules of correspondence (or operational definitions) for the theoretical notions of general relativity will differ in their specific empirical reference as the theory is applied to different types of physical systems. It would not be possible to retain the same operational definition for a given theoretical notion without curtailing the domain of invariance and the range of application of the theory. A simple illustration of such differences in correspondence rules for the same theoretical notion is found in the circumstance that the term 'point' in general relativity theory is sometimes coordinated with a small area of the earth's surface, sometimes with the entire volume of the earth, sometimes with a planet, sometimes with a galaxy. However, the fact that no unique rule of correspondence is laid down once for all for a given theoretical notion does not mean that a coordinating definition with a specific empirical reference is not provided for the notion when the theory is applied to a concrete problem.[22]

2. Einstein's general theory has been further criticized because it employs a geometry with variable curvature as the framework for a system of mechanics. However, the criticism under consideration is not based on an a priori commitment to Euclidean geometry. The criticism rests on the claim that *uniform* spatiotemporal relations must be adopted as the framework for a physical theory if the contingent and heterogeneous events of nature are to be systematically analyzed and related. This is the reason A. N. Whitehead advanced for developing an alternative general theory of relativity that employs a geometry with constant rather than variable curvature. Whitehead declared that

[22] Cf. A. S. Eddington, *The Mathematical Theory of Relativity*, 2nd ed., London, 1924, pp. 85ff.

our experience requires and exhibits a basis of uniformity, and . . . in the case of nature this basis exhibits itself as the uniformity of spatiotemporal relations. This conclusion entirely cuts away the casual heterogeneity of these relations which is the essential of Einstein's later theory. . . . It is inherent in my theory to maintain the old division between physics and geometry. Physics is the science of the contingent relations of nature and geometry expresses its uniform relatedness.[23]

According to Whitehead we must assume a geometry of constant curvature for expressing the contingent facts of nature, whether that geometry be Euclidean, Lobachewskian, or Riemannian. For unless "relations of systematic uniformity" are postulated to extend beyond those isolated cases in which direct observation is possible, we are doomed to "know nothing until we know everything." [24]

Whitehead's dissenting views, though not stated very clearly, appear to raise two issues, one of fact and one of logic.

a.   The first question is whether a system of mechanics can be constructed within the framework of a "nonuniform" geometry, characterized by a variable curvature. The answer to this question is obviously in the affirmative, since Einstein has indeed constructed just such a system of mechanics. No weight can therefore be attached to the contention that, unless a geometry assuming "uniform relatedness" is adopted, we are precluded from knowing anything beyond the isolated physical occurrences that fall under direct observation.

It is not clear, in any event, why there is more "casual heterogeneity" in a theory like Einstein's, which employs a geometry with variable curvature, than in one that adopts Euclidean geometry as the framework for mechanics. It is of course true that in Einstein's theory the spatiotemporal structure of a region is determined by the (contingent) distribution of matter in that region, so that in consequence this structure can be ascertained only on the basis of specific empirical evidence. Nevertheless, that theory does provide general rules within an inclusive conceptual framework which prescribe in precisely what manner the geometry of a region is a function of the matter distributed in it. In this respect the situation confronting an alternative theory, based on a geometry with constant curvature, is essentially no different. For even if the geometry is adopted as an a priori system of conventions for classifying and naming the spatial properties of bodies, only experimental observation can decide just what geometrical properties the bodies in a given region actually possess. Moreover, though the experimental facts

[23] A. N. Whitehead, *The Principle of Relativity,* Cambridge, England, 1922, pp. v-vi.
[24] *Ibid.,* p. 29; see also p. 64.

may warrant the assumption that these properties are Euclidean, additional assumptions would have to be made (e.g., concerning the "local" and contingent distribution of matter, as well as concerning the contingent laws of motion) if the actual trajectories of bodies are to fall within the province of the analysis. Accordingly, just how we systematize our geometrical and physical knowledge—whether we systematize mechanics as an integral part of a comprehensive "geometry" or whether we retain the traditional distinction between geometry and mechanics—does not determine the possibility of our achieving physical knowledge.[25]

b. The second issue raised by Whitehead in effect reopens the question concerning the logical status of geometry. The question in the present context can be stated as follows. We are confronted with two alternative physical theories: Einstein's general theory of relativity formulated in terms of a Riemannian geometry with variable curvature, and Whitehead's theory based on Euclidean geometry. The theories are not equivalent mathematically, although thus far a decision between them on experimental grounds does not seem possible. How are we to construe the differences between the theories insofar as they employ different geometries? Are the geometries in each case simply alternative conventions for interpreting and ordering spatial relations, and hence not subject to empirical tests?

Whitehead's compressed exposition of his own version of relativity theory makes it difficult to answer the question. However, though no confident answer can be given, a discussion of the question will at any rate provide an opportunity for restating and reinforcing some conclusions we have already reached concerning the logical status of geometry. It is essential to note, in the first place, that the word "geometry" is used in a more inclusive sense in Einstein's relativity theory than in Whitehead's usage. In Einstein's but not in Whitehead's context, the word signifies a theory of mechanics as well as a theory of spatial relations. In examining the question before us we must therefore compare the "geometry" of Einstein with the combined "geometry" *and* physics of Whitehead. Moreover, we must ascertain whether, and if so what, correspondence rules are employed in each of the two systems, especially for the term 'straight line.' As has already been noted, the Einsteinian theory does have such rules when it is applied to concrete physical problems; and in fact paths of light rays and of freely moving bodies are designated as the geodesics of the theory. When judged on the basis of empirical evidence, these configurations are in general Riemannian straight lines. Accordingly, given the coordinating definitions of the Einstein

[25] Cf. Bertrand Russell, *Analysis of Matter*, London, 1927, p. 79.

theory, the claim that the spatial structures of a region satisfy the requirements of a Riemannian geometry with variable curvature is not a "concealed definition" but is warranted only because of the nature of the factual evidence.

On the other hand, it is not entirely clear what coordinating definitions are employed by Whitehead for his geometrical terms. The motivations controlling his theoretical construction appear to be philosophical rather than physical. He develops his "relational theory of space" as a system of relations between *immediately sensed events*, not between *physical objects*, since on his view the latter are simply complexes of such sensed events. He therefore maintains that the motions of bodies can be referred to coordinate systems fixed in "homogeneous" or "uniform" space, defined in terms of directly apprehended relations between sensory events. It nevertheless remains obscure which configurations of immediately sensed events are to be identified according to Whitehead as "straight lines"; and it is difficult to escape the impression that for him Euclidean geometry functions as a set of implicit definitions for systematizing the spatial qualities of immediately experienced events. However, if this impression is sound, no conflict is possible between the Einsteinian contention that the configurations his theory specifies as geodesics are Riemannian straight lines, on the one hand, and the Whiteheadian claim that a configuration is a geodesic only if it is a Euclidean straight line, on the other hand. For the Einsteinian contention is a factual thesis, while the Whiteheadian claim is a proposed convention. Accordingly, although it is possible that a given configuration (e.g., an optical path in a force-free field) is characterized as a geodesic by both systems, it is equally possible that some other given configuration (e.g., an optical path in a strong gravitational field) is characterized differently by them. Nevertheless, while geometry appears to have the status of a set of conventions in the Whiteheadian system but not in the Einsteinian one, the latter has conventional components of its own that do not correspond to similar components in the former. For example, it is by stipulation that in the Einsteinian theory only those force fields are to count as gravitational force fields which satisfy certain equations prescribed by the theory. In short, though the institution of conventions is an essential phase in the construction of a theory, the locus of such conventions is in general variable.

3.   Something must finally be said about certain misconstructions often placed on the fact that in general relativity theory the fundamental equations of motion are invariant under a very comprehensive class of transformations from one coordinate system to another. The advantages of

such formulations of the laws of nature are manifest. Such formulations make it possible to subsume an extensive variety of specialized laws under a common formula; they make explicit just what conditions are indispensable for the occurrence of processes, and so help us to discriminate between what is essential and what is irrelevant to the continuance of those processes; and they serve as powerful guides in the conduct of inquiry and the solution of concrete problems. Recognizing the great theoretical and practical importance of invariant formulations, many writers have therefore identified *invariance* with *objectivity*, so that according to these thinkers only what is expressible in such invariant form merits the title of "genuine reality."

The proposed identification of objectivity with invariance is unexceptionable, if it is proposed as *one* explication of the many senses associated with the word "objective" in science and elsewhere.[26] However, this does not appear to be the intent of most of those who offer this identification, since they often proceed, on the basis of this conception of objectivity, to deny the "reality" of concretely existing physical systems, and in particular even of systems *embodying* the structures that receive invariant formulations in a physical theory. It therefore seems useful to note briefly some considerations often neglected in such denials, especially when the denials supposedly rest on an analysis of general relativity theory.

The equations of motion of relativity theory are indeed invariant under a broad class of transformations. Nevertheless, the equations are not invariant under all possible transformations, but only under the restricted class of those that are both continuous and differentiable. Accordingly, on the proposed identification of objectivity with invariance, the objectivity of the structures formulated by general relativity theory is relative to a selected set of transformations. But since there are an indefinite number of sets of transformations that could be selected to serve for defining invariance, there is no compelling a priori reason for maintaining that the set employed in relativity theory is intrinsically superior to, and philosophically more fundamental than, some other set.

It is often overlooked, furthermore, that the requirement that equations of motion possess an invariant form does not, by itself, impose serious restrictions on the forms which laws of nature may take. Indeed, if no limitations are placed on complexity of formulation, any law can be made to satisfy this requirement.[27] Accordingly, it is not the mere invariance of the relativity equations which is the source of their importance; other

---

[26] Cf. the list of alternative criteria for applying the predicate 'physically real,' discussed in Chapter 6 above.

[27] Cf. P. W. Bridgman, *The Nature of Physical Theory*, p. 81; and L. Silberstein, *The Theory of Relativity*, 2nd ed., London, 1924, pp. 296ff.

considerations, not excluding the pragmatic one of comparative simplicity, also enter as determinants of their value.

But in any event, are there any cogent reasons for denying that those features which differentiate motions when the latter are referred to particular frames of reference (even though these features are ignored in the invariant formulations of the equations of motion) are as much a part of nature as are the pervasive structures stated by the equations? Whenever the equations are applied to a concrete physical system, their invariant formalism must be supplemented with statements of detail that are not invariant. Accordingly, why should a special instance of the equations be counted as less "real" than the invariant structure embodied in that instance? Let us press this point in terms of a simple example. The general algebraic equation in two variables can be interpreted as the general equation of a conic section. However, when special numerical values are assigned to the "arbitrary constants" in the general equation, the various equations thus obtained will represent special conics differing from one another in type, size, or position relative to an assumed reference frame. Although the individual conics differ in the indicated ways, they possess a common structure formulated by the general equation for conics. But it would be preposterous to claim that the equation represents a "general conic" which is neither an ellipse nor a circle nor a hyperbola nor a parabola, and which alone is "objectively real" while its specializations are not.

Similarly, the general Newtonian equations of motion do not, as such, distinguish between the different trajectories that a body falling freely toward the center of a gravitational field can be described to follow, when the motion of the body is referred to different inertial frames of reference. In relation to one such reference frame, the trajectory may have the form of a parabola, while in another reference frame it may be a straight line. But it would be absurd to deny that there are such differences in trajectories, even if the generalized formulation of the Newtonian equations does not mention them explicitly. Nor is there any warrant for claiming that only those features of trajectories that are common to all of them have an objective standing in nature, unless this claim is simply the consequence of a special terminology. The situation is in principle no different in general relativity theory. Certainly, no reasons can be found in this theory for denying that the special characteristics motions exhibit when analyzed in different reference frames are as much features of the world explored by physics as are the common structure of motions codified in the invariant formulations of the theory.

Causality and Indeterminism
in Physical Theory

10 Recent developments in physical science have made evident the limitations of the theories of classical physics as universally adequate systems of explanation. These developments have also brought under scrutiny the validity of many time-honored principles of scientific inquiry. The view that has been challenged with especial vigor is the conception that the events of nature occur in fixed causal orders which it is the business of the sciences to discover. It is frequently maintained that current findings in physics no longer warrant the assumption of such orders, and that the ideal of a science of physics with strictly deterministic theories must be relinquished as inherently unrealizable. It is with the issues arising from these claims that we must now deal.

The problem which the advance of physics has made acute is not the traditional one concerning the correct analysis of the meaning of "cause" in the various usages this word may have. Whether the causal relations that are affirmed in ordinary practical affairs, for example, are further analyzable, whether at bottom they indicate some kind of necessity or identity, or whether they can be rendered in terms of regular though contingent sequences of events, are questions that are irrelevant to the debate stimulated by quantum mechanics. The current problem is generated by the dominant position in one sector of physical inquiry of a comprehensive theory which appears to differ from the theories of classical physics in having a "noncausal" or "indeterministic" structure.

The primary question that is thus raised is the precise sense in which the theories of classical physics are deterministic while current subatomic theory is not; and it is with this problem that we shall be initially concerned. However, less specialized though more vague issues have also been precipitated by recent innovations in physical theory—issues concerning the meaning and cognitive status of the so-called "principle of causality," concerning the alleged occurrence of "pure chance" events, and concerning the import of recent theoretical innovations for an adequate view of the nature and aims of science; and we shall accordingly pay some heed to these questions as well.

I. *The Deterministic Structure of Classical Mechanics*

Classical mechanics is the generally acknowledged paradigm of a deterministic theory, and current discussions of determinism are heavily indebted to mechanics for many of their distinctions and much of their language. It is therefore desirable to be clear as to what features of classical mechanics characterize it as a deterministic theory.

Viewed quite generally, mechanics is a set of equations that formulate the dependence of certain traits of bodies on other physical properties. In their Newtonian form, the equations of motion assert that the time-rate of change in the momentum of each mass-point belonging to a given physical system is dependent upon a definite set of other factors. Although the word "cause" does not occur in these equations, they are nevertheless sometimes said to express "causal relations" simply because they assert such a functional dependence of the time-rate of change in one magnitude (i.e., the momentum) upon other magnitudes. However, the characterization of mechanics as "causal" on this basis alone does not adequately clarify the sense in which quantum mechanics is allegedly noncausal, since on this criterion the equations of quantum mechanics also formulate causal relations.[1]

When the equations of motion are stated quite generally they contain, as we have seen, an unspecified function, the force-function. As we have also seen, a special structure must be assigned to this function, and definite values must be given to any arbitrary constants that may appear in it, if the equations are to serve as means for analyzing concrete physical problems. Moreover, the equations of motion are linear differential

---

[1] For similar reasons, a similar locution is frequently employed in other branches of physical science. Thus, the equations of electromagnetic field theory are also said to be causal because they too connect time-rate of changes in the electric and magnetic field vectors with other magnitudes. On the other hand, equations of geometry or of thermodynamics (such as the Boyle-Charles' law for ideal gases, according to which $pV = kRT$) are not designated as causal, since they do not relate any time-rate of change in some magnitude with something else.

equations of the second order, and they must be integrated in order to obtain a solution for a given problem. Accordingly, two constants of integration finally appear for each equation that is employed: the components of the position and momentum at some indicated initial time of the point-mass under consideration, where positions and velocities are assumed to be specifiable with respect to some appropriate frame of reference.

The position and momentum of a point-mass at a given time are said to constitute the "mechanical state" of that point-mass at that time, and the variables defining the mechanical state are called the "state variables." Since each point-mass has three components of position and three of velocity, there are six parameters or coordinates that fix the mechanical state of a point-mass at a given time. Accordingly, the mechanical state at any instant of a system consisting of n point-masses is known, when the values at that instant of the corresponding $6n$ state variables are given.[2] In terms of this nomenclature, we can now formulate an important feature of classical mechanics. Given the force-function for a physical system, the mechanical state of the system at any time is completely and uniquely determined by the mechanical state at some arbitrary initial time. It is this feature of the equations of motion that distinguishes classical mechanics as a deterministic theory.[3]

Since the notion of the mechanical state of a system is crucial for explicating the sense in which classical mechanics is a deterministic

[2] In order to avoid inessential complications, the above explanation of "mechanical state" is restricted to the mechanics of point-masses. The mechanical state of a system of bodies whose relative dimensions cannot be ignored and which, in addition to translatory motions, may exhibit rotations, can be defined in an analogous manner.

[3] A simple illustration will help make this clear. Let us use the equations of motion to analyze the motion of a freely falling body near the surface of the earth. If a system of spatial coordinates 'x,' 'y,' and 'z' is fixed in the earth with z-axis perpendicular to the earth's surface, the equations of motion take the form

$$m\frac{d^2x}{dt^2} = F_x = 0 \quad m\frac{d^2y}{dt^2} = F_y = 0 \quad m\frac{d^2z}{dt^2} = F_z = mg$$

On integrating, we obtain

$$m\frac{dx}{dt} = mv_x = a_1 \quad m\frac{dy}{dt} = mv_y = a_2 \quad m\frac{dz}{dt} = mv_z = mgt + a_3$$

and finally

$$mx = a_1t + b_1 \qquad my = a_2t + b_2 \qquad mz = \frac{mgt^2}{2} + a_3t + b_3$$

where the 'a_i' and the 'b_i,' are the coordinates of state. Accordingly, if we know their values for any one instant, we can calculate the values of the state from the integrated equations for any other instant.

theory, it is worthwhile to dwell on it further. Suppose S is a system of bodies, completely isolated from the influence of all other systems. Assume further that members of S have certain properties (such as mass, velocity, distribution in space, etc.) falling into a definite class K of properties, and that the magnitudes of these properties are represented by values of a set of numerical variables 'v_1,' 'v_2,' 'v_3,' etc. Members of S may have traits other than those in K, but we ignore these. Nor are we concerned whether K includes "observable" as well as "theoretical" properties, or just how K is delimited from other classes of traits; we simply assume that in some fashion K is adequately specified. Let us now agree that the numerical values of traits in K possessed by members of S at any given time define the state of S for that time. We assume next that at time t_0, S is in the state (v_1^0, v_2^0, v_3^0, . . .), that the state of S changes with time, and that at time t_1 the system is in the state (v_1^1, v_2^1, v_3^1, . . .). Now imagine that S is brought back into the state it possessed at time t_0, that it again changes its state of its own accord, and that after an interval of time ($t_1 - t_0$) it is once more in the state it had at time t_1. Let us assume, finally, that S always behaves in the indicated manner, for every initial time and for every time interval. Since the state of S at any given time then uniquely determines its state at any other time, we shall say that S is a deterministic system with respect to the properties in K. We are not assuming, however, that when S is in the same state at two different times, members of S also possess at those times identical values of properties *not* belonging to K. We are defining what it is for S to be a deterministic system *relative* to a stated class of properties K.

This abstract model illustrates in a general way the sense in which mechanics is a deterministic theory. However, the illustration is not entirely satisfactory. It is at least potentially misleading in suggesting that it is a *system of bodies*, rather than a *theory* about certain properties of a system of bodies, which is said to be deterministic. Moreover, since no mention of any theory has been made in stating the model, the discussion fails to illustrate fully the sense in which mechanics as a theory is said to be deterministic. We must therefore introduce some complications into the account thus far given.

Let us assume that a set of general statements L has been established such that, given the state of S at some initial time, a unique state of S for any other time can be deduced with the help of L. Accordingly, it is in principle possible to calculate the state of S for any time, given L and the state of S for some initial time. This suggests that the set of laws L be called a deterministic set of laws for S relative to K. However, one additional complication needs to be introduced. If the number of variables needed to specify the state of S is very large, it will not be practically feasible to describe the state; and in that eventuality, it is unlikely also that a set of laws L can be established. We therefore assume that the total set of predicates designating the properties in K are definable in some manner in terms of a relatively small number of mutually independent predicates belonging to the set—for the sake of definiteness, let us suppose that all the variables representing magnitudes of properties in K can be defined in terms of the independent variables 'v_1' and 'v_2.' On this hypothesis, if we know the values of these latter variables for S at a given time, we also know the state of S (as originally defined) at that time. In consequence, we now emend this original definition, so that the variables of state for S are just the variables in the small subclass of mutually independent variables in terms of which the remaining ones can be defined. Accordingly, the set of laws L constitute a deterministic set of laws for S relative to K if, given the state of S at any initial time, the laws L logically determine a unique state of S for any other time.

This discussion can be directly applied to mechanics. Mechanics studies the relations between a large number of properties belonging to a certain type or class. However, if all these properties have to be taken into consideration when the mechanical state of a system is described, it is doubtful whether a practically effective theory of motion would ever be achieved. Fortunately, not all these properties need be explicitly noted, since there is a small set of variables (consisting of the coordinates of position and momentum of a point-mass) in terms of which variables for other mechanical properties can be defined, so that coordinates of position and momentum constitute the variables of state in mechanics. For example, if the position and momentum of a particle are known, its kinetic and potential energies can be calculated. Accordingly, when the force-function and the mechanical state for a system at some initial instant are given, the equations of motion determine a unique mechanical state for the system for any other time, and hence also the magnitudes of all other "mechanical properties" of the system at this time.

In a frequently quoted passage, Laplace declared that for an intelligence acquainted with the positions of all material particles and with the forces acting between them, "the future as well as the past would be present to its eyes." [4] It will be evident that Laplace was here simply

---

[4] The entire passage is as follows: "We ought to regard the present state of the universe as the effect of its antecedent state and as the cause of the state that is to follow. An intelligence knowing all the forces acting in nature at a given instant, as well as the momentary positions of all things in the universe, would be able to comprehend in one single formula the motions of the largest bodies as well as of the lightest atoms in the world, provided that its intellect were sufficiently powerful to subject all data to analysis; to it nothing would be uncertain, the future as well as the past would be present to its eyes. The perfection that the human mind has

expounding the feature of mechanics that makes it a deterministic theory. Moreover, when nineteenth-century physicists subscribed to determinism as an article of scientific faith, most of them took for their ideal of a deterministic theory one that defines the state of a physical system in the manner of particle mechanics. As we shall see, this ideal continues to control in considerable measure current discussions of causality and determinism in physics. However, before examining the relevance of the notion of the "state of a physical system" for branches of physics other than mechanics, we must try to eliminate sources of possible misconceptions concerning the sense in which mechanics itself is a deter-ministic theory.

1. Only passing mention need be made of the fact that, like any branch of inquiry, classical mechanics deals with but a limited set of properties and relations of bodies. It is therefore essential to remember that, though mechanics is a deterministic theory, it is deterministic *exclusively* with respect to the "mechanical properties" of physical systems, and more particularly with respect to the mechanical states of systems. Thus, to fix our ideas, if the force-function is known, but only the initial positions of a system of particles (and *not* their initial velocities) are given, mechanics does not enable us to calculate the positions of the particles or their kinetic energy for some other time. Again, even though both the force-function and the state of a system at some initial instant are given, classical mechanics does not enable us to predict variations in the thermal or electromagnetic properties of a system—indeed, quite obviously not, if mechanics satisfies the requirements for what we called a "pure mechanical theory" in Chapter 7.

Laplace was therefore guilty of a serious *non sequitur* when he declared that "nothing would be uncertain" for an intelligence possessing complete knowledge of the mechanical states of particles at some instant and of the forces acting upon them. Such a claim would be warranted only if, in addition to knowing these things, Laplace's divine intelligence would be able to analyze *all* traits of physical objects whatsoever (such as their optical, thermal, chemical, and electromagnetic properties) as definable in terms of the variables that constitute the mechanical state of a system. However, mechanics does not rest on the assumption that such an analysis is in fact possible. Nor does the determinism of mechanics exclude the possibility that alterations in the

been able to give to astronomy affords a feeble outline of such an intelligence. Discoveries in mechanics and geometry, coupled with those in universal gravitation, have brought the mind within reach of comprehending in the same analytical formula the past and the future state of the system of the world. All the mind's efforts in the search for truth tend to approximate to the intelligence we have just imagined, although it will forever remain infinitely remote from such an intelligence."—*Théorie analytique des probabilités,* Paris, 1820, Preface.

mechanical state of a system may be the consequences of changes in properties of a system (e.g., chemical changes) not analyzable in this way. Accordingly, if such alterations should occur, mechanics would not be able to predict future states of a system on the basis of some given initial state. In brief, the determinism of classical mechanics is severely limited to a determinism with respect to mechanical states.

2. It is also advisable not to overlook the obvious and therefore easily ignored point that the theory of mechanics does not provide a summary account of the sequential or concomitant order in which concrete events actually occur. For as we have repeatedly noted, the theory of mechanics formulates in general terms only pervasive patterns of relationships, and codifies those patterns with the help of "ideal" or "limiting" notions (such as instantaneous position and instantaneous velocity) rather than by way of experimental concepts. Accordingly, the determinism of mechanics holds strictly only for the *theoretical* mechanical states of systems, whose state variables are *instantaneous* positions and moments. It does not follow, however, that, given the initial positions and momenta of a system of bodies as ascertained by *actual measurement,* the theory of mechanics makes it possible to predict a unique set of *similarly measured* positions and momenta of the bodies for any later time. Whether mechanics does in fact enable us to make such predictions is a separate issue, one that cannot be settled by analyzing only the formal structure of mechanical theory.

This important point deserves emphasis and expansion. The mechanical coordinates of state as stipulated by the theory are not defined in terms of any statistical conceptions or procedures. On the other hand, experimentally measured values of positions and momenta are never instantaneous values, and are in fact average values during some interval of time. Thus, when the velocity of a body is ascertained by measuring the distance it moves during one second, the numerical value so obtained is simply a statistical mean of the velocities that, from the perspective of the theory, the body possesses at the various "instants" during that second. Although the second may be cut down to a smaller duration, the interval cannot be diminished without limit in any experimental measurement of velocity. Accordingly, the theoretical variables of mechanical state can be made to correspond to experimentally measured magnitudes that are in effect only statistical coefficients, and that are therefore associated with a nonvanishing "spread" of experimentally detectable magnitudes. In consequence, the experimentally discriminable positions and momenta that constitute the actual starting point and terminus in any predictive procedure conducted with the aid of mechanics are not the theoretically unique mechanical states of a system. What we can successfully predict is at best only a *class* of values for positions and momenta which are a good approximation to a theoretical state of a system, and not a unique set of values.

3.   Considerations such as those just mentioned have sometimes been used to support the conclusion that classical mechanics is after all not actually a deterministic theory, but only approximates a deterministic theory. It has been argued, for example, that if we understand by the "mechanical state of a system" not the set of theoretical variables of state, but the set of experimentally measurable values of positions and momenta, the theory of mechanics asserts no more than a high statistical correlation (or "relation of probability") between experimentally defined mechanical states of systems at different times. Accordingly, though the laws of mechanics are customarily formulated as strictly universal statements, on this view such formulations must be construed as idealized schematizations of the true state of affairs. There are actually no strictly universal relations of dependence between experimentally defined mechanical states, so the argument runs, but only relations of probability. These probability relations are codified in terms of the schematism of strictly universal statements because the coefficient of probability is close to the maximum value of 1; and such a codification is justified because the discrepancy between the actual value of the probability and the maximum value is so small that the discrepancy can be neglected in practice.[5]

However, the argument for this conclusion is not entirely cogent. In the first place, the argument appears to assume that a theory is simply a generalized description of the sequential order of observable events. If this assumption is granted, then indeed a theory may be plausibly construed as asserting what are at best only relations of high degrees of probability between classes of events. However, we have found reasons for questioning that assumption, so that if the argument does really depend upon it, the conclusion is itself questionable.

But in the second place, the argument also appears to confound two questions that must be distinguished. On the one hand, there is a question of logical analysis, bearing on the internal structure of a theory, and concerned with identifying the theoretical state variables that stand in relations of logical determination to one another. On the other hand, there is a question of empirical fact, bearing on the adequacy of a theory to its subject matter, and dealing with the problem of the precision with which predictions of the theory are actually confirmed by experimental

[5] Cf. Hans Reichenbach, *Philosophic Foundations of Quantum Mechanics*, Berkeley, Calif., 1944, p. 2; also the same author's *Theory of Probability*, Berkeley, Calif., 1949, pp. 435-36.

data. Both questions are clearly important, though they are nonetheless different questions. Our earlier discussion, which maintained that mechanics is a deterministic theory, is obviously an attempt to answer the logical question. The contention that mechanics is not quite a deterministic theory is a proposed answer to the second question. Although the two answers seem to be at variance, they are patently not contradictory.

Moreover, it is almost truistic to claim that classical mechanics is not a deterministic theory, if the claim simply means that actual measurements confirm the predictions of the theory only approximately or only within certain statistically expressed limits. Any theory formulated, as is classical mechanics, in terms of magnitudes capable of mathematically continuous variation must, in the nature of the case, be statistical and not quite deterministic in this sense. For the numerical values of physical magnitudes (such as velocity) obtained by experimental measurement never form a mathematically continuous series; and any set of values so obtained will show some dispersion around values calculated from the theory. Nevertheless, a theory is labeled properly as a deterministic one if analysis of its internal structure shows that the theoretical state of a system at one instant logically determines a unique state of that system for any other instant. In this sense, and with respect to the theoretically defined mechanical states of systems, mechanics is unquestionably a deterministic theory.[6]

II. *Alternative Descriptions of Physical State*

Mechanics is not the only branch of physics, and it is not the only theory possessing a deterministic structure. However, even a cursory examination of the other theories makes clear that not all of them employ definitions of physical state identical with the one used in mechanics.

Although the analytical mechanics of point-masses dominated the minds of physicists for two centuries as the most qualified candidate for the role of a universal science of nature, only in astronomy was this theory employed with all strictness and practical success. The Laplacian ideal of a rigorously deterministic science, in which the mechanical definition of state is an essential feature, proved to be either unrealizable or too difficult to realize in most other domains. Physicists continued to pay lip service to that ideal. But in actual practice they found unavoid-

---

[6] In point of fact, it is on this basis that in current discussions theories are labeled as deterministic or indeterministic, and not on the basis of an examination of the experimental data which support them. This is also true for Reichenbach's own account of quantum theory; indeed, his analysis of quantum mechanics would lose its point if he were not concerned with showing that quantum theory is indeterministic because of its internal structure.

able the adoption of different (or at least modified) definitions of physical state in most branches of their science—even in parts of physics, such as hydrodynamics and elasticity, which were believed to fall without question into the province of mechanics. For example, physicists did not find it generally feasible to analyze the motions of liquids on the assumption of Newtonian forces acting between point-masses. The mathematical difficulties in such an approach were far too great to be overcome by mortal men, and only a Laplacian divine intelligence would be able to resolve them. Accordingly, instead of employing coordinates of position and momentum as state variables, physicists introduced for this purpose certain other parameters (such as the density of a liquid at a point) that could be interpreted as *average* values of the mechanical state variables. Analogous modifications of the mechanical definition of state were required in the study of the elastic properties of substances and in the kinetic theory of gases. Moreover, after decades of unsuccessful effort to develop a theory of electromagnetism within the framework of the requirements for a pure mechanical theory, Maxwell constructed a fully adequate theory of the subject by employing a form of state-description different from the mechanical one.

It is nevertheless a striking testimony to the hold of the notion of mechanical state upon the imaginations of scientists and laymen alike that "determinism" is so often identified with "mechanism." Indeed, it has been frequently assumed that the mark of a deterministic theory is its use of the mechanical definition of state. Innovations in physical theory that involve modes of state-description diverging from the canonical one in the mechanics of point-masses have therefore been widely signalized as the "bankruptcy" of "deterministic" physics. Something like this reception has been accorded by many writers to the advent of electromagnetic field theory, of statistical mechanics, of general relativity theory, and more recently of quantum theory. Nevertheless, the identification of determinism with mechanism is a mistaken one. We must now show that there are genuine alternatives to the mechanical definition of state, and that a physical theory may be rigorously deterministic even if it employs one of these alternative ways of specifying the state of a physical system.

It would take us too far afield to examine in detail even a partial list of theories that are deterministic but do not use the mechanical state-description. We can, however, indicate briefly one systematic way of classifying types of state-descriptions alternative to the mechanical definition of state, and illustrate some of them. With this end in view, let us note some generic features of the mechanical state-description. Observe, first, that the mechanical state of a system is specified by two state variables. If a point-mass is referred to some Cartesian frame of reference, its mechanical state will be defined by six coordinates of state—one for each of the three components of position, and one for each of the components of velocity. Accordingly, since a physical system to which the techniques of analysis of particle mechanics are directly applicable contains only a finite (though possibly very large) number of point-masses, the mechanical state of any system is specified by a *finite* number of values of the state variables. Secondly, each coordinate is an instantaneous value of state variable, so that the mechanical state is an *instantaneous* state. Finally, each coordinate represents a property or relation ascribed to an individual point-mass. Accordingly, the mechanical state of a system represents what we shall call an *individual* property —that is, a property that can be significantly predicated only of a single point-mass, or of a set of such individuals taken distributively rather than collectively.

However, each of these three features of a mechanical state-description is just one in a family of alternative (or contrary) features. In consequence, there are alternative ways of defining state-descriptions, in which each type is obtained by using a feature contrary to the one characterizing the mechanical state description. Let us examine some of these alternative possibilities.

1. A state-description might be defined in terms of an *infinite* rather than a finite number of values of some set of state variables. Indeed, a state-description of this type is employed in the "field theories" of physics, such as Maxwell's electromagnetic theory. The state of an electromagnetic field at an instant is fixed by the values of two vectors— the electric and magnetic field vectors—at each point (infinite in number) of the field. Although the state of a system in this case is specified by a finite number (two) of *state variables,* these variables are in fact functions defined for every point of a region. In consequence, the state of an electromagnetic field at a given time is known, only if in principle the *infinite* number of values of the two state variables at that time are known.

Field theories were first developed in physics in the study of continuous media, for whose analysis partial (as distinct from ordinary) differential equations were required. However, field theories came into special prominence in inquiries into processes involving the transmission of disturbances with finite speeds, with mechanisms that cannot be effectively analyzed in terms of Newtonian forces acting instantaneously "at a distance." Electrical and magnetic waves, for example, are propagated with a finite velocity. Moreover, the force that a moving electrically charged particle exerts on a magnetic pole depends not only on the distance between them, but also on their relative velocities and on the character of the medium in which they are embedded. Furthermore, the acceleration that a magnetic pole suffers because of the motion of an electric charge is not along the straight line joining the pole and the charge—as in the case of the acceleration of a body induced by a Newtonian force, like gravitation, with its source in another body—but in a direction perpendicular to this line. Maxwell's electromagnetic field theory offered a coherent schema of explanation for the experimental findings of Coulomb, Ampère, and Faraday; and it also supplied a satisfactory mathematical tool for dealing with the distinctive formal features of electromagnetic phenomena. Maxwell's approach encountered some resistance at first from those who were reluctant to abandon the mechanical conception of state as the basis for electromagnetic theory. Eventually, however, the theory took its place alongside the particle mechanics of Newton as a well-established system of ideas for understanding an extensive domain of experimental fact. Indeed, serious attempts were soon made to exhibit mechanics itself as simply a special branch of electromagnetic field theory, so that mechanics lost its traditional preeminence as the universal science of nature.

But the main burden of the present discussion is that classical electromagnetic theory possesses a deterministic structure, despite the fact that the electromagnetic description of the state of a system is defined differently than is the mechanical state. Thus, if the values of the electromagnetic vectors are given for every point of some region at an initial instant, then, provided that the boundary conditions of the problem remain unaltered, the values of these vectors for that region at any other time are uniquely determined by Maxwell's equations.[7] Quite analogous conclusions hold for other examples of field theories in physics, such as the Fourier theory of heat flow or the general theory of relativity.

2. A state-description might be defined in terms of the values of variables *at several different times,* or in terms of their values *during some temporal interval,* rather than the instantaneous values of variables. As a matter of fact, the mechanical state-description can be regarded as in a sense belonging to this type. For instead of defining the mechanical state in terms of the simultaneous positions and momenta of particles at *one* instant, an essentially equivalent state-description is obtained when the state is defined in terms of the positions alone at *two* distinct instants.

---

[7] It is clear that the infinite set of coordinates required to specify the electromagnetic state of a system cannot be ascertained by overt measurement at each point of a region. Special laws must be assumed, based on an empirical study of the boundary conditions in a given problem, which formulate the values of the field vectors as functions of the positions.

There are, however, better and more interesting examples of this type of state-description, which are not merely equivalent forms of the standard one in particle mechanics. Thus, it is common knowledge that in the biological sciences, and notably in medical practice, forecasts concerning the behavior of an organism usually require information about the *history* of the organism, and not simply about its momentary state. But even within physics there are areas of inquiry in which such historical knowledge is required, at any rate on certain levels of theoretical analysis. For example, in the study of elastic fatigue of metals, or of magnetic and electric hysteresis, it is not sufficient to specify the instantaneous values of certain variables in order to predict successfully the subsequent behavior of the physical systems under discussion. Thus, when an elastic wire is twisted, the forces brought to bear on it may leave permanent deformations, so that the wire will not in general return to its initial position of equilibrium. The subsequent motions of the wire—its twistings and untwistings—cannot therefore be predicted if we know only the angular torsion and the angular velocity of the wire at some *one* instant. In this case we must have information about the values of these magnitudes throughout the history of the wire since the deforming forces were first impressed on it. The study of this class of problems has led to the development of what is sometimes called "hereditary mechanics"; and in this branch of physics the state of a physical system is defined in terms of the sums of instantaneous values of certain functions during an *interval* of time.[8]

The use of noninstantaneous variables of state is sometimes regarded by mathematical physicists as only a makeshift, until hereditary phenomena can be explained by a theory employing instantaneous state-descriptions. It has been argued, for example, that molecular theory (or some other form of microscopic theory) is in principle capable of accounting for the macroscopic phenomena associated with metal fatigue. Accordingly, though our present technical inability to ascertain the instantaneous states of the molecules is admitted, it has been claimed that we cannot accept as final a theory of hereditary phenomena that employs noninstantaneous variables of state. It has even been asserted by Painlevé in this connection that "The notion according to which one must know the whole past of a physical system in order to predict its future, is the very denial of science."[9] Nevertheless, such rejection of noninstantaneous state-descriptions on grounds of general principle appears to have no firmer foundation than the dubious assumption that only state-descriptions of the type used in classical mechanics can be "ulti-

[8] Vito Volterra, *Theory of Functionals*, London, 1930, pp. 147ff.
[9] Paul Painlevé, *Les axiomes de la mécanique*, Paris, 1922, p. 40.

mate.'" In consequence, the rejection rests on the postulate that if macroscopic phenomena cannot be explained by macroscopic theories using such state-descriptions, those phenomena *must* be explained by a microscopic theory that does employ a type of state-description used in mechanics. Now it is of course abstractly possible that the Laplacian ideal of science will some day be realized, even if the present direction of scientific development does not make this likely; and it is not inherently absurd to pursue that ideal, even though it may be quixotic to do so. On the other hand, the Laplacian ideal does not represent an indispensable logical condition that all physical theories must satisfy. There is therefore no a priori reason for maintaining that a theory not employing the mechanical state-description cannot be as "ultimate" as one that does.

But in any event—and this is the point of chief concern to us at present—a theory may be deterministic with respect to its mode of specifying the state of a system, even though the state-description is defined in terms of noninstantaneous variables of state.

3.   One further type of state variable requires our attention. A state-description might be defined in terms of values of a variable that represents a *statistical* property of some class of elements, rather than a property that can be significantly predicated only of individuals. State-descriptions of this type occur in the mechanics of continuous media, to the extent that the theoretical analyses employ state variables (e.g., density functions and strain vectors) representing *average values* of magnitudes associated with properties of point-masses. However, state variables of this kind are especially distinctive of theories with a more pronounced statistical content, such as classical statistical mechanics and modern quantum theory.

Since we shall have occasion to refer repeatedly to such statistical theories, a general familiarity with their character is desirable. We shall therefore sketch in broad outline the distinctive features of classical statistical mechanics. This theory was initially developed to account for the properties of gases, though eventually its range of application was extended so that even astrophysical questions fell into its province. But in its original form the theory assumed that a gas is an aggregate of a very large number of minute particles or molecules, whose motions can be analyzed in terms of the Newtonian equations of mechanics. On the other hand, it is not actually possible to ascertain the mechanical state of such a system of molecules. Moreover, even if we could do so, we would be unable to predict future mechanical states of the system because of the grave mathematical difficulties presented by the problem of solving the enormous number of simultaneous differential equations of motion. In order to outflank these difficulties a statistical approach was adopted, so that, even though the individual motions of the molecules could not be predicted, it nevertheless became feasible to predict certain *average* values of magnitudes associated with those individual motions.

Accordingly, an additional statistical hypothesis was added to the usual nonstatistical Newtonian assumptions about the motions of the molecules. This new hypothesis stipulated that during any small interval of time the molecules of a gas are in various mechanical states with specified *degrees of probability* (or relative frequency). It can then be shown that the probability of the molecules' being in various mechanical states is a certain function of the mean kinetic energy of the molecules. It also follows that there is an overwhelmingly large probability that molecules will be in mechanical states which fall into a restricted subclass of the set of all possible mechanical states. In short, though statistical mechanics does not predict the individual motions of the molecules, it can characterize a steady condition of equilibrium for the system in terms of certain statistical properties of the individual motions of the molecules. These statistical properties are represented by statistical parameters; and it turns out that a number of these parameters are associated with magnitudes of observable macroscopic properties of the gas. Thus far, however, the analysis deals only with conditions of equilibrium. But the analysis can be extended so as to apply to systems of molecules whose states change with time, as in problems dealing with the diffusion of gases or with Brownian motions. To achieve this, additional statistical assumptions must be made, concerning the probability with which molecules in one set of mechanical states move into another set of mechanical states with lapse of time. The statistical parameters employed in this analysis are the variables of state in the theory, and it is possible to estimate the values of the parameters from experimental data. Accordingly, given the values of these statistical state variables for any initial time, the theory uniquely determines the values of the state variables for any other time.

Although statistical mechanics does not predict the individual mechanical states of the molecules of a gas, it would be a mistake to conclude that statistical mechanics is not a deterministic theory. For in the first place, statistical mechanics includes the assumptions of classical particle mechanics, so that at least in theory the initial mechanical state of the individual molecules uniquely determines the mechanical state for any other time. But what is more to the point is that the statistical-mechanical state-description is defined in terms of *statistical* state variables, not in terms of the state variables of particle mechanics. With respect to its *own* mode of specifying the state of a system, statistical mechanics is a strictly deterministic theory.

There are thus at least three pairs of contrary generic features that may characterize a state-description. The state of a system may be specified by either a finite or an infinite number of values of state variables; the values of state variables may either be instantaneous or may be measures representing traits of a system during a nonvanishing temporal period; and the state variables may be either individual or statistical parameters. Since the alternatives belonging to any one of these pairs are logically independent of the alternatives belonging to another pair, there are at least *eight* logically possible types of state-descriptions. The definition of the state of a system employed in classical particle mechanics belongs to one of them, and examples of three other types have been mentioned. On the other hand, the remaining types do not seem to have been used thus far in modern physical science.

This brief survey of possible alternatives to the mechanical definition of state is schematic and incomplete. It nevertheless suffices to make evident that classical mechanics is not the only deterministic theory in modern physics; and the discussion suggests a general definition of "determinism" that covers theories other than classical particle mechanics. According to this definition, a theory is deterministic if, and only if, given the values of its state variables for some initial period, the theory logically determines a unique set of values for those variables for any other period. If this definition is adopted, it is incorrect to deny that a theory is deterministic on either of the following two grounds: on the ground that a theory does not establish such unique one-to-one correspondences between the values at different times of *every* set of magnitudes mentioned by the theory; or on the ground that experimentally measured values of theoretical state variables are not in precise agreement with their theoretically calculated ones.

One final point of considerable importance should be noted. A definition of the 'state of a system,' suitable for a given empirical subject matter, cannot be supplied *in advance* of an adequate "causal" theory for that subject matter.[10] It will be recalled that, in explicating the notion of a deterministic system earlier in the present chapter, we first defined the state of a system S in terms of properties belonging to a certain class K. Enough was said at that time to make it obvious that K does not consist of an *arbitrarily selected* set of properties possessed by S. The discussion also made clear that K cannot be the set of *all* properties of S, if only because such a definition of state would be practically useless. Nor can K be generally identified with the set of all observable properties of S. For it is unsafe to assume, as the history of

[10] "Causal" in the sense explained at the outset of the present chapter, so that a theory is causal if it relates time-rate of changes in some set of magnitudes with other magnitudes.

science has repeatedly shown, that, if S exhibits the same observable properties at two different times, S is in the same state at those times. Thus, a system may manifest identical observable traits at two distinct instants, and yet differ in its theoretical properties at those times. Accordingly, it is only on the basis of some accepted "causal" theory that we can decide which variables are to count as the state variables.

It follows that one is uttering a truism in saying that a causal theory is deterministic relative to the state-description used by that theory. For as we have seen, a set of variables counts as the class of state variables for a system only if there is a theory that is deterministic with respect to a state-description defined by those state variables.[11] Nevertheless, though it is truistic to say this, it is not trivial to do so. On the contrary, the statement that every causal theory is deterministic relative to its own specification of the state of a system calls attention to the important point that, if a causal theory is nevertheless characterized as being in some sense "indeterministic," the alleged indeterminism must be explicated in terms of some special features distinguishing the state-description which the theory employs. This point will guide us in examining the characterization of modern quantum theory as indeterministic, as well as in considering the logical status of the so-called "principle of causality."

Meanwhile, however, we can sum up our discussion thus far by saying that there are genuine alternatives to the mechanical definition of the state of a physical system, and that the possibility of developing deterministic theories of physics is not contingent on the use of the mechanical variables of state.

III. *The Language of Quantum Mechanics*

What light does the preceding discussion throw on the alleged indeterminism of modern quantum mechanics? Let us first recall the customary grounds upon which this imputation is made.

1. Quantum theory was first introduced to account for a number of experimental laws concerning phenomena of thermal radiation and spectroscopy, phenomena which were apparently inexplicable on the basis of classical radiation theory. Eventually, however, quantum theory was modified and expanded so as to cover facts encountered in physical optics, crystallography, chemistry, and many other special domains of in-

---

[11] To say that $v_1, v_2, \ldots, v_k$ are variables of state, or that the set $[v_1, \ldots, v_k]$ constitutes a state-description is to say that there are functions $f_i(v_1, \ldots, v_k)$ such that $\frac{dv_i}{dt} = f_i(v_1, \ldots, v_k)$, where $i = 1, 2, \ldots, k$, and that the functions formulate the relations postulated by the theory. Cf. Philipp Frank, *Das Kausalgesetz*, Vienna, 1932, pp. 145ff.

quiry. In its most recent form, quantum theory can be developed in two mathematically equivalent ways—in terms of either the matrix algebra first introduced by Heisenberg or the formalism associated with the Schrödinger wave equation. We shall use the latter formulation as the basis for discussion, although practically all the technical details of the theory, as well as the experimental evidence for it, will be ignored. The theory is usually stated in terms of a model, and ostensibly postulates several distinct kinds of subatomic "particles" and "processes." As in the case of all theories, and especially microscopic theories, the empirical evidence for the postulates of quantum theory is logically incomplete, and is connected with the fundamental assumptions by long chains of deductions and many subsidiary hypotheses. Moreover, the empirical evidence is not in absolutely precise agreement with the numerical laws deduced from the theory, although the discrepancies are in general well within the limits of experimental error. In these respects there is nothing novel in quantum theory.

However, the standard interpretation of the experimental evidence for the theory yields the conclusion that in certain situations some of the postulated subatomic elements (such as electrons) have properties characteristic of *particles*, while in other situations they exhibit properties characteristic of *waves*. This apparently "dual nature" of its fundamental elements is a distinctive mark of the theory and has been the source of much puzzlement and speculation. But the feature of quantum mechanics that has precipitated the current discussion on determinism in physics, and is the usual ground for regarding quantum mechanics as an "indeterministic" theory, is the set of formulas logically derivable from the assumptions of the theory and known as the "Heisenberg uncertainty relations." One of these relations is expressed by the formula $\Delta p \, \Delta q \geqq h/4\pi$. In this formula, the variables 'p' and 'q' are commonly read as the instantaneous coordinates of the "momentum" and "position," respectively, of an electron or other subatomic element, and 'h' as the universal Planck's constant. On the other hand, 'Δp' is interpreted as the coefficient of dispersion (or deviation, sometimes also called the "uncertainty") from the mean value of the momentum at a given instant; and similarly for 'Δq.' The formula therefore asserts that at any given instant the product of the dispersions of the momentum and position, respectively, of a subatomic "particle" is never less than $h/4\pi$. Accordingly, this form of the Heisenberg uncertainty relation can be construed as saying that, if one of these coordinates is measured with great precision, it is not possible to obtain *simultaneously* an *arbitrarily precise* value for the conjugate coordinate. For example, if q is vanishingly small, p must be enormously large and for practical purposes "infinite." In consequence, if a measurement enables us to ascertain with great accuracy the position of an electron at a given time, no measurement can assign a precise value to the momentum (and hence to the velocity) of the particle at that time.

The argument leading to the conclusion that, because of the uncertainty relations, quantum theory is indeterministic usually takes the following form. It is in principle impossible to ascertain with unlimited precision the simultaneous positions and momenta of elementary physical particles. Indeed, the uncertainty relations assert that the position and momentum of a particle at any given time are not independent of one another, but are so related that a sharply delimited spatial location is incompatible with a sharply delimited velocity for the particle. The equations of quantum mechanics cannot, therefore, establish a unique correspondence between precise positions and momenta at one time, and precise positions and momenta at other times. Nevertheless, quantum theory is capable of calculating the *probability* with which a particle has a specified momentum when it has a given position, and vice versa. Accordingly, quantum theory is not deterministic in its structure, but is inherently statistical in content; and the unquestionably great successes of the theory must be taken as an indication that the "principle of causality" is inapplicable in the domain of subatomic processes.[12]

Before examining this argument and its conclusion, it will be desirable to mention briefly some of the comments physicists have offered on both the uncertainty relations and the "dual nature" of subatomic elements. A widely held, and *prima facie* plausible, interpretation of the uncertainty relations is that they formulate the relatively large but inherently unpredictable variations in certain features of subatomic particles and processes, produced by the interaction of the latter with the instruments used in measuring those features. For example, Heisenberg declared that when measuring large-scale objects, the effects generated in those objects by the processes of measuring them can be neglected, since the magnitudes of the disturbances thus produced are relatively small. In subatomic physics, on the other hand,

> the interaction between observer and object causes uncontrollable and large changes in the system being observed, because of the discontinuous character of atomic processes. The immediate consequence of this circumstance is that in general every experiment performed to determine some numerical quantity renders the knowledge of others illusory, since the uncontrollable perturbations of the observed system alters the values of previously determined quantities.[18]

[12] Cf. Richard C. Tolman, *The Principles of Statistical Mechanics*, Oxford, 1938, p. 187; also P. W. Bridgman, *Reflections of a Physicist*, New York, 1950, p. 135.

[18] Werner Heisenberg, *The Physical Principles of the Quantum Theory*, Chicago, 1930, p. 3. See also Niels Bohr: "Now, the quantum postulate implies that any

On the other hand, the wave-particle duality ascribed to such elements as electrons is often taken to indicate that there are limits to the interpretation of the quantum mechanical formalism in terms of traditional notions of space and time. It has been maintained, for example, that we must surrender as a universal schema of analysis the long-familiar practice of describing nature by way of specifying the properties and relations of spatiotemporally located individuals; and we are advised to abandon the hope of explaining "all phenomena as relations between objects in space and time." Indeed, the inapplicability of the principle of causality to subatomic processes, so it has been urged, stems entirely from the fact that, although subatomic processes cannot be described in this way, every application of the principle assumes the possibility of such description. But if the traditional mode of description and analysis is abandoned in the subatomic domain, the claim continues, we can avoid ascribing a wave-particle duality to electrons and can at the same time retain the principle of causality. Thus, again according to Heisenberg, description of atomic processes in spatiotemporal terms, on the one hand, and the exact validity of the principle of causality for atomic processes, on the other hand,

> represent complementary and mutually exclusive aspects of atomic phenomena. This situation is clearly reflected in the theory which has been developed. There exists a body of exact mathematical laws, but these cannot be interpreted as expressing simple relationships between objects existing in space and time. The observable predictions of this theory can be approximately described in such terms, but not uniquely—the wave and the corpuscular pictures both possess the same approximate validity. This indeterminateness of the picture of the process is a direct result of the indeterminateness of the concept 'observation'—it is not possible to decide, other than arbitrarily, what objects are to be considered as part of the observed system and what as part of the observer's apparatus. In the formulas of the theory this arbitrariness often makes it possible to use quite different analytical methods for the treatment of a single physical experiment.[14]

Heisenberg therefore proposed the following dilemma. We may interpret the equations of quantum theory as descriptions of subatomic processes in customary spatiotemporal terms, but (in view of the uncertainty relations) at the price of abandoning deterministic explanations for those

observation of atomic phenomena will involve an interaction with the agency of observation not to be neglected. Accordingly, an independent reality in the ordinary sense can neither be ascribed to the phenomena nor to the agencies of observation"—*Atomic Theory and the Description of Nature*, London, 1934, p. 54.

[14] Heisenberg, *op. cit.* Similarly, Bohr declares that the Schrödinger representation of atomic processes by the wave equation entails that "in the interpretation of observation a fundamental renunciation regarding the spatio-temporal description is unavoidable."—*Op. cit.*, p. 77.

processes. On the other hand, we may retain such explanations, but at the price of surrendering the possibility of interpreting the equations of the theory as referring to individuals and processes localized in space and time. Both horns of the dilemma thus involve radical readjustments in traditional ways of studying physical processes.

2. However, despite the high authority for these interpretations of the uncertainty relations and of the source of the wave-particle duality ascribed to electrons, the comments as stated are neither entirely clear nor persuasive.

a. Consider first the claim that the uncertainty relations express the "uncertainties" produced by the interactions of objects measured and measuring instruments, and that in consequence the classical distinction between "observed" and "observer" cannot be maintained in subatomic physics except in an arbitrary manner. This claim is sometimes put forth as if the uncertainty relations were the conclusions of a purely factual examination of the overt laboratory measurements conducted for testing quantum theory—and hence as if they were warranted on purely inductive grounds, independently of whether or not one accepts quantum theory. In point of fact, however, such a contention puts the cart before the horse. For the "uncontrollable" and "unpredictable" alterations that electrons are said to suffer when they interact with instruments of measurement do not constitute the *evidence* for the uncertainty relations, but are part of the *consequences* drawn from the relations. This will be clear if we ask ourselves what ground we have for claiming that the alterations are uncontrollable and unpredictable, and if we remind ourselves that perturbations produced by measuring instruments in the objects measured were fully recognized in classical physics. In classical physics, however, the extent of such perturbations can in principle be precisely evaluated with the help of established physical laws, so that the mere fact of such perturbations does not lead to the uncertainty relations. According to the Heisenberg uncertainty relations, however, the alterations produced in electrons by measurements on them cannot be calculated even in principle, because in this case electrons undergo "uncontrollable changes." The claim that the alleged changes are indeed unpredictable cannot, therefore, be merely an inductive conclusion from the facts of laboratory measurement. It is a conclusion based on the uncertainty relations, and in consequence on the assumptions of quantum theory from which those relations are logically derived.

It is well to note, moreover, that the Heisenberg relations set no limits upon the precision with which, for example, the coordinate of position of an electron may be measured. These relations simply set limits to the precision with which the simultaneous values of *both* the position and momentum coordinates can be ascertained by measurement. Accordingly, despite the assumed interaction between an electron and the apparatus used to measure it, either coordinate of the electron taken singly can in principle be measured with absolute precision. In consequence, it is hardly cogent to argue that the simultaneous positions and momenta of electrons cannot be ascertained with unlimited precision, on the ground that perturbations are produced in electrons when they are measured. In short, the impossibility of such unlimited precision follows from the uncertainty relations and not, as is sometimes maintained, simply from the familiar experimental facts concerning the effects produced by the "observing" instruments upon the "observed" objects of measurement.

b. Let us now turn to Heisenberg's comments on the source for the "dual nature" commonly imputed to electrons, protons, and other subatomic elements. On the face of it, his contention that traditional spatiotemporal notions are not suitable for "describing" subatomic processes is puzzling. For how else are *processes* to be described, one is led to ask, if not in spatiotemporal terms? It is nevertheless possible that the burden of his remarks is only concealed by this obscure formulation. Indeed, there is reason to suspect that the real point of his comments is that when electrons and the like are said to be "particles" or "waves," these characterizations are employed largely under the control of certain formal analogies and are not to be construed literally. Is it possible that the postulated elements of subatomic physics cannot be described in spatiotemporal terms, not because spatiotemporal notions are inadequate but because electrons, protons, and so on are not particles or waves in the familiar senses of these terms as established in classical physics for macroscopic objects? This suggestion seems worth pursuing, whatever may have been the actual intent of Heisenberg's comments. If the suggestion has merit, not only does the usual interpretation of the uncertainty relations require some emendation, but also the view that quantum mechanics is not a deterministic theory needs to be qualified. We shall therefore examine the language of quantum theory more closely, with the intent of placing these matters in clearer light.

i. The mathematical formalism of quantum mechanics is an outgrowth and adaptation of the formalism and notation first developed in classical physics. In consequence, the subatomic elements postulated by quantum theory are frequently described in a language customarily used for describing point-masses in classical mechanics. In particular, certain coordinates associated in quantum theory with electrons are called coordinates of "position" and of "momentum." It is quite natural, therefore, that use of the language of classical physics for formulating the assumptions of quantum mechanics should often produce the impression that, according to the quantum mechanical conception of electrons, an electron possesses both a determinate position and a determinate momentum at every instant. In view of the uncertainty relations, on the other hand, those using this language frequently feel compelled to add that it is nevertheless impossible to determine with unlimited precision the simultaneous position and momentum of a subatomic "particle." It is but a short step from this to the conclusion, apparently implied by the language employed, that, although subatomic particles do possess both a determinate position and a determinate momentum at all times, it is inherently impossible to discover the precise simultaneous values of these coordinates. In any event, this conclusion often serves as part of the ground for the claim that quantum mechanics is indeterministic.

Nevertheless, were this conclusion really necessitated by quantum mechanics, the situation would be even more perplexing than the one presented by the assumption of absolute space in Newtonian mechanics. Although Newtonian theory precludes the possibility of distinguishing on the basis of any *mechanical* experiment between rest and uniform velocity with respect to absolute space, the theory does provide an alleged criterion for identifying accelerated motions relative to absolute space. Moreover, Newtonian theory does not in principle exclude the possibility that some *nonmechanical* experiment (e.g., an optical one) could be devised for distinguishing between absolute rest and absolute uniform velocity. On the other hand, if quantum theory is construed in accordance with the above conclusion, we are required to assert that, although an electron does in theory have determinate positions and momenta at all times, when the *precise* position is ascertained for a given instant then no experiment whatever can discover even the *approximate* value of the momentum. Small wonder, therefore, that commentators who accept the above conclusion as soundly based often maintain that quantum mechanics requires at least the partial abandonment of the ideal of verifiability, which has controlled the development of so much of modern physics.

However, this perplexing situation seems in large measure to be the result of characterizing electrons and other postulated elements as "particles" but of overlooking the fact that this characterization is based on what are at best only *partial analogies* between the mathematical formalisms of classical and quantum mechanics. Indeed, the language of electrons as "particles" is replaced in certain contexts by the language of electrons as "waves," because each of these analogies is only partial and fails at various points. But, conversely, the characterization of electrons as "waves" also rests on such partial analogies between the symbolic structures of classical and quantum mechanics. Many presentations of quantum theory are in consequence formulated by way of a not always smoothly blended mixture of two modes of speech, neither of which is uniformly appropriate or wholly free from misleading associations. There is, of course, no question whatever that the terminology of "particles" and "waves" is suggestive and heuristically valuable. Nevertheless, the usefulness of this terminology must not hide from us the fact that it is employed analogically and is not to be construed literally.

ii. Let us examine this point more closely. The fundamental assumptions of quantum theory are expressed with the help of a highly complex mathematical symbolism. In the Schrödinger version of the theory a differential equation having the general form of the classical "wave equation" plays a central role. As in the case of all theories, coordinating definitions must be supplied for a certain number of the nonlogical terms occurring in this mathematical formalism, so that experimentally testable statements may then be derived from the theory. Once such coordinating definitions are stated, the empirical content of the theory is fixed for the time being. On the other hand, as has been argued in an earlier chapter, it is not essential in point of strict logic to provide a "model" for the theory which will illustrate the structural content of the theory in a more or less "pictorial" manner. Nevertheless, there are great psychological advantages in having such models for a theory. In consequence, with such models as their objective, physicists frequently formulate the content of quantum mechanics in the language of classically conceived particles and waves, because of certain analogies between the formal structures of classical and quantum mechanics.[15]

However, any model for quantum theory must satisfy the formal equations of the theory. These equations *implicitly define* the subatomic elements and processes postulated by any model for the theory. Accordingly, whatever further traits those postulated elements may be conceived to possess, those elements must at least have the structural characteristics stipulated by the equations. In consequence, any formulas logically entailed by the fundamental postulates of quantum theory—for example, the Heisenberg uncertainty relations—are in effect also implicit definitions that impose restrictions on the component elements in any model for the theory. In short, no hypothetical physical system can be a fully adequate model for quantum theory if certain features of the system do not satisfy the uncertainty relations.

It thus follows that if the variables 'p' and 'q,' which must satisfy the

---

[15] This point is made very clearly by Linus Pauling and E. Bright Wilson in the passage from their *Introduction to Quantum Mechanics* quoted above in footnote 4 on page 112.

uncertainty relations, are interpreted as the measures, respectively, of the "momentum" and "position" of an electron, then despite the names used for these measurable traits of electrons, these traits cannot be identified with characteristics of particles denoted by the words "momentum" and "position" as they are used in classical physics. For it is evident that though 'p' and 'q' in quantum mechanics are *called* coordinates of "momentum" and "position," the words are now being employed in an unusual sense. In classical mechanics these words are so used that a particle always must have a determinate position and simultaneously a determinate momentum, and in theory both the position and momentum can be ascertained with unlimited precision. In this usage it is nonsense to say that a particle has a determinate position but not a determinate momentum, or that it is logically impossible to discover the precise value of one but not the other. But in quantum mechanics the uses legislated for these words are patently different. Accordingly, if in conformity with the assumptions of quantum theory an electron is said to be a "particle," possessing magnitudes represented by the symbols 'p' and 'q' whose simultaneous values cannot be ascertained with unlimited precision even in principle, then either the word "particle" is being used in some Pickwickian sense or these symbols cannot represent momenta and positions in the familiar classical meanings of the words.

iii. A similar conclusion has been often urged by Niels Bohr, although on the basis of quite different considerations.[16] We shall state his argument in brief outline. Let us call an interpretation for a set of postulates a "uniformly complete" one, if (a) an interpretation is assigned to every nonlogical term employed in the postulates, and (b) the interpretation remains fixed for every context of the application of the postulates. In the case of a uniformly complete interpretation it is thus never the case that a nonlogical term receives either no interpretation in some context

---

[16] But other outstanding physicists have also argued for it. For example, Heisenberg notes that the "uncertainty relation specifies the limits within which the particle picture can be applied. Any use of the words 'position' and 'velocity' with an accuracy exceeding that given by the uncertainty equation is just as meaningless as the use of words whose sense is not defined."—*Op. cit.*, p. 6. Again, von Neumann remarks that it would be entirely meaningless to distinguish between a term '$p.q$.' and a term '$q.p$.,' as is done in quantum physics, if these are construed in the sense specified by classical physics. [J. von Neumann, *Mathematische Grundlagen der Quantenmechanik*, Berlin, 1932, p. 6 (English translation, *Mathematical Foundations of Quantum Mechanics*, Princeton, 1955, p. 9).] And Schrödinger declared that "the object referred to by quantum mechanics . . . is not a material point in the old sense of the word. . . . It should neither be disputed nor passed over in tactful silence (as is done in certain quarters) that the concept of the material point undergoes a considerable change which as yet we fail thoroughly to understand."—Erwin Schrödinger, *Science and the Human Temperament*, New York, 1935, pp. 71-72.

or different interpretations in different contexts. Now according to Bohr, a uniformly complete interpretation of the formalism of quantum mechanics, in terms of a subatomic model whose elements possess the customary traits of macroscopic objects (such as precise positions and velocities), results inevitably in the imputation of a paradoxical "dual nature" to those subatomic entities, so that these will possess both corpuscular and wavelike attributes. Attempts at such an interpretation must therefore be abandoned, if this paradox is to be avoided. On the other hand, the reason for assigning both corpuscular and wavelike attributes to electrons is that the *empirical evidence* for quantum theory is most conveniently described by using the language developed for talking about *classical* particles and waves. Indeed, the empirical evidence for any theory is inevitably drawn from the macroscopic domain; and in describing any experimental arrangements or observations, we have no alternative to using the common language of gross experience, suitably supplemented by the terminology of classical physics. Nevertheless, in Bohr's judgment:

> . . . the evidence obtained under different experimental conditions cannot be comprehended within a single picture, but must be regarded as *complementary* in the sense that only the totality of the phenomena exhausts the possible information about the objects.[17]

Accordingly, although a satisfactory uniformly complete interpretation of quantum mechanics based on a *single* model cannot be given, the theory can be satisfactorily interpreted for *each concrete experimental situation* to which the theory is applied.

More specifically, Bohr's view is that there are experimental situations within which the expression 'position of an electron' can be assigned a definite meaning; there are other experimental situations within which the expression 'momentum of an electron' can be meaningfully employed; but there are no experimental situations within which the expression 'position and momentum of an electron' can be given an experimental sense. On Bohr's analysis, therefore, the impossibility of assigning precise simultaneous values to conjugate coordinates (e.g., to those called the coordinates of "position" and "momentum") is simply a consequence of two facts: the fact that each coordinate can be interpreted, not uniformly and once for all for every context, but differently for each type of experimental situation to which quantum theory can be applied; and the fact that there are no contexts in which an experimentally significant sense can be assigned to both coordinates simultaneously. A type of ex-

---

[17] Niels Bohr, "Discussions with Einstein on Epistemological Problems in Atomic Physics," in *Albert Einstein, Philosopher-Physicist* (ed. by Paul A. Schilpp), Evanston, Ill., 1949, p. 210.

perimental arrangement suitable for measuring what is called the "position" of an electron thus fixes the meaning of the phrase 'the position of an electron' within a limited set of contexts; and similarly for the phrase 'the momentum of an electron.' However, the two types of experimental situations do not overlap, and must therefore be distinguished. In short, since no experimental arrangements can be instituted within which both phrases can be simultaneously interpreted, it follows trivially that no measurement can ever assign precise values to both conjugate coordinates simultaneously. But it also follows that the words 'particle,' 'position,' and 'momentum' as used in quantum theory cannot be construed in the senses assigned to these words in classical physics.[18]

iv.   This conclusion can be placed in a clarifying perspective if we recall another historically important adaptation of familiar language to new uses: the gradual extension of the word 'number' from its original context as a name for cardinal and ordinal integers to its current use as a label for a much more inclusive domain of mathematical entities. As is well known, the operations of addition and multiplication, and their inverses, were first developed in connection with the cardinal numbers and were eventually employed to define various properties of cardinals (such as odd and even, prime, perfect square, etc.). Subsequently, however, the word 'number' came to be applied to the ratios of cardinals (usually represented as fractions), in large measure because certain operations can be defined for ratios that are closely analogous to the familiar operations with cardinals. Thus, ratios can be "added" and "multiplied"; and these *distinctive operations* with ratios exhibit patterns of relations that, *up to a point*, are abstractly similar to the patterns exhibited by cardinal addition and multiplication. For example, addition and multiplication are commutative as well as associative operations, for cardinals as well as for ratios; e.g., $a + b = b + a$, and $a + (b + c) = (a + b) + c$. On the other hand, the multiplication of ratios always has an inverse; i.e., the division of one ratio by a second always yields a third ratio, with the usual restriction on division by "zero" ratios. This is not the case for cardinal multiplication; i.e., the division of one cardinal by a second does not always yield a third cardinal. Moreover, although certain properties can be defined for ratios that are formally analogous to certain properties of cardinals, cardinals have various defined properties for which there are no analogues in ratios. For example, ratios as well as cardinals may be perfect squares. However, though it makes good sense to ask whether a given cardinal number is odd, it makes no sense to raise a similar question about a given ratio, simply because the predicate

[18] *Ibid.*, pp. 232-35.

'being odd' happens not to be defined for ratios. It is worth noting in this connection that our inability to answer the question whether, say, ⅔ is odd, has its source neither in a temporary insufficiency of our knowledge nor in some alleged inherently unknowable nature of ratios; our inability stems from the undramatic fact that for ratios the question has no defined sense.

These brief observations about the rationale for extending the word "number" to cover the ratios as well as cardinals obviously apply to further extensions of the "number concept" to cover still different kinds of mathematical entities, such as the irrationals, the imaginaries, and the so-called "signed numbers." Moreover, these comments are also relevant for an appreciation of the reasons why certain mathematical operations have been given names familiar from arithmetic, even though the operations are neither upon numbers in the extended sense of the word, nor formally similar in many respects to the like-named arithmetical operations. For example, an operation called "multiplication" has been defined for certain types of ordered sets of numbers known as "matrices." This operation is associative but in general not commutative, so that in some respects it is like arithmetical multiplication while in other respects it is unlike it. The bald statement that multiplication is therefore not always commutative may indeed have the appearance of a profound paradox. But if a puzzle is generated by such a statement, the puzzle arises only if one overlooks the fact that, although in its original sense the word "multiplication" does denote a commutative operation, the word has been adapted to new uses. The operation named by the word in the new context is not the operation it named in the older one. If, nevertheless, the word is retained as a label for both operations, the reason is that despite important differences between the operations there are also important analogies.

The words 'position,' 'momentum,' 'particle,' and 'wave' in quantum theory must similarly be recognized as borrowings from classical physics. Their introduction into quantum mechanics has been guided by important formal analogies between the older and the newer theories; and their extension to this fresh domain has facilitated quantum mechanical formulations as well as suggested fresh lines of inquiry. Nevertheless, when these words are employed in the new context, they must be understood in terms of the restrictions placed upon their use by the postulates of quantum theory, and not in terms of the senses established for them in classical physics. Accordingly, since the rules governing the use of the words are not identical in the two contexts, what they mean in quantum mechanics cannot be the same as their more familiar historical meanings. It is therefore a blunder to suppose, as some commentators on quantum mechanics have done, that by improving our experimental techniques we may perhaps be able to ascertain the precise simultaneous values of the positions and momenta of electrons—in the senses of 'position' and 'momentum' fixed by current quantum theory. Such a supposition is on par with the conjecture that by more intense study we may eventually be able to discover whether or not the ratio ⅔ is odd. The supposition overlooks the crucial point that in virtue of the uncertainty relations the expression 'the precise simultaneous values of the position and momentum of an electron' has no defined sense in quantum mechanics.

Although Heisenberg admits this point, and indeed even insists on it, he also succeeds in ignoring it when he declares, in the passage cited above (page 295), that if an experiment determines the value of some numerical quantity (e.g., the precise position of an electron) "it renders the knowledge of others illusory" (e.g., of the value of the electron's momentum). For if the expression 'the precise simultaneous values of the position and momentum of an electron' is not defined, then there is no momentum to be known under the assumed circumstances. It is therefore difficult to understand how knowledge of an electron's alleged momentum can be "illusory" if, as the analysis indicates, there is no such thing as an electron's momentum to be the object of knowledge.[19]

IV. *The Indeterminism of Quantum Theory*

Quantum mechanics cannot therefore be validly characterized as indeterministic *merely* on the ground that the uncertainty relations exclude the possibility of precise values for the simultaneous "positions" and "momenta" of electrons and other subatomic "particles." If the above considerations have merit, these words have different senses in quantum theory from those they have in classical physics. In consequence, it is a *non sequitur* to conclude that the "positions" and "momenta" which the uncertainty relations assert to be conjugately "uncertain" are the self-same traits of particles which in classical mechanics are subject to precise numerical determination, so that while classical mechanics has a deterministic structure quantum mechanics does not.

1. There is, moreover, a further point to be noted in the customary argument for the indeterministic structure of quantum mechanics. The argument tacitly assumes that, like classical particle mechanics, quantum

[19] Similar comments are in order on Heisenberg's claim that subatomic physics has made questionable the classical distinction between "observer" and "observed," or "subject" and "object." Such a claim is intelligible only on the assumption that the terms of this distinction have a defined sense in quantum physics, and that this sense is the same as in classical physics. But we now have sufficient grounds for challenging such an assumption.

theory defines the state of a system as the set of instantaneous values of the position and momentum for each particle belonging to the system. Were this assumption sound, it would undoubtedly establish what the argument sets out to prove. For, since the state of a system so defined could never be specified for any time, it would be obviously impossible even in principle to calculate the state of the system for any other time. In point of fact, however, quantum mechanics does not define the state of a system in this manner. Accordingly, though it must be granted that quantum theory is not deterministic with respect to a state-description assumed to be defined in terms of positions and momenta as the state variables, it does not follow that the theory is not deterministic with respect to a state-description differently defined.

Indeed, an examination of the fundamental equations of quantum mechanics shows that the theory employs a definition of state quite unlike that of classical mechanics, but that relative to its own form of state-description quantum theory is deterministic in the same sense that classical mechanics is deterministic with respect to the mechanical description of state. However, the state-description employed in quantum theory is extraordinarily abstract; and, although its formal structure can be readily analyzed, it does not lend itself to an intuitively satisfactory nontechnical exposition. In any event, in the Schrödinger or wave mechanical formulation, quantum theory employs as the state-description of a system a certain function, the so-called "Psi-function." The arguments to this function are in general coordinates of "position" and "time." The function must satisfy the fundamental wave equation for the system under study; and it must be continuous, single-valued, and finite throughout the region for which the function is defined. But the feature of the Psi-function of special importance for the present discussion is that, given the values of the function for each point of the region at some initial instant, the Schrödinger wave equation determines a unique set of values for the function at any other instant. Quantum mechanics is therefore a fully deterministic theory with respect to the quantum mechanical state-description defined by the Psi-function.

But what does the Psi-function represent, and how may it be interpreted? It cannot be interpreted in terms of some visualizable physical model, whose moving parts are classical particles or waves. As has already been noted, all attempts at such interpretations of quantum theory yield models whose component elements have the "dual nature" of being both corpuscular and wavelike. Nevertheless, the unavailability of a uniformly complete interpretation in terms of a classical model is not fatal to the effective use of quantum mechanics. Like many other theories in physics, quantum mechanics formulates its assumptions with the help of a variety of variables and functions, most of which are associated neither with a pictorial image nor with identifiable experimental notions. Moreover, as in the case of other physical theories, coordinating definitions in terms of experimentally observable phenomena are in general supplied, not for the primitive variables and functions of quantum mechanics taken singly, but for certain combinations of them. In particular, it is not for the Psi-function itself, but for a certain mathematical construct into which the function enters, that an interpretation is given.

In broad outlines, the standard interpretation for the Psi-function is as follows. The function is in general complex, in the technical mathematical sense of 'complex'; but a mathematical construct can be formed out of it (the square of its absolute value) that is real. The square of the absolute value of Psi is then interpreted as the *probability* that the elementary constituents of the system for which it is defined (e.g., the system consisting of the nucleus and the single electron of a hydrogen atom) are at various points in space.[20] However, this interpretation of the Psi-function is still quite formal, especially in the light of our earlier discussion which argued that the word 'position' in such expressions of quantum theory as 'the position of an electron' is used in a somewhat Pickwickian sense. Let us therefore pursue the matter into some details.

Although the Psi-function is the quantum mechanical definition of state, both Psi and the probabilities associated with the square of its amplitude are at bottom only auxiliary parameters and play an intermediary role in the theory. They are important because they make it possible to calculate various *other* probabilities. For example, the postulates of the theory specify that atoms can occur only in certain states of energy; and the possible energy levels of atoms can be deduced from the fundamental wave equations for physical systems consisting of atoms. With the help of the Psi-function, however, we can also calculate the probabilities with which atoms in given energy states have certain mean diameters. Furthermore, the theory specifies that, when an atom emits or absorbs radiation with a given wave length, the atom passes from one energy level to another. With the help of the Psi-function, however, it is possible to calculate the probabilities of such transitions, and thereby to deduce the distribution of energies in the spectra of the radiations emitted by atoms. On the other hand, coordinating definitions in terms of experimentally significant concepts can be supplied for such theoretical expressions as 'the mean diameter of atoms' and 'the probability of transition from one energy level to another.' In consequence, deductions

[20] More exactly, if $q_1, \ldots, q_k$ are the coordinates of position of a system, so that at a given instant the Psi-function can be written as '$\Psi(q_1, \ldots, q_k)$,' and if $a_1, \ldots, a_k$ is a definite point, then the square of the absolute value of $\Psi(a_1, \ldots, a_k)$ is the probability that the elementary constituents which are in the state $\Psi(a_1, \ldots, a_k)$ are at the point $a_1, \ldots, a_k$.

from the theory such as those mentioned can be put to an experimental test.

This brief exposition will perhaps make evident that the theoretical state-description defined by the Psi-function is related to matters of observation by a circuitous route. The Psi-function itself receives no interpretation in terms of a subatomic model; the square of the amplitude of Psi is interpreted as a probability distribution function for elementary constituents in a subatomic model; these probabilities associated with Psi then enter into the calculation of various other probabilities; and some of these latter probabilities are finally coordinated by rules of correspondence with certain experimental concepts.

Let us now ask how the quantum mechanical state of a system may be ascertained. It will be obvious that we cannot do this directly by experimental observation, but that the procedure just outlined must in a sense be reversed. In assigning a Psi-function to a given system we must in any event adopt a number of intermediary assumptions about probability distributions that are confirmed only indirectly by empirical evidence. Accordingly, while in classical mechanics the variables of state are associated with properties of the *individuals* postulated by the theory, in quantum mechanics the state variable is associated with a *statistical* property of the postulated elements. In consequence, the fact that actual observations on a system are only in approximate agreement with the predictions of theory is interpreted in different ways in the two cases. In classical mechanics the discrepancy is attributed to a lack of precise knowledge concerning the *initial* state of the system. In quantum mechanics the discrepancy is also explained in part in terms of experimental errors, but an additional part of the explanation consists in noting that the assumptions and rules which coordinate the theoretical state of a system with experimental data contain an uneliminable statistical component.[21]

Despite the fact that quantum mechanics is deterministic with respect to the state-description defined by the Psi-function, it is for this reason that outstanding physicists maintain that quantum theory is "in the nature of the case indeterministic, and therefore the affair of statistics."[22] This characterization is unquestionably appropriate. It expresses succinctly the fundamental point that quantum theory is "indeterministic" in the important sense that its state-description is associated with a statistical interpretation and that its predictions are based on statistical assumptions. On the other hand, it is imperative not to misconstrue the characterization nor to draw unwarranted inferences from it. Let us therefore remind ourselves briefly of some essential facts.

[21] Cf. Max Planck, *The Philosophy of Physics*, New York, 1936, pp. 65-66.
[22] Max Born, *Atomic Physics*, London, 1935, p. 90.

In the first place, it is not the Psi-function itself, but only the square of its amplitude, that is interpreted as a probability distribution function. The Psi-function is no more a probability function than are the state-descriptions in Fourier's theory of heat conduction or in Maxwell's theory of electromagnetism. The Psi-function "represents" an abstract feature of physical systems, a feature that rigorously determines certain probabilities associated with those systems. However, since the Psi-function plays a role in quantum theory only through the function that is the square of the absolute magnitude of Psi, and hence through the theoretical probabilities determined by this derived function, Psi can be conveniently regarded as a quasi-statistical state variable.

In the second place, the interpretation of the square of Psi's absolute magnitude as a *probability* function is intelligible only on the assumption that certain subatomic processes form statistical aggregates, to which the notion of probability as a relative frequency is applicable. The Psi-function must therefore be taken to characterize those processes *only* with respect to some of their *statistical* properties. Accordingly, when a given property is ostensibly predicated of *individual* elementary constituents in those subatomic aggregates (for example, when an electron is said to possess a 'momentum' whose magnitude falls within a specified interval), such statements must be understood as *elliptic* formulations. Suitably expanded and made explicit, the illustrative statement about an electron's momentum in fact asserts either (a) that a momentum of the indicated magnitude occurs with a certain *relative frequency in a large class of electrons,* or (b) that a momentum of the indicated magnitude is exhibited by a given electron with a certain *relative frequency during a fairly long temporal period.* In short, if the interpretation associated with the Psi-function is a statistical one, then all predictions based exclusively on that interpretation must also be statistical, and cannot be predications of nonstatistical properties to individuals. There is therefore no warrant for the conclusion that because quantum theory does not predict the detailed individual behaviors of electrons and other subatomic elements, the behavior of such elements is "inherently indeterminate" and the manifestation of "absolute chance." It is of course true that quantum mechanics in its current formulation does not describe such detailed behavior of individual electrons or predict their individual trajectories. However, if the fundamental assumptions of quantum theory have only a statistical content, as indeed they do have on the standard interpretation given to them, it is neither surprising nor paradoxical that all conclusions derivable from those assumptions exclusively should likewise have only a statistical import. It would be surprising and paradoxical if the outcome were otherwise, unless indeed those assumptions are supplemented by additional stipulations or rules so as to permit the deduction of nonstatistical conclusions from the augmented set of assumptions.

On the other hand, quantum mechanics is commonly characterized as an "essentially statistical" theory, since its state variables, unlike the state variables of classical statistical mechanics, cannot be analyzed in terms of any available deterministic theory employing only nonstatistical state-descriptions. In consequence, despite the brilliant successes of quantum theory in explaining, coordinating, and predicting systematically vast areas of experimental fact, some distinguished physicists (including Planck, Einstein, and De Broglie) have expressed serious dissatisfaction with it, on the ground that in its current form quantum theory is "an incomplete representation of real things." For example, Einstein expressed his reservations as follows:

> The Psi-function does not in any way describe a condition which could be that of a single system; it relates rather to many systems, to 'an ensemble of systems' in the sense of statistical mechanics. If, except for certain special cases, the Psi-function furnishes only *statistical* data concerning measurable magnitudes, the reason lies not only in the fact that the *operation of measurement* introduces unknown elements, which can be grasped only statistically, but because of the very fact that the Psi-function does not, in any sense, describe the condition of *one* single system. The Schrödinger equation determines the time variations which are expressed by the ensemble of systems which may exist with or without external action on the single system. . . . But now I ask: Is there really any physicist who believes that we shall never get any inside view of these important alterations in the single system, in their structure and their causal connections, and this regardless of the fact that these single happenings have been brought so close to us, thanks to the marvelous inventions of the Wilson chamber and the Geiger counter? To believe this is logically possible without contradiction; but it is so very contrary to my scientific instinct that I cannot forego the search for a more complete conception.[23]

It is evident, however, that the preference Einstein manifested for a type of theory different from quantum mechanics is not arguable. Nor is it possible to produce cogent evidence, whether pro or con, for his belief that the type of theory he did favor will eventually triumph. In this respect, at any rate, the future is inscrutable. It is worth noting, moreover, that the characterization of quantum mechanics as "incomplete" is based on an assumption that is by no means self-evident. The assumption

[23] Albert Einstein, "Physik und Realität," *Journal of the Franklin Institute,* Vol. 221 (1936), reprinted in translation in *Out of My Later Years,* New York, 1950, pp. 89-91. Einstein's more technical statement of his claim that quantum mechanics is "incomplete" is contained in the paper by A. Einstein, B. Podolsky, and N. Rosen, "Can Quantum-Mechanical Description of Physical Reality Be Considered Complete?" *Physical Review,* Vol. 47 (1935), pp. 777-80.

is that it is always possible to construct an otherwise satisfactory theory for which a uniformly complete interpretation can be given. The component elements of the model used in this interpretation can in principle be described in a manner analogous to those employed in various theories of classical physics, with the help of *individual* rather than statistical variables of state. The alleged incompleteness of current quantum theory apparently consists in the fact that the theory formulates only certain *statistical* properties of subatomic processes but says nothing about the detailed behavior of the "individual" elements in those processes. The imputation of incompleteness thus seems to be made from the perspective of some *other* theory, envisaged in a general way as employing state variables different from those of current quantum mechanics and more like those of classical physics. But there can be no guaranty that such an alternative theory is bound to be developed and will eventually replace the present quantum theory; nearly all physicists today are frankly skeptical that this possibility will be realized in the foreseeable future.

Nevertheless, there are no conclusive reasons for maintaining, as many leading physicists seem to maintain, that the type of "indeterministic" theory illustrated by current quantum mechanics is here to stay permanently. One argument for this claim is based on an important theorem established by John von Neumann. According to this theorem, current quantum theory cannot be supplemented by introducing additional "hidden parameters" for defining the state of a system, so as to convert the theory into a nonstatistical one, without obtaining consequences from the emended theory incompatible with the vast quantities of experimental data that confirm so impressively the current theory. But von Neumann's theorem shows only that, as long as we remain within the basic framework of ideas of the present quantum theory and interpret the data of experiment in terms of its rules, an emendation of theory in the indicated manner is impossible. Von Neumann did not prove, and in the nature of the case he could not prove, that a satisfactory nonstatistical theory, having the scope of current quantum theory but constructed on quite different lines, is logically excluded. To be sure, no such alternative theory is now available, and the difficulties involved in the task of constructing one are formidable. At the same time, the experimental discovery of a variety of strange and partly unexpected "elementary particles" possessing high energies, for which present quantum theory does not provide an adequate explanation, has called attention to serious limitations in the theory. Stimulated by this new "crisis" in physics, physicists have in fact recently attempted to develop nonstatistical theories that escape the interdicts of the von Neumann theorem. These attempts deal with technical matters concerning which only professional physicists can have a competent opinion. But the fact that such attempts are being made by trained students of physics does indicate that the "essentially statistical" form of current subatomic theory is not necessarily the final word on the subject.[24]

2. Many physicists have become thoroughly convinced that quantum theory is the logically fundamental part of physics, in terms of whose basic ideas results achieved elsewhere in the science are to be understood. In consequence, the view has become widespread that all laws whatever (even those about macroscopic objects and events) are at bottom statistical, and that in the end all natural processes are "acausal."

The conception that all the laws of physics represent merely average or statistical regularities was vigorously championed by Charles Peirce, long before the advent of quantum mechanics.[25] The work of Boltzmann on the statistical interpretation of the second law of thermodynamics certainly appeared to confirm this thesis. Peirce's idea was revived in independent form by the Viennese physicist Exner,[26] who in turn stimulated Schrödinger to develop it in the light of more recent physical discoveries.[27] But in any event, the view that all physical laws are basically statistical and acausal has been affirmed by Eddington, among others, as a direct consequence of modern quantum theory. "There is no strict causal behavior anywhere," he declared. "It is impossible to trap modern physics into predicting anything with perfect determinism because it deals with probabilities from the outset."[28]

What is the argument for this claim? It appears to be the following. Macroscopic objects are complex structures of subatomic ones. The properties and relations of the former therefore occur under conditions that can be formulated in terms of the arrangements and interactions of the latter. But the established theory concerning subatomic objects is statis-

[24] The proof of the von Neumann theorem is given in his *Mathematische Grundlagen der Quantenmechanik*, Berlin, 1932 (English translation, *Mathematical Foundations of Quantum Mechanics*, Princeton, 1955), Chap. 4, Sec. 2, pp. 167-73. Discussions of the theorem, as well as of other matters touched upon in the text above, will be found in David Bohm, *Causality and Chance in Modern Physics*, London, 1957; Louis de Broglie, *The Revolution in Physics*, New York, 1953, Chap. 10; and *Observation and Interpretation, A Symposium of Philosophers and Physicists* (ed. by S. Körner), New York and London, 1957.

[25] Charles S. Peirce, "The Doctrine of Necessity Examined," *The Monist*, Vol. 2 (1892), reprinted in *Collected Papers of Charles S. Peirce*, Cambridge, Mass., 1935, Vol. 6, pp. 28-45.

[26] Franz Exner, *Vorlesungen über die Physikalischen Grundlagen der Naturwissenschaften*, Vienna, 1919, pp. 657ff., 696ff.

[27] Erwin Schrödinger, "What Is a Law of Nature?" in *Science and the Human Temperament*, New York, 1935, pp. 133-47.

[28] Arthur S. Eddington, *The Nature of the Physical World*, New York, 1928, p. 309; *New Pathways in Science*, Cambridge, 1935, p. 105. See also J. von Neumann, *op. cit.*, p. 172 (English translation, pp. 326ff.).

tical and indeterministic: to the best of our knowledge, the behavior of subatomic objects exhibits only statistical regularities. Accordingly, so the argument concludes, since the behavior of macroscopic objects is compounded out of the behavior of their subatomic constituents, the regularities manifested by the former are also only statistical.

But the argument is something less than cogent, even if one ignores the ambiguity in the characterization of quantum theory as "indeterministic." Since large philosophic issues concerning human freedom and responsibility are often made to hinge on the conclusion asserted by it, let us examine the argument with some care. However, the conclusion that all physical theories and laws are "statistical" is trivial though true, if it is understood in the sense that quantitative data obtained by experimental measurement confirm numerical laws only approximately, and not with absolute precision. We have already discussed this question, and need not consider it further. But we must recall the distinction we then made between what a statement actually asserts, and the precision with which empirical evidence agrees with what the statement asserts.[29] The claim we now want to examine is that what is asserted by all physical laws is statistical in content.

A tacit assumption in the argument for this claim is that if a theory (e.g., quantum mechanics) is statistical, then every conclusion derived from the theory must also be statistical. Although this assumption is in general sound, there are exceptions to it. Such exceptions may occur, for example, when the coordinating definitions for various statistical parameters in the theory associate with those parameters *nonstatistical* experimental concepts, so that what is *prima facie* a nonstatistical experimental law can then be deduced.

An example will make this clear. Planck's radiation law formulates the distribution of energy in the experimentally continuous spectrum of a black body, and asserts that the energy associated with emitted rays of a given wave length is a certain function of that wave length and of the temperature of the black body.[30] On the face of it, the law makes no statistical assertion. It can be put to an experimental test by measuring the energies at various places in the spectrum (for example, by placing a sensitive bolometer at some position in the spectrum, noting the temperature, and then calculating the energy with the help of other laws), and so ascertaining whether the magnitude of the energy at each place

---

[29] See page 284 above.

[30] If E_λ is the energy associated with a ray whose wave length is λ, T the absolute temperature of the radiating black body, h Planck's constant, c the velocity of light, and k Boltzmann's gas constant, then Planck's radiation law is given by

$$E_\lambda = \frac{hc^2}{\lambda^5}\left(\frac{1}{e^{hc/kT\lambda} - 1}\right)$$

has the value required by the law. But the law can be derived from a complicated set of assumptions, including postulates of quantum mechanics as well as of statistical mechanics and electrodynamics, applied to the physical system consisting of radiations from a black body. The derivation of the experimental law depends, among other things, on several coordinating definitions. One of these definitions, for example, associates the *nonstatistical* experimental concept of temperature with the theoretical *statistical* notion of the mean kinetic energy of the black-body oscillators. Another coordinating definition associates the *nonstatistical* experimental notion of energy with the theoretical notion of the *statistically determined number* of oscillators having a certain wave length.

The point illustrated by this example merits further discussion. Like other theories, a statistical microscopic theory is introduced in order to explain the occurrence of experimentally identifiable properties (often called "macrostates") of macroscopic objects. Such a theory postulates a set of microscopic elements, capable of standing to each other in various stipulated relations. Let us call each theoretically possible and distinguishable "arrangement" of the microscopic constituents of a system a "microstate" of the system. The theory explains the occurrence of the macrostates of a system in terms of assumptions concerning changes in the microstates, so that the explanation depends upon the institution of correspondences between macro- and microstates. However, the correspondences are usually so specified that to a given macrostate there corresponds not one but a large number of distinct microstates. For example, in the kinetic theory of gases, the temperature of a gas (the macrostate) corresponds to the mean kinetic energy of the molecules of the gas; but a given value for the mean kinetic energy is compatible with a large number of distinct microstates (where each microstate is described by a particular set of values for the positions and velocities of the molecules), so that a given macrostate corresponds to many microstates.[81] Let us assume that each macrostate M_i of the system corresponds to a *class* of microstates m_i, and that these classes m_i do not overlap. Suppose, further, that the occurrence at a given time t of a microstate

---

[81] Thus, suppose that there are just four molecules each with unit mass, each of which can occupy any one of eight positions, and each of which may have a velocity of either 1 or 2 feet per second. Then the total number of distinct microstates is $4^{10} = 1,048,576$. If the mean kinetic energy of the four molecules is

$$\frac{(1^2 + 2^2 + 2^2 + 1^2)}{(2 \times 4)} = \frac{5}{4},$$

this is compatible with any one of $6 \times 4^8 = 393,212$ microstates. Thus a single macrostate, the temperature of the gas, corresponds to almost 400,000 distinct microstates.

belonging to m_i does not determine the occurrence at some later time t' of a *unique* microstate, but that it *does* determine the occurrence of a microstate belonging to some class m_j, where the precise relation between i and j is specified by the microscopic theory. The theory is then a statistical theory with respect to microstates, and microstates succeed each other only with a statistical regularity. But it by no means follows that the succession of *macrostates* will also exhibit only a statistical regularity; on the contrary, the macrostates of the system may be related to each other in accordance with a strictly universal, nonstatistical law. It is therefore a *non sequitur* to conclude that, because quantum mechanics is the foundation for all other parts of physics but is a statistical theory, all physical laws deducible from quantum mechanics must also be statistical.

There is, however, another assumption, though of a vaguer sort, that appears to be a tacit premise in the argument for the claim that all physical laws are statistical. According to this assumption, if a system is analyzable into a structure of elementary constituents (whether these are "absolutely" or only "relatively" simple), the constituents are in some unclear sense more "ultimate" than, or "metaphysically prior" to, the complex system. What is perhaps meant is that no property or character has an indisputable place in an account of any complex thing, unless that property can also be predicated of the "ultimate" elements out of which the thing is constituted. In particular, though a law about macroscopic objects may *appear* to have a nonstatistical content, its content is "really" statistical if the law can be deduced from an essentially statistical theory about the fundamental elements of all natural processes.

.But if this is what this widely held assumption means, it is difficult to take it seriously. Indeed, were the assumption sound, it would be pointless to develop theoretical explanations for the behavior of macroscopic objects in terms of their elementary parts. For on that assumption macroscopic objects would possess indisputable properties only if those properties also characterized the elementary components of the objects. But, since the question of whether the hypothetical microscopic constituents of macroscopic objects do have certain traits cannot be settled except by making observations of macroscopic objects and their properties, the vicious circle thus generated cannot be avoided. Moreover, on that assumption the elementary constituents of macroscopic objects would simply be diminutive duplicates of the macroscopic objects, and would possess all the traits whose explanation is being sought. In point of fact, when a theory explains the behavior of macroscopic objects in terms of microscopic elements, special laws must be assumed that connect the manifest traits of the former with certain other traits of the latter. It would be absurdly fatuous to assume such laws if those manifest traits, though not characterizing the elementary constituents of things, were not as indisputably features of the world as the traits of the elements are presumed to be.

It should be noted, finally, that, even if we accept without question the most extreme claims for the indeterministic behavior of the subatomic elements postulated by quantum theory, this indeterminacy is not exhibited in any experimentally detectable behavior of *macroscopic* objects. Indeed, the theoretical indeterminism as calculated from quantum mechanics in the motions even of molecules, to say nothing of bodies with larger masses, is far smaller than the limits of experimental accuracy. As De Broglie observed, the theoretical indeterminacy of subatomic processes by no means contradicts the "apparent Determinism" of large-scale phenomena. For that indeterminacy "is completely masked by the errors introduced in the course of experiment, and everything happens therefore as though it did not exist at all. . . . In practice, as also in experiment, everything happens as though . . . there were a strict Determinism." [32]

In consequence, the statistical content of quantum mechanics does not annul the deterministic and nonstatistical structure of other physical laws. It also follows that conclusions concerning human freedom and moral responsibility, when based on the alleged "acausal" and "indeterministic" behavior of subatomic processes, are built on sand. Neither the analysis of physical theory, nor the study of the subject matter of physics, yields the conclusion that "There is no strict causal behavior anywhere."

V. *The Principle of Causality*

The extraordinarily successful career of quantum mechanics has been widely hailed as demonstrating the inapplicability to subatomic processes of the so-called "law of causality," and as marking its decline as a universally valid principle.[33] A brief discussion is therefore in order as to what this "law" asserts, what its logical status is, and whether the announcements of its general collapse are indeed warranted.

The law, or principle, of causality is usually distinguished from various special causal laws or theories, such as the theory of classical mechanics. But there is no generally accepted standard formulation of it, nor is there general agreement as to what it affirms. The principle is usually understood to have a wider scope than any special causal law. On the other hand, some writers take it to be a statement on par with particular

[32] Louis de Broglie, *Matter and Light*, New York, 1939, p. 230.
[33] Cf. W. Heisenberg, *op. cit.*, p. 63.

causal assertions, though affirming something about a trait pervasive throughout nature and not simply about features of a limited subject matter. Others understand by it a principle of higher rank than are the specialized causal laws; and they maintain that the principle asserts something about laws and theories rather than about the subject matters of laws and theories. Still other writers take it to be a regulative principle for inquiry rather than a formulation of connections between events and processes. Some regard it as an inductive generalization, some believe it to be a priori and necessary, and others maintain that it is a convenient maxim and the expression of a resolution. Since so many different notions are subsumed under the label of the "principle of causality," it is small wonder that the current claims concerning its "breakdown" have provoked discussions as ambiguous and indecisive as the claims themselves.

1. It would not be profitable to examine in any detail even the major formulations proposed for the principle during several centuries of debate. Moreover, although several statements of the principle by contemporary writers have the merit of possessing relatively great clarity, these formulations have been proposed primarily in connection with the problem of validating (or "justifying") inductive inferences; [34] and it would not be relevant to the present context of discussion to consider them. It will be useful, however, to examine briefly the view that the principle is an empirical generalization about the constitution of nature.

A familiar and influential statement of this conception was given by John Stuart Mill. According to him, the principle of the uniformity of nature (Mill's name for the principle of causality) asserts that "There are such things in nature as parallel cases; that what happens once, will, under a sufficient degree of similarity of circumstances, happen again." [35] Although Mill undoubtedly believed this statement to have an empirical content, whether it does or not hinges on how one understands the expression "a sufficient degree of similarity of circumstances." When are circumstances sufficiently similar? A superficial resemblance between circumstances is obviously not enough. Moreover, two sets of circumstances may be judged to be alike even by trained and discriminating observers, and yet an effect may follow upon one set but not upon the other. For example, two solutions of sugar and water may exhibit no sensible differences even under careful scrutiny, although one solution may rotate the plane of polarization of transmitted light in a clockwise

---

[34] Cf. J. M. Keynes, *A Treatise on Probability*, London, 1921, Part 3; Rudolf Carnap, *Logical Foundations of Probability*, Chicago, 1950, pp. 178ff.; Bertrand Russell, *Human Knowledge*, New York, 1948, Part 6.

[35] J. S. Mill, *A System of Logic*, London, 1879, Book 3, Chap. 3, Sec. 1.

direction while the other rotates it counterclockwise. Must a defender of the principle of causality therefore surrender it as invalid? Not at all. He could claim that the two solutions are in fact not really similar and that the sugars differ in their atomic structures, even if no independent evidence for such an alleged difference should be available. But in that eventuality, it is patent that the expression "a sufficient degree of similarity of circumstances" is being so used that two sets of circumstances are said to be sufficiently similar only if they have similar consequences. On that assumption, Mill's formulation of the principle does not possess an empirical content but has the status of a stipulative definition.

But could not a sense be assigned to the phrase in question so that the principle would then be a genuine factual assertion about the "order of nature"? Attempts to fix such a sense, without giving the principle a less general form than Mill's version of it, have not been successful. A typical example of a more specialized formulation of the principle is the one proposed by Laplace in the passage cited earlier in this chapter. Laplace assumed classical mechanics to be the universal science of nature; and he therefore adopted the mechanical definition of state in his formulation of the circumstances in which things must be similar if they are to have similar consequences. Accordingly, Laplace's version of the principle of causality asserts that, if a physical system is in the same mechanical state at any two distinct times, the system will go through the same evolutions subsequent to those times and will possess all properties in common at corresponding instants in that evolution.

Nevertheless, the principle of causality encounters difficulties even on this formulation. In the first place, as is evident from the preceding discussions in this chapter, it is erroneous to maintain that the mechanical state of a system determines *all* the properties of the system. In the second place, this formulation of the principle is almost as empty of empirical content as is Mill's version, and like the latter it appears to be compatible with every possible state of affairs. Suppose, for example, that a system is judged to be in the same mechanical state at two different times, but that nonetheless the system does not exhibit the same properties at corresponding subsequent moments. But despite its apparent incompatibility with the facts, the principle of causality need not be abandoned as false. It could be retained as perfectly valid, simply by assuming that the system has concealed constituents that were not in the same mechanical state at the two initial times. And finally, though the principle appears irrefutable by any empirical evidence, it has in fact been abandoned in the construction of theories in many fields of physical inquiry. It has been abandoned in these areas because those traits of things (i.e., the mechanical state) upon which this version of the principle places exclusive stress have not turned out to be suitable as the basis for advancing our theoretical understanding of many physical processes. Accordingly, if the principle of causality is construed in the sense stated by Laplace's version of it, the claim that the principle is inapplicable in subatomic physics is obviously warranted.

2. For these reasons it is extremely difficult if not hopeless to regard the principle of causality as a universally valid inductive truth concerning the pervasive order of events and processes. Let us therefore consider whether the principle does not stand in a more favorable light if it is formulated as a regulative or methodological rule of inquiry.

Suppose, for example, that Newtonian mechanics is applied to the study of the relative motions of the sun and earth, on the assumption that the force-function is the familiar inverse-square law, which does not explicitly mention the time of motion and depends only on the masses of the two bodies and the distances between them. As is well known, the theoretical orbit of the earth is then an ellipse, with the center of mass of the two bodies at one of the foci. However, the actual positions and velocities of the earth, as established by observation at different times, differ from the theoretical values of these coordinates of state by more than the margin of experimental error. Indeed, on the assumed hypothesis, the earth appears to behave as if the force varied with the time in some irregular manner: at certain times the earth has positions and velocities differing from its theoretical ones to a greater extent than at other times, the variations in the discrepancies exhibiting no obvious pattern. It seems, therefore, that Newtonian theory is not quite satisfactory, and it is conceivable that physicists might in consequence reject the theory.

As everyone knows, however, physicists do not do so. They account for the discrepancies by attributing them to the fact that the sun-earth system is not an "isolated" one, and that in fact there are celestial bodies (e.g., the known planets) which produce "perturbations" in the earth's motion. The procedure physicists adopt is to *enlarge* the initial system which appeared at first to behave in a manner disagreeing with Newtonian theory. More specifically, physicists enlarge the initial system by including in it further bodies, until the force on the earth in the enlarged system no longer appears to vary with the time in some unaccountable manner.

This example illustrates a standard scientific procedure that has yielded much valuable fruit in the past. Thus it was that by using this procedure Adams and Leverrier postulated the existence of the previously unknown planet Neptune, subsequently identified telescopically, in order to account for the "irregularities" in the motion of the planet Uranus. But the tacit rule governing this procedure is that version of the principle of causality according to which the principle is a *maxim* for inquiry rather than a statement with a definite empirical content. Construed as a maxim, the principle bids us analyze physical processes in such a way that their evolution can be shown to be independent of the particular times and places at which those processes occur. More generally, the maxim enjoins us to look for laws and theories that contain no explicit reference to the times and places at which events and processes occur. In point of fact, Maxwell did state the principle of causality as follows: "The difference between one event and another does not depend on the mere difference of the times or the places at which they occur, but only on differences in the nature, configuration, or motion of the bodies concerned." [86] Although this formulation does not make fully explicit the sense of the principle as a methodological rule, and is stated with the special requirements of classical mechanics in mind, this sense of the principle is not far beneath the surface of Maxwell's words. The interpretation of the principle as a maxim has been expressed with great vigor in a more recent formulation: "Whenever you come across an *incomplete or disturbed system,* try to the best of your ability to *amplify it to one undisturbed whole,* looking for the *supplement* first among things known, near and far. If the desired supplement is not found among them, search for it among *things unknown.*" [87]

The principle of causality so construed is thus a generalized recommendation. It bids us construct theories and find appropriate systems to which those theories can be successfully applied, with no restrictions upon the detailed form of the theories, except for the requirement that, when the state of a system is given for some initial time (in whatever manner the state may be defined), the theory for it must determine a unique state of the system for any other time. When formulated in this general way, however, the principle is admittedly vague, and supplies no specific directives for achieving the objectives it recommends. Indeed, unless the formulation is understood in the light of certain additional though usually tacit stipulations, the principle is well-nigh reduced to triviality. To see this, consider Maxwell's version of the principle. Suppose that the processes in some domain of inquiry exhibit no obvious regularity, and that they depend on the time of their occurrence in such a manner that no explanations for this dependence can be found which refer only to the "nature, configuration, or motion of the bodies concerned." Nevertheless, it is demonstrable that there must be a mathe-

[86] J. C. Maxwell, *Matter in Motion,* New York, 1920, p. 13. But recent developments in physical cosmology suggest that Maxwell's formulation of the principle of causality may require modification.

[87] L. Silberstein, *Causality,* New York, 1933, p. 71. See also Ernst Cassirer, *Determinism and Indeterminism in Modern Physics,* New Haven, 1956, Part 2.

matical function relating the processes to the time of their occurrence; and with luck we may even chance upon this function.[38] Moreover, if the function satisfies certain very broad mathematical conditions, it is even possible to eliminate from the function all explicit references to the specific times and places at which the processes occur (thus satisfying Maxwell's requirement), without troubling ourselves to enlarge the system of processes in the manner indicated above—provided that we are prepared to employ in our theory differential equations of arbitrarily high order and any degree of complexity.[39]

In point of fact, however, most physicists would very likely demur at granting this proviso. They would decline to admit it on the ground that a law or theory cannot be regarded as satisfactory if its mathematical form is so complex that it cannot be used conveniently for purposes of calculation and prediction, or if its basic notions are so opaque that it can be applied to concrete situations only with the greatest difficulty. Accordingly, although the task enjoined by the principle of causality when it is quite generally formulated may in many cases be executed almost trivially, tacit restrictions are in fact placed upon the complexity and the character of theories acceptable as satisfying the principle's "real" intent. Such restrictions, couched, when made explicit, in such terms as "simplicity," "convenience," and "naturalness," prevent the principle from being trivially satisfied; but because these terms are vague and cannot be assigned fixed precise meanings, the content of the principle's recommendations is itself imprecise. Nevertheless, there is usually at least a rough consensus among scientists of a given period concerning the broad limits within which suitable theories are to be sought, even if these limits are flexible, depend on the current state of a science, and

[38] The basis for the claim that there must be such a function is simply that, if some magnitude x takes definite values for different times t, this correspondence between values of x and t "defines" the function.

[39] For example, in classical mechanics the differential equations for forced vibrations take the form:

$$\frac{d^2x}{dt^2} = \alpha \frac{dx}{dt} + \beta x + \gamma \cos \omega t$$

where 'α,' 'β,' 'γ' are certain constants. By differentiating this equation twice with respect to the time, we obtain:

$$\frac{d^4x}{dt^4} = \alpha \frac{d^3x}{dt^3} + \beta \frac{d^2x}{dt^2} - \gamma \omega^2 \cos \omega t$$

and hence by eliminating the time variable, finally:

$$\frac{d^4x}{dt^4} = \alpha \frac{d^3x}{dt^3} + (\beta - \omega^2) \frac{d^2x}{dt^2} + \alpha \omega^2 \frac{dx}{dt} + \beta \omega^2 x$$

which does not contain 't' explicitly.

may change with the development of mathematical and experimental techniques.[40]

But if the principle of causality is construed as a maxim of the sort that has been suggested, it is clear that, contrary to the views of J. S. Mill and others, the principle is not an empirical generalization about the pervasive structure of the world and does not appear as the "ultimate major premise" in any explanation. The function of the principle so interpreted is to make explicit a generalized goal of inquiry and to formulate in general terms a *condition* that premises proposed as explanations are required to satisfy. Moreover, it is also clear why on this interpretation the principle cannot be disproved by any experiment or series of experiments, although *special forms* of the principle may be abandoned as ill-advised in the light of experience. For the principle is a *directive*, instructing us to search for explanations possessing certain broadly delimited features; and even repeated failure to find such explanations for any given domain of events is no logical bar to further search.

On the other hand, when the directives stated by the principle assume particular forms, it may be wise strategy to ignore them in the face of

[40] Maxwell's formulation of the principle of causality has been criticized by Moritz Schlick on the ground that it is too restrictive and that it states a sufficient but not a necessary condition under which a law would be called a "causal" one. Schlick maintained that a world is conceivable in which all laws would contain the time explicitly, and yet in which such laws might be regarded as fully determinate. For example, in such a world the elementary electric charge might not be constant throughout time but might increase or diminish by 5% of its present value at certain stated intervals (say, after 7 hours, then after 7 hours, then again after 5 hours, etc.), though no further explanations of such a fluctuation might be possible. Schlick therefore concluded that the necessary and sufficient condition for a law to be causal is that it makes prediction possible, and he formulated the principle of causality as the injunction to look for laws in accordance with the dictum "All events are in principle predictable."—Moritz Schlick, "Die Kausalität in der gegenwärtigen Physik," in his *Gesammelte Aufsätze*, Vienna, 1938.

But neither Schlick's criticism of Maxwell's formulation nor his own proposed substitute for it are entirely satisfactory. As we have seen in the text, a theory (in the form of differential equations) which contains the time explicitly can in general be transformed in such a way that the time variable does not so appear. Accordingly, if it is at all possible to formulate a theory which establishes some relation between the variations in a set of magnitudes and the time, Maxwell's requirements will be satisfied—unless some further tacit stipulations are made concerning the "simplicity" of the theory. Moreover, Schlick's proposed criterion in terms of predictability, if taken strictly, leads to the conclusion that no theory or law is strictly causal. For as we have seen, all actual predictions made with the help of a theory are at best only approximate. Moreover, a law or theory enables us to predict only if we are in the position to ascertain the requisite initial conditions; and in many cases we may not be in the position to do so without thereby refusing to characterize a theory as a deterministic one. Schlick recognizes this by his qualification that the prediction be possible "in principle." But this qualification in effect transposes the issue from a question of predictability to a discussion of the structure of theories. On this general point concerning predictability as a criterion of causal laws, cf. Max Planck, *The Philosophy of Physics*, New York, 1936, pp. 56-57, 64ff.

repeated failures to achieve their objectives. Thus, if the principle is understood, as it often has been, as an injunction for every domain of inquiry to develop only such theories that employ a special type of state-description (e.g., the state-description of classical mechanics), unswerving adherence to the principle may eventually become an obstacle to theoretical invention and discovery. It is also abstractly conceivable that nothing short of the entire cosmos is an isolated system with respect to certain types of occurrences for which explanations are sought. In consequence, it might not be possible to devise explanations for such occurrences involving reference to only a limited set of objects and properties, in conformity with the principle. In that eventuality, the pursuit of scientific knowledge in connection with events of that type would be impossible, and the principle would be a useless guide. For both theoretical and experimental science proceed on the assumption that everything is not relevant to everything else, and that occurrences in one part of the world are not dependent upon what happens everywhere else. It is a fact of history that the search for isolated systems (or enlarged systems, in the sense previously indicated) not coinciding with the total cosmos has thus far been successful. Indeed, our generally unhesitant readiness to conduct our inquiries in accordance with the principle is undoubtedly based on the high frequency of successes that have rewarded our past actions guided by it.

In sum, therefore, the principle as a maxim expresses the general objective of theoretical science to achieve *deterministic* explanations, in the now familiar sense of "determinism" according to which, given the state of a system for some initial time, the explanatory theory logically establishes a unique state for the system for any other time. In its most general formulation, the principle does not prescribe a particular definition for the state-description (such as the state-description of classical mechanics), nor does it postulate as the goal of science the development of theories possessing some special logical form (such as that of being expressible by differential equations). It does not proscribe the use of statistical or quasi-statistical variables of state, and recent developments in subatomic physics are therefore not in conflict with its directives. The current claim that the principle of causality is inapplicable to the subject matter of quantum mechanics is tenable only if it is construed to legislate the use of special types of state-descriptions, and only if the use of statistical state variables by a theory is taken to mark the theory as lacking a deterministic structure.

3. What is the upshot of this discussion of the logical status of the principle of causality? Is the principle an empirical generalization, an a priori truth, a concealed definition, a convention that may be accepted or not as one pleases?

The view that the principle is an empirical generalization, it has been argued, is difficult to maintain. For when the principle is formulated in a fully general way, without mention of which factors determine the occurrences of things and processes, the principle excludes nothing whatever from the logically possible orders of events in the world; and in effect the principle collapses into an implicit definition of what it is to be a causal or determining factor in natural processes. On the other hand, if the principle is formulated in a more limiting manner, so that it does mention which traits of things are the causally determining ones in natural processes, the principle turns out not to be universally true, and can therefore be asserted as sound only for certain special subject matters.

But if the principle is a maxim, is it a rule that may be followed or ignored at will? Is it merely an arbitrary matter what general goals are pursued by theoretical science in its development? It is undoubtedly only a contingent historical fact that the enterprise known as "science" does aim at achieving the type of explanations prescribed by the principle; for it is logically possible that in their efforts at mastering their environments men might have aimed at something quite different. Accordingly, the goals men adopt in the pursuit of knowledge are *logically* arbitrary.

Nevertheless, the actual pursuit of theoretical science in modern times is directed toward certain goals, one of which is formulated by the principle of causality. Indeed, the phrase "theoretical science" appears to be so generally used that an enterprise not controlled by those objectives would presumably not be subsumed under this label. It is at least plausible to claim, therefore, that the acceptance of the principle of causality as a maxim of inquiry (whether the acceptance is explicit or only illustrated in the overt actions of scientists, and whether the principle is formulated with some precision or only vaguely) is an *analytical consequence* of what is commonly meant by "theoretical science." In any event, one can readily grant that, when the principle assumes a special form, so that it prescribes the adoption of a particular type of state-description by every theory, the principle might be abandoned in various areas of investigation. But it is difficult to understand how it would be possible for modern theoretical science to surrender the general ideal expressed by the principle without becoming thereby transformed into something incomparably different from what that enterprise actually is.

VI.   *Chance and Indeterminism*

Thus far, the discussion of determinism in physics has examined the issues almost entirely in terms of the logical structure of physical *theories* and of the *concepts* employed in the latter. We have deliberately avoided the question, prominent in current debates on the foundations of physics as well as in the historical literature of philosophy, whether the *actual events* of nature are not themselves, in part or in whole, "undetermined" or "chance" occurrences, and whether the use of essentially statistical variables of state is not a consequence of the fact that certain physical processes are in the domain of fortuitous happenings. This is the question that will now occupy us.

The word "chance" is notoriously ambiguous as well as vague. Our first task must therefore be to distinguish between various senses of this word, in order to decide whether in any of these meanings of the word the characterization of an event as a chance occurrence is incompatible with also characterizing that event as caused or determined.

1. Perhaps the most familiar and widespread use of the word 'chance' occurs in contexts in which something unexpected happens rather than as the consequence of a deliberate plan. Thus, two friends out for a walk who meet without prior arrangement are said to have a chance encounter. Again, a gardener who finds a gold coin when he is turning up the soil in preparation for planting is said to have found the treasure by chance or accident. However, it is usually not the mere unexpectedness of the event that suffices for the application of the label "chance" in these cases. If in the first example one of the friends passes a complete stranger just five minutes after he starts his walk, or if in the second example the gardener turns up a pebble after digging for ten seconds, neither occurrence would normally be described as a chance happening—even though neither the event of meeting the stranger after five minutes, nor the event of finding a pebble after just ten seconds of spading may have been literally expected. To be described as a chance event, the event must usually have some striking features, and its occurrence must be felt to intrude into a fairly definite plan of action. But in this usage the word "chance" is quite vague, and no clear limits for its application can be fixed. On the other hand, an event that is said to be a chance occurrence in this sense is usually not assumed to be "uncaused" or to lack determinate conditions for its occurrence. For example, the gardener mentioned above need not necessarily withdraw his characterization of the happening as a chance occurrence if he should learn that the coin he found was buried by some ancestor; and yet he might withdraw the label if he should discover that the coin was deliberately buried by a friend, so that his apparently chance discovery was part of a definite plan. But in any event, this sense of "chance" is hardly relevant to discussions on the foundations of physics.

2. The word 'chance' is predicated of an event in another sense, either when there is practically complete ignorance concerning the determining conditions for the event, or when these conditions are known

to belong to some *class* of alternative types of conditions but are not known to belong to a particular type in that class. Until fairly recently the weather was commonly regarded to be a matter of chance in this sense, as the expression "The wind bloweth where it listeth" indicates; and whether a symmetrically constructed coin falls with its head or with its tail uppermost is still the stock example of a chance event. In the first instance, the causes of the weather were not known; and in the second, it is generally assumed that, though the initial position of the coin and the forces acting upon it to determine its final state can be analyzed into an exhaustive set of possibilities, we do not in fact know which of these possibilities will be realized.

It is usually not regarded as sufficient to characterize an event as "due to chance" in the present sense of the word, merely because events of the kind in question occur under certain conditions with a relative frequency that is less than one. For example, suppose that a "fair" coin is tossed a large number of times in some standard manner, and that the head turns up in about half the cases. However, the outcome of the tosses would not be credited to chance if the sequence of heads (H) and tails (T) were to be of the following kinds: $HTHTHTHT \ldots$; or $HHTTHHHTTTHHHHTTTT \ldots$; or even $HHTTHTTTHHHHTT$ $HTTTHH. \ldots$ For in each of these series (and many others that can be constructed) a little inspection shows that head and tail succeed each other with an easily formulable regularity. In order to be counted as chance occurrences, the outcome of the tosses must exhibit a certain "fortuitous" or "random" character. A variety of definitions have been proposed for the "randomness" intended, though not all are satisfactory and some are more stringent than others. A definition that has considerable merit is an implicit one. According to it, a linearly ordered set of events is random if and only if the set satisfies certain postulates of the calculus of probability. But further details must be omitted at this place.[41] The essential point to be noted is that, when a given type of event is said to be "due to chance" in the present sense of the word, some *definition* of "fortuitous" or "random" occurrence is taken for granted. It is also essential to note, moreover, that to say that an event occurs by chance in this sense is not incompatible with maintaining that it is caused; for admission of ignorance concerning the specific determining conditions for an event obviously does not entail a denial of the existence of such conditions.

3. In historical and sociological discussions an event is commonly said to be a chance happening if it occurs "at the intersection of two independent causal series." Suppose, for example, that a man leaves his

[41] A few further comments on randomness are added on page 333.

home in order to purchase some tobacco, but that on the way he is felled by a brick displaced from the roof of a building. The man's misfortune is then said to be a chance occurrence not because it is "uncaused" (indeed, the description of the event indicates the cause), but because it occurs at the "juncture" of two independent causal sequences, one that terminates in the man's being beside the building at a given time, and the other that terminates in the brick's motion at that time. These causal series are said to be "independent" in the sense that the events in one do not determine the events in the other: had the brick not fallen the man would have proceeded on his journey to the tobacconist, and had the man not been at a particular spot the brick would have struck the ground. Accordingly, the man's injury is alleged to be fortuitous or accidental, since, however complete may be our knowledge of the circumstances leading to the man's journey or of the conditions making for the brick's passage, neither body of knowledge by itself suffices to foretell the accident.

The sense of "chance" here intended requires further clarification. The notion needing special attention is that associated with the phrase "independent causal chains." The notion is obviously based on the image of two distinct lines (or chains) intersecting in a common point. The sequence of points (or links) in each line is supposedly determined by the "inherent" character of the line, but not determined by the "nature" of the other line; and the fact that the lines do have a point in common is determined by the nature of neither line taken singly. But the supposition that concrete events are analogous to points on the line, that they are self-contained occurrences whose "natures" are exhausted by their "positions" in some one specified linear sequence of events, and that the occurrence of an event in such a sequence is determined by the "nature" of the preceding parts of the sequence, is at best a suggestive but loose metaphor and at worst a barely intelligible phantasy. Concrete happenings do not appear to possess such inherent, self-contained natures; for a given event manifests an indefinite number of characters, and if we can credit current physical theories there are thus an indefinite number of distinct causal determinants for the occurrence of any specific event. Accordingly, if the image of a line or chain is adopted for describing the causal relations of events, an event is more appropriately described as being the common intersection of an indefinite (if not an infinite) number of lines. But if this more complex image is employed, it no longer is even apparently clear just what we are to understand by "independent causal lines," since now every event is the node of very many causal influences.

Greater clarity concerning the sense of 'chance' under discussion can be obtained if we reformulate the distinction in terms of relations between *statements* rather than events or happenings. Let S_1 be a statement asserting the occurrence of some event, for example, a statement of the form 'x is injured by a falling brick at time t and place y,' and, more generally, is of the form 'x is in the relation R to z at time t and place y.' Suppose T_1 is a theory or law which states in a general way the conditions and manner of occurrence of some one factor manifested in that event, but states them without reference to the presence or absence of certain other factors which are also involved in it; and suppose further that T_2 performs a similar task for these other factors. Moreover, we shall make the explicit assumption that T_1 and T_2 cannot be deduced from each other. To fix our ideas and to make the discussion specifically relevant to the general form assumed for S_1, let T_1 assert that, if under conditions C_1, x is in a state P at time t and place y, then x is in the state P' at time t' and place y'. And analogously, let T_2 assert that, if under conditions C_2, z is in the state Q at time t and place y, then z is in the state Q' at t' and y'. Accordingly, given T_1 and some appropriate initial data D_x concerning x, it is possible to calculate the state of x for other times and places; and similarly for z with the help of T_2 and initial data D_z. Moreover, in virtue of the assumptions made concerning T_1 and T_2, the state of x at any time *cannot* be calculated from T_2 and D_z, nor the state of z from T_1 and D_x. The sequence of states of x and z, respectively, may therefore be called "independent chains," this independence being a consequence of the assumed logical independence of T_1 and T_2. Now it is clear that S_1 itself is derivable neither from T_1 and D_x alone, nor from T_2 and D_z alone, nor even from the conjunction of T_1 and T_2. The first two cases are excluded because S_1 involves reference to *both* individuals x and y and a certain relation between them, while the suggested premises of the deduction do not do so; and the last case is excluded because S_1 is a simple singular statement, while T_1 and T_2 are both universal conditionals. S_1 is therefore logically independent of T_1 as well as of T_2, whether these are taken singly or in conjunction; and it is also logically independent of T_1 and D_x, as well as of T_2 and D_z. We may therefore say that the event expressed by S_1 is a "chance occurrence"—*relative* to the sequence of states determined by T_1 and D_x, and also *relative* to the sequence of states determined by T_2 and D_z.

On the other hand, if the conditions C_1 and C_2 mentioned in T_1 and T_2, respectively, are physically compatible, and if the relation R mentioned in S_1 is correctly analyzable in terms of the states P and Q also mentioned in T_1 and in T_2, then in general S_1 is deducible from the complex conjunction of T_1 and D_x and T_2 and D_z. It follows that the event mentioned by S_1 is *not* a chance occurrence with respect to the sequence of states of x and y determined by this complex formula. It also follows that the characterization of an event as a chance happening in the present sense of the word does *not* entail that the event is uncaused, or even that we are ignorant of the conditions which determine its occurrence; nor is the attribute that is thus predicated of an event something "subjective," and so merely the expression of the state of mind of one who is predicating the attribute. When such a judgment is made fully explicit and not expressed in elliptic language, it requires the use of a *relational* though "objective" predicate—in the same sense in which the assertion that one side of a street is the "other side" requires the use of a relational though objective characterization.[42]

4. A fourth sense of "chance" is intimately related to the one just discussed, but merits special attention. An event is said to be a "chance" or "contingent" occurrence in this sense of the word, if in a given context of inquiry the statement asserting its occurrence is not derived from anything else. Thus, if we wish to predict some future position and velocity of the planet Mars with the help of Newtonian gravitational theory, we must supply some initial position and velocity for the planet; and the fact that at such an initial time Mars is at a certain place and is moving with a certain velocity is thus a chance occurrence. It is of course not denied that an event which is characterized as a chance happening in some context may be shown to be the consequence of some other happening, or, more precisely, that the statement affirming the first occurrence may be derived from other statements about different happenings, with the help of appropriate assumptions (in the above example, a given initial state of Mars can certainly be derived by means of Newtonian theory from some other initial state). However, contrary to frequent claims, the fact that it can be so derived does not wipe out the distinction between a chance and a non-chance event in the sense here intended. For in the first place, an event is said to be a chance happening in a *given context;* and the fact that it is not a chance occurrence in some *other* context does not preclude its being such an occurrence in the given one. It is clear, therefore, that there is no incompatibility between saying that an event is a chance happening (in the present sense) and saying that nevertheless there are determinate conditions or causes for its occurrence. And in the second place, though an event which is a

---

[42] There is an obvious similarity between the above discussion of "chance" and Aristotle's analysis of "accident." Aristotle also took the view that whether a predicate represented an accidental property of a subject is relative to the definition of the latter. However, he adopted an "absolutistic" (or "essentialist") view of definitions, since he maintained that a definition states the fixed "essence" or "nature" of a substance. Since this latter claim rests on assumptions that are unwarranted in the light of current knowledge and are incompatible with much that is said in the present book, there is therefore a fundamental disagreement between Aristotle's account of accidents and the above discussion of "chance."

chance happening in one context may not be such in a second context, some *other* event must be recognized which *is* a chance occurrence in the latter context. For the occurrence of an event is formulated as a simple singular statement; and such statements can be deduced from theories or laws only if appropriate initial conditions are supplied for the latter.

However, not only are events characterized as chance occurrences in the present sense of the expression, but by a natural extension the expression is sometimes used to characterize laws and theories. There is nevertheless a slight ambiguity in this more extended usage. Sometimes a law or theory is said to be "contingent" or to hold "by chance" if in a given context the law or theory is not derived from any other premises; and here there is a fairly strict parallelism to the use of the word in connection with events. For example, the law of linear thermal expansion of solids was once said to be merely a contingent truth, because no explanation of it in terms of an accepted physical theory was then available. For this reason the law was also commonly labeled as merely an "empirical" formula, because its acceptance was based only on a mass of direct experimental evidence. On the other hand, though the Boyle-Charles' law for ideal gases was at one time held to be simply a fortuitous empirical truth, it is generally not so characterized now, since its derivation from the assumptions of the kinetic theory of gases. Again, a theory such as the kinetic theory of gases or the electromagnetic theory is said to be a contingent body of assumptions because it is (at any rate at present) not explainable by any more inclusive theory and because it is not accepted on the ground that it is the logical consequence of other well-established premises. Since at a given state of scientific development the process of explanation cannot continue indefinitely, it is clear that there must always be some theories which are contingent in the present sense of this word. Scientists and philosophers who maintain that "ultimately" or "in the last analysis" the sciences do not provide explanations for anything often have only something like this in mind; and they are to be understood as saying that the grounds for accepting the premises in any alleged explanation are in the end not purely demonstrative.

Sometimes, however, a theory or law is said to be a contingent truth, whether or not the theory or law is derivable from some other assumptions, simply because the theory or law is not itself a logically necessary truth and can be established only on the basis of empirical evidence. It is of course tacitly assumed that there are some statements which are logically necessary and can be certified as true by considering only the meanings of their terms, and some which are not. Statements such as 'No spiders are insects,' 'The angle sum of a Euclidean triangle is equal to two right angles,' and 'All prime numbers greater than two are odd' are typical examples of the first class, while 'No mammals have gills,' 'Under electrolysis, water is decomposed into hydrogen and oxygen,' and 'A moving electrically charged body generates a magnetic field' are common illustrations of the second class. Those who reject the distinction between logically necessary (or "analytic") statements and logically indeterminate (or "synthetic") ones—whether because they believe that all true statements are "in the end" logically necessary, or because they hold that even the statements of formal logic and arithmetic are simply well-attested empirical generalizations, or because they maintain that at bottom the distinction is one of degree rather than of kind—doubtless see no point in characterizing a special group of statements as "contingent" truths.[48] But in actual scientific practice the distinction is generally observed and appears to be firmly based on the differences of procedure employed in establishing statements in various branches of inquiry. Accordingly, since scientific theories and laws are assumed to be only contingently true at best, no individual occurrence in nature and no patterns of coexistence or change that theories or laws formulate are logically necessary. If a "completely rational" explanation is identified—as it frequently has been—with an explanation whose premises are necessary truths, then no completely rational account of the world or any occurrence in it can be given.

5.   One remaining sense of 'chance' needs attention; in this sense the word refers to some "absolute" rather than relational character of events. An event of which 'chance' is predicated in this sense is sometimes held to be "uncaused," so that not only do we not know the determining conditions for its occurrence but, so it is assumed, there are no such conditions. The ancient Epicurean conception of atoms swerving spontaneously from their normal paths is one illustration of this use of 'chance.' More recently, as already noted, Charles Peirce developed a speculative evolutionary cosmogony on the assumption of a radical tychism. Trace the causes of irregular departure from any accepted physical law as far back as you please, Peirce maintained, "and you will be forced to admit they are always due to arbitrary determination, or chance." [44] According to him, diversification is always taking place, and, by admitting "pure spontaneity" as a character of the universe "acting always and everywhere though restrained within narrow bounds by law, producing in-

[48] Recently some philosophers have challenged the "dualism" of the analytic-synthetic distinction. Cf. W. V. O. Quine, *From a Logical Point of View*, Cambridge, Mass., 1953, Chap. 2; Morton White, *Toward Reunion in Philosophy*, Cambridge, Mass., 1956, Chap. 8.

[44] Charles S. Peirce, *Collected Papers*, Cambridge, Mass., 1935, Vol. 6, p. 37.

finitesimal departures from law continually," Peirce believed he was able to account "for all the variety and diversity of the universe." [45] And many contemporary physicists also appear to maintain that subatomic processes, at any rate, are characterized by absolute chance, so that, for example, the emission of particles by radioactive substances is regarded as "a process due to the spontaneous decomposition of its atoms." [46]

Sometimes, however, an event is said to be an "absolutely chance event," not because there are no determinate conditions for its occurrence but because, though there are such conditions, the event exhibits certain "novel features" incomparably different from those which the conditions manifest. Accordingly, so the present sense of 'chance' is sometimes explained, even if the conditions for the occurrence of a chance event were known with utmost precision, the event could not be predicted from them—unless, indeed, events of that type had already been actually observed to be regularly associated with such conditions. Thus, it is often claimed that, when sulfuric acid was added to table salt for the first time, the gas with its peculiar properties that was then produced could not have been actually predicted; and the generation of the gas under the indicated conditions is said to be a chance event. It is also possible that Peirce's tychism contains this notion of chance as one component. However, this special sense of 'absolute chance' plays an essential role in current doctrines of "emergent evolution"; and we shall therefore postpone further discussion of it until these doctrines are examined in the next chapter.

Let us return to the first sense of 'absolute chance' as absence of determining conditions for the occurrence of an event. This notion of chance is free from internal contradictions, except for reservations to be noted below, and claims to the contrary such as Bradley's [47] are certainly mistaken. Nor are there any other a priori reasons for ruling out the possibility of chance events in this sense. On the other hand, there appear to be no unquestionably authentic cases of such events. Indeed, it is impossible in the nature of the case to establish beyond question that any event is an absolutely chance occurrence. For to show beyond all possible doubt that a given happening (e.g., the decomposition of an atom) is spontaneous and without determining circumstances, it would

[45] *Ibid.*, p. 41. Like Epicurus and many current writers, Peirce postulated his radical tychism in order to allow for human free will. Thus he declared, "By supposing the rigid exactitude of causation to yield, I care not how little—be it but by a strictly infinitesimal amount—we gain room to insert mind into our scheme, and to put it into the place where it is needed, into the position which, as the sole self-intelligible thing, it is entitled to occupy, that of the fountain of existence; and in so doing we resolve the problem of the connection of soul and body."—*Ibid.*, pp. 42-43.

[46] Max Planck, *The Philosophy of Physics*, New York, 1936, p. 52.

[47] F. H. Bradley, *Appearance and Reality*, London, 1920, pp. 387-94.

be necessary to show that there is nothing whatever upon which its occurrence depends. But this would be tantamount to showing that no satisfactory theory could ever be devised to explain what present theories already explain, and in addition account for the allegedly spontaneous event. However, though any amount of evidence might be produced to show that the given event's occurrence does not depend on a specified set of factors, the possibility would not be thereby excluded that other factors may eventually be found which do determine the event's occurrence, and that in consequence a theory might yet be constructed which will do what our present theories fail to do.

Accordingly, the assumption that events of a certain type are absolutely chance occurrences cannot be conclusively validated, even if the available evidence may make that assumption plausible. It must be admitted, in any case, that the radioactive disintegration of atoms is at present not known to occur only under specific determining conditions. It may therefore well be that those events are 'absolutely chance' occurrences. On the other hand, though current physical theory is not incompatible with the assumption that atomic disintegrations occur by absolute chance, it makes no specific use of this assumption in its formulations. Accordingly, current theory is also compatible with the weaker assumption that these events are *relatively* chance occurrences, in one of the senses of 'chance' previously distinguished.

Moreover, there is a serious difficulty associated with the notion of chance that makes the assumption of 'absolute chance' a gratuitous hypothesis. The reason that events are usually said to occur in a totally fortuitous manner is that no "order" whatsoever appears in the *sequence* of their occurrences, and that in consequence no functional relations can be stated between the events and the times at which they happen. However, the claim that any sequence of events exhibits an absolute disorder is tenable only if the terms 'order' and 'disorder' are used in some selective or qualified sense, and only if 'functional relation' is understood to mean some limited class of mathematical functions.

To fix our ideas, consider the atoms in a given piece of radium, and suppose that the time at which each atom disintegrates is recorded. Now there will doubtless be no obvious formula connecting the number of disintegrations with the times at which they occur. Nevertheless, since by hypothesis there is a correspondence between the disintegrations and the times, a mathematical function relating the former with the latter is thereby *defined* in extension. It is therefore not logically impossible that a general formula can be constructed which states this correspondence, even if the formula should turn out to be forbiddingly complex. Accordingly, there is no "absolute" disorder in the distribution of the atomic disintegrations in time, since there clearly is *some* order in their arrangement. In short, the notion of an absolute, unqualified disorder is self-contradictory. This does not mean that *each* event in a series could not possibly occur in an absolutely chance manner. It *does* mean, however, that the disorder predicated of the *distribution of these events in time* must be understood as being relative to some (perhaps only vaguely delimited) type of order or class of mathematical functions.[48] The logically incoherent assumption of an absolutely random distribution must therefore be replaced by the coherent hypothesis of *relative disorder* (or relative randomness), according to which a sequence of events is a random or disordered sequence, if the events occur in an order that cannot be deduced from any law belonging to some specified class of laws. On the other hand, though the occurrence of events of a certain type may be random relative to one class of laws, their occurrence may not be random relative to some other class of laws. Accordingly, if the thesis that an event is "uncaused" or completely fortuitous is based on the claim that the *sequence* of events of that type exhibits no order in the occurrence of the events, the strength of this claim as a support for the thesis must be appraised in the light of the fact that the alleged disorder is only a *relative* disorder.

The main outcome of this discussion is that saying an event "happens by chance" is not in general incompatible with asserting the event to be determined, except when "happening by chance" is understood to mean that the event has no determining conditions for its occurrence. However, we do not in fact know the precise conditions for the occurrence of many kinds of events, even though we may be confident that there are such conditions. In lieu of such knowledge, we can often establish relations of dependence between statistical properties of events, rather than between individual events or their individual properties. Indeed, the use of statistical variables of state in modern physical theories is based on the assumption that, although we do not know the detailed behavior of the "individual" microscopic elements postulated by theory,

---

[48] In the literature of probability, attempts have been made to define precisely the notion of "randomness" or "disorder." Thus, Richard von Mises proposed the following definition: Let x_1, x_2, x_3, . . . be a nonterminating series of elements, and let Q be some property which characterizes some of these elements. Then the occurrence of Q in that series is random, if two conditions are satisfied: (1) the relative frequency of Q converges to some limit p; and (2) the relative frequency of Q in *all* partial sequences that may be selected from the original series converges to the same limit p. Cf. Richard von Mises, *Probability, Statistics and Truth*, New York, 1939, pp. 32ff. But it has been shown that the second condition, requiring the invariance of p in all partial sequences (and hence in a *nondenumerably infinite* number of partial sequences) leads to a contradiction. The requirement must be modified so that p is invariant only under a *denumerably* infinite class of partial sequences. Cf. A. Wald, "Die Widerspruchsfreiheit des Kollektivbegriffes der Wahrscheinlichkeitsrechnung," in *Ergebnisse eines mathematischen Kolloquiums* (ed. by Karl Menger), Heft 8.

we can significantly reduce our ignorance by considering various statistical properties of those elements.

In classical physics (e.g., classical statistical mechanics), however, the "random" behavior assumed for the postulated individuals is not taken to be the manifestation of some radically "acausal" or "intrinsically fortuitous" character of the motions of those individuals. On the contrary, the sense in which the individual motions are said to occur "by chance" is explicitly asserted to be the sense of 'chance' as relative, which is the second meaning of the word we identified earlier. In quantum mechanics, on the other hand, the use of a statistical state-description is widely believed to reflect the inherently undetermined or absolutely chance nature of certain subatomic processes. Nevertheless, the question whether or not these processes are absolutely fortuitous is not an issue of scientific moment, for, as has been noted, quantum theory is compatible with either alternative. Physicists who maintain that quantum mechanics requires only the notion of relative chance and who have "scientific instincts" hostile to the notion of absolute chance,[49] may one day develop an essentially nonstatistical theory to replace the present quantum theory. Should their hopes be realized, the current belief will undoubtedly be reversed that physics has established the completely fortuitous character of subatomic processes. But until such an alternative theory becomes available, the question of absolute chance will remain the subject of inconclusive controversy.

[49] In a letter to Born, Einstein declared, "You believe in God playing dice and I in perfect laws in the world of things existing as real objects, which I try to grasp in a wildly speculative way."—*Albert Einstein, Philosopher-Scientist* (ed. by Paul A. Schilpp), Evanston, Ill., 1949, p. 176; also Max Born, *Natural Philosophy of Cause and Chance*, New York, 1949, p. 122.

The Reduction of Theories

11 It is a commonplace that classical mechanics is no longer
regarded as the universal and fundamental science of
nature. Its brilliant successes in explaining and bringing
into systematic relations a large variety of phenomena
were at one time indeed unprecedented. And the belief,
once widely held by physicists and philosophers, that
all the processes of nature must eventually fall within the scope of its
principles was repeatedly confirmed by the absorption of various sectors
of physics into it. Nevertheless, the period of the "imperialism" of mechanics was practically over by the latter part of the nineteenth century.
The difficulties which faced the extension of mechanics into still uncon-
quered territory, and particularly into the domain of electromagnetic
phenomena, came to be acknowledged as insuperable.

However, new candidates for the office of a universal physical science
were proposed, sometimes with the backing of a priori arguments analogous to those once used to support the claims of mechanics. To be sure,
with some few doubtful exceptions no serious student of the sciences today believes that any physical theory can be warranted on a priori
grounds, or that such arguments can establish a theory in that high office. Moreover, many outstanding physicists are frankly skeptical whether
it is possible to realize the ideal of a comprehensive theory which will
integrate all domains of natural science in terms of a common set of
principles and will serve as the foundation for all less inclusive theories.
Nevertheless, that ideal continues to leaven current scientific speculation;
and, in any case, the phenomenon of a relatively autonomous theory be-

coming absorbed by, or reduced to, some other more inclusive theory is an undeniable and recurrent feature of the history of modern science. There is every reason to suppose that such reduction will continue to take place in the future.

The present chapter is concerned with this phenomenon, and with some of the broader issues associated with it. Scientists as well as philosophers have exploited both successful and unsuccessful reductions of one theory to another as occasions for developing far-reaching interpretations of science, of the limits of human knowledge, and of the ultimate constitution of things in general. These interpretations have taken various forms, but only a few typical ones need be mentioned here.

Discoveries relating to the physics and physiology of perception are often used to support the claim that the findings of physics are radically incompatible with so-called "common sense"—with customary beliefs that the familiar things of everyday experience possess the traits they manifest even to carefully controlled observation. Again, the successful reduction of thermodynamics to statistical mechanics in the nineteenth century was taken to prove that spatial displacements are the only form of intelligible change, or that the diverse qualities of things and events which men encounter in their daily lives are not "ultimate" traits of the world and are perhaps not even "real." But, conversely, the difficulty in finding consistent visualizable models for the mathematical formalism of quantum mechanics has been taken as evidence for the "mysterious" character of subatomic processes and for the claim that behind the opaque symbolism of the "world of physics" there is a pervasive "spiritual reality" that is not indifferent or alien to human values. On the other hand, the failure to explain electromagnetic phenomena in terms of mechanics, and the general decline of mechanics from its earlier position as the universal science of nature, have been construed as evidence for the "bankruptcy" of classical physics, for the necessity of introducing "organismic" categories of explanation in the study of all natural phenomena, and for a variety of sweeping doctrines concerning levels of being, emergence, and creative novelty.

We shall not examine the detailed arguments that culminate in these and similar contentions. However, one broad comment is relevant to most of the claims. As has been repeatedly noted in previous chapters, expressions associated with certain established habits or rules of usage in one context of inquiry are frequently adopted in the exploration of fresh fields of study because of assumed analogies between the several domains. Nevertheless, its users do not always note that, when the range of application of a given expression is thus extended, the expression often undergoes a critical change in its established meaning. Serious misunderstandings and spurious problems are then bound to be generated unless care is taken to understand the expression in the sense relevant to, and required by, the special context in which the expression has acquired a fresh use. Such alterations are particularly prone to occur when one theory is explained by, or reduced to, another theory; and the shifts in the meanings of familiar expressions that often result as a consequence of the reduction are not always accompanied by a clear awareness of the logical and experimental conditions under which the reduction has been effected. In consequence, both successful and unsuccessful attempts at reduction have been occasions for comprehensive philosophical reinterpretations of the import and nature of physical science, such as those cited in the preceding paragraph. These interpretations are in the main highly dubious because they are commonly undertaken with little appreciation for the conditions that must be fulfilled if a successful reduction is to be achieved. It is therefore of some importance to state with care what these conditions are, both for the light that the discussion of those conditions throws on the structure of scientific explanation and for the help which the discussion can provide toward an adequate appraisal of a number of widely held philosophies of science. An examination of the conditions for reduction and of their bearing on some moot issues in the philosophy of science is the central task of the present chapter.

I. *The Reduction of Thermodynamics to Statistical Mechanics*

Reduction, in the sense in which the word is here employed, is the explanation of a theory or a set of experimental laws established in one area of inquiry, by a theory usually though not invariably formulated for some other domain. For the sake of brevity, we shall call the set of theories or experimental laws that is reduced to another theory the "secondary science," and the theory to which the reduction is effected or proposed the "primary science." However, many cases of reduction seem to be normal steps in the progressive expansion of a scientific theory and rarely generate serious perplexities or misunderstandings. It will therefore be convenient to distinguish, with the help of some examples, between the two types of reduction, the first of which is commonly regarded as being quite unproblematic and which we shall ignore in consequence, while the second is often felt to be a source of intellectual discomfort.

1. A theory may be formulated initially for a type of phenomenon exhibited by a somewhat restricted class of bodies, though subsequently the theory may be extended to cover that phenomenon even when manifested by a more inclusive class of things. For example, the theory of mechanics was first developed for the motions of point-masses (i.e., for

the motions of bodies whose dimensions are negligibly small when compared with the distances between the bodies) and was eventually extended to the motions of rigid as well as deformable bodies. In such cases, if laws have already been established within the more inclusive domain (perhaps on a purely experimental basis, and before the development of the theory), these laws are then reduced to the theory. However, in these cases there is a marked qualitative similarity between the phenomena occurring in the initial and the enlarged provinces of the theory. Thus, the motions of point-masses are quite like those of rigid bodies, since the motions in both cases involve only changes in spatial position, even though rigid bodies can exhibit a form of motion (rotation) that point-masses do not. Such reductions usually raise no serious questions as to what has been effected by them.

Analogously, the range of application of a macroscopic theory may be extended from one domain to another homogeneous with it in the features under study, so that substantially the same concepts are employed in formulating the laws in both domains. For example, Galileo's *Two New Sciences* was a contribution to the physics of free-falling terrestrial bodies, a discipline which in his day was considered to be distinct from the science of celestial motions. Galileo's laws were eventually absorbed into Newtonian mechanics and gravitational theory, which was formulated to cover both terrestrial and celestial motions. Although the two classes of motions are clearly distinct, no concepts are required for describing motions in one area that are not also employed in the other. Accordingly, the reduction of the laws of terrestrial and celestial motions to a single set of theoretical principles has for its outcome simply the incorporation of two classes of qualitatively similar phenomena into a more inclusive class whose members are likewise qualitatively homogeneous. Because of this circumstance, the reduction again generates no special logical puzzles, although it did in point of fact produce a revolution in men's outlook upon the world.

In reductions of the sort so far mentioned, the laws of the secondary science employ no descriptive terms that are not also used with approximately the same meanings in the primary science. Reductions of this type can therefore be regarded as establishing deductive relations between two sets of statements that employ a homogeneous vocabulary. Since such "homogeneous" reductions are commonly accepted as phases in the normal development of a science and give rise to few misconceptions as to what a scientific theory achieves, we shall pay no further attention to them.

2. The situation is usually different in the case of a second type of reduction. Difficulties are frequently experienced in comprehending the import of a reduction as a consequence of which, a set of distinctive

traits of some subject matter is assimilated to what is patently a set of quite dissimilar traits. In such cases, the distinctive traits that are the subject matter of the secondary science fall into the province of a theory that may have been initially designed for handling qualitatively different materials and that does not even include some of the characteristic descriptive terms of the secondary science in its own set of basic theoretical distinctions. The primary science thus seems to wipe out familiar distinctions as spurious, and appears to maintain that what are *prima facie* indisputably different traits of things are really identical. The acute sense of mystification that is thereby engendered is especially frequent when the secondary science deals with macroscopic phenomena, while the primary science postulates a microscopic constitution for those macroscopic processes. An example will show the sort of puzzle that is generated.

Most adults in our society know how to measure temperatures with an ordinary mercury thermometer. If provided with such an instrument, they know how to determine with reasonable accuracy the temperature of various bodies; and, in terms of operations that are performed with the instrument, they understand what is meant by such statements as that the temperature of a glass of milk is 10° C. A good fraction of these adults would doubtless be unable to explicate the meaning of the word 'temperature' to the satisfaction of someone schooled in thermodynamics; and these same adults would probably also be unable to state explicitly the tacit rules governing their use of the word. Nevertheless, most adults do know how to use the word, even if only within certain limited contexts.

Let us now assume that some person has come to understand what is meant by 'temperature' exclusively in terms of manipulating a mercury thermometer. If that individual were told that there is a substance which melts at a temperature of fifteen thousand degrees, he would probably be at a loss to make sense of this statement, and he might even claim that what has been told him is quite meaningless. In support of this claim he might maintain that, since a temperature can be assigned to bodies only on the basis of employing a mercury thermometer, and since such thermometers are vaporized when brought into the proximity of bodies whose temperatures (as specified by a mercury thermometer) are a little above 350° C, the phrase "temperature of fifteen thousand degrees" has no defined sense and is therefore meaningless. However, his puzzlement over the information given him would be quickly removed by a little study of elementary physics. For he would then discover that the word 'temperature' is associated in physics with a more embracing set of rules of usage than the rules that controlled his own use of the word. In particular, he would learn that laboratory scientists employ the word to refer to a certain state of physical bodies, and that variations in this state are often manifested in other ways than by the volume expansion of mercury—for example, in changes in the electrical resistance of a body, or in the generation of electrical currents under specified conditions. Accordingly, once the laws are explained that formulate the relations between the behaviors of instruments such as thermocouples, which are sometimes used to record changes in the physical state of bodies called their 'temperature,' the person understands how the word can be meaningfully employed in situations other than those in which a mercury thermometer can be used. The enlargement of the word's range of application then appears no more puzzling or mysterious than does the extension of the word 'length,' from its primitive meaning as fixed by using the human foot for determining lengths, to contexts in which a standard metal bar replaces the human organism as a measuring instrument.

Suppose, however, that the layman for whom 'temperature' thus acquires a more generalized meaning now pursues his study of physics into the kinetic theory of gases. Here he discovers that the temperature of a gas is the mean kinetic energy of the molecules which by hypothesis constitute the gas. This new information may then generate a fresh perplexity, and indeed in an acute form. On the one hand, the layman has not forgotten his earlier lesson, according to which the temperature of a body is specified in terms of various overtly performed instrumental operations. But on the other hand, he is also assured by some authorities he now consults that the individual molecules of a gas cannot be said to possess a temperature, and that the meaning of the word is identical "by definition" with the meaning of 'the mean kinetic energy of molecules.' [1] Confronted by such apparently conflicting ideas, he may therefore find a host of typically "philosophical" questions both relevant and inescapable.

If the meaning of 'temperature' is indeed the same as that of 'the mean kinetic energy of molecules,' what is the plain man in the street talking about when he says that milk has a temperature of 10° C? Most consumers of milk who might make such statements are surely not asserting anything about the energies of molecules; for even though they understand and know how to use such statements, they are generally uninstructed in physics, and know nothing about the molecular composition of milk. Accordingly, when the man in the street learns about molecules in milk, he may come to believe that he is confronted with a serious issue as to what is genuine "reality" and what is only "appearance." He may then perhaps be persuaded by a traditional philosophical argu-

[1] Cf. Bernhard Bavink, *The Anatomy of Science,* London, 1932, p. 99.

ment that the familiar distinctions between hot and cold (indeed, even the distinctions between various temperatures of bodies as specified in terms of instrumental operations), refer to matters that are "subjective" manifestations of an underlying but mysterious physical reality, a reality which cannot properly be said to possess temperatures in the common-sense meaning of the word. Or he may accept the view, supported by a different mode of reasoning, that it is temperature as defined by procedures involving the use of thermometers and other such instruments which is the genuine reality, and that the molecular energies in terms of which the kinetic theory of matter "defines" temperature are just a fiction. Alternatively, if the layman adopts a somewhat more sophisticated line of thought, he may perhaps come to regard temperature as an "emergent" trait, manifested at certain "higher levels" of the organization of nature but not at the "lower levels" of physical reality; and he may then question whether the kinetic theory, which ostensibly is concerned only with those lower levels, does after all "really explain" the occurrence of emergent traits such as temperature.

Perplexities of this sort are frequently generated by reductions of the type of which the above example is an instance. In such reductions, the subject matter of the primary science appears to be qualitatively discontinuous with the materials studied by the secondary science. Put somewhat more precisely, in reductions of this type the secondary science employs in its formulations of laws and theories a number of distinctive descriptive predicates that are not included in the basic theoretical terms or in the associated rules of correspondence of the primary science. Reductions of the first or "homogeneous" type can be regarded as a special case of reductions of the second or "heterogeneous" type. But it is with reductions of the second type that we shall be concerned in what follows.

3. To fix our ideas, let us consider a definite example of a reduction of this variety. The incorporation of thermodynamics within mechanics—more exactly, within statistical mechanics and the kinetic theory of matter—is a classic and generally familiar instance of such a reduction. We shall therefore outline one small fragment of the argument by which the reduction is effected, on the assumption that this part of the argument is sufficiently representative of reductions of this type to serve as a basis for a generalized discussion of the logic of reduction in theoretical science.

Let us first briefly recall some historical facts. The study of thermal phenomena goes back in modern times to Galileo and his circle. During the subsequent three centuries a large number of laws were established dealing with special phases of the thermal behavior of bodies; and it was eventually shown with the help of a small number of general principles that these laws have certain systematic interrelations. Thermodynamics, as this science came to be called, uses concepts, distinctions, and general laws that are also employed in mechanics; for example, it makes free use of the notions of volume, weight, and pressure and of laws such as Hooke's law and the laws of the lever. In addition, however, thermodynamics employs a number of distinctive notions such as temperature, heat, and entropy, as well as general assumptions that are not corollaries to the fundamental principles of mechanics. Accordingly, though many laws of mechanics are constantly used in the explorations and explanations of thermal phenomena, thermodynamics was regarded for a long time as a special discipline, plainly distinguishable from mechanics and not simply a chapter in the latter. Indeed, thermodynamics is still usually expounded as a relatively autonomous physical theory; and its concepts, principles, and laws can be understood and verified without introducing any reference to some postulated microscopic structure of thermal systems and without assuming that thermodynamics can be reduced to some other theory such as mechanics. However, experimental work early in the nineteenth century on the mechanical equivalent of heat stimulated theoretical inquiry to find a more intimate connection between thermal and mechanical phenomena than the customary formulation of thermal laws seems to assert. Bernoulli's earlier attempts in this direction were revived, and Maxwell and Boltzmann were able to give a more satisfactory derivation of the Boyle-Charles' law from assumptions, statable in terms of the fundamental notions of mechanics, concerning the molecular constitution of ideal gases. Other thermal laws were similarly derived; and Boltzmann was able to interpret the entropy principle—perhaps the most characteristic assumption of thermodynamics and one which appears to differentiate the latter from mechanics most definitely—as an expression of the statistical regularity that characterizes the aggregate mechanical behavior of molecules. In consequence, thermodynamics was held to have lost its autonomy with respect to mechanics, and to have been reduced to the latter branch of physics.

Just how is this reduction effected? By what reasoning is it apparently possible to derive statements containing such terms as 'temperature,' 'heat,' and 'entropy' from a set of theoretical assumptions in which those terms do not appear? It is not possible to exhibit the complete argument without reproducing a treatise on the subject. Let us therefore fix our attention on but a small part of the complicated analysis, the derivation of the Boyle-Charles' law for ideal gases from the assumptions of the kinetic theory of matter.

If we suppress most of the details that do not contribute to the clarification of the main issue, a simplified form of the derivation is in outline as follows. Suppose that an ideal gas occupies a container whose volume is V. The gas is assumed to be composed of a large number of perfectly elastic spherical molecules possessing equal masses and volumes but with dimensions that are negligible when compared with the average distances between them. The molecules are further assumed to be in constant relative motions, subject only to forces of impact between themselves and the perfectly elastic walls of the container. Thus the molecules within their container constitute by postulation an isolated or conservative system, and the molecular motions are analyzable in terms of the principles of Newtonian mechanics. The problem now is to calculate the relation of other features of their motion to the pressure (or force per unit area) exerted by the molecules on the walls of the container because of their constant bombardments.

However, since the instantaneous coordinates of state of the individual molecules are not actually ascertainable, the usual mathematical procedures of classical mechanics cannot be applied. In order to make headway, a further assumption must be introduced—a statistical one concerning the positions and momenta of the molecules. This statistical assumption takes the following form: Let the volume V of the gas be subdivided into a very large number of smaller volumes, whose dimensions are equal among themselves and yet large when compared with the diameters of the molecules; and also divide the maximum range of the velocities that the molecules may possess into a large number of equal intervals. Now associate with each small volume all possible velocity-intervals, and call each complex obtained by associating a volume with a velocity-interval a "phase-cell." The statistical assumption then is that the probability of a molecule's occupying an assigned phase-cell is the same for all molecules and is equal to the probability of a molecule's occupying any other phase-cell, and that (subject to certain qualifications involving among other things the total energy of the system) the probability that one molecule occupies a phase-cell is independent of the occupation of that cell by any other molecule.

If in addition to these various assumptions it is stipulated that the pressure p exerted at any instant by the molecules on the walls of the container is the average of the instantaneous momenta transferred from the molecules to the walls, it is possible to deduce that the pressure p is related in a very definite way to the mean kinetic energy E of the molecules, and that in fact $p = 2E/3V$, or $pV = 2E/3$. But a comparison of this equation with the Boyle-Charles' law (according to which $pV = kT$, where k is a constant for a given mass of gas, and T its absolute temperature) suggests that the law could be deduced from the assumptions mentioned *if* the temperature were in some way related to the mean kinetic energy of the molecular motions. Let us therefore introduce the postulate that $2E/3 = kT$, that is, that the absolute temperature of an ideal gas is proportional to the mean kinetic energy of the molecules assumed to constitute it. Just what the character of this postulate is we shall for the moment not inquire. But our final result is that the Boyle-Charles' law is a logical consequence of the principles of mechanics, when these are supplemented by a hypothesis about the molecular constitution of a gas, a statistical assumption concerning the motions of the molecules, and a postulate connecting the (experimental) notion of temperature with the mean kinetic energy of the molecules.[2]

II. *Formal Conditions for Reduction*

Although the derivation of the Boyle-Charles' law from the kinetic theory of gases has only been sketched, the outline can nevertheless serve as a basis for stating the general conditions that must be satisfied if one science is to be reduced to another. It is convenient to divide the discussion into two parts, the first dealing with matters that are primarily of a formal nature, the second with questions of a factual or empirical character. We consider the formal matters first.

1. It is an obvious requirement that the axioms, special hypotheses, and experimental laws of the sciences involved in a reduction must be available as explicitly formulated statements, whose various constituent terms have meanings unambiguously fixed by codified rules of usage or by established procedures appropriate to each discipline. To the extent that this elementary requirement is not satisfied, it is hardly possible to decide with assurance whether one science (or branch of science) has in fact been reduced to another. It must be acknowledged, moreover, that in few if any of the various scientific disciplines in active development is this requirement of maximum explicitness fully realized, since in the normal practice of science it is rarely necessary to spell out in detail all the assumptions that may be involved in attacking a concrete problem. This requirement of explicitness is thus an ideal demand, rather than a description of the actual state of affairs that obtains at a given time. Nevertheless, the statements within each specialized discipline can be classified into distinct groups, based on the logical roles of the statements in the discipline. The following schematic catalogue of such groups of statements is not intended to be exhaustive, but to list the more important types of statements that are relevant to the present discussion.

---

[2] For a detailed exposition of the deduction, see, for example, James Rice, *Introduction to Statistical Mechanics*, New York, 1930, Chap. 4, or J. H. Jeans, *The Dynamical Theory of Gases*, Cambridge, England, 1925, Chap. 6.

a. In the highly developed science S (such as mechanics, electrodynamics, or thermodynamics) there is a class T of statements consisting of the fundamental theoretical postulates of the discipline. These postulates appear as premises (or partial premises) in all deductions within S. They are not derived from other assumptions in a given codification of the science, although in an alternative exposition of S a different set of logically primitive statements may be employed. Since T is adopted in order to account for, and to direct further inquiry into, experimental laws and observable events, there will also be a class R of coordinating definitions (or rules of correspondence) for a sufficient number of theoretical notions occurring in T or in statements formally derivable from those in T. Moreover, T will presumably satisfy the usual requirements for an adequate scientific theory. In particular, T will be capable of explaining systematically a large class of experimental laws belonging to S; it will not contain any assumptions whose inclusion does not significantly augment the explanatory power of T but serves merely to account for perhaps only one or two experimental laws; it will be "compendent" (in the sense that any pair of postulates in T will have at least one theoretical term in common); and the postulates in T will be "simple" and not too numerous. As noted in Chapter 6, it is sometimes convenient to use the assumptions T not as premises but as leading principles or as methodological rules of analysis. However, the issues that arise from stressing the role of theories as guiding principles rather than as premises have already been discussed, and those issues are in any case of no moment in the present context.

It is often possible to establish a hierarchy among the statements of T in respect to their generality (in the sense of "generality" examined in Chapter 3). When this can be done it is then useful to distinguish the subclass T_1, containing the most general theoretical assumptions in T, from the remaining subclass T_2, of more specialized ones. The most general postulates T_1 normally have a scope of application that is more inclusive than the scope of the theory T taken as a whole. Accordingly, the postulates T_1 are comprehensive postulates of which T is but a special case, while the assumptions T_2 are hypotheses concerning some *special type* of physical systems. For example, the most general theoretical assumptions in the kinetic theory of gases are the Newtonian axioms of motion, so that they belong to T_1; and their scope is clearly more embracing than is the scope of the kinetic theory. On the other hand, the postulate that every gas is a system of perfectly elastic molecules whose dimensions are negligible, or the postulate that all the molecules have the same probability of occupying a given phase-cell, are less general than the Newtonian axioms, and belong to T_2. The assumptions T_2 can thus be regarded as variable supplements to those in T_1, for they can be varied without altering the content of those in T_1, since the latter are applied to different types of systems. For example, the Newtonian axioms are supplemented by distinctive assumptions concerning the molecular structures of gases, liquids, and solids, when these axioms are used in theories about the properties of different states of aggregation of matter. Again, although the kinetic theory of gases retains the fundamental assumptions of Newtonian mechanics when it deals with various types of gases, the theory does not always postulate that gas molecules have negligible dimensions; moreover, the forces assumed by the theory to be acting between the molecules depend on whether or not the gas is far removed from its point of liquefaction.

Although it may not always be possible to distinguish sharply between the more general postulates T_1 of a theory and the less general variable supplements to them, some such distinction is commonly recognized. Thus, despite the fact that the primary science to which thermodynamics has been reduced contains other postulates than those of classical mechanics, thermodynamics is often said (even if only loosely) to be reducible to *mechanics,* presumably because the Newtonian axioms of motion are the most general assumptions of the kinetic theory of gases, so that they formulate the basic framework of ideas within which the special conclusions of the theory are embedded. Moreover, were the kinetic theory of gases able to account for some of the experimental laws of thermodynamics only by modifying one or more of its less general assumptions T_2, it is unlikely that anyone would therefore dispute the reducibility of thermodynamics to mechanics, provided that the principles of mechanics are retained as the most general explanatory premises of the revised theory.

b. A science S possessing a fundamental theory T will also have a class of theorems that are logical consequences of T. Some of the theorems will be formally derivable exclusively from T (indeed, often from the most general postulates T_1) without any help from the correspondence rules R, while others can be obtained only by using R as well. For example, a familiar theorem of the first kind in planetary theory is that, if a point-mass is moving under the action of a single central force, its orbit is a conic section; a theorem of the second kind is that, if a planet is moving under the action of the sun's gravitational force alone, its areal velocity is constant.

But whether or not S has a comprehensive theory, it will in general contain a class L of experimental laws that are conventionally regarded as falling into the special province of S. Thus, the various laws dealing with the reflection, refraction, and diffraction of light constitute part of the experimental content of the science of optics. Although at any given stage of development of S the class of its experimental laws *L* is in principle unambiguously determinable, this class is frequently augmented (and sometimes even diminished) with the progress of inquiry. Nor is there a permanently fixed demarcation between the experimental laws *L* that are grouped together as belonging to one branch of science S and the laws that are considered to fall into a different branch. Thus, it was not always understood that electrical and magnetic phenomena are intimately related; and in older books on physics, though not in most of the recent ones, experimental laws about *prima facie* different phenomena are classified as belonging to distinct departments of experimental inquiry. Indeed, the limits assumed for the domain of a given science, and the rationale operative in classifying experimental laws under different scientific disciplines, are often based on the explanatory scope of currently held theories.

   c. Every positive science contains a large class of singular statements that either formulate the outcome of observations on the subject matter regarded as the province of the science or describe the overt procedures instituted in conducting some actual inquiry within that discipline. We shall call such singular statements "observation statements," but with the understanding that in using this label we are not committed to any special psychological or philosophical theory as to what are the "real" data of observation. In particular, observation statements are not to be identified with statements about "sense data" sometimes alleged to be exclusive objects of "direct experience." Thus, 'There was a total eclipse of the sun at Sabral in North Brazil on May 29, 1919,' and 'The switch was turned on yesterday in my office when the temperature of the room dropped to 50° F,' both count as observation statements in the present use of this designation. Observation statements may on occasion formulate initial and boundary conditions for a theory or law; they may also be employed to confirm or refute theories and laws.

   d. Many observation statements of a given science S describe the arrangement and behavior of apparatus required for conducting experiments in S or for testing various assumptions adopted in S. Accordingly, the assertion of such observation statements may tacitly involve the use of laws concerning characteristics of different sorts of instruments; some of these laws may not fall into the generally acknowledged province of S and may not be explained by any theory of S. For example, photographic equipment attached to telescopes is commonly employed in testing Newtonian gravitational theory, so that the construction of such apparatus, as well as the interpretation of data obtained with its help, takes for granted theories and experimental laws both of optics and chemistry. However, the general assumptions thus taken for granted do not belong to the science of mechanics; and Newtonian gravitational theory does not pretend to explain or to warrant optical and chemical laws. When cameras and telescopes are used in inquiries into mechanical phenomena, distinctions and laws are therefore "borrowed" from other special disciplines. We shall refer to such laws, which are *used* in a science S but are not established or explained within S itself, as "borrowed laws" of S.

Most sciences will also contain statements that are certifiable as logically true, such as those of logic and mathematics. Even if we ignore these, we have identified four major classes of statements that may occur in a science S, whether or not any degree of autonomy is claimed for it relative to other special disciplines: (a) the theoretical postulates of S, the theorems derivable from them, and the coordinating definitions associated with theoretical notions in the postulates or theorems; (b) the experimental laws of S; (c) the observation statements of S; and (d) the borrowed laws of S.

2. We come to the second formal point. Every statement of a science S can be analyzed as a linguistic structure, compounded out of more elementary expressions in accordance with tacit or explicit rules of construction. It will be assumed that, though these elementary expressions may be vague in varying degrees, they are employed unambiguously in S, with meanings fixed either by habitual usage or explicitly formulated rules. Some of the expressions will be locutions of formal logic, arithmetic, and other branches of mathematical analysis. We shall, however, be primarily concerned with the so-called "descriptive expressions," signifying what are generally regarded as "empirical" objects, traits, relations, or processes, rather than purely formal or logical entities. Although there are difficulties in developing a precise distinction between logical and descriptive expressions, these difficulties do not impinge upon the present discussion. Let us in any case consider the class D of descriptive expressions in S that do not occur in borrowed laws of S.

Many of the descriptive expressions of a science are simply taken over from the language of ordinary affairs, and retain their everyday meanings. This is frequently true for expressions occurring in observation statements, since a large fraction of the overt procedures employed even in carefully devised laboratory experiments can be described in the language of gross experience. On the other hand, other descriptive expressions may be specific to a given science; they may have a use restricted to highly specialized technical contexts; and the meanings assigned to them in that science may even preclude their being employed to describe matters identifiable either by direct or indirect observation. Descriptive expressions of this latter sort occur typically in the theoretical assumptions of a science.

It is often possible to explicate the meaning of an expression in D with the aid of other expressions in D supplemented by logical ones. Such explications can sometimes be supplied in the form of conventional *explicit* definitions, though usually more complicated techniques for fixing the meanings of terms are required. But whatever formal techniques of explication may be used, let us call the set of expressions in D which, with the help of purely logical locutions, suffice to explicate the meanings of all other expressions in D, the "primitive expressions" of S. There will always be at least one set P of primitive expressions, since, in the least favorable cases, when no descriptive expression can be explicated in terms of others, the set P will be identical with the class D. On the other hand, there may be more than one such set P, for, as is well known, expressions that are primitive in one context of analysis may lose their primitive status in another context; but this possibility does not affect the present discussion.

However, if S has a comprehensive theory as well as observation statements and experimental laws, the explication of an expression may proceed in either one of two directions that must be noted, since in general each direction involves the use of a distinctive set of primitives.

a. Let us designate as "observation expressions" those expressions in D that refer to things, properties, relations, and processes capable of being observed. The distinction between observation expressions and other descriptive ones is admittedly vague, especially since different degrees of stringency may be used in different contexts in deciding what matters are to count as observable ones. But, despite its vagueness, the distinction is useful and is unavoidable in both scientific inquiry and everyday practice. In any event, many explications aim at specifying the meanings of descriptive expressions in terms of observable ones. The program (advocated by Peirce and Bridgman among others) of fixing the meanings of terms by giving what are currently known as "operational definitions" for them appears to have explications of this sort for its objective. Let us call the set P_1 of observation expressions needed for explicating in this manner the maximum number of expressions in D, the "observation primitives" of S. For example, the meaning of 'temperature' is frequently explained in physics in terms of the volume expansions of liquids and gases or in terms of other observable behaviors of bodies; in such cases the explication of 'temperature' is given by way of observable primitives.

b. Let us suppose that S has a theory capable of explaining all the experimental laws of the science; and let us designate as the "theoretical expressions" of S the descriptive expressions employed in the theoretical postulates (exclusive of the coordinating definitions) and the theorems formally derivable from them. Many explications aim at specifying the meanings of expressions by way of theoretical ones; and we shall call the set P_2 of theoretical expressions needed for explicating in this way the maximum number of expressions in D, the "theoretical primitives" of S. For example, the meaning of 'temperature' is given a theoretical explication in the science of heat with the help of statements describing the Carnot cycle of heat transformations, and therefore in terms of such theoretical primitives as 'perfect nonconductors,' 'infinite heat reservoirs,' and 'infinitely slow volume expansions.'

As we have seen in Chapter 6, the question whether theoretical expressions are explicitly definable in terms of observable ones has been much debated. If theoretical expressions were always so definable, they could be eliminated in favor of observable ones, so that the distinction would have little point. However, although a negative answer to the question has not been demonstrably established, all the available evidence supports that answer. Indeed, there are good reasons for maintaining the stronger claim that theoretical expressions cannot in general be adequately explicated with the help of observation ones alone, even when forms of explication other than explicit definitions are employed. It is not necessary to adopt a position on these questions for the purposes of the present discussion. We must nevertheless not assume as a matter of course that the set of observation primitives P_1 is sufficient to explicate all the descriptive expressions D; and we must allow for the possibility that the class P of primitive expressions of S does not in general coincide with the class P_1. Accordingly, although 'temperature' is explicated in the science of heat both in terms of theoretical and of observation primitives, it does not follow that the word understood in the sense of the first explication is synonymous with 'temperature' construed in the sense of the second.

3. We can now turn to the third formal consideration on reduction. The primary and secondary sciences involved in a reduction generally have in common a large number of expressions (including statements) that are associated with the same meanings in both sciences. Statements certifiable in formal logic and mathematics are obvious illustrations of such common expressions, but there usually are many other descriptive ones as well. For example, many laws belonging to the science of mechanics, such as Hooke's law or the laws of the lever, also appear in the science of heat, if only as borrowed laws; and the latter science employs in its own experimental laws such expressions as 'volume,' 'pressure,' and 'work' in senses that coincide with the meanings of these words in mechanics. On the other hand, before its reduction the secondary sci-

ence generally uses expressions and asserts experimental laws formulated with their help which do not occur in the primary science, except possibly in the latter's classes of observation statements and borrowed laws. For example, the science of mechanics in its classical form does not count the Boyle-Charles' law as one of its experimental laws; nor does the term 'temperature' occur in the theoretical assumptions of mechanics, though the word may sometimes be employed in its experimental inquiries to describe the circumstances under which some law of the science is being used.

It is, however, of utmost importance to note that expressions belonging to a science possess meanings that are fixed by its *own* procedures of explication. In particular, expressions distinctive of a given science (such as the word 'temperature' as employed in the science of heat) are intelligible in terms of the rules or habits of usage of that branch of inquiry; and when those expressions are used in that branch of study, they must be understood in the senses associated with them in that branch, whether or not the science has been reduced to some other discipline. Sometimes, to be sure, the meaning of an expression in a science can be explicated with the help of the primitives (whether theoretical or observational) of some other science. For example, there are firm grounds for the assumption that the word 'pressure' as understood in thermodynamics is synonymous with the term 'pressure' as explicated by way of the theoretical primitives of mechanics. It nevertheless does not follow that in general every expression employed in a given science, in the sense specified by its own distinctive rules or procedures, is explicable in terms of the primitives of some other discipline.

With these preliminaries out of the way, we must now state the formal requirements that must be satisfied for the reduction of one science to another. As has already been indicated in this chapter, a reduction is effected when the experimental laws of the secondary science (and if it has an adequate theory, its theory as well) are shown to be the logical consequences of the theoretical assumptions (inclusive of the coordinating definitions) of the primary science. It should be observed that we are not stipulating that the borrowed laws of the secondary science must also be derivable from the theory of the primary science. However, if the laws of the secondary science contain terms that do not occur in the theoretical assumptions of the primary discipline (and this is the type of reduction to which we agreed earlier to confine the discussion), the logical derivation of the former from the latter is *prima facie* impossible. The claim that the derivation is impossible is based on the familiar logical canon that, save for some essentially irrelevant exceptions, no term can appear in the conclusion of a formal demonstration unless the term also appears in the premises.[3] Accordingly, when the laws of the secondary science do contain some term '*A*' that is absent from the theoretical assumptions of the primary science, there are two necessary formal conditions for the reduction of the former to the latter: (1) Assumptions of some kind must be introduced which postulate suit-

[3] Possible objections to this logical canon are based for the most part on the fact that, in view of some theorems in modern formal logic, a valid deductive argument can have a conclusion containing terms not occurring in the premises.

There are at least two laws in the sentential calculus (or logic of unanalyzed propositions) that permit the deduction of such conclusions. According to one of them, any statement of the form 'If S_1, then S_1 or S_2,' where S_1 and S_2 are any statements, is logically true, so that S_1 or S_2 is derivable from S_1. But since S_2 can be chosen arbitrarily, S_1 or S_2 can be made to contain terms not occurring in S_1. According to a second logical law, any statement of the form 'S_1, if and only if S_1 and (S_2 or not-S_2)' is logically true; hence 'S_1 and (S_2 or not-S_2)' is derivable from S_1, with the same general outcome as in the first case. However, it is clear that neither type of deductive step can yield the Boyle-Charles' law from the kinetic theory of gases. If it could (for example, by way of substituting this law for S_2 in the first of the two logical laws mentioned), then, since S_2 is entirely arbitrary, the deduction would also yield the contradictory of this law; and this cannot happen, unless the kinetic theory itself is self-contradictory. This argument is quite general and applies to other examples of reduction. Accordingly, insofar as reductions make use only of the logical laws of the sentential calculus in deducing statements of the secondary science from the theory of the primary science, it is sufficient to meet the objection to the logical canon mentioned in the text by emending the latter to read: In a valid deduction no term appears in the conclusion that does not occur in the premises, unless a term enters into the conclusion via logical laws of the sentential calculus, which permit the introduction of any *arbitrary* term into the conclusion.

However, there are other logical laws, developed in other parts of formal logic, that also sanction conclusions with terms not in the premises. Substitution for variables expressing universality is a familiar type of such inference. For example, although the premise "For any x, if x is a planet then x shines by reflected light" does not contain the term 'Mars,' the statement "If Mars is a planet, then Mars shines by reflected light" can be validly deduced from it. Another type of such inference is illustrated by the derivation from "All men are mortal" of the conclusion "All hungry men are hungry mortals." Nevertheless, an examination of the derivation of the Boyle-Charles' law reveals that the term 'temperature,' contained in this law but not in the kinetic theory, is not introduced into the derivation by way of any such universally valid deductive steps; and an argument, analogous to the one presented in the preceding paragraph of this note for the case of deductions in the sentential calculus, can be constructed to show that this must also be the case in the deduction of other laws, containing distinctive terms, of a secondary science that is reducible to some primary one. Accordingly, these various exceptions to the logical canon of the text can be ignored as not relevant to the matters under discussion.

A different objection to this canon is that, formal logic aside, we often do recognize arguments as valid even though they ostensibly violate the canon. Thus, 'John is a cousin of Mary' is said to follow from 'The uncle of John is the father of Mary,' and 'Smith's shirt is colored' is said to follow from 'Smith's shirt is red,' despite the fact that a term appears in each of the conclusions that is absent from the corresponding premise. However, these examples and others like them are essentially enthymematic inferences, with a tacit assumption either in the form of an explicit definition or some other kind of a priori statement. When these suppressed assumptions are made explicit, the examples no longer appear to be exceptions to the logical canon under examination.

able relations between whatever is signified by 'A' and traits represented by theoretical terms already present in the primary science. The nature of such assumptions remains to be examined; but without prejudging the outcome of further discussion, it will be convenient to refer to this condition as the "condition of connectability." (2) With the help of these additional assumptions, all the laws of the secondary science, including those containing the term 'A,' must be logically derivable from the theoretical premises and their associated coordinating definitions in the primary discipline. Let us call this the "condition of derivability."[4]

There appear to be just three possibilities as to the nature of the linkages postulated by these additional assumptions: (1) The first is that the links are *logical connections* between established meanings of expressions. The assumptions then assert 'A' to be logically related (presumably by synonymy or by some form of one-way analytical entailment) to a theoretical expression 'B' in the primary science. On this alternative, the meaning of 'A' as fixed by the rules or habits of usage of the secondary science must be explicable in terms of the established meanings of theoretical primitives in the primary discipline. (2) The second possibility is that the linkages are *conventions*, created by deliberate fiat. The assumptions are then coordinating definitions, which institute a correspondence between 'A' and a certain theoretical primitive, or some construct formed out of the theoretical primitives, of the primary science. On this alternative, unlike the preceding one, the meaning of 'A' is not being explicated or analyzed in terms of the meanings of theoretical primitives. On the contrary, if 'A' is an observation term of the secondary science, the assumptions in this case *assign* an experimental significance to a certain theoretical expression of the primary science, consistent with other such assignments that may have been previously made. (3) The third possibility is that the linkages are *factual* or *material*. The assumptions then are physical hypotheses, asserting that the occurrence of the state of affairs signified by a certain theoretical expression 'B' in the primary science is a sufficient (or necessary and sufficient) condition for the state of affairs designated by 'A.' It will be evident that in this case independent evidence must in principle be

---

[4] The condition of connectability requires that *theoretical* terms of the primary science appear in the statement of these additional assumptions. It would not suffice, for example, if these assumptions formulated an explication of 'A' by way of observation primitives of the primary science, even if the theoretical primitives could also be explicated by way of the observation primitives. For it would not thereby follow that 'A' could be explicated by way of the theoretical primitives. Thus, although 'uncle' and 'grandfather' are each definable in terms of 'male' and 'parent,' 'uncle' is not definable in terms of 'grandfather.' In consequence, the additional assumption would not contribute toward the fulfillment of the condition of derivability.

obtainable for the occurrence of each of the two states of affairs, so that the expressions designating the two states must have identifiably different meanings. On this alternative, therefore, the meaning of 'A' is not related analytically to the meaning of 'B.' Accordingly, the additional assumptions cannot be certified as true by logical analysis alone, and the hypothesis they formulate must be supported by empirical evidence.[5]

In the light of this discussion, let us now examine the derivation of the Boyle-Charles' law from the kinetic theory of gases. For the sake of simplicity let us also assume that the word 'temperature' is the only term in this law that does not occur in the postulates of that theory. However, as already noted, the deduction of the law from the theory depends on the additional postulate that the temperature of a gas is proportional to the mean kinetic energy of its molecules. Our problem is to decide on the status of this postulate and to determine which if any of the three types of linkage we have been discussing is asserted by the postulate.

For reasons mentioned in the first section of the present chapter, it is safe to conclude that 'temperature,' in the sense the word is employed in classical thermodynamics, is not synonymous with 'mean kinetic energy of molecules,' nor can its meaning be extracted from the meaning of the latter expression. Certainly no standard exposition of the kinetic theory of gases pretends to establish the postulate by analyzing the meanings of the terms occurring in it. The linkage stipulated by the postulate cannot therefore be plausibly regarded as a logical one.

But it is far more difficult to decide which of the remaining two types

---

[5] It follows that the condition of connectability is in general not sufficient for reduction and must be supplemented by the condition of derivability. Connectability would indeed assure derivability if, as has been rightly argued by John G. Kemeny and Paul Oppenheim ["On Reduction," *Philosophical Studies*, Vol. 7 (1956), p. 10], for every term 'A' in the secondary science but not in the primary one there is a theoretical term 'B' in the primary science such that A and B are linked by a biconditional: A if and only if B. If the linkage has this form, 'A' can be replaced by 'B' in any law L of the secondary science in which 'A' occurs, and so yield a warranted theoretical postulate L'. If L' is not itself derivable from the available theory of the primary science, the theory need only be augmented by L' to become a modified theory, but nonetheless a theory of the primary science. In any event, L will be deducible from a theory of the primary science with the help of the biconditionals. However, the linkage between A and B is not necessarily biconditional in form, and may for example be only a one-way conditional: If B, then A. But in this eventuality 'A' is not replaceable by 'B,' and hence the secondary science will not in general be deducible from a theory of the primary discipline. Accordingly, even if we waive the question whether a reduction is satisfactory when achieved by augmenting the theory of the primary science by a new postulate L' which is empirically confirmed but may contribute next to nothing to the explanatory power of the initial theory, connectability does not in general suffice to assure derivability. On the other hand, the condition of derivability is both necessary and sufficient for reduction, since derivability obviously entails connectivity. The condition of connectability is nevertheless stated separately, because of its importance in the analysis of reduction.

of linkage is asserted by the postulate, for there are plausible reasons favoring each of these alternatives. The argument in support of the claim that the postulate is simply a coordinating definition is essentially as follows: The kinetic theory of gases cannot be put to experimental test, unless rules of correspondence first associate some of its theoretical notions with experimental concepts. But the postulate itself cannot be subjected to experimental control. For, although the temperature of a gas can be determined by familiar laboratory procedures, there is apparently no way of ascertaining the mean kinetic energy of the hypothetical gas molecules—unless, indeed, the temperature is stipulated by fiat to be a measure of this energy. Accordingly, the postulate can be nothing other than one of the correspondence rules which institute an association between theoretical and experimental concepts.[6] On the other hand, the claim that the postulate is a physical hypothesis is also not an unfounded one; and, indeed, it is in this fashion that the postulate is introduced in many technical presentations of the subject. The major reason advanced for this claim is that, although the postulate cannot be tested by direct measurements on the mean kinetic energy of gas molecules, the value of this energy can nevertheless be ascertained indirectly, by calculation from experimental data on gases other than data obtained by measuring temperatures. In consequence, it does seem possible to determine experimentally whether the temperature of a gas is proportional to the mean kinetic energy of its molecules.

Despite appearances to the contrary, these alternative claims and supporting reasons for them are not necessarily incompatible. Indeed, the alternatives illustrate what is by now a familiar point—that the cognitive status of an assumption often depends on the mode adopted for articulating a theory in a particular context. The reduction of thermodynamics to mechanics can undoubtedly be so expounded that the additional postulates about the proportionality of temperature to the mean kinetic energy of gas molecules institutes what is at first the sole link between the theoretical notions of the primary science and experimental concepts of the secondary one. In such a context of exposition, the postulate cannot be subjected to experimental test but functions as a coordinating definition. However, different modes of exposition are also possible, in which coordinating definitions are introduced for other pairs of theoretical and experimental concepts. For example, one theoretical notion can be made to correspond to the experimental idea of viscosity, and another can be associated with the experimental concept of heat flow. In consequence, since the mean kinetic energy of gas molecules is related, by virtue of the assumptions of the kinetic theory, to these other

[6] Cf. Norman R. Campbell, *Physics, the Elements*, Cambridge, England, 1920, pp. 126ff.

theoretical notions, a connection may thus be indirectly established between temperature and kinetic energy. Accordingly, in such a context of exposition, it would make good sense to ask whether the temperature of a gas is proportional to the value of the mean kinetic energy of the gas molecules, where this value is calculated in some indirect fashion from experimental data other than that obtained by measuring the temperature of the gas. In this case the postulate would have the status of a physical hypothesis.

It is therefore not possible to decide in general whether the postulate is a coordinating definition or a factual assumption, except in some given context in which the reduction of thermodynamics to mechanics is being developed. This circumstance does not, however, wipe out the distinction between rules of correspondence and material hypotheses, nor does it destroy the importance of the distinction. But in any event, the present discussion does not require that a decision be made between these alternative interpretations of the postulate. The essential point in this discussion is that in the reduction of thermodynamics to mechanics a postulate connecting temperature and mean kinetic energy of gas molecules must be introduced, and that this postulate cannot be warranted by simply explicating the meanings of the expressions contained in it.

One objection to this central contention must be briefly considered. The redefinition of expressions with the development of inquiry, so the objection notes, is a recurrent feature in this history of science. Accordingly, though it must be admitted that in an earlier use the word 'temperature' had a meaning specified exclusively by the rules and procedures of thermometry and classical thermodynamics, it is *now* so used that temperature is "identical by definition" with molecular energy. The deduction of the Boyle-Charles' law does not therefore require the introduction of a further postulate, whether in the form of a coordinating definition or a special empirical hypothesis, but simply makes use of this definitional identity. This objection illustrates the unwitting double talk into which it is so easy to fall. It is certainly possible to redefine the word 'temperature' so that it becomes synonymous with 'mean kinetic energy of molecules.' But it is equally certain that on this redefined usage the word has a different meaning from the one associated with it in the classical science of heat, and therefore a meaning different from the one associated with the word in the statement of the Boyle-Charles' law. However, if thermodynamics is to be reduced to mechanics, it is temperature in the sense of the term in the classical science of heat which must be asserted to be proportional to the mean kinetic energy of gas molecules. Accordingly, if the word 'temperature' is redefined as suggested by the objection, the hypothesis must be invoked that the state of bodies described as 'temperature' (in the classical thermodynamical sense) is also characterized by 'temperature' in the redefined sense of the term. *This* hypothesis, however, will then be one that does not hold as a matter of definition, and will not be one for which logical necessity can be rightly claimed. Unless the hypothesis is adopted, it is not the Boyle-Charles' law which can be derived from the assumptions of the kinetic theory of gases. What is derivable without the hypothesis is a sentence similar in syntactical structure to the standard formulation of that law, but possessing a sense that is unmistakably different from what the law asserts.

III.   *Nonformal Conditions for Reduction*

We must now turn to features of reduction that are not primarily formal, though some of them have already been touched upon in passing.

1.   The two formal conditions for reduction discussed in the previous section do not suffice to distinguish trivial from noteworthy scientific achievements. If the sole requirement for reduction were that the secondary science is logically deducible from arbitrarily chosen premises, the requirement could be satisfied with relatively little difficulty. In the history of significant reductions, however, the premises of the primary science are not *ad hoc* assumptions. Accordingly, although it would be a far too strong condition that the premises must be known to be true, it does seem reasonable to impose as a nonformal requirement that the theoretical assumptions of the primary science be supported by empirical evidence possessing some degree of probative force. The problems connected with the logic of weighing evidence are difficult and at many crucial points still unsettled. However, the issues raised by these problems are not exclusively relevant to the analysis of reduction; and, except for some brief comments especially pertinent to the reduction of thermodynamics to mechanics, we shall not at this place examine the notion of adequate evidential support.

The evidence for the several assumptions of the kinetic theory of gases comes from a variety of inquiries, only a fraction of which fall into the domain of thermodynamics. Thus, the hypothesis of the molecular constitution of matter was supported by quantitative relations exhibited in chemical interactions even before thermodynamics was reduced to mechanics; and it was also confirmed by a number of laws in molar physics not primarily about thermal properties of bodies. The adoption of this hypothesis for the new task of accounting for the thermal behavior of gases was therefore in line with the normal strategy of the science to exploit on a new front ideas and analogies found to be fruitful elsewhere. Similarly, the axioms of mechanics, constituting the most general parts of the premises in the primary science to which thermodynamics is reduced, are supported by evidence from many fields quite distinct from the study of gases. The assumption that these axioms also hold for the hypothetical molecular components of gases thus involved the extrapolation of a theory from domains in which it was already well confirmed into another domain postulated to be homogeneous in important respects with the former ones. But the point having greatest weight in this connection is that the combined assumptions of the primary science to which the science of heat was reduced have made it possible to incorporate into a unified system many apparently unrelated laws of the science of heat as well as of other parts of physics. A number of gas laws had of course been established before the reduction. However, some of these laws were only approximately valid for gases not satisfying certain narrowly restrictive conditions; and most of the laws, moreover, could be affirmed only as so many independent facts about gases. The reduction of thermodynamics to mechanics altered this state of affairs in significant ways. It paved the way for a reformulation of gas laws so as to bring them into accord with the behaviors of gases satisfying less restrictive conditions; it provided leads to the discovery of new laws; and it supplied a basis for exhibiting relations of systematic dependence among gas laws themselves, as well as between gas laws and laws about bodies in other states of aggregation.

This last point deserves a brief elaboration. If the Boyle-Charles' law were the sole experimental law deducible from the kinetic theory of gases, it is unlikely that this result would be counted by most physicists as weighty evidence for the theory. They would probably take the view that nothing of significance is achieved by the deduction of only this one law. For prior to its deduction, so they might maintain, this law was known to be in good agreement with the behavior of only "ideal" gases, that is, those at temperatures far above the points at which the gases liquefy; and by hypothesis, nothing further follows from the theory as to the behavior of gases at lower temperatures. Moreover, physicists would doubtless call attention to the telling point that even the deduction of this law can be effected only with the help of a special postulate connecting temperature with the energy of gas molecules—a postulate that, under the circumstances envisaged, has the status of an *ad hoc* assumption, supported by no evidence other than the evidence warranting the Boyle-Charles' law itself. In short, if this law were the sole experimental consequence of the kinetic theory, the latter would be dead wood from which only artificially suspended fruit could be gathered.

In actual fact, however, the reduction of thermodynamics to the kinetic theory of gases achieves much more than the deduction of the Boyle-Charles' law. There is available other evidence that counts heavily with most physicists as support for the theory and that removes from the special postulate connecting temperatures and molecular energy even the appearance of arbitrariness. Indeed, two related sets of considerations make the reduction a significant scientific accomplishment. One set consists of experimental laws, deduced from the theory, which have not been previously established or which are in better agreement with a wider range of facts than are laws previously accepted. For example, the Boyle-Charles' law holds only for ideal gases and is deducible from the kinetic theory when some of the less general assumptions of the kinetic theory have the limiting form corresponding to a gas being an ideal gas. However, these special assumptions can be replaced by others without modifying the fundamental ideas of the theory, and in particular by assumptions less simple than those introduced for ideal gases. Thus, instead of the stipulations with the aid of which the Boyle-Charles' law is derivable from the theory, we can assume that the dimensions of gas molecules are not negligible when compared to the mean distance between them, and that in addition to forces of impact there are also cohesive forces acting upon them. It is then possible to deduce from the theory employing these more complex special assumptions the van der Waals' law for gases, which formulates more adequately than does the Boyle-Charles' law the behavior of both ideal and nonideal gases. In general, therefore, for a reduction to mark a significant intellectual advance, it is not enough that previously established laws of the secondary science be represented within the theory of the primary discipline. The theory must also be fertile in usable suggestions for developing the secondary science, and must yield theorems referring to the latter's subject matter which augment or correct its currently accepted body of laws.

The second set of considerations in virtue of which the reduction of thermodynamics to mechanics is generally regarded as an important achievement consists of the intimate and frequently surprising relations of dependence that can thereby be shown to obtain between various experimental laws. An obvious type of such dependence is illustrated when laws, hitherto asserted on independent evidential grounds, are deducible from an integrated theory as a consequence of the reduction. Thus, both the second law of thermodynamics (according to which the entropy of a closed physical system never diminishes) as well as the Boyle-Charles' law are derivable from statistical mechanics, although in classical thermodynamics these laws are stated as independent primitive assumptions. In some ways a more impressive and subtler type of dependence is illustrated when some numerical constant appearing in different experimental laws of the secondary science is exhibited as a definite function of theoretical parameters in the primary discipline—an outcome that is particularly striking when congruous numerical values can be calculated for those parameters from experimental data obtained in independent lines of inquiry. Thus, one of the postulates of the kinetic theory is that, under standard conditions of temperature and pressure, equal volumes of a gas contain an equal number of molecules, irrespective of the chemical nature of the gas. The number of molecules in a liter of gas under standard conditions is thus the same for all gases, and is known as Avogadro's number. Moreover, a certain constant appearing in several gas laws (among others, in the Boyle-Charles' law and the laws of specific heats) can be shown to be a function of this number and other theoretical parameters. On the other hand, Avogadro's number can be calculated in alternative ways from experimental data gathered in different kinds of inquiry, e.g., from measurements in the study of thermal phenomena, of Brownian movements, or of crystal structure; and the values obtained for the number from each of these diverse sets of data are in good agreement with one another. Accordingly, apparently independent experimental laws (including thermal ones) are shown to involve a common invariant component, represented by a theoretical parameter that in turn becomes firmly tied to several kinds of experimental data. In consequence, the reduction of thermodynamics to kinetic theory not only supplies a unified explanation for the laws of the former discipline; it also integrates these laws so that directly relevant evidence for any one of them can serve as indirect evidence for the others, and so that the available evidence for any of the laws cumulatively supports various theoretical postulates of the primary science.

2. These general comments on the considerations that determine whether a reduction is a significant advance in the organization of knowledge or only a formal exercise, and on the character of the evidence that actually supports the kinetic theory, direct attention to an important feature of sciences in active development. As has already been suggested, different branches of science may sometimes be delimited on the basis of the theories used as explanatory premises and leading principles in their respective domains. Nevertheless, theories do not as a rule remain unaltered with the progress of inquiry; and the history of science provides many examples of special branches of knowledge becoming reorganized around new types of theory. Moreover, even if a discipline continues to retain the most general postulates of some theoretical system, the less general ones are often modified or are augmented by others as fresh problems arise.

Accordingly, the question whether a given science is reducible to another cannot in the abstract be usefully raised without reference to some particular stage of development of the two disciplines. Questions about reducibility can be profitably discussed only if they are made definite by specifying the established content at a given date of the sciences under consideration. Thus, no practicing physicist is likely to take seriously the claim that the contemporary science of nuclear physics is reducible to some variant of classical mechanics—even if the claim should be accompanied by a formal deduction of the laws of nuclear physics from admittedly purely mechanical assumptions—unless these assumptions are supported by adequate evidence available at the time the claim is made, and also possess at that time the heuristic advantages normally expected of the theory belonging to a proposed primary science. Again, it is one thing to say that thermodynamics is reducible to mechanics when the latter counts among its recognized postulates assumptions (including statistical ones) about molecules and their modes of action; it is quite a different thing to claim that thermodynamics is reducible to a science of mechanics that does not countenance such assumptions. In particular, though contemporary thermodynamics is undoubtedly reducible to a statistical mechanics postdating 1866 (the year in which Boltzmann succeeded in giving a statistical interpretation for the second law of thermodynamics with the help of certain statistical hypotheses), that secondary science is not reducible to the mechanics of 1700. Similarly, certain parts of nineteenth-century chemistry (and perhaps the whole of this science) is reducible to post-1925 physics, but not to the physics of a hundred years ago.

Moreover, the possibility should not be ignored that little if any new knowledge or increased power for significant research may actually be gained from reducing one science to another at certain periods of their development, however great may be the potential advantages of such reduction at some later time. Thus, a discipline may be at a stage of active growth in which the imperative task is to survey and classify the extensive and diversified materials of its domain. Attempts to reduce the discipline to another (perhaps theoretically more advanced) science, even if successful, may then divert needed energies from what are the crucial problems at this period of the discipline's expansion, without being compensated by effective guidance from the primary science in the conduct of further research. For example, at a time when the prime need of botany is to establish a systematic typology of existing plant life, the discipline may reap little advantage from adopting a physicochemical theory of living organisms. Again, although one science may be reducible to another, the secondary discipline may be progressively solving its own special class of problems with the help of a theory expressly devised for dealing with the subject matter of that discipline. As a basis for attacking these problems, this less inclusive theory may well be more satisfactory than the more general theory of the primary science—perhaps because the primary science requires the use of techniques too refined and cumbersome for the subjects under study in the secondary science, or because the initial conditions needed for applying it to these subjects are not available, or simply because its structure does not suggest fruitful analogies for handling these problems. For example, even if biology were reducible to the physics of current quantum mechanics, at the present stage of biological science the gene theory of heredity may be a more satisfactory instrument for exploring the problems of biological inheritance than would be the quantum theory. An integrated system of explanation by some inclusive theory of a primary science may be an eventually realizable intellectual ideal. But it does not follow that this ideal is best achieved by reducing one science to another with an admittedly comprehensive and powerful theory, if the secondary science at that stage of its development is not prepared to operate effectively with this theory.

Much controversy over the interrelations of the special sciences, and over the limits of the explanatory power of their theories, neglects these elementary considerations. The irreducibility of one science to another (for example, of biology to physics) is sometimes asserted absolutely, and without temporal qualifications. In any event, arguments for such claims often appear to forget that the sciences have a history, and that the reducibility (or irreducibility) of one science to another is contingent upon the specific theory employed by the latter discipline at some stated time. On the other hand, converse claims maintaining that some particular science is reducible to a favored discipline also do not always give sufficient heed to the fact that the sciences under consideration must be at appropriately mature levels of development if the reduction is to be of scientific importance. Such claims and counterclaims are perhaps most charitably construed as debates over what is the most promising direction systematic research should take at some given stage of a science. Thus biologists who insist on the "autonomy" of their science and who reject *in toto* so-called "mechanistic theories" of biological phenomena sometimes appear to adopt these positions because they believe that in the present state of physical and biological theory biology stands to gain more by carrying on its investigations in terms of distinctively biological categories than by abandoning them in favor of modes of analysis typical of modern physics. Analogously, mechanists in biology can often be understood as recommending the reduction of biology to physics, because in their view biological problems can now be handled more effectively within the framework of current physical theories than with the help of any purely biological ones. As we shall see in the following chapter, however, this is not the way that the issues are usually stated by those taking sides in such debates. On the contrary, largely because of a failure to note that claims concerning the reducibility or irreducibility of a science must be temporally qualified, questions that at bottom relate to the strategy of research, or to the logical relations between sciences as constituted at a certain time, are commonly discussed as if they were about some ultimate and immutable structure of the universe.

3. Throughout the present discussion stress has been placed on conceiving the reduction of one science to another as the deduction of one set of empirically confirmable *statements* from another such set. However, the issues of reduction are frequently discussed on the supposition that reduction is the derivation of the *properties* of one subject matter from the properties of another. Thus a contemporary writer maintains that psychology is demonstrably an autonomous discipline with respect to physics and physiology, because "a headache is not an arrangement or rearrangement of particles in one's cranium," and "our sensation of violet *is* not a change in the optic nerve." Accordingly, though the mind is said to be "connected mysteriously" with the physical processes, "it cannot be reduced to those processes, nor can it be explained by the laws of those processes." [7] Another recent writer, in presenting the case for the occurrence of "genuine novelties" in inorganic nature, declares that "it is an error to assume that *all* the properties of a compound can be deduced solely from the nature of its elements." In a similar vein, a third contemporary author asserts that the characteristic behavior of a chemical compound, such as water, "*could* not, even in theory, be deduced from the most complete knowledge of the behavior of its components, taken separately or in other combinations, and of their properties and arrangements in this whole." [8] We must now briefly indicate that the conception of reduction as the deduction of *properties* from other properties is potentially misleading and generates spurious problems.

The conception is misleading because it suggests that the question of whether one science is reducible to another is to be settled by inspecting the "properties" or alleged "natures" of things rather than by investigating the logical consequences of certain explicitly formulated *theories* (that is, systems of statements). For the conception ignores the crucial point that the "natures" of things, and in particular of the "elementary constituents" of things, are not accessible to direct inspection and that we cannot read off by simple inspection what it is they do or do not imply. Such "natures" must be stated as a theory and are not the objects of observation; and the range of the possible "natures" which

---

[7] Brand Blanshard, "Fact, Value and Science," in *Science and Man* (ed. by Ruth N. Anshen), New York, 1942, p. 203.

[8] C. D. Broad, *The Mind and Its Place in Nature*, London, 1925, p. 59.

chemical elements may possess is as varied as the different theories about atomic structures that we can devise. Just as the "fundamental nature" of electricity used to be stated by Maxwell's equations, so the fundamental nature of molecules and atoms must be stated as an explicitly articulated theory about them and their structures. Accordingly, the supposition that, in order to reduce one science to another, some properties must be deduced from certain other properties or "natures" converts what is eminently a logical and empirical question into a hopelessly irresolvable speculative one. For how can we discover the "essential natures" of the chemical elements (or of anything else) except by constructing theories which postulate definite characteristics for these elements, and then controlling the theories in the usual way by confronting consequences deduced from the theories with the outcome of appropriate experiments? And how can we know in advance that no such theory can ever be constructed which will permit the various laws of chemistry to be derived systematically from it?

Accordingly, whether a given set of "properties" or "behavioral traits" of macroscopic objects can be explained by, or reduced to, the "properties" or "behavioral traits" of atoms and molecules is a function of whatever theory is adopted for specifying the "natures" of these elements. The deduction of the "properties" studied by one science from the "properties" studied by another may be impossible if the latter science postulates these properties in terms of one theory, but the reduction may be quite feasible if a different set of theoretical postulates is adopted. For example, the deduction of the laws of chemistry (e.g., of the law that under certain conditions hydrogen and oxygen combine to form a stable compound commonly known as water, which in turn exhibits certain definite modes of behavior in the presence of other substances) from the physical theories of the atom accepted fifty years ago was rightly held to be impossible. But what was impossible relative to one theory need not be impossible relative to another physical theory. The reduction of various parts of chemistry to the quantum theory of atomic structure now appears to be making steady if slow headway; and only the stupendous mathematical difficulties involved in making the relevant deductions from the quantum theoretical assumptions seem to stand in the way of carrying the task much further along. Again, to repeat in the present context a point already made in another, if the "nature" of molecules is stipulated in terms of the theoretical primitives of classical statistical mechanics, the reduction of thermodynamics is possible only if an additional postulate is introduced that connects temperature and kinetic energy. However, the impossibility of the reduction without such special hypothesis follows from purely formal considerations, and not from some alleged ontological hiatus between the mechanical and the thermodynamical. Laplace was thus demonstrably in error when he believed that a Divine Intelligence could foretell the future in every detail, given the instantaneous positions and momenta of all material particles as well as the magnitudes and directions of the forces acting between them. At any rate, Laplace was in error if his Divine Intelligence is assumed to draw inferences in accordance with the canons of logic, and is therefore assumed to be incapable of the blunder of asserting a statement as a conclusion of an inference when the statement contains terms not occurring in the premises.

However this may be, the reduction of one science to a second—e.g., thermodynamics to statistical mechanics, or chemistry to contemporary physical theory—does not wipe out or transform into something insubstantial or "merely apparent" the distinctions and types of behavior which the secondary discipline recognizes. Thus, if and when the detailed physical, chemical, and physiological conditions for the occurrence of headaches are ascertained, headaches will not thereby be shown to be illusory. On the contrary, if in consequence of such discoveries a portion of psychology will be reduced to another science or to a combination of other sciences, all that will have happened is that an explanation will have been found for the occurrence of headaches. But the explanation that will thus become available will be of essentially the same sort as those obtainable in other areas of positive science. It will not establish a logically necessary connection between the occurrence of headaches and the occurrence of certain events or processes specified by physics, chemistry, and physiology. Nor will it consist in establishing the synonymy of the term 'headache' with some expression defined by means of the theoretical primitives of these disciplines. It will consist in stating the conditions, formulated by means of these primitives, under which, and as a matter of sheer contingent fact, a determinate psychological phenomenon takes place.

IV. *The Doctrine of Emergence*

The analysis of reduction is intimately relevant to a number of currently debated theses in general philosophy, especially the doctrine known as "emergent evolution" or "holism." Indeed, some results of that analysis have already been applied in the preceding section of this chapter to some of the issues raised by the doctrine of emergence. We shall now examine this doctrine more explicitly, in the light supplied by the discussion of reduction.

The doctrine of emergence is sometimes formulated as a thesis about the hierarchical organization of things and processes, and the consequent occurrence of properties at "higher" levels of organization which are not predictable from properties found at "lower" levels. On the other hand, the doctrine is sometimes stated as part of an evolutionary cosmogony, according to which the simpler properties and forms of organization already in existence make contributions to the "creative advance" of nature by giving birth to more complex and "irreducibly novel" traits and structures. In one of its forms, at any rate, emergent evolution is the thesis that the present variety of things in the universe is the outcome of a progressive development from a primitive stage of the cosmos containing only undifferentiated and isolated elements (such as electrons, protons, and the like), and that the future will continue to bring forth unpredictable novelties. This evolutionary version of the emergence doctrine is not entailed by the conception of emergence as irreducible hierarchical organization, and the two forms of the doctrine must be distinguished. We shall first consider emergence as a thesis about the nonpredictability of certain characteristics of things, and subsequently examine briefly emergence as a temporal, cosmogonic process.

1.   Although emergence has been invoked as an explanatory category most frequently in connection with social, psychological, and biological phenomena, the notion can be formulated in a general way so as to apply to the inorganic as well. Thus, let O be some object that is constituted out of certain elements $a_1, \ldots, a_n$ standing to each other in some complex relation R; and suppose that O possesses a definite class of properties P, while the elements of O possess properties belonging to the classes $A_1, \ldots, A_n$ respectively. Although the elements are numerically distinct, they may not all be distinct in kind; moreover, they may enter into relations with one another (or with other elements not parts of O) that are different from R, to form complex wholes different from O. However, the occurrence of the elements $a_1, \ldots, a_n$ in the relation R is by hypothesis the necessary and sufficient condition for the occurrence of O characterized by the properties P.

Let us next assume what proponents of the doctrine of emergence call "complete knowledge" concerning the elements of O: we know *all* the properties the elements possess when they exist "in isolation" from one another; and we also know *all* the properties exhibited by complexes other than O that are formed when some or all of these elements stand to each other (or to additional elements) in relations other than R, as well as *all* the properties of the elements in these complexes. According to the doctrine of emergence, two cases must be distinguished. In the first case, it is possible to predict (that is, deduce) from such complete knowledge that, if the elements $a_1, \ldots, a_n$ occur in the relation R, then the object O will be formed and will possess the properties P. In the second case, there is at least one property P_e in the class P such that, despite complete knowledge of the elements, it is impossible to predict from this knowledge that, if the elements stand to each other in relation R, then an object O possessing P_e will be formed. In the latter case, the object O is an "emergent object" and P_e an "emergent property" of O.

It is this form of the emergence doctrine that underlies the passage from Broad cited in the preceding section of this chapter (page 364). Broad illustrates this version of emergence as follows:

> Oxygen has certain properties and Hydrogen has certain other properties. They combine to form water, and the proportions in which they do this is fixed. Nothing that we know about Oxygen itself or in its combination with anything but Hydrogen could give us the least reason to suppose that it could combine with Hydrogen at all. Nothing that we know about Hydrogen by itself or in its combination with anything but Oxygen, could give us the least reason to expect that it would combine with Oxygen at all. And most of the chemical and physical properties of water have no known connexion, either quantitative or qualitative, with those of Oxygen and Hydrogen. Here we have a clear instance where, so far as we can tell, the properties of a whole composed of two constituents could not have been predicted from a knowledge of those properties taken separately, or from this combined with a knowledge of the properties of other wholes which contain these constituents.[9]

There are several issues raised by the present version of the doctrine of emergence, though most of them have already been touched upon in the preceding discussion of reduction and can be settled on the basis of considerations which were introduced there.

a.   The supposition underlying the notion of emergence is that, although it is possible in some cases to deduce the properties of a whole from the properties of its constituents, in other cases it is not possible to do so. We have seen, however, that both the affirmative and negative parts of this claim rest upon incomplete and misleading formulations of the actual facts. It is indeed impossible to deduce the properties of water (such as viscosity or translucency) from the properties of hydrogen alone (such as that it is in a gaseous state under certain conditions of pressure and temperature) or of oxygen alone, or of other compounds containing these elements as constituents (such as that hydrofluoric acid dissolves glass). But frequent claims to the contrary notwithstanding, it is also impossible to deduce the behavior of a clock merely from the properties and organization of its constituent parts. However, the deduction is impossible for the same reasons in both cases. It is not *properties*, but *statements* (or propositions) which can be deduced. Moreover, statements about properties of complex wholes can be deduced from statements about their constituents only if the premises contain a suitable *theory* concerning these constituents—one which makes it possible to analyze the behavior of such wholes as "resultants" of the assumed behaviors of the constituents. Accordingly, all descriptive expressions occurring in a statement that is allegedly deducible from the theory must also occur among the expressions used to formulate the theory or the assumptions adjoined to the theory when it is applied to specialized circumstances. Thus a statement like 'Water is translucent' cannot indeed be deduced from any set of statements about hydrogen and oxygen which do not contain the expressions 'water' and 'translucent'; but this impossibility derives entirely from purely formal considerations and is relative to the special set of statements adopted as premises in the case under consideration.

b. It is clear, therefore, that to say of a given property that it is an "emergent" is to attribute to it a character which the property may possess relative to one theory or body of assumptions but may not possess relative to some other theory. Accordingly, the doctrine of emergence (in the sense now under discussion) must be understood as stating certain *logical* facts about formal relations between statements rather than any experimental or even "metaphysical" facts about some allegedly "inherent" traits of *properties* of objects.

It is worth repeating in this connection, and particularly when the constituents of complex wholes are assumed to be submicroscopic particles and processes, that the "properties" of such constituents cannot be ascertained by inspection and their "structure" cannot be learned by any form of "direct perception." What these properties and structures are can be formulated only by way of some *theory*, which postulates the existence of those constituents and assumes various characteristics for them. It is patent, moreover, that the theory is subject to indefinite modifications in the light of macroscopic evidence. Accordingly, the question whether a given property of compounds can be predicted from the properties of their atomic constituents cannot be settled by considerations concerning alleged "inherent natures" that atoms are antecedently known to possess. For while one theory of atomic structure may be unequal to the task of predicting a given property, another theory postulating a different structure for atoms may make it possible to do so.

This view of the question is supported by the history of atomic theory. The ancient atomic theory of matter was revived by Dalton in the first quarter of the nineteenth century in order to account systematically for a limited range of chemical facts—initially, facts about constancies in the ratios of combining weights of substances participating in chemical reactions. Dalton's form of the theory postulated relatively few properties for atoms, and his theory was incapable of explaining many features of chemical transformations; for example, it did not account for chemical valence or for thermal changes manifested in chemical transformations. Eventually, however, Dalton's theory was modified, so that an increasing number and variety of laws, dealing with optical, thermal, and electromagnetic as well as chemical phenomena, could be explained by its later variants. But with this series of modifications of the theory, the conception of the "intrinsic nature" of atoms was also transformed; for each variant of the theory—more precisely, each theory in a certain series of theoretical constructions having a number of broad assumptions in common—postulated (or "defined") distinctive kinds of submicroscopic components for macroscopic objects, with distinctive "natures" for the components in each case. Accordingly, the "atoms" of Democritus, the "atoms" of Dalton, and the "atoms" of modern physicochemical theory are quite different sorts of particles; and they can be subsumed under the common name of "atom" chiefly because there are important analogies between the various theories that define them.

We must therefore not be misled by the convenient habit of thinking of the various atomic theories as representing a progress in our knowledge concerning a fixed set of submicroscopic objects. This way of describing the historical succession of atomic theories easily generates the belief that atoms can be said to exist and to have ascertainable "inherent natures," independent of any particular theory that postulates the existence of atoms and prescribes what properties they possess. In point of fact, however, to maintain that there are atoms having some definite set of characteristics is to claim that a certain theory about the constitution of physical objects is warranted by experimental evidence. The succession of atomic theories propounded in the history of science may indeed represent not only advances in knowledge concerning the order and connection of macroscopic phenomena, but also a progressively more adequate understanding of the atomic constitution of physical things. It nevertheless does not follow that, apart from some particular atomic theory, it is possible to assert just what can or cannot be predicted from the "natures" of atomic particles.

In any event, it is certainly the case that properties of compounds not predictable from certain older theories of atomic structure (e.g., the chemical and optical properties of the stable substance formed when hydrogen and oxygen combine under certain conditions) *can* be predicted from the current electronic theory of the composition of atoms. It therefore follows that an elliptic formulation is being employed when it is claimed that a given property of a compound is an "emergent" one. For, although a property may indeed be an emergent trait relative to some given theory, it need not be emergent relative to some different theory.

c. However, while it is an error to claim that a given property is "inherently" or "absolutely" an emergent trait, it is equally an error to maintain that in characterizing a trait as an emergent we are only baptizing our ignorance. It has been argued, for example, that

it may be that no physical-chemist could have predicted all the properties of H_2O before having studied it, and yet it seems probable that this incapacity to predict is only an expression of ignorance of the nature of H and O. If, on their combination, H and O yield water, presumably they contain in some sense the potentiality of forming water. In fact it is of the essence of Emergent Evolution that nothing new is added from without, that 'emergence' is the consequence of new kinds of relatedness between existents. The presumption is, then, that with sufficient knowledge of the components, highly probable predictions of the properties of water could have been made. In fact, chemists have successfully predicted the properties of compounds they have never observed and have proceeded to produce these 'emergents.' They have even predicted the existence and the properties of elements which had not been observed.[10]

Objections of this sort miss the force of the doctrine of emergence and appear to deny even what is demonstrably sound in it. In the first place, the doctrine employs the phrase 'to predict' in the sense of 'to deduce with strict logical rigor.' A proponent of emergence could readily admit that an allegedly emergent property might be foretold, whether invariably or only occasionally, by some happy insight or fortunate guess, but he would not thereby be compelled to surrender his claim that the property in question cannot be *predicted*. In the second place, it is possible to show that in some cases a given property cannot be predicted from certain other properties—more strictly, that a given statement about the occurrence of a designated property cannot be deduced from a specified set of other statements. For it may be possible to demonstrate with the help of established logical techniques that the statement about the first property is *not entailed* by the statements about the other properties; and such a demonstration is easily produced, especially when the former statement contains expressions that do not appear in the latter class of statements. Third and finally, our alleged "ignorance" or "incomplete knowledge" concerning the "natures" of atoms is entirely irrelevant to the issue at stake. For that issue is the simple one whether a given statement is deducible from a *given* set of statements, and not whether the statement is deducible from some *other* set of statements. As we have

[10] William McDougall, *Modern Materialism and Emergent Evolution*, New York, 1929, p. 129.

already seen, when we are said to improve or enlarge our knowledge concerning the "nature of H and O," what we are doing in effect is replacing one theory about H and O with another theory; and the fact that H and O combine to form water can be deduced from the second theory does not contradict the fact that the statement cannot be deduced from the initial set of premises. As was noted in discussing the reduction of thermodynamics to mechanics, the Boyle-Charles' law cannot be deduced from the assumptions of statistical mechanics unless a postulate is added relating the term 'temperature' to the expression 'mean kinetic energy of molecules.' This postulate cannot itself be deduced from statistical mechanics in its classical form; and this fact—that a postulate (or something equivalent to it) must be added to statistical mechanics as an independent assumption if the gas law is to be deduced —illustrates what is perhaps the central thesis in the doctrine of emergence as we have been interpreting it.

d. We have thus admitted the essential correctness of the doctrine of emergence when construed as a thesis concerning the *logical relation* between certain statements. It should be noted, however, that the doctrine so understood has a far wider range of application than proponents of emergence usually maintain. The doctrine has been urged for the most part in connection with chemical, biological, and psychological properties because these properties characterize systems at "higher levels" of organization and are allegedly "emergents" relative to properties occurring at "lower levels." Indeed, the doctrine is often advanced in opposition to the supposedly universalistic claims of "mechanical explanations," since, if some properties are in fact emergents, their occurrence is held to be inexplicable in "mechanical" terms. The truth of the doctrine of emergence is therefore sometimes believed to set limits to the science of mechanics, in which the principle of composition of forces is a warranted principle of analysis, and to differentiate from mechanics other systems of explanation in which that principle does not hold.[11] Accordingly, proponents of the doctrine often seem to suggest, if they do not explicitly maintain, that there are no emergent properties within the province usually assigned to mechanics or possibly even within the domain of physics; and the commonly cited example of a nonemergent property is the behavior of a clock, which is supposedly predictable from a knowledge of the properties and organization of its constituent cogs and springs.

But the logical point constituting the core of the doctrine of emer-

[11] Cf. the distinction drawn by Mill between the "mechanical" and the "chemical" modes of the "conjunct action of causes," which is the classical source of the doctrine of emergence. J. S. Mill, *A System of Logic*, London, 1879, Book 3, Chap. 6.

gence is applicable to all areas of inquiry and is as relevant to the analysis of explanations within mechanics and physics generally as it is to discussions of the laws of other sciences. The above discussion of the reduction of thermodynamics to mechanics makes this quite evident. But for the sake of additional clarity and emphasis, consider the clock example. It is well to note that the "behavior" of the clock which is predictable on the basis of mechanics is only that phase of its behavior which can be characterized entirely in terms of the primitive ideas of mechanics —for example, the behavior constituted by the motion of the clock's hands. Any phase of its behavior that cannot be brought within the scope of those ideas—for example, behavior consisting in variation in the clock's temperature or in changes in magnetic forces that may be generated by the relative motions of the parts of the clock—is not explained or predicted by mechanical theory. However, it appears that nothing but arbitrary custom stands in the way of calling these "nonmechanical" features of the clock's behavior "emergent properties" relative to mechanics. On the other hand, such nonmechanical features are certainly explicable with the help of theories of heat and magnetism, so that, relative to a wider class of theoretical assumptions, the clock may display no emergent traits.

Proponents of the doctrine of emergence are sometimes inclined to make a special point of the fact that the occurrence of so-called "secondary qualities" cannot be predicted by physical theories. For example, it has been argued that, from a complete knowledge of the microscopic structure of atoms, a mathematical archangel might be able to predict that nitrogen and hydrogen would combine when an electric discharge is passed through a mixture of the two, and would form water-soluble ammonia gas. However, though the archangel might be able to deduce what the exact microscopic structure of ammonia must be,

> he would be totally unable to predict that a substance with this structure must smell as ammonia does when it gets into the human nose. The utmost that he could predict on this subject would be that certain changes would take place in the mucous membrane, the olfactory nerves and so on. But he could not possibly know that these changes would be accompanied by the appearance of a smell in general or of the peculiar smell of ammonia in particular, unless someone told him so or he had smelled it for himself.[12]

But this claim is at best a truistic one, and can be affirmed with the same warrant for physical (or "primary") qualities of things as it can for secondary qualities. It is undoubtedly the case that a theory of chemistry that in its formulations makes no use of expressions referring to olfactory properties of substances cannot predict the occurrence of

[12] Broad, *op. cit.*, p. 71.

smells. But it cannot do so for the same reason that mechanics cannot account for optical or electrical properties of matter—namely that, when a deduction is made formally explicit, no statement employing a given expression can be logically derived from premises that do not also contain the expression. Accordingly, if a mathematical archangel is indeed incapable of predicting smells from a knowledge of the microscopic structure of atoms, this limitation in his powers is simply a consequence of the fact that the logical conditions for deducibility are the same for archangels as they are for men.

2.   Let us now briefly consider the doctrine of emergence as an evolutionary cosmogony, whose primary stress is upon the alleged 'novelty' of emergent qualities. The doctrine of emergent evolution thus maintains that the variety of individuals and their properties that existed in the past or occur in the present is not complete, and that qualities, structures, and modes of behavior come into existence from time to time the like of which has never been previously manifested anywhere in the universe. Thus, according to one formulation of the doctrine, an emergent evolution is said to have taken place if, when the present state of the world (called "Ph.N.") is compared with any prior phase (called "Ph.A."), one or more of the following features lacking in Ph.A. can be shown to be present in Ph.N.:

> (1) Instances of some general type of change . . . common to both phases (e.g., relative motion of particles), of which instances the manner or condition of occurrence could not be described in terms of, nor predicted from, the laws which would have been sufficient for the description and . . . the prediction of all changes of that type occurring in Ph.A. Of this evolutionary emergence of laws one, though not the only conceivable, occasion would be the production, in accordance with one set of laws, of new local integrations in matter, the motions of which, and therefore of their component particles, would therefore conform to vector, i.e., directional, laws emergent in the sense defined. . . . (2) New qualities . . . attachable to entities already present, though without those accidents in Ph.A. (3) Particular entities *not* possessing all the essential attributes characteristic of those found in Ph.A., and having distinctive types of attributes (not merely configurational) of their own. (4) Some type or types of event or process irreducibly different in kind from any occurring in Ph.A. (5) A greater quantity, or number of instances, not explicable by transfer from outside the system, of any one or more types of prime entity common to both phases.[18]

[18] Arthur O. Lovejoy, "The Meanings of 'Emergence' and Its Modes," in *Proceedings of the Sixth International Congress of Philosophy* (ed. by Edgar S. Brightman), New York, 1927, pp. 26-27.

Emergent evolution as a doctrine of unceasing "creative novelty" is therefore commonly placed in opposition to the preformationist view, attributed especially to seventeenth-century science, that all the events of nature are simply the spatial rearrangements of a set of ultimate, simple "entities," whose total number, qualities, and laws of behavior remain invariant throughout the various juxtapositions into which they enter. However, some writers have gone beyond the assertion of such "creative novelty" and have outlined what they believe to be the successive stages of creative evolution; but we shall not concern ourselves with the details of these cosmic speculations.

a. It should be noted in the first place that the doctrine of creative evolution appears neither to entail nor to be entailed by the conception of emergence as the unpredictability of various properties. For it may very well be the case that a property is an emergent relative to a given theory but is not novel in a *temporal* sense. To take an extreme example, the property that bodies possess weight is not deducible from the classical theory of physical geometry; however, there is no reason to believe that bodies came to exhibit gravitational properties *after* they acquired spatial ones. On the other hand, it might be possible to deduce from some theory of atomic structure that nitrogen and oxygen could combine to form a water-soluble ammonia gas, although, because the prevailing physical conditions did not permit the formation of water in liquid state—say, before the time when the earth became sufficiently cool —no actual instance of ammonia dissolving in water had ever occurred. A subsequent formation of water with the dissolution of some ammonia gas in it would then be a temporally novel event. Accordingly, the question whether any properties are "emergents" in the sense of being temporally novel is a problem of a different order from the issue whether any properties are "emergents" in the sense of being unpredictable. The latter is an issue largely though not exclusively concerned with the *logical relations* of statements; the former is primarily a question that can be settled only by empirical *historical* inquiry.

b. Accordingly, the question whether a property, process, or mode of behavior is a case of emergent evolution is a straightforward empirical problem and can be resolved at least in principle by recourse to historical inquiry. Nevertheless, there are some difficulties facing attempts to answer it which deserve brief mention. One of these difficulties is a practical one, and arises from the circumstance that to answer the question conclusively we must possess a detailed knowledge of all the past occurrences in the universe (or in some portion of it), so as to be able to decide whether an alleged emergent trait or process is really such. But our knowledge of the past is seriously incomplete, and we possess fairly reliable evidence only in a limited class of cases to show that certain properties and processes could not have occurred before a given time. Thus we do not possess a sufficient basis for deciding beyond a reasonable doubt whether various processes on the atomic and sub-atomic levels which are believed to occur at present have always taken place, or whether they are characteristic of the current cosmic epoch. On the other hand, if we take for granted the dependence of living organisms upon favorable temperature conditions, and if we also assume that at one time the temperature of the earth was far too great for the functioning of such organisms, it becomes practically certain that living forms did not appear on the earth (or perhaps anywhere in the universe) before a certain age.

A second difficulty has its source in the vagueness of such words as 'property' and 'process' and in the lack of precise criteria for judging whether two properties or processes are to be counted as "the same" or as "different." Thus, the "mere" spatial rearrangement of a set of objects is apparently not to be regarded as an instance of an emergent property, even when that specific rearrangement has not previously occurred. Nevertheless, it is pertinent to ask whether every spatial redistribution of things is not always associated with some "qualitative" changes, so that spatial changes are *ipso facto* also alterations in the "properties" of the things redistributed. For example, the pattern formed by a square resting on one base certainly "looks different" from the pattern formed when the square is rotated so as to stand on one of its vertices. If the second pattern had not existed before, would its occurrence count as the appearance of a novel property? If it would not, what is the mark of a new trait? But if it would count as something novel, then almost any change must also be regarded as an illustration of emergent evolution. For a given state of affairs may be analyzable into a set of traits, each of which has occurred in the past. On the other hand, in their present manifestation the traits occur in a determinate context of relations; and, although the specific *pattern* of these relations is a repeatable one, those traits may in fact never have been previously exemplified in just that pattern. Accordingly, the given state of affairs would in that eventuality illustrate an emergent property; and, since every situation may very well exhibit such novel patterns, especially if no limits are placed on the spatiotemporal extent of a situation, the doctrine of emergence barely escapes collapse into the trivial thesis that things change.

Furthermore, just what is to be understood by the stipulation contained in the above quotation that a particular entity is to count as an instance of emergent evolution, if it does not possess "all the essential attributes" of entities in previous phases of evolution? In general, whether or not an attribute is to be regarded as an "essential" one depends on the context of the question and on the problem under consideration. But if this is so, then in view of that stipulation the distinction between an emergent trait and a nonemergent one would shift with changes in interest and with the purposes of an inquiry. These difficulties are not cited as being fatal to the doctrine of emergence. They do indicate, however, that, unless the doctrine is formulated with greater care than is customary, it can easily be construed as simply a truism.

c. The claim that there are emergent properties in the sense of emergent evolution is entirely compatible with the belief in the universality of the causal principle, at any rate in the form that there are determinate conditions for the occurrence of all events. Some proponents of emergent evolution do indeed combine the doctrine with versions of radical indeterminism; others invariably associate emergence with so-called "teleological" causation, thus attributing the appearance of novel qualities and processes to the operation of purposive agents. However, neither a belief in indeterminism nor in teleological causation is essential to emergent evolution. There are in fact many emergent evolutionists who maintain that the occurrence of a new chemical compound, for example, is always contingent upon the formation of definite though unique configurations of certain chemical elements; and they hold, furthermore, that, whenever these elements are conjoined in that special manner, whether through the agency of purposive creatures or adventitious circumstances, a compound of the same type is invariably formed.

d. It is also worth noting that, despite widespread opinion to the contrary, the assumptions and procedures of classical physics (and of mechanics in particular) neither imply nor contradict the thesis of emergent evolution. To be sure, there are philosophical interpretations of physics, according to which the properties of things are "ultimately" those distinctive of mechanics, and according to which also the only "real" changes in nature are spatial ones. However, such interpretations are of doubtful validity and cannot be assumed to be adequate accounts of the nature of physical theory. As we have seen, the science of mechanics does indeed operate with a limited and selected set of theoretical notions. However, this fact does not entail the requirement that the science deny either the actual existence or the possible emergence of traits of things other than those with which mechanics is primarily concerned. Such a denial would be unwarranted, even if earlier hopes of physicists had been realized and mechanics had continued to retain its one-time eminence as the universal science of nature. For a mechanical explanation of an event or process consists simply in stating the conditions for its occurrence in *mechanical terms*. But such explanations would clearly be impossible (on pain of making the enterprise of giving explanations for things self-defeating) if the event or process were not first identified by observing its characteristics—whether or not the characteristics are purely mechanical properties, and whether or not they are novel. In short, when the structure of mechanics or of any other theory of classical physics is analyzed, it becomes evident that the operative efficacy of the theory does not depend on acceptance or denial of the *historical* thesis that in the course of time novel traits and individuals appear in the universe.

e. Perhaps the most intriguing suggestion contained in the doctrine of emergent evolution is that the "laws of nature" may themselves change, and that novel patterns of dependence between events are manifested during different cosmic epochs. It will of course be clear that what is intended is not simply that our *knowledge* or our *formulation* of the structures of events and processes may be undergoing development, but that these *structures themselves* are altering with time. Thus, the Boyle-Charles' law is not as adequate a formulation of the behavior of gases as is the van der Waals' equation; but the fact that we have replaced the former with the latter is not taken to signify that the pattern of behavior of gases has undergone a change. Moreover, the suggestion does not consist merely in the supposal that the mode of behavior of some specific physical system is evolving. For example, there is evidence to indicate that the period of the earth's axial rotation is diminishing. However, this special fact is explained not by the assumption that the laws of mechanics are being altered, but in terms of such factors as the "braking" effect of the tides, produced by the sun and the moon in accordance with presumably unchanging laws. Accordingly, what the suggestion contemplates is the possibility that pervasive *types of structure* are changing, or that novel relational patterns are manifested by things; for example, instead of permanently remaining inversely proportional to the square of the distance, the gravitational force between all pairs of particles may be slowly changing so that this latter exponent is increasing with time; or various chemical elements may exhibit progressively new properties and new modes of combination with one another. However, the suggestion is not without serious difficulties, some of which must now be noted.

Perhaps the most obvious and crucial of these stems from the fact that we cannot be sure whether an apparent change in a law is really such, or whether it merely indicates that our knowledge was incomplete concerning the conditions under which some type of structure prevails.

Suppose, for example, that evidence were available which seems to show that some universal constant (such as the velocity of light *in vacuo*) is changing, so that its value during the present century is smaller than it was during prehistoric times. However, other things have also changed in the interim: the relative positions of the galaxies are no longer the same; there have been internal changes in the stars and in the quantity of radiation they emit; and possibly even some hitherto undetected trait of physical bodies has varied (some trait comparable to the electric properties of matter, which have been discovered by men only relatively recently). It is therefore at least conceivable that the hitherto asserted law of the constancy of the velocity of light is simply erroneous, and that this velocity varies with some such factors as have been mentioned. It would certainly not be a simple task to eliminate this alternative interpretation of the evidence; and in fact most scientists would doubtless be more inclined to regard the hitherto accepted law as correct only when certain antecedent conditions are satisfied—and therefore to regard it as simply a limiting case of a more inclusive law—rather than to assume that the pervasive structure of physical occurrences is undergoing evolution. In any event, whether such an assumption will ever be widely accepted will most likely depend on how effective and convenient it proves to be in establishing a thoroughly inclusive and integrated system of knowledge. Accordingly, although the suggestion that some laws may be evolving does not fall outside the bounds of possibility, it is at best a highly speculative one for which it is not easy to supply reasonably conclusive evidence.

There is an additional difficulty of a different order which faces the doctrine that *all* laws are changing with time.[14] For how is evidence obtained for the claim that a law is undergoing change? A pervasive pattern of relations cannot be literally "seen" to evolve, and the basis for such a conclusion must be obtained from comparisons of the present with the past. However, the past is not accessible to direct inspection. It can only be reconstructed from data available in the present, with the help of laws which must be assumed to be unchanging at least during the epoch which includes that past and the present. For example, suppose that the gravitational force between bodies is alleged to be slowly diminishing, on the ground that in the past the tides were generally higher than in the present, even though the number and relative position of celestial bodies were the same as at present. But how can we know that the past was indeed like this, unless we use laws that have not altered in order to infer those past facts from present data? Thus, we might

[14] Cf. Henri Poincaré, "L'Evolution des Lois," in *Dernières Pensées*, Paris, 1926; Pascual Jordan, *Die Herkunft der Sterne*, Stuttgart, 1947.

find deposits of sea salt at altitudes now out of the reach of the tides. However, even if we waive the question whether the land had not been elevated by geological action rather than because of a diminution in the height of tides, the conclusion that the salt was deposited by the ocean takes for granted various laws concerning the motions of tidal water and the evaporation of liquids. Accordingly, the assumption that all laws are simultaneously involved in a process of change is self-annihilating, for, since the past would then be completely inaccessible to knowledge we would be unable to produce any evidence for that assumption.

The form in which the suggestion of emergent laws appears most plausible is that new types of behavior conforming to novel modes of dependence arise when hitherto nonexistent combinations and integrations of matter occur. For example, chemists have produced substances in the laboratory which, as far as we can tell, have never existed before, and which possess properties and ways of interacting with other substances that are distinctive and novel. What has thus occasionally happened in the laboratory of chemists has undoubtedly happened more frequently in the older and vaster laboratory of nature. It might of course be said that such novel types of dependence are not "really novel" but are only the realizations of "potentialities" that have always been present in "the natures of things"; and it might also be said that, with "sufficient knowledge" of these "natures," anyone having the requisite mathematical skills could predict the novelties in advance of their realization. We have already commented sufficiently on the latter part of this rejoinder, and can therefore discount it without further ado as both invalid and irrelevant. As for the first part of the objection, it must be admitted that it is irrefutable; but it will also be clear that what the objection asserts has no factual content, and that its irrefutability is that of a definitional truism.

V.   *Wholes, Sums, and Organic Unities*

Before leaving the subjects of reduction and emergence, it will be convenient to discuss a familiar thesis frequently associated with these themes. According to this conception, there is an important type of individual wholes (physical, biological, psychological, as well as social) distinguished from others by the fact that they are "organic unities," and not simply "aggregates" of independent parts or members. Wholes of this type are often characterized by the dictum that they possess an organization in virtue of which "the whole is more than the sum of its parts." Since the existence of organic wholes is sometimes taken to place fixed limits on the possibility of effecting reductions in the sciences, as well as on the scope of the methods of physics, it is desirable to examine such wholes with care.

A preliminary point must first be noted. As commonly employed, the words 'whole,' 'sum,' and their derivatives are unusually ambiguous, metaphorical, and vague. It is therefore frequently impossible to assess the cognitive worth as well as the meaning of statements containing them, so that some of the many senses of those words must be distinguished and clarified. Some examples will make evident the need for such clarification. A quadrilateral encloses an area, and either one of its two diameters divides the figure into two partial areas whose sum is equal to the area of the initial figure. In this geometrical context, and in many analogous ones as well, the statement 'The whole is equal to the sum of its parts' is normally accepted as true. Indeed, the statement in this context is frequently acknowledged to be not only true but necessarily true, so that its denial is regarded as self-contradictory. On the other hand, in discussing the taste of sugar of lead as compared with the tastes of its chemical components, some writers have maintained that in this case the whole is *not* equal to the sum of its parts. This claim is obviously intended to be informative about the matters discussed, and it would be high-handed to reject it outright as simply a logical absurdity. It is clear, nevertheless, that in the context in which this claim is made the words 'whole,' 'part,' 'sum,' and perhaps even 'equal,' are being employed in senses different from those associated with them in the geometrical context. We must therefore assume the task of distinguishing between a number of senses of these words that appear to play a role in various inquiries.

1. The words 'whole' and 'part' are normally used for correlative distinctions, so that x is said to be a whole in relation to something y which is a component or part of x in some sense or other. It will be convenient, therefore, to have before us a brief list of certain familiar "kinds" of wholes and corresponding parts.

a. The word 'whole' is used to refer to something with a spatial extension, and anything is then called a 'part' of such a whole which is spatially included in it. However, several special senses of 'whole' and 'part' fall under this head. In the first place, the terms may refer to specifically spatial properties, so that the whole is then some length, area, or volume containing as parts lengths, areas, or volumes. In this sense, neither wholes nor parts need be spatially continuous; thus, the United States and its territorial possessions are not a spatially continuous whole, and continental United States contains as one of its spatial parts desert regions which are also not spatially continuous. In the second place, 'whole' may refer to a nonspatial property or state of a spatially extended thing, and 'part' designates an identical property of some spatial part of the thing. Thus, the electric charge on a body is said to have for its parts the electric charges on spatial parts of the body. In the third place, though sometimes the only spatial properties counted as parts of a spatial whole are those that have the same spatial dimensions as the whole, at other times the usage of the terms is more liberal. Thus the surface of a sphere is frequently said to be a part of the sphere, but on other occasions only volumes in the sphere's interior are so designated.

b. The word 'whole' refers to some temporal period whose parts are temporal intervals in it. As in the case of spatial wholes and parts, temporal ones need not be continuous.

c. The word 'whole' refers to any class, set, or aggregate of elements, and 'part' may then designate either any proper subclass of the initial set, or any element in the set. Thus, by a part of the whole consisting of all the books printed in the United States during a given year may be understood either all the novels printed that year, or some particular copy of a novel.

d. The word 'whole' sometimes refers to a property of an object or process, and 'part' to some analogous property standing to the first in certain specified relations. Thus, a force in physics is commonly said to have for its parts or components other forces into which the first can be analyzed according to a familiar rule. Similarly, the physical brightness of a surface illuminated by two sources of light is sometimes said to have for one of its parts the brightness associated with one of the sources. In the present sense of the words, a part is not a spatial part of the whole.

e. The word 'whole' may refer to a pattern of relations between certain specified kinds of objects or events, the pattern being capable of embodiment on various occasions and with various modifications. However, 'part' may then designate different things in different contexts. It may refer to any one of the elements which are related in that pattern on some occasion of its embodiment. Thus, if a melody (say "Auld Lang Syne") is such a whole, one of its parts is then the first tone that is sounded when the melody is sung on a particular date. Or it may refer to a class of elements that occupy corresponding positions in the pattern in some specified mode of its embodiment. Thus, one of the parts of the melody will then be the class of first notes when "Auld Lang Syne" is sung in the key of E flat. Or the word 'part' may refer to a subordinate phrase in the total pattern. In this case, a part of the melody may be the pattern of tones that occurs in its first four bars.

f. The word 'whole' may refer to a process, one of its parts being another process that is some discriminated phase of the more inclusive one. Thus, the process of swallowing is part of the process of eating.

g. The word 'whole' may refer to any concrete object, and 'part' to any of its properties. In this sense, the character of being cylindrical in shape or being malleable is a part of a given piece of copper wire.

h. Finally, the word 'whole' is often used to refer to any system whose spatial parts stand to each other in various relations of dynamical dependence. Many of the so-called "organic unities" appear to be systems of this type. However, in the present sense of 'whole' a variety of things are customarily designated as its parts. Thus, a system consisting of a mixture of two gases inside a container is frequently, though not always in the same context, said to have for its parts one or more of the following: its spatially extended constituents, such as the two gases and the container; the properties or states of the system or of its spatial parts, such as the mass of the system or the specific heats of one of the gases; the processes which the system undergoes in reaching or maintaining thermodynamical equilibrium; and the spatial or dynamical organization to which its spatial parts are subject.

This list of senses of 'whole' and 'part,' though by no means complete, will suffice to indicate the ambiguity of these words. But what is more important, it also suggests that, since the word 'sum' is used in a number of contexts in which these words occur, it suffers from an analogous ambiguity. Let us therefore examine several of its typical senses.

2. We shall not inquire whether the word 'sum' actually is employed in connection with each of the senses of 'whole' and 'part' that have been distinguished, and if so just what meaning is to be associated with it. In point of fact, it is not easy to specify a clear sense for the word in many contexts in which people do use it. We shall accordingly confine ourselves to noting only a small number of the well-established uses of 'sum' and to suggesting interpretations for it in a few contexts in which its meaning is unclear and its use misleading.

a. It is hardly surprising that the most carefully defined uses of 'sum' and 'addition' occur in mathematics and formal logic. But even in these contexts the word has a variety of special meanings, depending on what type of mathematical and logical "objects" are being added. Thus, there is a familiar operation of addition for the natural integers; and there are also identically named but really distinct operations for ratios, real numbers, complex numbers, matrices, classes, relations, and other mathematical or logical "entities." It is not altogether evident why all these operations have the common name of 'addition,' though there are at least certain formal analogies between many of them; for example, most of them are commutative and associative. However, there are some important exceptions to the general rule implicit in this example, for the addition of *ordered* sets is not uniformly commutative, though it is associative. On the other hand, the sum of two entities in mathematics is invariably some unique entity which is of the same type as the summands;—thus the sum of two integers is an integer, of two matrices a matrix, and so on. Moreover, though the word 'part' is not always defined or used in connection with mathematical "objects," whenever both it and 'sum' are employed they are so used that the statement 'The whole is equal to the sum of its parts' is an analytic or necessary truth.

However, it is easy to construct an apparent counter-instance to this last claim. Let K^* be the *ordered* set of the integers, ordered in the following manner: first the odd integers in order of increasing magnitude, and then the even integers in that order. K^* may then be represented by the notation: $(1, 3, 5, \ldots, 2, 4, 6, \ldots)$. Next let K_1 be the class of odd integers and K_2 the class of even ones, neither class being an ordered set. Now let K be the class-sum of K_1 and K_2, so that K contains all the integers as members; K also is not an ordered class. But the membership of K is the same as that of K^*, although quite clearly K and K^* are not identical. Accordingly, so it might be argued, in this case the whole (namely K^*) is not equal to the sum (i.e., K) of its parts.

This example is instructive on three counts. It shows the possibility of defining in a precise manner the words 'whole,' 'part,' and 'sum' so that 'The whole is unequal to the sum of its parts' is not only not logically absurd but is in fact logically true. There is thus no a priori reason for dismissing such statements as inevitable nonsense; and the real issue is to determine, when such an assertion is made, in what sense if any the crucial words in it are being used in the given context. But the example also shows that, though such a sentence may be true on one specified usage of 'part' and 'sum,' it may be possible to assign other senses to these words so that the whole *is* equal to the sum of its parts in this redefined sense of the words. Indeed, it is not standard usage in mathematics to call either K_1 or K_2 a part of K^*. On the contrary, it is customary to count as a part of K^* only an *ordered* segment. Thus, let K_1^* be the ordered set of odd integers arranged according to increasing magnitude, and K_2^* the corresponding ordered set of even integers. K_1^* and

K_2* are then parts of K*. [K* has other parts as well, for example, the ordered segments indicated by the following: (1, 3, 5, 7), (9, 11, . . . , 2, 4), and (6, 8, . . .).] Now form the *ordered sum* of K_1* and K_2*. But *this* sum yields the ordered set K*, so that in the specified senses of 'part' and 'sum' the whole *is* equal to the sum of its parts. It is thus clear that, when a given system has a special type of organization or structure, a *useful* definition of 'addition,' if such can be given, must take into account that mode of organization. Any number of operations could be selected for the label 'summation,' but not all of them are relevant or appropriate for advancing a given domain of inquiry.

Finally, the example suggests that, though a system has a distinctive structure, it is not in principle impossible to specify that structure in terms of relations between its elementary constituents, and moreover in such a manner that the structure can be correctly characterized as a 'sum' whose 'parts' are themselves specified in terms of those elements and relations. As we shall see, many students deny, or appear to deny, this possibility in connection with certain kinds of organized systems (such as living things). The present example therefore shows that, though we may not be able *as a matter of fact* to analyze certain highly complex "dynamic" (or "organic") unities in terms of some given theory concerning their ultimate constituents, such inability cannot be established as a matter of *inherent logical necessity*.

b. If we now turn to the positive sciences, we find that here too are a large number of well-defined operations called 'addition.' The major distinction that needs to be drawn is between scalar and vector sums. Let us consider each in turn. Examples of the former are the addition of the numerosity of groups of things, of spatial properties (length, area, and volume), of temporal periods, of weights, of electrical resistance, electric charge, and thermal capacity. They illustrate the first three senses of 'whole' and 'part' which we distinguished above; and in each of them (and in many other cases that could be mentioned) 'sum' is so specified that the whole is the sum of appropriately chosen parts.

On the other hand, there are many magnitudes, such as density or elasticity, for which no operation of addition is defined or seems capable of being defined in any useful manner; most of these cases fall under the last four of the above distinctions concerning 'whole' and 'part.' Moreover, there are some properties for which addition is specified only under highly specialized circumstances; for example, the sum of the brightness of two sources of light is defined only when the light emitted is monochromatic. It makes no sense, therefore, to say that the density (or the shape) of a body is, or is not, the sum of the densities (or shapes) of its parts, simply because there are neither explicitly formulated rules nor ascertainable habits of procedure which associate a usage with the word 'sum' in such a context.

The addition of vector properties, such as forces, velocities, and accelerations, conforms to the familiar rule of parallelogram composition. Thus, if a body is acted on by a force of 3 poundals in a direction due north, and also by a force of 4 poundals in a direction due east, the body will behave as if it were acted on by a single force of 5 poundals in a northeasterly direction. This single force is said to be the 'sum' or 'resultant' of the other two forces, which are called its 'components'; and, conversely, any force can be analyzed as the sum of an arbitrary number of components. This sense of 'sum' is commonly associated with the fourth of the above distinctions concerning 'whole' and 'part'; and it is evident that here the sense of 'sum' is quite different from the sense of the word in such contexts as 'the sum of two lengths.'

It has been argued by Bertrand Russell that a force cannot rightly be said to be the sum of its components. Thus he declared:

> Let there be three particles A, B, C. We may say that B and C both cause accelerations in A, and we compound these accelerations by the parallelogram law. But this composition is not truly addition, for the components are not *parts* of the resultant. The resultant is a new term, as simple as their components, and not by any means their sum. Thus the effects attributed to B and C are never produced, but a third term different from either is produced. This, we may say, is produced by B and C together, taken as a whole. But the effect which they produce as a whole can only be discovered by supposing each to produce a separate effect: if this were not supposed, it would be impossible to obtain the two accelerations whose resultant is the actual acceleration. Thus we seem to reach an antinomy: the whole has no effect except what results from the effects of the parts, but the effects of the parts are nonexistent.[15]

However, all this argument shows is that by the component of a force (or of an acceleration) we do not mean anything like what we understand by a component or part of a length—the components of forces are not *spatial parts* of forces. It does not establish the claim that the addition of forces "is not truly addition," unless, indeed, the word 'addition' is being used so restrictively that no operation is to be so designated which does not involve a juxtaposition of spatial (or possibly temporal) parts of the whole said to be their sum. But in this latter event many other operations that are called 'addition' in physics, such as the addition of electrical capacities, would also have to receive different labels. Moreover, no antinomy arises from the supposition that, on the one hand,

15 Bertrand Russell, *The Principles of Mathematics*, Cambridge, England, 1903, p. 477.

the effect of each component force acting alone does not exist, while on the other hand the actual effect produced by the joint action of the components is the resultant of their partial effects. For the supposition simply expresses what is the case, in a language conforming to the antecedent *definition* of the addition and resolution of forces.

The issue raised by Russell is thus terminological at best. His objection is nevertheless instructive. For it calls needed attention to the fact that, when the matter is viewed abstractly, the 'sum' of a given set of elements is simply an element that is *uniquely determined* by some *function* (in the mathematical sense) of the given set. This function may be assigned a relatively simple and familiar form in certain cases, and a more complex and strange form in others; and in any event, the question whether such a function is to be introduced into a given domain of inquiry, and if so what special form is to be assigned to it, cannot be settled a priori. The heart of the matter is that when such a function is specified, and if a set of elements satisfies whatever conditions are prescribed by the function, it becomes possible to *deduce* from these premises a class of statements about some structural complex of those elements.[16]

c. We must now consider a use of 'sum' associated with the fifth sense of 'whole' and 'part' distinguished above—a use also frequently associated with the dictum that the whole is more than, or at any rate not merely, the sum of its parts. Let us assume that the following statement is typical of such usage: "Although a melody may be produced by sounding a series of individual tones on a piano, the melody is not the sum of its individual notes." The obvious question that needs to be asked is: "In what sense is 'sum' being employed here?" It is evident that the statement can be informative only if there *is* such a thing as the sum of the individual tones of melody. For the statement can be established as true or false only if it is possible to compare such a sum with the whole that is the melody.

However, most people who are inclined to assert such a statement do

---

[16] An issue similar to the one raised by Russell has been raised in connection with the addition of velocities in relativity theory. Let A, B, C be three bodies, so that the velocity of A with respect to B is v_{AB}, that of B with respect to C is v_{BC} (where the direction of v_{BC} is parallel to the direction of v_{AB}), and of A with respect to C is v_{AC}. Then according to classical mechanics, $v_{AC} = v_{AB} + v_{BC}$. But according to the special relativity theory,

$$v_{AC} = \frac{v_{AB} + v_{BC}}{1 + \dfrac{v_{AB}v_{BC}}{c^2}}$$

where c is the velocity of light. It has been argued that in the latter we are not "really adding" velocities. However, this objection can be disposed of in essentially the same manner as can Russell's argument.

not specify what that sum is supposed to be; and there is thus a basis for the supposition that they either are not clear about what they mean or do not mean anything whatever. In the latter case the most charitable view that can be taken of such pronouncements is to regard them as simply misleading expressions of the possibly valid claim that the notion of summation is *inapplicable* to the constituent tones of melodies. On the other hand, some writers apparently understand by 'sum' in this context the *unordered class* of individual tones; and what they are therefore asserting is that this class is not the melody. But this is hardly news, though conceivably there may have been some persons who believed otherwise. In any event, there appears to be no meaning, other than this one, which is normally associated with the phrase 'sum of tones' or similar phrases. Accordingly, if the word 'sum' is used in this sense in contexts in which the word 'whole' refers to a pattern or configuration formed by elements standing to each other in certain relations, it is perfectly true though trivial to say that the whole is more than the sum of its parts.

As has already been noted, however, this fact does not preclude the possibility of *analyzing* such wholes into a set of elements related to one another in definite ways; nor does it exclude the possibility of assigning a different sense to 'sum' so that a melody might then be construed as a sum of appropriately selected parts. It is evident that at least a partial analysis of a melody is effected when it is represented in the customary musical notation; and the analysis could obviously be made more complete and explicit, and even expressed with formal precision.[17]

But it is sometimes maintained in this connection that it is a fundamental mistake to regard the constituent tones of a melody as independent parts, out of which the melody can be reconstituted. On the contrary, it has been argued that what we "experience at each place in the melody is a *part* which is itself determined by the character of the whole. . . . The flesh and blood of a tone depends from the start upon its role in the melody: a *b* as leading tone to *c* is something radically different from the *b* as tonic." [18] And as we shall see, similar views have been advanced in connection with other cases and types of Gestalts and "organic" wholes.

Now it may be quite true that the *effect* produced by a given tone depends on its position in a context of other tones, just as the effect produced by a given pressure upon a body is in general contingent upon

---

[17] For an interesting sketch of a generalized formal analysis of Gestalts such as melodies, cf. Kurt Grelling and Paul Oppenheim, "Der Gestaltbegriff in Lichte der neuen Logik," *Erkenntnis*, Vol. 7 (1938), pp. 211-25.

[18] Max Wertheimer, "Gestalt Theory," in *A Source Book of Gestalt Psychology* (ed. by Willis D. Ellis), New York, 1950, p. 5.

what other pressures are operative. But this supposed fact does not imply that a melody cannot rightly be viewed as a relational complex whose component tones are identifiable independently of their occurrence in that complex. For if the implication did hold, it would be impossible to describe how a melody is constituted out of individual tones, and therefore impossible to prescribe how it is to be played. Indeed, it would then be self-contradictory to say that "a *b* as leading tone to *c* is something radically different from the *b* as tonic." For the name '*b*' in the expression '*b* as leading tone to *c*' could then not refer to the same tone to which the name '*b*' refers in the expression '*b* as tonic'; and the presumable intent of the statement could then not be expressed. In short, the fact that, in connection with wholes that are patterns or Gestalts of occurrences, the word 'sum' is either undefined or defined in such a way that the whole is unequal to the sum of its parts, constitutes no inherently insuperable obstacle to analyzing such wholes into elements standing to each other in specified relations.

d. We must finally examine the use of 'sum' in connection with wholes that are organized systems of dynamically interrelated parts. Let us assume as typical of such usage the statement 'Although the mass of a body is equal to the sum of the masses of its spatial parts, a body also has properties which are not the sums of properties possessed by its parts.' The comments that have just been made about 'sum' in connection with patterns of occurrences such as melodies can be extended to the present context of usage of the word; and we shall not repeat them. In the present instance, however, an additional interpretation of 'sum' can be suggested.

When the behavior of a machine like a clock is sometimes said to be the sum of the behavior of its spatial parts, what is the presumptive content of the assertion? It is reasonable to assume that the word 'sum' does not here signify an unordered class of elements, for neither the clock nor its behavior is such a class. It is therefore plausible to construe the assertion as maintaining that, from the theory of mechanics, coupled with suitable information about the actual arrangements of the parts of the machine, it is possible to deduce statements about the consequent properties and behaviors of the entire system. Accordingly, it seems also plausible to construe in a similar fashion statements such as that of J. S. Mill: "The different actions of a chemical compound will never be found to be the sums of actions of its separate parts." [19] More explicitly, this statement can be understood to assert that from some assumed theory concerning the constituents of chemical compounds, even when it

[19] J. S. Mill, *A System of Logic*, London, 1879, Book 3, Chap. 6, § 2 (Vol. 1, p. 432).

is conjoined with appropriate data on the organization of these constituents within the compounds, it is not in fact possible to deduce statements about many of the properties of these compounds.

If we adopt this suggestion, we obtain an interpretation for 'sum' that is particularly appropriate for the use of the word in contexts in which the wholes under discussion are organized systems of interdependent parts. Let T be a theory that is in general able to explain the occurrence and modes of interdependence of a set of properties $P_1, P_2, \ldots, P_k$. More specifically, suppose it is known that, when one or more individuals belonging to a set K of individuals occur in an environment E_1 and stand to each other in some relation belonging to a class of relations R_1, the theory T can explain the behavior of such a system with respect to its manifesting some or all of the properties P. Now assume that some or all of the individuals belonging to K form a relational complex R_2 not belonging to R_1 in an environment E_2, which may be different from E_1, and that the system exhibits certain modes of behavior that are formulated in a set of laws L. Two cases may then be distinguished: from T, together with statements concerning the organization of the individuals in R_2, it is possible to deduce the laws L; or secondly, not all the laws L can be so deduced. In the first case, the behavior of the system R_2 may be said to be the 'sum' of the behaviors of its component individuals; in the second case, the behavior of R_2 is *not* such a sum. It is evident that in the terminology and distinctions of the present chapter, both conditions for the reducibility of L to T are satisfied in the first case; in the second case, however, although the condition of connectability may be satisfied, the condition of derivability is not.

If this interpretation of 'sum' is adopted for the indicated contexts of its usage (let us call this the "reducibility sense" of the word), it follows that the distinction between wholes that are sums of their parts and those that are not is *relative to some assumed theory T* in terms of which the analysis of a system is undertaken. Thus, as we have seen, the kinetic theory of matter as developed during the nineteenth century was able to explain certain thermal properties of gases, including certain relations between the specific heats of gases. However, that theory was unable to account for these relations between specific heats when the state of aggregation of molecules is that of a solid rather than a gas. On the other hand, modern quantum theory is capable of explaining the facts concerning the specific heats of solids, and presumably also all other thermal properties of solids. Accordingly, although relative to classical kinetic theory the thermal properties of solids are not sums of the properties of their parts, relative to quantum theory those properties are such sums.

3. We must now briefly consider the distinctive feature of those systems that are commonly said to be "organic unities" and that exhibit a mode of organization often claimed to be incapable of analysis in terms of an "additive point of view." However, although living bodies are the most frequently cited examples of organic wholes, we shall not be now concerned specifically with such systems. For it is generally admitted that living bodies constitute only a special class of systems possessing a structure of internally related parts; and it will be an advantage to ignore for the present special issues connected with the analysis of vital phenomena.

Organic or "functional" wholes have been defined as systems "the behavior of which is not determined by that of their individual elements, but where the part-processes are themselves determined by the intrinsic nature of the whole." [20] What is distinctive of such systems, therefore, is that their parts do not act, and do not possess characteristics, *independently* of one another. On the contrary, their parts are supposed to be so related that any alteration in one of them causes a change in *all* the other parts. [21] In consequence, functional wholes are also said to be systems which cannot be built up out of elements by combining these latter seriatim without producing changes in all those elements. Moreover, such wholes cannot have any part removed without altering both that part and the remaining parts of the system. [22] Accordingly, it is often claimed that a functional whole cannot be properly analyzed from an "additive point of view"; that is, the characteristic modes of functioning of its constituents must be studied *in situ,* and the structure of activities of the whole cannot be inferred from properties displayed by its constituents in isolation from the whole.

A purely physical example of such functional wholes has been made familiar by Köhler. Consider a well-insulated electric conductor of arbitrary shape, for example, one having the form of an ellipsoid; and assume that electric charges are brought to it successively. The charges will immediately distribute themselves over the surface of the conductor

[20] Max Wertheimer, *op. cit.,* p. 2. Cf. also Koffka's statement: "Analysis if it wants to reveal the universe in its completeness has to stop at the wholes, whatever their size, which possess functional reality. . . . Instead of starting with the elements and deriving the properties of the wholes from them a reverse process is necessary, i.e., to try to understand the properties of parts from the properties of wholes. The chief content of Gestalt as a category is this view of the relation of parts and wholes involving the recognition of intrinsic real dynamic whole-properties."—K. Koffka, "Gestalt," in *Encyclopedia of the Social Sciences,* New York, 1931, Vol. 6, p. 645, quoted by kind permission of the publishers, The Macmillan Company.
[21] Cf. Kurt Lewin, *Principles of Topological Psychology,* New York, 1936, p. 218.
[22] W. Köhler, *Die physischen Gestalten in Ruhe und in stationären Zustand,* Braunschweig, 1924, p. 42; also Ellis, *op. cit.,* p. 25.

in such a way that the electric potential will be the same throughout the surface. However, the density of the charge (i.e., the quantity of charge per unit surface) will not in general be uniform at all points of the surface. Thus, in the ellipsoidal conductor, the density of the charge will be greatest at the points of greatest curvature and will be smallest at the points of least curvature.[23] In brief, the distribution of the charges will exhibit a characteristic pattern or organization—a pattern which depends on the shape of the conductor but is independent of the special materials of its construction or of the total quantity of charge placed upon it.

It is, however, not possible to build up this pattern of distribution bit by bit, for example, by bringing charges first to one part of the conductor and then to another so as to have the pattern emerge only after all the charges are placed on the conductor. For when a charge is placed on one portion of the surface, the charge will not remain there but will distribute itself in the manner indicated; and in consequence, the charge density at one point is not independent of the densities at all other points. Similarly, it is not possible to remove some part of the charge from one portion of the surface without altering the charge densities at all other points. Accordingly, although the total charge on a conductor is the sum of separable partial charges, the configuration of charge densities cannot be regarded as composed from independent parts. Köhler thus declares:

> The natural structure assumed by the total charge is not described if one says: at this point the charge-density is this much 'and' at that point the density is that much, etc.; but one might attempt a description by saying: the density is so much at this point, so much at that point, all mutually interdependent, and such that the occurrence of a certain density at one point determines the densities at all other points.[24]

Many other examples—physical, chemical, biological, and psychological—could be cited which have the same intent as this one. Thus

---

[23] More generally, the charge density on the ellipsoid is proportional to the fourth root of the curvature at a point.

[24] Köhler, *op. cit.*, p. 58, and cf. also p. 166. Many other physical examples of such "functional" wholes could be cited. The surfaces assumed by soap films provide an intuitively evident illustration. The general principle underlying the analysis of such surfaces is that, subject to the boundary conditions imposed on the surface, its area is a minimum. Thus, neglecting gravity, a soap film bounded by a plane loop of wire will assume a plane surface; a soap bubble will assume the shape of a sphere, a figure which has the minimum surface for a given volume. Now consider a part of the surface of a soap bubble bounded by a circle. If this part were removable from the spherical surface, it would no longer retain its convex shape, but would become a plane. Thus, the shape assumed by a part of the film depends on the whole of which it is a part. Cf. the accounts of soap film experiments in Richard Courant and Herbert Robbins, *What Is Mathematics?* New York, 1941, pp. 386ff.

there is no doubt that in many systems the constituent parts and processes are "internally" related, in the sense that these constituents stand to each other in relations of mutual causal interdependence. Indeed, some writers have found it difficult to distinguish sharply between systems which are of this sort and systems which allegedly are not; and they have argued that all systems whatever ought to be characterized as wholes which are "organic" or "functional" in some degree or other.[25] In point of fact, many who claim that there is a fundamental difference between functional and nonfunctional (or "summative") wholes tacitly admit that the distinction is based on *practical decisions* concerning what causal influences may be ignored for certain purposes. Thus, Köhler cites as an example of a "summative" whole a system of three stones, one each in Africa, Australia, and the United States. The system is held to be a summative grouping of its parts, because displacement of one stone has no effect on the others or on their mutual relations.[26] However, if current theories of physics are accepted, such a displacement is not without *some* effects on the other stones, even if the effects are so minute that they cannot be detected with present experimental techniques and can therefore be practically ignored. Again, Köhler regards the total charge on a conductor as a summative whole of independent parts, though it is not at all evident that the electronic constituents of the charge undergo no alterations when parts of the charge are removed from it. Accordingly, although the occurrence of systems possessing distinctive structures of interdependent parts is undeniable, no general criterion has yet been proposed which makes it possible to identify in an absolute way systems that are "genuinely functional" as distinct from systems that are "merely summative."[27]

[25] This is the contention of A. N. Whitehead's philosophy of organism. Cf. his *Process and Reality*, New York, 1929, esp. Part 2, Chaps. 3 and 4.

[26] Köhler, *op. cit.*, p. 47.

[27] This suggestion that the distinction between functional and nonfunctional wholes is not a sharp one, is borne out by an attempt to state more formally the character of an "organic" whole. Let S be some system and K a class of properties $P_1, \ldots, P_n$ which S may exhibit. Assume, for the sake of simplicity of exposition, that these properties are measurable in some sense, so that the specific forms of these properties can be associated with the values of numerical variables; and assume, also for the sake of simplicity, that statements about these properties have the form 'At time t the property P_i of S has the value x,' or, more compactly, '$P_i(S,t) = x$.' We now define a property in K, say P_1, to be "dependent" on the remaining properties in K when P_1 has the same value at different times if the remaining properties have equal values at those times; that is, when for every property P_i in K, if $P_i(S,t_1) = P_i(S,t_2)$ then $P_1(S,t_1) = P_1(S,t_2)$. Moreover we shall say that the class K of properties is "interdependent" if *each* property in the class is dependent on the remaining properties in K, that is, when for every P_i and P_j in K, if $P_i(S,t_1) = P_i(S,t_2)$ then $P_j(S,t_1) = P_j(S,t_2)$. On the other hand, we can define the class K to be an "independent" class if no property in K is dependent on the remaining properties of K. To fix our ideas, let S be a gas, V its volume, p its pressure, and T its absolute temperature. Then according to the Boyle-Charles' law, V is depend-

Moreover, it is essential to distinguish in this connection between the question whether a given system can be *overtly constructed* in a piecemeal fashion by a seriatim juxtaposition of parts, and the question whether the system can be *analyzed in terms of a theory* concerning its assumed constituents and their interrelations. There undoubtedly are wholes for which the answer to the first question is affirmative—for example, a clock, a salt crystal, or a molecule of water; and there are wholes for which the answer is negative—for example, the solar system, a carbon atom, or a living body. However, this difference between systems does not correspond to the intended distinction between functional and summative wholes; and our inability to construct effectively a system out of its parts, which in some cases may only be a consequence of temporary technological limitations, cannot be taken as evidence for deciding negatively the second of the above two questions.

But let us turn to this second question, for it raises what appears to be the fundamental issue in the present context. That issue is whether the analysis of "organic unities" necessarily involves the adoption of irreducible laws for such systems, and whether their mode of organization precludes the possibility of analyzing them from the so-called "additive point of view." The main difficulty in this connection is that of ascertaining in what way an "additive" analysis differs from one which is not. The contrast seems to hinge on the claim that the parts of a functional whole do not act independently of one another, so that any laws which may hold for such parts when they are not members of a functional whole cannot be assumed to hold for them when they actually are members. An "additive" analysis therefore appears to be one which

ent on p and T; and also this class of properties is an interdependent class of properties. Again, if S is an insulated conductor possessing a definite shape, R the curvature at any point, s the charge density at any region, and p the pressure at any region, then p is not dependent on R and s, and the properties p, R, and s do not form an interdependent class, though they do not form an independent class either. For this analysis, and further details involved in its elaboration, see the papers by Kurt Grelling, "A Logical Theory of Dependence," and Kurt Grelling and Paul Oppenheim, "Logical Analysis of 'Gestalt' and 'Functional Whole,'" reprinted for members of the Fifth International Congress for the Unity of Science held in Cambridge, Mass., 1939, from the *Journal of Unified Science*, Vol. 9. This volume of the Journal was a casualty of World War II and has never been published.

However, if now we define a system S to be a "functional whole" with respect to a class K of properties if K is an interdependent class, and also define S to be a "summative whole" if K is an independent class, two points should be noted. In the first place, whether a property will be said to depend on certain others will be affected in part by the degree of experimental precision with which values of the properties in question can be established. This is the point already made in the text. In the second place, though S may not be a functional whole in the sense defined, it need not therefore be a summative whole; for some properties in K may be dependent on the remaining ones, though not all are. Accordingly, there may be various "degrees" of interdependence of parts of a system.

accounts for the properties of a system in terms of assumptions about its constituents, where these assumptions are not formulated with specific reference to the characteristics of the constituents as elements in the system. A "nonadditive" analysis, on the other hand, seems to be one which formulates the characteristics of a system in terms of relations between certain of its parts as functioning elements in the system.

However, if this is indeed the distinction between these allegedly different modes of analysis, the difference is not one of fundamental principle. We have already noted that it does not seem possible to distinguish sharply between systems that are said to be "organic unities" and those which are not. Accordingly, since even the parts of summative wholes stand in relations of causal interdependence, an additive analysis of such wholes must include special assumptions about the actual organization of parts in those wholes when it attempts to apply some fundamental theory to them. There are certainly many physical systems, such as the solar system, a carbon atom, or a calcium fluoride crystal, which despite their complex form of organization lend themselves to an "additive" analysis; but it is equally certain that current explanations of such systems in terms of theories about their constituent parts cannot avoid supplementing these theories with statements about the special circumstances under which the constituents occur as elements in the systems. In any event, the mere fact that the parts of a system stand in relations of causal interdependence does not exclude the possibility of an additive analysis of the system.

The distinction between additive and nonadditive analysis is sometimes supported by the contrast commonly drawn between the particle physics of classical mechanics and the field approach of electrodynamics. It will therefore be instructive to dwell for a moment on this contrast. According to Newtonian mechanics, the acceleration induced in a particle by the action of other bodies is the vector sum of the accelerations which would be produced by each of these bodies were they acting singly; and the assumption underlying this principle is that the force exerted by one such body is independent of the force exerted by any other. In consequence, a mechanical system such as the solar system can be analyzed additively. In order to account for the characteristic behavior of the solar system as a whole, we need to know only the force (as a function of the distance) that each body in the system exerts separately on the other bodies.

But in electrodynamics the situation is different. For the action of an electrically charged body on another depends not only on their distances but also on their relative motions. Moreover, the effect of a change in motion is not propagated instantaneously, but with a finite velocity. Accordingly, the force on a charged body due to the presence of other such bodies is not determined by the positions and velocities of the latter but by the conditions of the electromagnetic "field" in the vicinity of the former. In consequence, since such a field cannot be regarded as a 'sum' of 'partial' fields, each due to a distinct charged particle, an electromagnetic system is commonly said to be incapable of an additive analysis. "The field can be treated adequately only as a unit," so it is claimed, "not as the sum total of the contributions of individual point charges." [28]

Two brief comments must be made on this contrast. In the first place, the notion of 'field' (as used in electromagnetic theory) undoubtedly represents a mathematical technique for analyzing phenomena that is different in many important respects from the mathematics employed in particle mechanics. The latter operates with discrete sets of state variables, so that the state of a system is specified by a finite number of coordinates; the field approach requires that the values of each of its state variables be specified for each point of a mathematically continuous space. And there are further corresponding differences in the kinds of differential equations, the variables that enter into them, and the limits between which mathematical integrations are performed.

But in the second place, though it is true that the electromagnetic field associated with a set of charged particles is not a 'sum' of partial fields associated with each particle separately, it is also true that the field is uniquely determined (i.e., the values of each state variable for each point of space are unequivocally fixed) by the set of charges, their velocities, and the initial and boundary conditions under which they occur. Indeed, in one technique employed within field theory, the electromagnetic field is simply an intermediary device for formulating the effects of electrically charged particles upon other such particles. [29] Accordingly, though it may be convenient to treat an electromagnetic field as a "unit," this convenience does not signify that the properties of the field cannot be analyzed in terms of assumptions concerning its constituents. And though the field may not be a 'sum' of partial fields in any customary sense, an electromagnetic system is a 'sum' in the special sense of the word proposed previously, namely, there is a theory about the constituents of these systems such that the relevant laws of the system can be deduced from the theory. In point of fact, if we take a final glance at the functional whole

[28] Peter G. Bergmann, *Introduction to the Theory of Relativity,* New York, 1942, p. 223. It would be pointless to ask in the present context whether any "physical reality" is to be assigned to electromagnetic fields or whether, as some writers maintain, electromagnetic fields are only a "mathematical fiction." It is sufficient to note that, whatever its "ultimate status," the field concept in physics represents a mode of analysis which can be distinguished from the particle approach.

[29] The technique to which reference is made is the device of retarded potentials. Cf. the remarks in Max Mason and Warren Weaver, *The Electromagnetic Field,* Chicago, 1929, Introduction.

illustrated by the charges on the insulated conductor, the law that formulates the distribution of charge densities can be deduced from assumptions concerning the behavior of charged particles.[30]

The upshot of this discussion of organic unities is that the question whether they can be analyzed from the additive point of view does not possess a general answer. Some functional wholes certainly can be analyzed in that manner, while for others (for example, living organisms) no fully satisfactory analysis of that type has yet been achieved. Accordingly, the mere fact that a system is a structure of dynamically interrelated parts does not suffice, by itself, to prove that the laws of such a system cannot be reduced to some theory developed initially for certain assumed constituents of the system. This conclusion may be meager; but it does show that the issue under discussion cannot be settled, as so much of extant literature on it assumes, in a wholesale and a priori fashion.

[30] Cf., for example, O. D. Kellogg, *Foundations of Potential Theory*, Berlin, 1929, Chap. 7.

Mechanistic Explanation and Organismic Biology

12 The analytic methods of the modern natural sciences are universally admitted to be competent for the study of all nonliving phenomena, even those which, like cosmic rays and the weather, are still not completely understood. Moreover, attempts at unifying special branches of physical science, by reducing their several systems of explanation to an inclusive theory, are generally encouraged and welcomed. During the past four centuries these methods have also been fruitfully employed in the study of living organisms; and many features of vital processes have been successfully explained in physicochemical terms. Outstanding biologists as well as physical scientists have therefore concluded that the methods of the physical sciences are fully adequate to the materials of biology, and many of these scientists have been confident that eventually the whole of biology would become simply a chapter of physics and chemistry.

But, despite the undeniable successes of physicochemical explanations in the study of living things, biologists of unquestioned competence continue to regard such explanations as not entirely adequate for the subject matter of biology. Most biologists are in general agreement that vital processes, like nonliving ones, occur only under determinate physicochemical conditions and form no exceptions to physicochemical laws. Some of them nevertheless maintain that the mode of analysis required for understanding living phenomena is fundamentally different from that which obtains in the physical sciences. Opposition to the systematic ab-

398

sorption of biology into physics and chemistry is sometimes based on the practical ground that it does not conform to the correct strategy of biological research. However, such opposition is often also supported by theoretical arguments which aim to show that the reduction of biology to physicochemistry is inherently impossible. Biology has long been an area in which crucial issues in the logic of explanation have been the subject of vigorous debate. In any event, it is instructive to examine some of the reasons biologists commonly advance for the claim that the logic of explanatory concepts in biology is distinctive of the science and that biology is an inherently autonomous discipline.

What are the chief supports for this claim?

1. Let us first dispose of two less weighty ones. Although it is difficult to formulate in precise terms the generic differences between the living and the nonliving, no one seriously doubts the obvious fact that there are such differences. Accordingly, the various "life sciences" are concerned with special questions that are patently unlike those with which physics and chemistry deal. In particular, biology studies the anatomy and physiology of living things, and investigates the modes and conditions of their reproduction, development, and decay. It classifies vital organisms into types or species; and it inquires into their geographic distribution, their lines of descent, and the modes and conditions of their evolutionary changes. Biology also analyzes organisms as structures of interrelated parts and seeks to discover what each part contributes to the maintenance of the organism as a whole. Physics and chemistry, on the other hand, are not specifically concerned with such problems, although the subject matter of biology also falls within the province of these sciences. Thus a stone and a cat when dropped from a height exhibit behaviors which receive a common formulation in the laws of mechanics; and cats as well as stones therefore belong to the subject matter of physics. Nevertheless, cats possess structural features and engage in processes in which physics and chemistry, at any rate in their current form, are not interested. Stated more formally, biology employs expressions referring to identifiable characteristics of living phenomena (such as 'sex,' 'cellular division,' 'heredity,' or 'adaptation') and asserts laws containing them (such as 'Hemophilia among humans is a sex-linked hereditary trait') that do not occur in the physical sciences and are not at present definable or derivable within these sciences. Accordingly, while the subject matter of biology and the physical sciences is not disparate, and though biology makes use of distinctions and laws borrowed from the physical sciences, the two sciences do not at present coincide.

It is no less evident that the techniques of observation and experimentation in biology are in general different from those current in the physical sciences. To be sure, some tools and techniques of observations, measurement, and calculation (such as lenses, balances, and algebra) are used in both groups of disciplines. But biology also requires special skills (such as those involved in the dissection of organic tissues) which serve no purpose in physics; and physics employs techniques (such as those needed for handling high-voltage currents) that are irrelevant in present-day biology. A physical scientist untrained in the special techniques of biological research is no more likely to perform a biological experiment successfully than is a pianist untutored in playing wind instruments likely to perform well on an oboe.

These differences between the special problems and techniques of the physical and biological sciences are sometimes cited as evidence for the inherent autonomy of biology, and for the claim that the analytical methods of physics are not fully adequate to the objectives of biological inquiry. However, though the differences are genuine, they certainly do not warrant such conclusions. Mechanics, electromagnetism, and chemistry, for example, are *prima facie* distinct branches of physical science, in each of which different special problems are pursued and different techniques are employed. But as we have seen, these are not sufficient reasons for maintaining that each of those divisions of physical science is an autonomous discipline. If there is a sound basis for the alleged absolute autonomy of biology, it must be sought elsewhere than in the differences between biology and the physical sciences which have been noted thus far.

2. What, then, are the weightier reasons which support that allegation? The main ones appear to be as follows. Vital processes have a *prima facie* purposive character; organisms are capable of self-regulation, self-maintenance, and self-reproduction, and their activities seem to be directed toward the attainment of goals that lie in the future. It is usually admitted that one can study and formulate the morphological characteristics of plants and animals in a manner comparable with the way physical sciences investigate the structural traits of nonliving things. Thus, the categories of analysis and explanation in physics are generally held to be adequate for studying the gross and minute anatomy of the human kidney, or the serial order of its development. But morphological studies are only one part of the biologist's task, since it also includes inquiry into the *functions* of structures in sustaining the activities of the organism as a whole. Thus, biology studies the role played by the kidney and its microscopic structure in preserving the chemical composition of the blood, and thereby in maintaining the whole body and its other parts in their characteristic activities. It is such manifestly "goal-directed" be-

havior of living things that is often regarded as requiring a distinctive category of explanation in biology.

Moreover, living things are organic wholes, not "additive systems" of independent parts, and the behavior of these parts cannot be properly understood if they are regarded as so many isolable mechanisms. The parts of an organism must be viewed as internally related members of an integrated whole. They mutually influence one another, and their behavior regulates and is regulated by the activities of the organism as a whole. Some biologists have argued that the coordinated, adaptive behavior of living organisms can be explained only by assuming a special vitalistic agent; others believe that an explanation is possible in terms of the hierarchical organization of internally related parts of the organism. But in either case, so it is frequently claimed, biology cannot dispense with the notion of organic unity; and in consequence it must use modes of analysis and formulation that are unmistakably *sui generis*.

Accordingly, two main features are commonly alleged to differentiate biology from the physical sciences in an essential way. One is the dominant place occupied by *teleological* explanations in biological inquiry. The other is the use of conceptual tools uniquely appropriate to the study of systems whose total behavior is not the resultant of the activities of independent components. We must now examine these claims in some detail.

I. *The Structure of Teleological Explanations*

Almost any biological treatise or monograph yields conclusive evidence that biologists are concerned with the functions of vital processes and organs in maintaining characteristic activities of living things. In consequence, if "teleological analysis" is understood to be an inquiry into such functions, and into processes directed toward attaining certain end-products, then undoubtedly teleological explanations are pervasive in biology. In this respect, certainly, there appears to be a marked difference between biology and the physical sciences. It would surely be an oddity on the part of a modern physicist were he to declare, for example, that atoms have outer shells of electrons in order to make chemical unions between themselves and other atoms possible. In ancient Aristotelian science, categories of explanations suggested by the study of living things and their activities (and in particular by human art) were made canonical for all inquiry. Since nonliving as well as living phenomena were thus analyzed in teleological terms—an analysis which made the notion of final cause focal—Greek science did not assume a fundamental cleavage between biology and other natural science. Modern science, on the other hand, regards final causes to be vestal virgins which bear no fruit in the study of physical and chemical phenomena; and, because of the association of teleological explanations with the doctrine that goals or ends of activity are dynamic agents in their own realizations, it tends to view such explanations as a species of obscurantism. But does the presence of teleological explanations in biology and their apparent absence from the physical sciences entail the absolute autonomy of the former? We shall try to show that it does not.

1.   Quite apart from their association with the doctrine of final causes, teleological explanations are sometimes suspect in modern natural science because they are assumed to invoke purposes or ends-in-view as causal factors in natural processes. Purposes and deliberate goals admittedly play important roles in human activities, but there is no basis whatever for assuming them in the study of physicochemical and most biological phenomena. However, as has already been noted, a great many explanations counted as teleological do not postulate any purposes or ends-in-view; for explanations are often said to be "teleological" only in the sense that they specify the *functions* which things or processes possess. Most contemporary biologists certainly do not impute purposes to the organic parts of living things whose functions are investigated; most of them would probably also deny that the means-ends relationships discovered in the organization of living creatures are the products of some deliberate plan on the part of a purposeful agent, whether divine or in some other manner supranatural. To be sure, there are biologists who postulate psychic states as concomitants and even as directive forces of all organic behavior. But such biologists are in a minority; and they usually support their views by special considerations that can be distinguished from the facts of functional or teleological dependencies which most biologists do not hesitate to accept. Since the word 'teleology' is ambiguous, confusion and misunderstandings would doubtless be prevented if the word were eliminated from the vocabulary of biology. But biologists do use it, and say they are giving a teleological explanation when, for example, they explain that the function of the alimentary canal in vertebrates is to prepare ingested materials for absorption into the bloodstream. The crucial point is that when biologists do employ teleological language they are not necessarily committing the pathetic fallacy or lapsing into anthropomorphism.

We shall therefore assume that teleological (or functional) statements in biology normally neither assert nor presuppose in the materials under discussion either manifest or latent purposes, aims, objectives, or goals. Indeed, it seems safe to suppose that biologists would generally deny they are postulating any conscious or implicit ends-in-view even when they employ such words as 'purpose' in their functional analyses—as when the 'purpose' (i.e., the function) of kidneys in the pig is said to be that of eliminating various waste products from the bloodstream of the organism. On the other hand, we shall adopt as the mark of a teleological statement in biology, and as the feature that distinguishes such statements from nonteleological ones, the occurrence in the former but not in the latter of such typical locutions as 'the function of,' 'the purpose of,' 'for the sake of,' 'in order that,' and the like—more generally, the occurrence of expressions signifying a means-ends nexus.

Nevertheless, despite the *prima facie* distinctive character of teleological (or functional) explanations, we shall first argue that they can be reformulated, without loss of asserted content, to take the form of nonteleological ones, so that in an important sense teleological and nonteleological explanations are equivalent. To this end, let us consider a typical teleological statement in biology, for example, 'The function of chlorophyll in plants is to enable plants to perform photosynthesis (i.e., to form starch from carbon dioxide and water in the presence of sunlight).' This statement accounts for the presence of chlorophyll (a certain substance A) in plants (in every member S of a class of systems, each of which has a certain organization C of component parts and processes). It does so by declaring that, when a plant is provided with water, carbon dioxide, and sunlight (when S is placed in a certain "internal" and "external" environment E), it manufactures starch (a certain process P takes place yielding a definite product or outcome) only if the plant contains chlorophyll. The statement usually carries with it the additional tacit assumption that without starch the plant cannot continue its characteristic activities, such as growth and reproduction (it cannot maintain itself in a certain state G); but for the present we shall ignore this further claim.

Accordingly, the teleological statement is a telescoped argument, so that when the content is unpacked it can be rendered approximately as follows: When supplied with water, carbon dioxide, and sunlight, plants produce starch; if plants have no chlorophyll, even though they have water, carbon dioxide, and sunlight, they do not manufacture starch; hence, plants contain chlorophyll. More generally, a teleological statement of the form 'The function of A in a system S with organization C is to enable S in environment E to engage in process P' can be formulated more explicitly by: Every system S with organization C and in environment E engages in process P; if S with organization C and in environment E does not have A, then S does not engage in P; hence, S with organization C must have A.

It is clearly not relevant in the present context to inquire whether the premises in this argument are adequately supported by competent evidence. However, because the issue is sometimes raised in discussions of teleological explanations, at least passing notice deserves to be given to the question of whether chlorophyll is really necessary to plants and whether they could not manufacture starch (or other substances essential for their maintenance) by some alternative process not requiring chlorophyll. For, if the presence of chlorophyll is not actually necessary for the production of starch (or if plants can maintain themselves without the mechanism of photosynthesis), so it has been urged, the second premise in the above argument is untenable. The premise would then have to be modified; and in its emended form it would assert that chlorophyll is an element in a set of conditions that is *sufficient* (but not necessary) for the production of starch. In that case, however, the new argument with the emended premise would be invalid, so that the proposed teleological explanation of the presence of chlorophyll in plants would apparently be unsatisfactory.

This objection is in part well-taken. It is certainly *logically* possible that plants might maintain themselves without manufacturing starch, or that processes in living organisms might produce starch without requiring chlorophyll. Indeed, there are plants (the funguses) that can flourish without chlorophyll; and in general, there is more than one way of skinning a cat. On the other hand, the above teleological explanation of the occurrence of chlorophyll in plants is presumably concerned with living organisms having certain determinate forms of organization and definite modes of behavior—in short, with the so-called "green plants." Accordingly, although living organisms (plants as well as animals) capable of maintaining themselves without processes involving the operation of chlorophyll are both abstractly and physically possible, there appears to be no evidence whatever that in view of the limited capacities green plants possess as a consequence of their *actual* mode of organization, these organisms can live without chlorophyll.

Two important complementary points thus emerge from these considerations. In the first place, teleological analyses in biology (or in other sciences in which such analyses are pursued) are not explorations of merely logical possibilities, but deal with the actual functions of definite components in concretely given living systems. In the second place, on pain of failure to recognize the possibility of alternative mechanisms for achieving some end-product, and of unwittingly (and perhaps mistakenly) assuming that a process known to be indispensable in a given class of systems is also indispensable in a more inclusive class, a teleological explanation must articulate with exactitude both the character of the end-product and the defining traits of the systems manifesting them, relative to which the indicated processes are supposedly indispensable.

In any event, however, the above teleological account of chlorophyll, in its expanded form, is simply an illustration of an explanation that conforms to the deductive model, and contains no locution distinctive of teleological statements. Accordingly, the initial, unexpanded statement about chlorophyll appears to assert nothing that is not asserted by 'Plants perform photosynthesis only if they contain chlorophyll,' or alternatively by 'A necessary condition for the occurrence of photosynthesis in plants is the presence of chlorophyll.' These latter statements do not explicitly ascribe a function to chlorophyll, and in that sense are therefore not teleological formulations. If this example is taken as a paradigm, it seems that, when a function is ascribed to a constituent element in an organism, the content of the teleological statement is fully conveyed by another statement that is not explicitly teleological and that simply asserts a necessary (or possibly a necessary and sufficient) condition for the occurrence of a certain trait or activity of the organism. In the light of this analysis, therefore, a teleological explanation in biology indicates the *consequences* for a given biological system of a constituent part or process; the equivalent nonteleological formulation of this explanation, on the other hand, states some of the *conditions* (sometimes, but not invariably, in physicochemical terms) under which the system persists in its characteristic organization and activities. The difference between a teleological explanation and its equivalent nonteleological formulation is thus comparable to the difference between saying that Y is an effect of X, and saying that X is a cause or condition of Y. In brief, the difference is one of selective attention, rather than of asserted content.

This point can be reinforced by another consideration. If a teleological explanation had an asserted content different from the content of every conceivable nonteleological statement, it would be possible to cite procedures and evidence employed for establishing the former that differ from the procedures and evidence required for warranting the latter. But in point of fact there appear to be no such procedures and evidence. Consider, for example, the teleological statement 'The function of the leucocytes in human blood is to defend the body against foreign microorganisms.' Now whatever may be the evidence that warrants this statement, that evidence also confirms the nonteleological statement 'Unless human blood contains a sufficient number of leucocytes, certain normal activities of the body are impaired,' and conversely. If this is so, however, there is a strong presumption that the two statements do not differ in factual content. More generally, if, as seems to be the case, the conceivable evidence for any given teleological explanation is identical with the conceivable evidence for a certain nonteleological one, the conclusion appears inescapable that those statements cannot be distinguished with respect to what they *assert*, even though they are distinguishable in other ways.

2. This proposed equation of teleological and nonteleological explanations must nevertheless face a fundamental objection. Many biologists would perhaps admit that a teleological statement *implies* a certain nonteleological one; but some of them, at any rate, are prepared to maintain that the latter statement generally does not in turn imply the former one, and that in consequence the alleged equivalence between the statements does not in fact hold.

The claim that there is indeed no such equivalence can be forcefully presented as follows. If there were such an equivalence, not only could a teleological explanation be replaced by a nonteleological one, but conversely a nonteleological explanation could also be replaced by a teleological one. In consequence, the customary statements of laws and theories in the physical sciences would be translatable without change in asserted content into teleological formulations. In point of fact, however, modern physical science does not appear to sanction such reformulations. Indeed, most physical scientists would doubtless resist the introduction of teleological statements into their disciplines as a misguided attempt to reinstate the point of view of Greek and medieval science. For example, the statement 'The volume of a gas at constant temperature varies inversely with its pressure' is a typical physical law, which is entirely free of teleological connotations. If it were equivalent to a teleological statement, its equivalent (constructed on the model of the example adopted above as paradigmatic) would presumably be 'The function of a varying pressure in a gas at constant temperature is to produce an inversely varying volume of the gas,' or perhaps 'Every gas at constant temperature under a variable pressure alters its volume in order to keep the product of the pressure and the volume constant.' But most physicists would undoubtedly regard these formulations as preposterous, and at best as misleading. Accordingly, if no teleological statement can correctly translate a law of physics, the contention that for every teleological statement a logically equivalent nonteleological one can be constructed seems hardly tenable. There must therefore be some important difference between teleological and nonteleological statements, so the objection concludes, that the discussion has thus far failed to make explicit.

The difficulty just expounded cannot be disposed of easily. To assess it adequately, we must consider the type of subject matter in which teleological analyses are currently undertaken, and in which teleological explanations are not rejected ostensibly as a matter of general principle.

a. The attitude of physical scientists toward teleological formulations in their disciplines is doubtless as alleged in this objection. Nevertheless, this fact is not completely decisive on the point at issue. Two comments are in order which tend to weaken its critical force.

In the first place, it is not entirely accurate to maintain that the physical sciences never employ formulations that have at least the appearance of teleological statements. As is well known, some physical laws and theories are often expressed in so-called "isoperimetric" or "variational" form, rather than in the more familiar manner of numerical or differential equations. When laws and theories are expressed in this fashion, they strongly resemble teleological formulations, and have in fact been frequently assumed to express a teleological ordering of events and processes. For example, an elementary law of optics states that the angle of incidence of a light ray reflected by a surface is equal to the angle of reflection. However, this law can also be rendered by the statement that a light ray travels in such a manner that the length of its actual path (from its source to reflecting surface to its terminus) is the minimum of all possible paths. More generally, a considerable part of classical as well as contemporary physical theory can be stated in the form of "extremal" principles. These principles assert that the actual development of a system proceeds in such a manner as to minimize or maximize some magnitude which represents the possible configurations of the system.[1]

The discovery that the principles of mechanics can be given such extremal formulations was once considered as evidence for the operation of a divine plan throughout nature. This view was made prominent by Maupertuis, an eighteenth-century thinker who was perhaps the first to state mechanics in variational form; and it was widely accepted in the eighteenth and nineteenth centuries. Such theological interpretations of extremal principles are now almost universally recognized to be entirely gratuitous; and with rare exceptions, physicists today do not accept the earlier claim that extremal principles entail the assumption of a plan or purpose animating physical processes. The use of such principles in physical science nevertheless does show that the dynamical structure of physical systems can be formulated so as to make focal the effect of constituent elements and subsidiary processes upon certain global properties of the system taken as a whole. If physical scientists dislike teleo-

---

[1] Cf. A. D'Abro, *The Decline of Mechanism in Modern Physics*, New York, 1939, Chap. 18; Adolf Kneser, *Das Prinzip der kleinsten Wirkung*, Leipzig, 1928; Wolfgang Yourgrau and Stanley Mandelstam, *Variational Principles in Dynamics and Quantum Theory*, London, 1955.

It can in fact be shown that, when certain very general conditions are satisfied, all quantitative laws can be given an "extremal" formulation.

logical language in their own disciplines, it is not because they regard teleological notions in this sense as foreign to their task. Their dislike stems in some measure from the fear that, except when such teleological language is made rigorously precise through the use of quantitative formulations, it is apt to be misunderstood as connoting the operation of purposes.

In the second place, the physical sciences, unlike biology, are in general not concerned with a relatively special class of organized bodies, and they do not investigate the conditions making for the persistence of some selected physical system rather than of others. When a biologist ascribes a function to the kidney, he tacitly assumes that it is the kidney's contribution to the maintenance of the living animal which is under discussion; and he ignores as irrelevant to his primary interest the kidney's contribution to the maintenance of any other system of which it may also be a constituent. On the other hand, a physicist generally attempts to discuss the effects of solar radiation upon a wide variety of things; and he is reluctant to ascribe a "function" to the sun's radiation, because no one physical system of which the sun is a part is of greater interest to him than is any other such system. And similarly for the law relating the pressure and volume of a gas: if a physicist views with suspicion the formulation of this law in functional or teleological language, it is because (in addition to the reasons which have been or will be discussed) he does not regard it as his business to assign special importance, even if only by vague suggestion, to one rather than another consequence of varying pressures in a gas.

b. However, the discussion thus far can be accused, with some justice, of naïveté if not of irrelevance, on the ground that it has ignored completely the fundamental point, namely, the "goal-directed" character of organic systems. It is because living things exhibit in varying degrees adaptive and regulative structures and activities, while the systems studied in the physical sciences do not—so it is frequently claimed—that teleological explanations are peculiarly appropriate for biological systems but not for physical systems. Thus, because the solar system, or any other system of which the sun is a part, does not tend to persist in some integrated pattern of activities in the face of environmental changes, and because the constituents of the system do not undergo mutual adjustments so as to maintain this pattern in relative independence from the environment, it is preposterous to ascribe any function to the sun or to solar radiation. Nor does the fact that physics can state some of its theories in the form of extremal principles, so the objection continues, minimize the differences between biological and purely physical systems. It is true that a physical system develops in such a way as to minimize or maximize a certain magnitude which represents a property of the system as a whole. But physical systems are not organized to maintain, in the face of considerable alterations in their environment, some *particular* extremal values of such magnitudes, or to develop under widely varying conditions in the direction of realizing some particular values of such magnitudes.

Biological systems, on the other hand, do possess such organization, as a single example (which could be matched by an indefinite number of others) makes quite clear. The human body maintains many of its characteristics in a relatively steady state (or homeostasis) by complicated but coordinated physiological processes. Thus, the internal temperature of the body must remain fairly constant if it is not to be fatally injured. In point of fact, the temperature of the normal human being varies during a day only from about 97.3° F to 99.1° F, and cannot fall much below 75° F or rise much above 110° F without permanent injury to the body. However, the temperature of the external environment can fluctuate much more widely than this; and it is clear from elementary physical considerations that the body's characteristic activities would be profoundly impaired or curtailed unless it were capable of compensating for such environmental changes. But the body is indeed capable of doing just this; and in consequence its normal activities can continue, in relative independence of the temperature of the environment—provided, of course, that the environmental temperature does not fall outside a certain interval of magnitudes. The body achieves this homeostasis by means of a number of mechanisms, which serve as a series of defenses against shifts in the internal temperature. Thus, the thyroid gland is one of several that control the body's basal metabolic rate (which is the measure of the heat produced by combustion in various cells and organs); the heat radiated or conducted through the skin depends on the quantity of blood flowing through peripheral vessels, a quantity which is regulated by dilation or contraction of these vessels; sweating and the respiration rate determine the quantity of moisture that is evaporated, and so affect the internal temperature; adrenaline in the blood also stimulates internal combustion, and its secretion is affected by changes in the external temperature; and automatic muscular contractions involved in shivering are an additional source of internal heat. There are thus physiological mechanisms in the body that automatically preserve its internal temperature, despite disturbing conditions in the body's internal and external environment.[2]

Three separate questions, frequently confounded, are raised by such facts of biological organization. (1) Is it possible to formulate in gen-

[2] Cf. Walter B. Cannon, *The Wisdom of the Body,* New York, 1932, Chap. 12.

eral but fairly precise terms the distinguishing structure of "goal-directed" systems, but in such a way that the analysis is neutral with respect to assumptions concerning the existence of purposes or the dynamic operation of goals as instruments in their own realization? (2) Does the fact, if it is a fact, that teleological explanations are customarily employed only in connection with "goal-directed" systems constitute relevant evidence for deciding the issue of whether a teleological explanation is equivalent to some nonteleological one? (3) Is it possible to explain in purely physicochemical terms—that is, exclusively in terms of the laws and theories of current physics and chemistry—the operations of biological systems? This third question will not concern us for the present, though we shall return to it later; but the other two require our immediate attention.

i. Since antiquity there have been many attempts at constructing machines and physical systems simulating the behavior of living organisms in one respect or another. None of these attempts has been entirely successful, for it has not been possible thus far to manufacture in the workshop and out of inorganic materials any device that acts fully like a living body. Nevertheless, it has been possible to construct physical systems that are self-maintaining and self-regulating in respect to certain of their features, and which therefore resemble living organisms in at least this one important characteristic. In an age in which servo-mechanisms (governors on engines, thermostats, automatic airplane pilots, electronic calculators, radar-controlled anti-aircraft firing devices, and the like) no longer excite wonder, and in which the language of cybernetics and "negative feedbacks" has become widely fashionable, the imputation of "goal-directed" behavior to purely physical systems certainly cannot be rejected as an absurdity. Whether "purposes" can also be imputed to such physical systems, as some expounders of cybernetics claim,[8] is perhaps doubtful, though the question is in large measure a semantic one; and in any event, this further issue is not relevant to the present context of discussion. Moreover, it is worth noting that the possibility of constructing self-regulating physical systems does not, by itself, constitute a proof that the activities of living organisms can be explained in exclusively physicochemical terms. Nevertheless, the fact that such systems have been constructed does suggest that there is no sharp demarcation setting off the teleological organizations, often assumed to be

[8] Cf. Arturo Rosenblueth, Norbert Wiener, Julian Bigelow, "Behavior, Purpose and Teleology," *Philosophy of Science*, Vol. 10 (1943); Norbert Wiener, *Cybernetics*, New York, 1948; A. M. Turing, "Computing Machines and Intelligence," *Mind*, Vol. 59 (1950); Richard Taylor, "Comments on a Mechanistic Conception of Purposefulness," *Philosophy of Science*, Vol. 17 (1950), and the reply by Rosenblueth and Wiener with a rejoinder by Taylor in the same volume.

distinctive of living things, from the goal-directed organizations of many physical systems. At the minimum, that fact offers strong support for the presumption that the teleologically organized activities of living organisms and of their parts can be analyzed without requiring the postulation of purposes or goals as dynamic agents.

With the homeostasis of the temperature of the human body as an exemplar, let us now state in general terms the formal structure of systems possessing a goal-directed organization.[4] The characteristic feature of such systems is that they continue to manifest a certain state or property G (or that they exhibit a persistence of development "in the direction" of attaining G) in the face of a relatively extensive class of changes in their external environments or in some of their internal parts —changes which, if not compensated for by internal modification in the system, would result in the disappearance of G (or in an altered direction of development of the systems). The abstract pattern of organization of such systems can be formulated with considerable precision, although only a schematic statement of that pattern can be presented in what follows.

Let S be some system, E its external environment, and G some state, property, or mode of behavior that S possesses or is capable of possessing under suitable conditions. Assume for the moment (this assumption will eventually be relaxed) that E remains constant in all relevant respects, so that its influence upon the occurrence of G in S may be ignored. Suppose also that S is analyzable into a structure of parts or processes, such that the activities of a certain number (possibly all) of them are causally relevant for the occurrence of G. For the sake of simplicity, assume that there are just three such parts, each capable of being in one of several distinct conditions or states. The state of each part at any given time will be represented by the predicates 'A_x,' 'B_y,' and 'C_z,' respectively, with numerical values of the subscripts to indicate the different particular states of the corresponding parts. Accordingly, 'A_x,' 'B_y,' and 'C_z' are state variables, though they are not necessarily numerical variables, since numerical measures may not be available for representing the states of the parts; and the state of S that is causally relevant to G at any given time will thus be expressed by a specialization of the matrix '$(A_x B_y C_z)$.' The state variables may, however, be quite complex in form—for example, 'A_x' may represent the state of the peripheral blood vessels in a human body at a given time—and they may be either individual or

4 The following discussion is heavily indebted to G. Sommerhoff, *Analytical Biology*, London, 1950. Cf. also Alfred J. Lotka, *Elements of Physical Biology*, New York, 1926, Chap. 25; W. Ross Ashby, *Design for a Brain*, London, 1953, and *An Introduction to Cybernetics*, London, 1956; and R. B. Braithwaite, *Scientific Explanation*, Cambridge, 1954, Chap. 10.

statistical coordinates. But in order to avoid inessential complications in exposition, we shall suppose that, whatever the nature of the state variables, in respect to the states they represent S is a deterministic system: the states of S change in such a way that, if S is in the same state at any two different moments, the corresponding states of S after equal lapses of time from those moments will also be the same.

One further important general assumption must also be made explicit. Each of the state variables can be assigned any particular "value" to characterize a state, provided the value is compatible with the known character of the part of S whose state the variable represents. In effect, therefore, the values of 'A_x' must fall into a certain restricted class K_A; and there are similar classes K_B and K_C for the permissible values of the other two state variables. The reason for these restrictions will be clear from an example. If S is the human body, and 'A_x' states the degree of dilation of peripheral blood vessels, it is obvious that this degree cannot exceed some maximum value; for it would be absurd to suppose that the blood vessels might have a mean diameter of, say, five feet. On the other hand, the possible values of one state variable *at a given time* will be assumed to be independent of the possible values of the other state variables *at that time*. This assumption must not be misunderstood. It does not assert that the value of a variable at one time is independent of the values of the other variables at some *other* time; it merely stipulates that the value of a variable at some specified instant is not a function of the values of the other variables *at that very same instant*. The assumption is the one normally made for state variables, and is introduced in part to avoid redundant coordinates of state. For example, the state variables in classical mechanics are the position and the momentum coordinates of a particle at an instant. Although the position of a particle at one instant will in general depend on its momentum (and position) at some *previous* time, the position at a given instant is not a function of the momentum *at that given instant*. If the position were such a function of the momentum, it is clear that the state of a particle in classical mechanics could be specified by just one state variable (the momentum), so that mention of the position would be redundant. In our present discussion we are similarly assuming that none of the state variables is dispensable, so that any combination of simultaneous values of the state variables yields a permissible specialization of the matrix '$(A_x B_y C_z)$,' provided that the values of the variables belong to the classes K_A, K_B, and K_C, respectively. This is tantamount to saying that, apart from the proviso, the state of S stipulated to be causally relevant to G must be so analyzed that the state variables employed for describing the state at a given time are mutually independent of one another.

Suppose now that if S is in the state $(A_0 B_0 C_0)$ at some initial time, then either S has the property G, or else a sequence of changes occurs in S as a consequence of which S will possess G at some subsequent time. Let us call such an initial state of S a "causally effective state with respect to G," or a "G-state" for short. Not every possible state of S need be a G-state, for one of the causally relevant parts of S may be in a certain state at a given time, such that *no* combination of possible states of the other parts will yield a G-state for S. Thus, suppose that S is the human body, G the property of having an internal temperature lying in the range 97° to 99° F, A_x again the state of the peripheral blood vessels, B_y the state of the thyroid glands, and C_z the state of the adrenal glands. It may happen that B_y assumes a value (e.g., corresponding to acute hyperactivity) such that for no possible values of A_x and C_z, respectively, will G be realized. It is of course also conceivable that no possible state of S is a G-state, so that in fact G is never realized in S. For example, if S is the human body and G the property of having an internal temperature lying in the range from 150° to 160°, then there is no G-state for S. On the other hand, more than one possible state of S may be a G-state. But if there is more than one possible G-state, then (since S has been assumed to be a deterministic system) the one that is realized at a given time is uniquely determined by the actual state of S at some previous time. The case in which there is more than one such possible G-state for S is of particular relevance to the present discussion, and we must now consider it more closely.

Assume once more that at some initial time t_0, the system S is in the G-state $(A_0B_0C_0)$. Suppose, however, that a change occurs in S so that in consequence A_0 is caused to vary, with the result that at time t_1 subsequent to t_0 the state variable 'A_x' has some other value. What value it will have at t_1 will in general depend on the particular changes that have taken place in S. We shall assume, however, that S will continue to be in a G-state at time t_1, provided that the values of 'A_x' at t_1 fall into a certain class K_A' (a subclass of K_A) containing more than one member, and provided also that certain further changes take place in the other state variables. To fix our ideas, suppose that A_1 and A_2 are the only possible members of K_A'; and assume also that neither $(A_1B_0C_0)$ nor $(A_2B_0C_0)$ is a G-state. In other words, if A_0 were changed into A_3 (a member of K_A but not of K_A'), S would no longer be in a G-state; but even though the new value of 'A_x' falls into K_A', if this were the only change in S the system would also no longer be in a G-state at time t_1. Let us assume, however, S to be so constituted that if A_0 is caused to vary so that the value of 'A_x' at time t_1 falls into K_A', there will be further compensatory changes in the values of some or all of the other state variables such that S continues to be in a G-state.

These further changes are stipulated to be of the following kind. If, as a concomitant of the change in A_0, the values of 'B_y' and 'C_z' at time t_1 fall into certain classes K_B' and K_C', respectively (where of course K_B' is a subclass, though not necessarily a proper subclass, of K_B, and K_C' is a subclass of K_C), then for each value in K_A' there is a unique pair of values, one member of the pair belonging to K_B' and the other to K_C', such that for those values S continues to be in a G-state at time t_1. These pairs of values can be taken to be elements in a certain class K_{BC}'. On the other hand, were the altered values of 'B_y' and 'C_z' not accompanied by the indicated changes in the value of 'A_x,' the system S would no longer be in a G-state at time t_1. In terms of the notation just introduced, accordingly, if at time t_1 the state variables of S have values such that two of them are members of a pair belonging to the class K_{BC}' while the value of the third variable is not the corresponding element in K_A', then S is not in a G-state. For example, suppose that, when A_0 changes into A_1, the initial G-state $(A_0B_0C_0)$ is changed into the G-state $(A_1B_1C_1)$, but that $(A_0B_1C_1)$ is not a G-state; and suppose also that when A_0 changes into A_2, the initial G-state is changed into the G-state $(A_2B_1C_2)$, with $(A_0B_1C_2)$ not a G-state. In this example, K_A' is the class (A_1, A_2); K_B' is the class (B_1); K_C' is the class (C_1, C_2); and K_{BC}' is the class of pairs $[(B_1, C_1), (B_1, C_2)]$, with A_1 corresponding to the pair (B_1, C_1) and A_2 to the pair (B_1, C_2).

Let us now bring together these various points, and introduce some definitions. Assume S to be a system satisfying the following conditions: (1) S can be analyzed into a set of related parts or processes, a certain number of which (say three, namely A, B, and C) are causally relevant to the occurrence in S of some property or mode of behavior G. At any time the state of S causally relevant to G can be specified by assigning values to a set of state variables 'A_x,' 'B_y,' and 'C_z.' The values of the state variables for any given time can be assigned independently of one another; but the possible values of each variable are restricted, in virtue of the nature of S, to certain classes of values K_A, K_B, and K_C, respectively. (2) If S is in a G-state at a given initial instant t_0 falling into some interval of time T, a change in any of the state variables will in general take S out of the G-state. Assume that a change is initiated in one of the state variables (say the parameter 'A'); and suppose that in fact the possible values of the parameter at time t_1 within the interval T but later than t_0 fall into a certain class K_A', with the proviso that if this were the sole change in the state of S the system would be taken out of its G-state. Let us call this initiating change a "primary variation" in S. (3) However, the parts A, B, and C of S are so related that, when the primary variation in S occurs, the remaining parameters also vary, and in point of fact their values at time t_1 fall into certain classes K_B' and K_C', respectively. These changes induced in B and C thus yield unique pairs of values for their parameters at time t_1, the pairs being elements of a class K_{BO}'. Were these latter changes the only ones in the initial G-state of S, unaccompanied by the indicated primary variation in S, the system would not be in a G-state at time t_1. (4) As a matter of fact, however, the elements of K_A' and K_{BO}' correspond to each other in a uniquely reciprocal manner, such that, when S is in a state specified by these corresponding values of the state variables, the system is in a G-state at time t_1. Let us call the changes in the state of S induced by the primary variation and represented by the pairs of values in K_{BO}' the "adaptive variations" of S with respect to the primary variation of S (i.e., with respect to possible values of the parameter 'A' in K_A'). Finally, when a system S satisfies all these assumptions for every pair of initial and subsequent instants in the interval T, the parts of S causally relevant to G will be said to be "directively organized during the interval of time T with respect to G"—or, more shortly, to be "directively organized," if the reference to G and T can be taken for granted.

This discussion of directively organized systems has been based on several simplifying assumptions. However, the analysis can be readily generalized for a system requiring the use of any number of state variables (including numerical ones), for changes in the state of a system that are initiated in more than one of the causally relevant parts of the system, and for continuous as well as discrete series of transitions from one G-state of a system to another.[5] Indeed, it is not difficult to develop

[5] When the state coordinates are assumed to be numerical, it is possible to formulate the conditions for a directively organized system as follows:

Let S be a system, G a trait of S, and 'x_1,' 'x_2,' . . . , 'x_n' the state variables for G. The variables are stipulated to be independent and continuous functions of the time; and superscripts will indicate their values at any given time t.

a) If S is a deterministic system with respect to G, the state of S at time t is uniquely determined by the state of S at some preceding time t_0. Hence

$$x_1{}^t = f_1(x_1{}^{t_0}, \ldots, x_n{}^{t_0}, t - t_0)$$
$$x_i{}^t = f_i(x_1{}^{t_0}, \ldots, x_n{}^{t_0}, t - t_0)$$
$$x_n{}^t = f_n(x_1{}^{t_0}, \ldots, x_n{}^{t_0}, t - t_0)$$

where the f_i's are single-valued functions of their arguments. Their first derivatives with respect to the time are also single-valued functions of their arguments and of no other functions of the time.

b) Since the special character of S imposes restrictions on the values of the state variables, the values of each variable 'x_i' will fall within an interval determined by a pair of numbers a_i and b_i. That is,

$$a_i \leq x_i \leq b_i$$

with $i \leq 1, 2, \ldots, n$, or alternately

$$x_i \in \Delta x_i$$

where Δx_i is some definite interval and 'ϵ' is the usual sign for class membership.

c) If S is in a G-state at a given time t falling into a given period of time T, the

within this framework of analysis the notion of a system exhibiting self-regulatory behaviors with respect to several G's at the same time, alternative (and even incompatible) G's at different times, a set of G's constituting a hierarchy on the basis of some postulated scale of "relative importance," or more generally a set of G's whose membership changes with time and circumstance. But apart from complexity, nothing immediately relevant would be gained by such extensions of the analysis; and the schematic and incompletely general definitions that have been presented will suffice for our purposes.

It will in any case be clear from the above account that if S is directively organized, the persistence of G is in an important sense independent of the variations in any one of the causally relevant parts of S, provided that these variations do not exceed certain limits. For although by hypothesis the occurrence of G in S depends upon S being in a G-state, and therefore upon the state of the causally relevant parts of S, an alteration in the state of one of those parts may be compensated by induced changes in one or more of the other causally relevant parts,

state variable must satisfy a set of conditions or equations. That S is in a G-state at time t can be expressed by requirement:

$$g_1(x_1{}^t, \ldots, x_n{}^t) = 0$$
$$\cdots\cdots\cdots\cdots\cdots$$
$$g_r(x_1{}^t, \ldots, x_n{}^t) = 0$$

where each g_j ($j = 1, 2, \ldots, r$) is a function differentiable with respect to each of the state variables, and $r < n$.

d) The values of each state variable '$x_i{}^t$' satisfying these equations defining a G-state of S fall into certain restricted intervals:

$$a_i \leq a_i{}^G \leq x_i{}^t \leq b_i{}^G \leq b_i$$

or alternately:

$$x_i{}^t \in \Delta x_i{}^G$$

where $\Delta x_i{}^G$ falls into the interval Δx_i.

e) Assume that S is in a G-state at the initial time t_0 during the period T, and that a change takes place in the value of some state variable 'x_k' so that at time t later than t_0 in T its value is $x_k{}^t$. The condition that this change is a G-preserving change (so that $x_k{}^t \in \Delta x_k{}^G$), is that for each function g_j:

$$\frac{\partial g_j}{\partial x_k{}^{t_0}} = \frac{\partial g_j}{\partial x_1{}^t} \frac{\partial x_1{}^t}{\partial x_k{}^{t_0}} + \frac{\partial g_j}{\partial x_2{}^t} \frac{\partial x_2{}^t}{\partial x_k{}^{t_0}} + \cdots + \frac{\partial g_j}{\partial x_n{}^t} \frac{\partial x_n{}^t}{\partial x_k{}^{t_0}} = 0$$

f) The system S is directively organized with respect to G during T if, when such G-preserving changes occur in any given state variable 'x_k,' there are compensating variations in one or more of the other state variables. Accordingly, there must be at least one function g_j such that in the partial differential equations just mentioned there are at least two nonvanishing summands. That is, there are at least two summands in one or more of these equations such that

$$\frac{\partial g_j}{\partial x_i{}^t} \frac{\partial x_i{}^t}{\partial x_k{}^{t_0}} \neq 0$$

so as to preserve S in its assumed G-state. The *prima facie* distinctive character of so-called "goal-directed" or teleological systems is thus formulated by the stated conditions for a directively organized system. The above analysis has therefore shown that the notion of a teleological system can be explicated in a manner not requiring the adoption of teleology as a fundamental or unanalyzable category. What may be called the "degree of directive organization" of a system, or perhaps the "degree of persistence" of some trait of a system, can also be made explicit in terms of the above analysis. For the property G is maintained in S (or S persists in its development, which eventuates in G) to the extent that the range K_A' of the possible primary variations is associated with the range of induced compensatory changes K_{BC}' (i.e., the adaptive variations) such that S is preserved in its G-state. The more inclusive the range K_A' that is associated with such compensatory changes, the more is the persistence of G independent of variations in the state of S. Accordingly, on the assumption that it is possible to specify a measure for the range K_A', the "degree of directive organization" of S with respect to variations in the state parameter 'A' could be defined as the measure of this range.

We may now relax the assumption that the external environment E has no influence upon S. But in dropping this assumption, we merely complicate the analysis, without introducing anything novel into it. For suppose that there is some factor in E which is causally relevant to the occurrence of G in S, and whose state at any time can be specified by some determinate form of the state variable 'F_w.' Then the state of the enlarged system S' (consisting of S together with E) which is causally relevant to the occurrence of G in S is specified by some determinate form of the matrix '$(A_x B_y C_z F_w)$,' and the discussion proceeds as before. However, it is generally not the case that a variation in any of the internal parts of S produces any significant variation in the environmental factors. What usually is the case is that the environmental factors vary quite independently of the internal parts; they do not undergo changes which compensate for changes in the state of S; and, while a limited range of changes in them may be compensated by changes in S so as to preserve S in some G-state, most of the states which environmental factors are capable of assuming cannot be so compensated by changes in S. It is customary, therefore, to talk of the "degree of plasticity" or the "degree of adaptability" of organic systems in relation to their environments, and not conversely. However, it is possible to define these notions without special reference to organic systems, in a manner analogous to the definition of the "degree of directive organization" of a system already suggested. Thus, suppose that the variations in the environmental state variable 'F,' assumed to be compensated by further changes in S so as to preserve S in some G-state, all fall into the class K_F'. If an appropriate measure for the magnitude of this class could be devised, the "degree of plasticity" of S with respect to the maintenance of some G in relation to F could then be defined as equal to the measure of K_F'.

This must suffice as an outline of the abstract structure of goal-directed or teleological systems. The account given deliberately leaves undiscussed the detailed mechanisms involved in the operation of particular teleological systems; and it simply assumes that all such systems can in principle be analyzed into parts which are causally relevant to the maintenance of some feature in those systems, and which stand to each other and to environmental factors in determinate relations capable of being formulated as general laws. The discovery and analysis of such detailed mechanisms is the task of specialized scientific inquiry. Accordingly, since the above account deals only with what is assumed to be the common distinctive structure of teleological systems, it is also entirely neutral on such substantive issues as to whether the operations of all teleological systems can be explained in exclusively physicochemical terms. On the other hand, if the account is at least approximately adequate, it requires a positive answer to the question whether the distinguishing features of goal-directed systems can be formulated without invoking purposes and goals as dynamic agents.

There is, however, one further matter that must be briefly discussed. The definition of directively organized systems has been so stated that it can be used to characterize both biological and nonvital systems. It is in fact easy to cite illustrations for the definition from either domain. The human body with respect to homeostasis of its internal temperature is an example from biology; a building equipped with a furnace and thermostat is an example from physicochemistry. Nevertheless, although the definition is not intended to distinguish between vital and nonvital teleological systems—for the differences between such systems must be stated in terms of the specific material composition, characteristics, and activities they manifest—it *is* intended to set off systems having a *prima facie* "goal-directed" character from systems usually not so characterized. The question therefore remains whether the definition does achieve this aim, or whether on the contrary it is so inclusive that almost *any* system (whether it is ordinarily judged to be goal-directed or not) satisfies it.

Now there are certainly many physicochemical systems that are not ordinarily regarded as being "goal-directed" but that nevertheless appear to conform to the definition of directively organized systems proposed above. Thus, a pendulum at rest, an elastic solid, a steady electric current flowing through a conductor, a chemical system in thermodynamic equilibrium, are obvious examples of such systems. It seems therefore that the definition of directive organization—and in consequence the proposed analysis of "goal-directed" or "teleological" systems—fails to attain its intended objective. However, two comments are in order on the point at issue. In the first place, though we admittedly do distinguish between systems that are goal-directed and those which are not, the distinction is highly vague, and there are many systems which cannot be classified definitely as of one kind rather than another. Thus, is the child's toy sometimes known as the "walking beetle"—which turns aside when it reaches the edge of a table and fails to fall off, because an idle wheel is then brought into play through the action of an "antenna"—a goal-directed system or not? Is a virus such a system? Is the system consisting of members of some biological species that has undergone evolutionary development in a steady direction (e.g., the development of gigantic antlers in the male Irish elk), a goal-directed one? Moreover, some systems have been classified as "teleological" at one time and in relation to one body of knowledge, only to be reclassified as "nonteleological" at a later time, as knowledge concerning the physics of mechanisms improved. "Nature does nothing in vain" was a maxim commonly accepted in pre-Newtonian physics, and on the basis of the doctrine of "natural places" even the descent of bodies and the ascent of smoke were regarded as goal-directed. Accordingly, it is at least an open question whether the current distinction between systems that are goal-directed and those that are not invariably has an identifiable objective basis (i.e., in terms of differences between the actual organizations of such systems), and whether the *same* system may not often be classified in alternative ways depending on the perspective from which it is viewed and on the antecedent assumptions adopted for analyzing its structure.

In the second place, it is by no means certain that physical systems such as the pendulum at rest, which is not usually regarded as goal-directed, really do conform to the definition of "directively organized" systems proposed above. Consider a simple pendulum that is initially at rest and is then given a small impulse (say by a sudden gust of wind); and assume that apart from the constraints of the system and the force of gravitation the only force that acts on the bob is the friction of the air. Then on the usual physical assumptions, the pendulum will perform harmonic oscillations with decreasing amplitudes, and finally assume its initial position of rest. The system here consists of the pendulum and the various forces acting on it, while the property G is the state of the pendulum when it is at rest at the lowest point of its path of oscillation. By hypothesis, its length and the mass of the bob are fixed, and so is the force of gravitation acting on it, as well as the coefficient of damping; the variables are the impulsive force of the gust of wind, and the restoring force which operates on the bob as a consequence of the constraints of the system and of the presence of the gravitational field. How-ever—and this is the crucial point—these two forces are *not* independent of one another. Thus, if the effective component of the former has a certain magnitude, the restoring force will have an equal magnitude with an opposite direction. Accordingly, if the state of the system at a given time were specified in terms of state variables which take these forces as values, these state variables would not satisfy one of the stipu-lated conditions for state variables of directively organized systems; for the value of one of them at a given time is uniquely determined by the value of the other at that same time. In short, the values of these pro-posed state variables at any given instant are not independent.[6] It there-fore follows that the simple pendulum is not a directively organized system in the sense of the definition presented. Moreover, it is also pos-sible to show in a similar manner that a number of other systems, gen-erally regarded as nonteleological ones, fail to satisfy that definition. Whether one could show this for all systems currently so regarded is admittedly an open question. However, since there are at least some systems not usually characterized as teleological which must also be so characterized on the basis of the definition, the label of 'directively or-ganized system' whose meaning the definition explicates does not apply to everything whatsoever, and it does not baptize a distinction without a difference. There are therefore some grounds for claiming that the defi-nition achieves what it was designed to achieve, and that it formulates

[6] This can be shown in greater detail by considering the usual mathematical dis-cussion of the simple pendulum. If l is the length of the pendulum, m the mass of its bob, g the constant force of gravity, k the coefficient of damping due to air re-sistance, t the time as measured from some fixed instant, and s the distance of the bob along its path of oscillation from the point of initial rest, the differential equa-tion of motion of the pendulum (on the assumption that the amplitude of vibration is small) is

$$m\frac{d^2s}{dt^2} + k\frac{ds}{dt} + \frac{mg}{l}s = 0$$

If at time t_0 the pendulum is at rest, both s_0 and $v_0 \left[= \left(\frac{ds}{dt}\right)_0 \right]$ are zero, so that

$$\left(m\frac{d^2s}{dt^2}\right)_0 = 0;$$

i.e., no unbalanced forces are acting on the bob. Suppose now that at time t_1 the bob is at s_1 with a velocity v_1; the restoring force will then be

$$\left(m\frac{d^2s}{dt^2}\right)_1 = -kv_1 - \frac{mg}{l}s_1$$

But an impulsive force F_1 communicated to the bob at time t_1 uniquely determines the velocity v_1 and the position s_1 of the bob at that time. Hence the restoring force can be calculated, so that it is uniquely determined by the impulsive force.

the abstract structure commonly held to be distinctive of "goal-directed" systems.

ii.   We can now settle quite briefly the second question, on page 410, we undertook to examine, namely, whether the fact that teleological explanations are usually advanced only in connection with "goal-directed" systems affects the claim that, in respect to its asserted content, every teleological explanation is translatable into an equivalent nonteleological one. The answer is clearly in the negative, if such systems are analyzable as directively organized ones in the sense of the above definition. For on the supposition that the notion of a goal-directed system can be explicated in the proposed manner, the characteristics that ostensibly distinguish such systems from those not goal-directed can be formulated entirely in nonteleological language. In consequence, every statement about the subject matter of a teleological explanation can in principle be rendered in nonteleological language, so that such explanations together with all assertions about the contexts of their use are translatable into logically equivalent nonteleological formulations.

Why, then, does it seem odd to render physical statements such as Boyle's law in teleological form? The answer is plain, if indeed teleological statements (and in particular, teleological explanations) are normally advanced only in connection with subject matters that are assumed to be directively organized. The oddity does not stem from any difference between the explicitly asserted content of a physical law and of its purported teleologically formulated equivalent. A teleological version of Boyle's law appears strange and unacceptable because such a formulation would usually be construed as resting on the assumption that a gas enclosed in a volume is a directively organized system, in contradiction to the normally accepted assumption that a volume of gas is not such a system. In a sense, therefore, a teleological explanation does connote more than does its *prima facie* equivalent nonteleological translation. For the former presupposes, while the latter normally does not, that the system under consideration in the explanation is directively organized. Nevertheless, if the above analysis is generally sound, this "surplus meaning" of teleological statements can always be expressed in nonteleological language.

3.   On the hypothesis that a teleological explanation can always be translated, with respect to what it explicitly asserts, into an equivalent nonteleological one, let us now make more explicit in what way two such explanations nevertheless do differ. The difference appears to be as follows: Teleological explanations focus attention on the culminations and products of specific processes, and in particular upon the contribu-

tions of various parts of a system to the maintenance of its global prop-erties or modes of behavior. They view the operations of things from the perspective of certain selected "wholes" or integrated systems to which the things belong; and they are therefore concerned with characteristics of the parts of such wholes, only insofar as those traits of the parts are relevant to the various complex features or activities assumed to be dis-tinctive of those wholes. Nonteleological explanations, on the other hand, direct attention primarily to the conditions under which specified proc-esses are initiated or persist, and to the factors upon which the contin-ued manifestations of certain inclusive traits of a system are contingent. They seek to exhibit the integrated behaviors of complex systems as the resultants of more elementary factors, frequently identified as constituent parts of those systems; and they are therefore concerned with traits of complex wholes almost exclusively to the extent that these traits are de-pendent on assumed characteristics of the elementary factors. In brief, the difference between teleological and nonteleological explanations, as has already been suggested, is one of emphasis and perspective in for-mulation.

If this account is sound, the use of teleological explanations in the study of directively organized systems is as congruent with the spirit of modern science as is the use of nonteleological ones. This conclusion is confirmed by an examination of two currently held assessments of teleological ex-planations, one suggesting a limit to the value of such explanations, the other objecting in principle to their use.

a.   The claim has been advanced that, although teleological explana-tions are in general legitimate, they are useful only when the knowledge we happen to possess of directively organized systems is of a certain kind.[7] Our available information about the range of environmental changes to which such a system can make adaptive responses (i.e., about what we have called the "plasticity" of goal-directed systems) may have two sources. It may have the status simply of an extrapolation to a given system of inductive generalizations obtained from a direct experimental study of quite similar systems. For example, the knowledge we have at present concerning the plasticity of a particular human organism in maintaining its internal temperature in the face of changes in tempera-ture of the environment is based on our familiarity with the adaptive responses of other human bodies. On the view under consideration, teleological explanations in such cases are valuable, since they enable us to predict certain future behaviors of a given system from our knowledge concerning the past behaviors of similar systems—future behaviors that

⁷ R. B. Braithwaite, *Scientific Explanation*, pp. 333ff.

would otherwise not be predictable in the assumed state of our knowledge On the other hand, our information about the plasticity of a given system may have the status of a body of deductions from previously established causal laws concerning the mechanisms embodied in the system. In such cases, adaptive responses of a given system to environmental changes can be calculated with the help of general assumptions, and can be predicted without any familiarity with the past behaviors of similar systems. In consequence, teleological explanations in such cases are said to have little if any value.

Although the distinction between these two types of sources of available knowledge concerning the plasticity of directively organized systems is patently sound, it is nevertheless not evident why the line between valuable and useless teleological explanations should be drawn in the indicated manner. Questions about the value of an explanation are not decided by reference to the logical source of the explanatory premises, and can be answered only by examining the effective role an explanation plays in inquiry and in the communication of ideas. It is in any event far from certain that teleological explanations for goal-directed systems concerning which we possess theoretically based knowledge are invariably or normally regarded as otiose. For there are in fact many artificial self-regulating systems (such as engines with governors regulating their speed) whose plasticity can be deduced from general theoretical assumptions. Teleological explanations for various features of such systems nevertheless continue to fill many pages in technical treatises about those systems, and there is no good reason to suppose that the explanations are commonly regarded as so much worthless lumber.

b. It is sometimes objected, however, that teleological explanations are inexcusably parochial. They are based, so it is argued, on a tacit assumption that a special set of complex systems have a privileged status; and in consequence such explanations make focal the role of things and processes in maintaining just those systems and no others. Processes have no inherent termini, the objection continues, and cannot rightly be supposed to contribute exclusively to the maintenance of some unique set of wholes. It is therefore misleading to say, for example, that *the* function of the white cells in the human blood is to defend the human body against foreign microorganisms. This is undoubtedly *a* function of the leucocytes; and this particular activity may even be said to be *the* function of these cells from the perspective of the human body. But leucocytes are elements in other systems as well; for example, they are parts of the blood stream considered in isolation from the rest of the body, of the system composed of some virus colony together with these white cells, or of the more inclusive and complex solar system. These other systems are also capable of persisting in their "normal" organization and activities only under definite conditions; and, from the perspective of the maintenance of these numerous other systems, the leucocytes possess other functions.

One obvious reply to this objection is in the form of a *tu quoque*. It is as legitimate to focus attention on consequences, culminations, and uses as it is on antecedents, starting points, and conditions. Processes do not have inherent termini, but neither do they have absolute beginnings. Things and processes are in general not elements engaged in maintaining some exclusively unique whole, but neither are wholes analyzable into an exclusively unique set of constituents. It is nevertheless intellectually profitable in causal inquiries to focus attention on certain earlier stages in the development of a process rather than on later ones, and on one set of constituents of a system rather than on another set. Similarly, it is illuminating to select as the point of departure for the investigation of some problems certain complex wholes rather than others. Moreover, as we have seen, some things are parts of directively organized systems, but do not appear to be parts of more than one such system. The study of the unique functions of parts in such unique directively organized systems is therefore not a preoccupation that assigns without warrant a special importance to certain particular systems. On the contrary, it is an inquiry that is sensitive to fundamental and objectively identifiable differences in subject matter.

There is nevertheless a point to the objection. For the refractive influence of provincial human interests on the construction of teleological explanations is perhaps more often overlooked than it is in the case of nonteleological analyses. In consequence, certain end-products of processes and certain directions of change are frequently assumed to be inherently "natural," "essential," or "proper," while all others are then labeled as "unnatural," "accidental," or even "monstrous." Thus, the development of corn seeds into corn plants is sometimes said to be natural, while their transformation into the flesh of birds or men is asserted to be merely accidental. In a given context of inquiry, and in the light of the problem which initiates it, there may be ample justification for ignoring all but one direction of possible changes and all but one system of activities to whose maintenance things and processes contribute. But such disregard of other functions that things may have, and of other wholes of which things may be parts, does not warrant the conclusion that what is ignored is less genuine or natural than what receives selective attention.

4. One final point in connection with teleological explanations in biology must be briefly noted. As has already been mentioned, some biologists maintain that the distinctive character of biological explana-

tions appears in physiological inquiries, in which the functions of organs and vital processes are under investigation, even though most biologists are quite prepared to admit that no special categories of explanation are required in morphology or the study of structural traits. Accordingly, great stress has been placed by some writers on the contrast between structure and function, and on the difficulties in assessing the relative importance of each as a determinant of living phenomena. It is generally conceded that "the development of functions goes hand in hand with the development of structure," that living activity does not occur apart from a material structure, nor does vital structure exist save as a product of protoplasmic activity. In this sense, structure and function are commonly regarded as "inseparable aspects" of biological organization. Nevertheless, eminent biologists believe it is still an unresolved and perhaps insoluble problem "to what extent structures may modify functions or functions structures"; they regard the contrast between structure and function as presenting a "dilemma." [8]

But what is this contrast, why do its terms raise an apparently irresolvable issue, and what does one of its terms cover which allegedly requires a mode of analysis and explanation specific to biology? Let us first remind ourselves in what way a morphological study of some biological organ, say the human eye, differs from the corresponding physiological investigation. A structural account of the eye usually consists in a description of its gross and minute anatomy. Such an account therefore specifies the various parts of the organ, their shapes and relative spatial arrangements with respect to each other and other parts of the body, and their cellular and physicochemical compositions. The phrase "structure of the eye" therefore ordinarily signifies the spatial organization of its parts, together with the physicochemical properties of each part. On the other hand, a physiological account of the organ specifies the activities in which its various parts can or do participate, and the role these parts play in vision. For example, the ciliary muscles are shown to be capable of contracting and slackening, so that because of their connection with the suspensory ligament the curvature of the lens can be accommodated to near and far vision; or the lachrymal glands are identified as the sources of fluids which lubricate and cleanse the conjunctival membranes. In general, therefore, physiology is concerned with the char-

[8] Cf. Edwin G. Conklin, *Heredity and Environment*, Princeton, 1922, p. 32, and Edmund B. Wilson, *The Cell*, New York, 1925, p. 670. In a later volume Conklin declared that "the relation of mechanism to finalism is not unlike that of structure to function—they are two aspects of organization. The mechanistic conception of life is in the main a structural aspect, the teleological view looks chiefly to ultimate function. These two aspects of life are not antagonistic, but complementary."—*Man: Real and Ideal*, New York, 1943, p. 117.

acter, the order, and the consequences of the activities in which the parts of the eye may be engaged.

If this example is typical of the way biologists employ the terms, the contrast between structure and function is evidently a contrast between the *spatial* organization of anatomically distinguishable parts of an organ and the *temporal* (or spatiotemporal) organization of changes in those parts. What is investigated under each term of the contrasting pair is a mode of organization or a type of order. In the one case the organization is primarily if not exclusively a spatial one, and the object of the investigation is to ascertain the spatial distribution of organic parts and the modes of their linkage. In the other case the organization has a temporal dimension, and the aim of the inquiry is to discover sequential and simultaneous orders of change in the spatially ordered and linked parts of organic bodies. It is evident, therefore, that structure and function (in the sense in which biologists appear to use these words) are indeed "inseparable." For it is difficult to make sense of any supposition that a system of activities having a temporal organization is not also a system of spatially structured parts manifesting these activities. In any event, there is obviously no antithesis between an inquiry directed to the discovery of the spatial organization of organic parts and an inquiry addressed to ascertaining the spatiotemporal structures that characterize the activities of those parts.

A comparable distinction between types of inquiries can also be introduced in the physical sciences. Descriptive physical geography, for example, is concerned primarily with the spatial distribution and spatial relations of mountains, plains, rivers, and oceans; historical geology and geophysics, on the other hand, investigate the temporal and dynamic orders of change in which such geographic features are involved. Accordingly, if inquiries into structure and function were antithetical in biology, a comparable antithesis would also occur within the nonbiological sciences. Every inquiry involves discriminating selection from the great variety of patterns of relations embodied in a subject matter; and it is both convenient and unavoidable to direct some inquiries to one kind of pattern and other inquiries to different kinds. There seems to be no reason for generating a fundamental puzzle from the fact that living organisms exhibit both a spatial and a spatiotemporal structure of their parts.

What, then, is the unsolved or irresolvable issue raised by the biological distinction between structure and function? Two questions can be distinguished in this connection. It may be asked, in the first place, what spatial structures are required for the exercise of specified functions, and whether a change in the pattern of activities of an organism or of its parts is associated with any change in the distribution and spatial organization of the constituents of that system. This is obviously a matter to be settled by detailed empirical inquiry, and, though there are innumerable unsettled problems in this connection, they do not raise issues of fundamental principle. One school of philosophers and biological theorists, for example, maintains that the development of certain comparable organs in markedly different species can be explained only on the assumption of a "vital impulse" that directs evolution toward the attainment of some future function. Thus the fact that the eyes of the octopus and of man are anatomically similar, though the evolution of each species from eyeless ancestors has followed different lines of development, has been offered as evidence for the claim that no explanation of this convergence is possible in terms of the mechanisms of chance variation and adaptations. That fact has in consequence been used to support the view that there is an "undivided original vital impulse" which so acts on inert matter as to create appropriate organs for the function of vision.[9] But even this hypothesis, however vague and otherwise unsatisfactory it may be, involves in part factual issues; and if most biologists reject it, they do so largely because the available factual evidence supports more adequately a different theory of evolutionary development.

In the second place, one may ask just why it is that a given structure is associated with a certain set of functions, or conversely. Now this question may be understood as a demand for an explanation, perhaps in physicochemical terms, for the fact that when a living body has a given spatial organization of its parts it exhibits certain patterns of activities. When the question is so understood, it is far from being a preposterous one. Although we may not possess answers to it in most cases, we do have reasonably adequate answers in at least a few others, so that we have some ground for the presumption that our ignorance is not necessarily permanent. However, such explanations must contain as premises not only statements about the physicochemical constitution of the parts of a living thing and about the spatial organization of these parts, but also statements of physicochemical laws or theories. Moreover, at least some of these latter premises must assert connections between the spatial organization of physicochemical systems and the temporal patterns of activities. But if the question continues to be pressed, and an explanation is demanded for these latter connections as well, an impasse is finally reached. For the demand then in effect assumes that the temporal or causal structure of physical processes is deducible simply from the spatial organization of physical systems, or conversely; and neither assumption is in fact tenable.

[9] Cf. H. Bergson, *Creative Evolution*, New York, 1911, Chap. 1, and the brief but incisive critique of views similar to those of Bergson in George G. Simpson, *The Meaning of Evolution*, New Haven, 1949, Chap. 12. See also Theodosius Dobzhansky, *Evolution, Genetics and Man*, New York and London, 1955, Chap. 14.

It is possible, analogously, to give a quite accurate account of the spatial relations in which the various parts of a clock stand to one another. We can specify the sizes of its cogwheels, the location of the mainspring and the escapement wheel, and so on. But although such knowledge of the clock's spatial structure is indispensable, it is not sufficient for understanding how the clock will operate. We must also know the laws of mechanics, which formulate the temporal structure of the clock's behavior by indicating how the spatial distribution of its parts at one time is related to the distribution at a later time. However, this temporal structure cannot be deduced simply from the clock's spatial structure (or its "anatomy"), any more than its spatial structure at any given time can be derived from the general laws of mechanics. Accordingly, the question why a given anatomical structure is associated with specific functions may be irresolvable, not because it is beyond our capacities to answer it, but simply because the question in the sense in which it is intended asks for what is *logically* impossible. In short, anatomical structure does not *logically* determine function, though *as a matter of contingent fact* the specific anatomical structure possessed by an organism does set bounds to the kinds of activities in which the organism can engage. And conversely, the pattern of behavior exhibited by an organism does not *logically* imply a unique anatomical structure, though *in point of fact* an organism manifests specific modes of activity only when its parts possess a determinate anatomical structure of a definite kind.

It follows from these various considerations that the distinction between structure and function covers nothing that distinguishes biology from the physical sciences, or that necessitates the use in biology of a distinctive logic of explanation. It has not been the aim of the present discussion to deny the patent differences between biology and other natural sciences with respect to the role played by functional analyses. Nor has it been its aim to cast doubt on the legitimacy of such explanations in any domain in which they are appropriate because of the special character of the systems investigated. The objective of the discussion has been to show only that the prevalence of teleological explanations in biology does not constitute a pattern of explanation incomparably different from those current in the physical sciences, and that the use of such explanations in biology is not a sufficient reason for maintaining that this discipline requires a radically distinctive logic of inquiry.

II. *The Standpoint of Organismic Biology*

Vitalism of the substantive type advocated by Driesch and other biologists during the preceding century and the earlier decades of the present one is now almost entirely a dead issue in the philosophy of biology.

The issue has ceased to be focal, perhaps less as a consequence of the methodological and philosophical criticisms to which vitalism has been subjected, than because of the sterility of vitalism as a guide in biological research and the superior heuristic value of other approaches to the study of vital phenomena. Nevertheless, the historically influential Cartesian conception of biology as simply a chapter of physics continues to meet resistance. Many outstanding biologists who find no merit in vitalism are equally dubious about the validity of the Cartesian program; and they sometimes advance what they believe are conclusive reasons for affirming the irreducibility of biology to physics and the intrinsic autonomy of biological method. The standpoint from which this antivitalistic and yet antimechanistic thesis is currently advanced commonly carries the label of "organismic biology." The label covers a variety of special biological doctrines that are not always mutually compatible. Nonetheless, the doctrines falling under it generally share the common premise that explanations of the "mechanistic" type are not appropriate for vital phenomena. We shall now examine the main contentions of organismic biology.

1. Although organismic biologists deny the suitability if not always the possibility of "mechanistic theories" for vital processes, it is frequently not clear what it is they are protesting against. But such unclarity can undoubtedly be matched by the ambiguity that often marks the statements of aims and programs by professed "mechanists" in biology. As we had occasion to note in an earlier chapter, the word "mechanism" has a variety of meanings, and "mechanists" in biology as well as their opponents take few pains to make explicit the sense in which they employ it. There are biologists who profess themselves to be mechanists simply in the broad sense that they believe that vital phenomena occur in determinate orders and that the conditions for their occurrence are spatiotemporal structures of bodies. But such a view is compatible with the outlook of all schools in biology, with the exception of the vitalists and radical indeterminists; and in any case, when mechanism in biology is so understood, no issue divides those who profess it from most organismic biologists. There have also been biologists who proclaimed themselves to be mechanists in the sense that they maintained that all vital phenomena were explicable exclusively in terms of the science of mechanics (more specifically, in terms of either pure or unitary mechanical theories in the sense of Chapter 7), and who therefore believed living things to be "machines" in the original meaning of this word. It is doubtful, however, whether any biologists today are mechanists in this sense. Physicists themselves have long since abandoned the seventeenth-century hope that a universal science of nature could be devel-

oped within the framework of the fundamental ideas of classical mechanics. And it is safe to say that no contemporary biologist subscribes literally to the Cartesian program of reducing biology to the science of mechanics, and especially to the mechanics of contact action.

In any event, most biologists today who call themselves mechanists profess a view that is at once much more specific than the general thesis of causal determinism, and much less restrictive than the one which identifies a mechanistic explanation with an explanation in terms of the science of mechanics. A mechanist in biology, we shall assume, is one who believes, as did Jacques Loeb, that all living processes "can be unequivocally explained in physicochemical terms," [10] that is, in terms of theories and laws which by common consent are classified as belonging to physics and chemistry. However, biological mechanism so understood must not be taken to deny that living bodies have highly complex organizations. On the contrary, most biologists who adopt such a standpoint usually note quite emphatically that the activities of living bodies are not explicable by analyzing "merely" their physical and chemical compositions without taking into account their "ordered structures or organization." Thus, Loeb's characterization of a living body as a "chemical machine" is an obvious recognition of such organization. It is recognized even more explicitly by E. B. Wilson, who declares, after defining the "development" of germ plasm as the totality of operations by which the germ gives rise to its typical product, that the particular course of this development

> is determined (given the normal conditions) by the specific 'organization' of the germ-cells which form its starting-point. As yet we have no adequate conception of this organization, though we know that a very important part of it is represented by the nucleus. . . . Its nature constitutes one of the major unsolved problems of nature. . . . Nevertheless the only available path toward its exploration lies in the mechanistic conception that somehow the organization of the germ-cell must be traceable to the physico-chemical properties of its component substances and the specific configurations which they may assume.[11]

If such is the content of current biological mechanism, and if organismic biologists, like mechanists, reject the postulation of nonmaterial "vitalistic" agents whose operations are to explain vital processes, in what way do the approach and content of organismic biology differ from those of mechanism? The main points of difference, as noted by organismic biologists themselves, appear to be the following:

[10] Jacques Loeb, *The Mechanistic Conception of Life*, Chicago, 1912.
[11] E. B. Wilson, *op. cit.*, p. 1037, quoted by kind permission of The Macmillan Company, New York.

a. It is a mistake to suppose that the sole alternative to vitalism is mechanism. There are sectors of biological inquiry in which physico-chemical explanations play little or no role at present, and a number of biological theories have been successfully exploited which are not physicochemical in character. For example, there is available an impressive body of experimental knowledge concerning embryological processes, though few of the regularities that have been discovered can be explained at present in exclusively physicochemical terms; and neither the theory of evolution even in its current forms, nor the gene theory of heredity, is based on any definite physicochemical assumptions concerning vital processes. It is certainly not inevitable that mechanistic explanations will eventually prevail in these domains; and, since in any event these domains are now being fruitfully explored without any necessary commitment to the mechanistic thesis, organismic biologists possess at least some ground for their doubts concerning the ultimate triumph of that thesis in all sectors of biology. For just as physicists may be warranted in holding that some branch of physics (e.g., electromagnetic theory) is not reducible to some other branch of the science (e.g., to mechanics), so an organismic biologist may be warranted in espousing an analogous view with respect to the relation of biology to the physical sciences. Thus there is a genuine alternative in biology to both vitalism and mechanism—namely, the development of systems of explanation that employ concepts and assert relations neither defined in nor derived from the physical sciences.

b. However, organismic biologists generally claim far more than this. Many of them also maintain that the analytic methods of the physico-chemical sciences are intrinsically unsuited to the study of living organisms; that the central problems connected with vital processes require a distinctive mode of approach; and that, since biology is inherently irreducible to the physical sciences, mechanistic explanations must be rejected as the ultimate goal of biological research. One reason commonly advanced for this more radical thesis is the "organic" nature of biological systems. Indeed, perhaps the dominant theme upon which the writings of organismic biologists play so many variations is the "integrated," "holistic," and "unified" character of a living thing and its activities. Living creatures, in contrast to nonliving systems, are not loosely jointed structures of independent and separable parts, are not assemblages of tissues and organs standing in merely external relations to one another. Living creatures are "wholes" and must be studied as "wholes"; they are not mere "sums" of isolable parts, and their activities cannot be understood or explained if they are assumed to be such "sums." But mechanistic explanations construe living organisms as "machines" possessing independent parts, and thereby adopt an "additive" point of view in analyzing vital phenomena. Accordingly, since the action of the whole organism "has a certain unifiedness and completeness" which is left out of account in the course of analyzing it into its elementary processes, E. S. Russell concludes that "the activities of the organism as a whole are to be regarded as of a different order from physico-chemical relations, both in themselves and for the purposes of our understanding." [12] Therefore biology must observe two "cardinal laws of method": "The activity of the whole cannot be fully explained in terms of the activities of the parts isolated by analysis"; and "No part of any living entity and no single process of any complex organic unity can be fully understood in isolation from the structure and activities of the organism as a whole." [18]

c.   An additional though closely related point which organismic biologists stress is the "hierarchical organization" of living bodies and processes. Thus, a cell is known to be a structure of various constituents, such as the nucleus, the Golgi bodies, and the membranes, each of which may be analyzable into other parts and these in turn into still others, so that the analysis presumably terminates in molecules, atoms, and their "ultimate" parts. But in multicellular organisms the cell is also only an element in the organization of a tissue, the tissue is a part of some organ, the organ a member of an organ system, and the organ system a constituent in the integrated organism. It is patent that these various "parts" do not occur at the same "level" of organization. In consequence, organismic biologists place great stress on the fact that an animate body is not a system of parts homogeneous in complexity of organization, but that on the contrary the "parts" into which an organism is analyzed must be distinguished according to the different levels of some particular type of hierarchical structure (there may be several such types) to which the parts belong. Now organismic biologists do not deny that physicochemical explanations are possible for the activities of parts on the "lower" levels of a hierarchy. Nor do they deny that the physico-chemical properties of the parts on lower levels "condition" or "limit" in various ways the occurrence and modes of action of higher levels of organization. They do deny, on the other hand, that the processes found

[12] E. S. Russell, *The Interpretation of Development and Heredity*, Oxford, 1930, pp. 171-72.
[18] *Ibid.*, pp. 146-47. Similar statements of the central tenet of organismic biology will be found in Russell's *Directiveness of Organic Activities*, Cambridge, England, 1945, esp. Chaps. 1 and 7; Ludwig von Bertalanffy, *Theoretische Biologie*, Berlin, 1932, Chap. 2; his *Modern Theories of Development*, Oxford, 1933, Chap. 2; and his *Problems of Life*, New York and London, 1952, Chaps. 1 and 2; and W. E. Agar, *The Theory of the Living Organism*, Melbourne and London, 1943.

at higher levels of a hierarchy are "caused" by, or are fully explicable in terms of, lower-level properties. Biochemistry is acknowledged to be the study of the "conditions" under which cells and organisms act the way they do. Organismic biology, on the other hand, investigates the activities of the whole organism "regarded as conditioned by, but irreducible to, the modes of action of lower unities." [14]

We must now examine these alleged differences between the organismic and the mechanistic approaches to biology, and attempt to assess the claim that the mechanistic approach is generally inadequate to biological subject matter.

2. At first blush, the sole issues raised by organismic biology are those we have already discussed in connection with the doctrine of emergence and the reduction of one science to another. In point of fact, other questions are also involved. But to the extent that the issues are those of reduction, we can dispose of them quite rapidly.

Let us first remind ourselves of the two formal conditions, examined at some length in the preceding chapter, that are necessary and sufficient for the reduction of one science to another. When stated with special reference to biology and physicochemistry, they are as follows:

a. *The condition of connectability.* All terms in a biological law that do not belong to the primary science (such as 'cell,' 'mitosis,' or 'heredity') must be "connected" with expressions constructed out of the theoretical vocabulary of physics and chemistry (out of terms such as 'length,' 'electric charge,' 'free energy,' and the like). These connections may be of several kinds. The meanings of the biological expressions may be analyzable, and perhaps even explicitly definable, in terms of physicochemical ones, so that in the limiting case the biological expressions are eliminable in favor of the physicochemical terms. An alternative mode of connection is that biological expressions are associated with physicochemical ones by some type of coordinating definition, so that the connections have the logical status of conventions. Finally, and this is the more frequent case, the biological terms may be connected with physicochemical ones on the strength of empirical assumptions, so that the suf-

[14] Russell, *The Interpretation of Development and Heredity*, p. 187. For an analogous view, cf. Ludwig von Bertalanffy and Alex B. Novikoff, "The Conception of Integrative Levels and Biology," *Science*, Vol. 101 (1945), pp. 209-15, and the discussion of this article in the same volume, pp. 582-85 and in Vol. 102 (1945), pp. 405-06. A careful and sober analysis of the nature of hierarchical organization in biology and of its import for the possibility of mechanistic explanation is given in J. H. Woodger, *Biological Principles*, New York, 1929, Chap. 6, and in Woodger's "The 'Concept of Organism' and the Relation between Embryology and Genetics," *Quarterly Review of Biology*, Vol. 5 (1930), and Vol. 6 (1931).

ficient conditions (and possibly the necessary ones as well) for the occurrence of whatever is designated by the biological terms can be stated by means of the physicochemical expressions. Thus, if the term 'chromosome' can be associated in neither of the first two ways with some expression constructed out of the theoretical vocabulary of physics and chemistry, then it must be possible to state in the light of an assumed law the truth-conditions for a sentence of the form 'x is a chromosome' entirely by means of a sentence constructed out of that vocabulary.

b. *The condition of derivability.* Every biological law, whether theoretical or experimental, must be logically derivable from a class of statements belonging to physics and chemistry. The premises in these deductions will contain an appropriate selection from the theoretical assumptions of the primary discipline, as well as statements formulating the associations between biological and physicochemical terms required by the condition of connectability. In general, some of the premises will state in the vocabulary of the primary science the boundary conditions or specialized spatiotemporal configurations under which the theoretical assumptions are being applied.

As was shown in the preceding chapter, the condition of derivability cannot be fulfilled unless the condition of connectability is satisfied. It is beyond dispute, however, that the task of satisfying the first of these conditions for biology is still far from completed. We do not know at present, for example, the detailed chemical composition of chromosomes in living cells. We are therefore unable to state in exclusively physicochemical terms the conditions for the occurrence of those organic parts, and hence to state in such terms the truth-conditions for the application of the word 'chromosome.' And a fortiori we are not able at present to formulate in physicochemical language the structure of any of the systems, such as cell nucleus, cell, or tissue, of which chromosomes are themselves parts. Accordingly, in the current state of biological knowledge it is logically impossible to deduce the totality of biological laws and theories from purely physicochemical assumptions. In short, biology is not at present simply a chapter of physics and chemistry.

Organismic biologists are therefore on firm ground in maintaining that mechanistic explanations of all biological phenomena are currently impossible, and will remain impossible until the descriptive and theoretical terms of biology can be shown to satisfy the first condition for the reduction of that science to physics and chemistry—that is, until the composition of every part or process of living things, and the distribution and arrangement of their parts at any time, can be exhaustively specified in physicochemical terms. Moreover, even if this condition were realized, the triumph of the mechanistic standpoint would not thereby be assured. For as we have already shown, the satisfaction of the condition of connectability is a necessary but in general not a sufficient requirement for the absorption of biology into physics and chemistry. Although the connectability condition might be fulfilled, there would still remain the question whether all biological laws are deducible from the current theoretical assumptions of these physical sciences. The answer to this question is conceivably in the negative, since physicochemical theory in its present form may be insufficiently powerful to permit the derivation of various biological laws, even if these laws were to contain only terms properly linked with expressions belonging to those primary disciplines. It should also be noted that, even if both formal conditions for the reducibility of biology were satisfied, the reduction might nevertheless have little if any scientific importance, for the reason that some of the conditions previously labeled "nonformal" might not be adequately realized.

On the other hand, the facts cited and the argument thus far examined do not warrant the conclusion that biology is *in principle* irreducible to the physical sciences. The task facing such a proposed reduction is admittedly a most difficult one; and it undoubtedly impresses many students as one which, if not utterly hopeless, is at present not worth pursuing. However, no *logical* contradiction has yet been exhibited to the supposition that both the formal and nonformal conditions for the reduction of biology may some day be fulfilled. We can therefore terminate this part of the discussion with the conclusion that the question whether biology is reducible to physicochemistry is an open one, that it cannot be settled by a priori argument, and that an answer to it can be provided only by further experimental and logical inquiry.

3. Let us next turn to the argument for the inherent "autonomy" of biology based on the fact that living systems are hierarchically organized. The burden of the argument, as we have seen, is that properties and modes of behavior occurring on a higher level of such a hierarchy cannot in general be explained as the resultants of properties and behaviors exhibited by isolable parts belonging to lower levels of an organism's structure.

There is no serious dispute among biologists over the thesis that the parts and processes into which living organisms are analyzable can be classified in terms of their respective loci into hierarchies of various types, such as the essentially spatial hierarchy mentioned earlier. Nor is there disagreement over the contention that the parts of an organism belonging to one level of a hierarchy frequently exhibit forms of relatedness and of activity not manifested by organic parts belonging to another level. Thus, a cat can stalk and catch mice; but though the continued beating of its heart is a necessary condition for these activities, the cat's heart cannot perform these feats. Again, the heart can pump blood by contracting and expanding its muscular tissues, although no single tissue can keep the blood in circulation; and no tissue is able to divide by fission, even though its constituent cells may have this property. Such examples suffice to establish the claim that modes of behavior appearing at higher levels of a hierarchically organized system are not explained by merely listing each of the various lower-level parts and processes of the system as an aggregate of isolated and unrelated elements. Organismic biologists do not deny that the occurrence of higher-level traits in hierarchically structured living organisms is contingent upon the occurrence, at different levels of the hierarchy, of various component parts related in definite ways. But they do deny, and with apparent good reason, that statements formulating the traits exhibited by components of an organism, when the components are not parts of an actually living organism, can adequately explain the behavior of the living system that contains those components as parts related in complex ways to other elements in a hierarchically structured whole.

But do these admitted facts establish the contention that mechanistic explanations are either impossible or unsuitable for biological subject matter? It should be noted that various forms of hierarchical organization are exhibited by the materials of physics and chemistry, and not only by those of biology Our current theories of matter assume atoms to be structures of electric charges, molecules to be organizations of atoms, solids and liquids to be complex systems of molecules. Moreover, the available evidence indicates that elements at different levels of this hierarchy exhibit traits which their component parts do not invariably possess. However, these facts have not stood in the way of establishing comprehensive theories for the more elementary physical particles and processes, in terms of which it has been possible to account for some if not all of the physicochemical properties exhibited by objects having a more complex organization. To be sure, we do not possess at present a comprehensive and unified theory competent to explain the full range even of purely physicochemical phenomena occurring on various levels of organization. Whether such a theory will ever be achieved is certainly an open question. It is also relevant to note in this connection that biological organisms are "open systems," never in a state of "true equilibrium" but at best only in a steady state of "dynamic equilibrium" with their environment, because they continually exchange material components and not only energies with the latter.[15] In this respect, living organisms are unlike the "closed systems" usually studied in current phys-

[15] L. von Bertalanffy, *Problems of Life*, Chap. 4.

ical science. Indeed, an adequate theory for physicochemical processes
in open systems—for example, a thermodynamics competent to deal with
systems in nonequilibrium as well as equilibrium states—is at present
only in an early stage of development. Nevertheless, the circumstance
remains that we can now account for some characteristics of fairly complex systems with the help of theories formulated in terms of relations
between relatively more simple ones, for example, the specific heats of
solids in terms of quantum theory, or the changes in phase of compounds
in terms of the thermodynamics of mixtures. This circumstance must
make us pause in accepting the conclusion that the hierarchical organization of living systems by itself precludes a mechanistic explanation for
their traits.

Let us, however, examine in greater detail some of the organismic
arguments on this issue. One of them has been persuasively stated by
J. H. Woodger, whose careful but sympathetic analyses of organismic
notions are important contributions to the philosophy of biology. Woodger maintains that it is essential to distinguish between chemical *entities*
and chemical *concepts;* he believes that, if the distinction is kept in
mind, it no longer appears plausible to assume that a thing can be satisfactorily described in terms of chemical concepts exclusively, merely because the thing is held to be composed of chemical entities. "A lump of
iron," Woodger declares, "is a chemical entity, and the word 'iron' stands
for a chemical concept. But suppose that the iron has the form of a poker
or a padlock, then although the iron is still chemically analyzable in the
same way as before it cannot still be fully described in terms of chemical
concepts. It now has an organization above the chemical level." [16]

Now there is no doubt that many of the uses to which iron pokers or
padlocks may be put are not, and may never be, described in purely
physicochemical terms. But does the fact that a piece of iron has the
form of a poker or of a padlock stand in the way of explaining an extensive class of its properties and modes of behavior in exclusively physicochemical terms? The rigidity, tensile strength, and thermal properties
of the poker, or the mechanism and the qualities of endurance of the
padlock, are certainly explicable in such terms, even if it may not be
necessary or convenient to invoke a microscopic physical theory to account for all these traits. Accordingly, the mere fact that a piece of iron
has a certain organization does not preclude the possibility of a physico-

[16] J. H. Woodger, *Biological Principles*, p. 263. Woodger continues, "In the same
way an organism is a physical entity in the sense that it is one of the things we become aware of by means of the senses, and is a chemical entity in the sense that
it is capable of chemical analysis just as is the case with any other physical entity,
but it does not follow from this that it can be fully and satisfactorily described in
chemical terms."

chemical explanation for some of the characteristics it exhibits as an organized object.

Some organismic biologists maintain that, even if we were able to describe in minute detail the physicochemical composition of a fertilized
egg, we would nevertheless still be unable to explain mechanistically the
fact that such an egg normally segments. In the view of E. S. Russell, for
example, we might be able on the stated supposition to formulate the
physicochemical conditions for segmentation, but we would be unable
to "explain the course which development takes." [17]

This claim raises some of the previously discussed issues associated
with the distinction between structure and function. But quite apart from
these issues, the claim appears to rest on a misunderstanding if not on a
confusion. It is cogent to maintain that a knowledge of the physicochemical composition of a biological organism does not suffice to explain mechanistically its modes of action—any more than an enumeration of the
parts of a clock together with a description of their spatial distribution
and arrangement suffices to explain or to predict the behavior of the
timepiece. To make such an explanation, we must also assume some theory or set of laws (in the case of the clock, the theory of mechanics)
which formulates the way certain elements act when they occur in some
initial distribution and arrangement, and which permits the calculation
(and hence the prediction) of the subsequent development of that organized system of elements. Moreover, it is conceivable that, despite our
assumed ability at some given stage of scientific knowledge to describe
in full detail the physicochemical composition of a living thing, we
might nevertheless be unable to deduce from the established physicochemical theories of the day the course of the organism's development. In
short, it is conceivable that the first but not the second formal condition
for the reducibility of one science to another is satisfied at a given time.
It is a misunderstanding, however, to suppose that a fully codified explanation in the natural sciences can consist only of instantial premises
formulating initial and boundary conditions but containing no statements
of law or theory. It is an elementary blunder to claim that, because some
one physicochemical theory (or some class of such theories) is not competent to explain certain vital phenomena, it is *in principle* impossible
to construct and establish a mechanistic theory that can do so.

On the other hand, it would be foolish to underestimate the enormity
of the task facing the mechanistic program in biology because of the in-
tricate hierarchical organization of living things. Nor should we dismiss
as pointless the protests of organismic biologists against versions of the
mechanistic thesis that appear to ignore the fact of such organization.

[17] E. S. Russell, *The Interpretation of Development and Heredity*, p. 186.

As biologists of all schools have often observed, there is no such thing as a homogeneous and structurally undifferentiated "living substance," analogous to "copper substance." There have nevertheless been mechanists who in their statements on biological method, if not in their actual practice as biological investigators, have in effect asserted the contrary. It is therefore worth stressing that the subject matters of their inquiry have compelled biologists to recognize not just a single type of hierarchical organization in living things but several types, and that a central problem in the analysis of organic developmental processes is the discovery of the precise interrelations between such hierarchies.

The hierarchy most frequently cited is generated by the relation of spatial inclusion, as in the case of cell parts, cells, organs, and organisms. However, on any reasonable criterion for distinguishing between various "levels" of such a hierarchy, it turns out that there are bodily parts in most organisms (such as the blood plasma) which cannot be fitted into it. Furthermore, there are types of hierarchy that are not primarily spatial. Thus, there is a "division hierarchy," with cells as elements, which is generated by the division of a zygote and of its cell descendants. Biologists also recognize a "hierarchy of processes": the hierarchy of physicochemical processes in a muscle, the contraction of the muscle, the reaction of a system of muscles, the reaction of the animal organism as a whole; and other types which could be added to this brief list. In any event, it should be noted that in embryological development the spatial hierarchy changes, since in this process new spatial parts are elaborated. This fact can be expressed by saying that, when the division hierarchy of an embryo is compared at different times, its spatial hierarchy at a later time contains elements that did not exist at earlier times. Accordingly, organismic biologists are obviously correct in claiming that to a large extent biological research is concerned with establishing relations of interdependence between various hierarchical structures in living bodies.[18]

Let us now, however, state briefly the schematic form of a hierarchical organization (not necessarily a spatial hierarchy), with a view to assessing in general terms one component in the organismic critique of biological mechanism. Suppose S is some biological system which is analyzable into three major constituents A, B, and C, so that S can be conceived as the relational complex $R(A,B,C)$, where R is some relation. Assume further that each major constituent is in turn analyzable into subordinate constituents $(a_1, a_2, \ldots, a_i)$, $(b_1, b_2, \ldots, b_j)$, and $(c_1, c_2, \ldots, c_k)$, respectively, so that the major constituents of S can be

[18] Cf. the writings of Woodger cited above, as well as his *Axiomatic Method in Biology*, Cambridge, Eng., 1937; also L. von Bertalanffy, *Problems of Life*, Chap. 2.

represented as the relational complexes $R_A(a_1, \ldots, a_i)$, $R_B(b_1, \ldots, b_j)$, and $R_C(c_1, \ldots, c_k)$. The a's, b's, and c's may be analyzable still further, but for the sake of simplicity we shall assume only two levels for the hierarchical organization of S. We also stipulate that some of the a's (and similarly some of the b's and c's) stand to each other in various special relations, subject to the condition that all of them are related by R_A to constitute A (with analogous conditions for the b's and c's). Moreover, we assume that some of the a's may stand in certain other special relations to some of the b's and c's, subject to the condition that the complexes A, B, and C are related by R to constitute S. If S is such a hierarchy, one aim of research on S will be to discover its various constituents, and to ascertain the regularities in the relations connecting them with S and with constituents on the same or on different levels.

The pursuit of this aim will in general require the resolution of many serious difficulties. To discover just what the presence of A, for example, contributes to the traits manifested by S taken as a whole, it may be necessary to establish what S would be like in the absence of A, as well as how A behaves when it is not a constitutive part of S. There may be grave experimental problems in attempting to isolate and identify such causal influences. But quite apart from these, the fundamental question must at some point be faced whether the study of A, when it is placed in an environment differing in various ways from the environment provided by S itself, can yield pertinent information about the behavior of A when it occurs as an actual constituent of S. Suppose, however, that we possess a theory T about the components a of A, such that if the a's are assumed to be in the relation R_A when they occur in an environment E, it is possible to show with the help of T just what traits characterize A in that environment. On this supposition it may not be necessary to experiment upon A in isolation from S. The above crucial question will nevertheless continue to be unresolved unless the theory T permits conclusions to be drawn not only when the a's are in the relation R_A in some artificial environment E, but also when they are in that relation in the particular environment that contains the b's and c's all jointly organized by the relations R_B, R_C, and R. Without such a theory, it will generally be the case that the only way of ascertaining just what role A plays in S is to study A as an actual component in the relational complex $R(A,B,C)$.

Accordingly, organismic biologists are right in insisting on the general principle that "an entity having the hierarchical type of organization such as we find in the organism requires investigation at all levels, and investigation of one level cannot replace the necessity for investigation of levels higher up in the hierarchy." [19] On the other hand, this prin-

[19] J. H. Woodger, *Biological Principles*, p. 316.

ciple does not entail the impossibility of mechanistic explanations for vital phenomena, though organismic biologists sometimes appear to believe that it does. In particular, if the *a*'s, *b*'s, and *c*'s in the above schematism are the submicroscopic entities of physics and chemistry, S is a biological organism, and T is a physicochemical theory, it is not impossible that the conditions for the occurrence of the relational complexes A, B, C, and S can be specified in terms of the fundamental concepts of T, and that furthermore the laws concerning the behaviors of A, B, C, and S can be deduced from T. But, as has been argued in the preceding chapter, whether in point of fact one science (such as biology) is reducible to some primary science (such as physicochemistry), is contingent on the character of the particular theory employed in the primary discipline at the time the question is put.

4. We must finally turn to what appears to be the main reason for the negative attitude of organismic biologists toward mechanistic explanations of vital phenomena, namely, the alleged "organic unity" of living things and the consequent impossibility of analyzing biological wholes as "sums" of independent parts. Whether there is merit in this reason obviously depends on what senses are attached to the crucial expressions 'organic unity' and 'sum.' Organismic biologists have done little to clarify the meanings of these terms, but at least a partial clarification has been attempted in the preceding and present chapters of this book. In the light of these earlier discussions the issue now under examination can be disposed of with relative brevity.

Let us assume, as do organismic biologists, that a living thing possesses an "organic unity," in the sense that it is a teleological system exhibiting a hierarchical organization of parts and processes, so that the various parts stand to each other in complex relations of causal interdependence. Suppose also that the particles and processes of physics and chemistry constitute the elements at the lowest level of this hierarchical system, and that T is the current body of physicochemical theory. Finally, let us associate with the word 'sum' in the statement 'A living organism is not the sum of its physicochemical parts,' the "reducibility" sense of the word distinguished in the preceding chapter. The statement will then be understood to assert that, even when suitable physicochemical initial and boundary conditions are supplied, it is not possible to deduce from T the class of laws and other statements about living things commonly regarded as belonging to the province of biology.

Subject to an important reservation, the statement construed in this manner may very well be true, and probably represents the opinion of most students of vital phenomena, whether or not they are organismic biologists. The statement is widely held, despite the fact that in many cases physicochemical conditions for biological processes have been ascertained. Thus, an unfertilized egg of the sea urchin does not normally develop into an embryo. However, experiments have shown that, if such eggs are first placed for about two minutes in sea water to which a certain quantity of acetic acid has been added and are then transferred to normal sea water, the eggs presently begin to segment and to develop into larvae. But, although this fact certainly counts as impressive evidence for the physicochemical character of biological processes, the fact has thus far not been fully explained, in the strict sense of 'explain,' in physicochemical terms. For no one has yet shown that the statement that sea urchin eggs are capable of artificial parthenogenesis under the indicated conditions is *deducible* from the purely physicochemical assumptions T. Accordingly, if organismic biologists are making only the *de facto* claim that no systems possessing the organic unity of living things have thus far been proved to be sums (in the reducibility sense) of their physicochemical constituents, the claim is undoubtedly well founded.

On the other hand, in the prevailing circumstances of our knowledge there should be no cause for surprise that the fact about the artificial parthenogenesis of sea urchin eggs is not deducible from T. The deduction is not possible, if only because the elementary logical requirements for performing it are currently not satisfied. No theory can explain the operations of any concretely given system unless a complete set of initial and boundary conditions for the application of the theory is stated in a manner consonant with the specific notions employed in the theory. For example, it is not possible to deduce the distribution of electric charges on a given insulated conductor merely from the fundamental equations of electrostatic theory. Additional instantial information must be supplied in a form prescribed by the character of the theory—in this instance, information about the shape and size of the conductor, the magnitudes and distribution of electric charges in the neighborhood of the conductor, and the value of the dielectric constant of the medium in which the conductor is embedded. In the case of the sea urchin eggs, however, although the physicochemical composition of the environment in which the unfertilized eggs develop into embryos is presumably known, the physicochemical composition of the eggs themselves is still unknown, and cannot be formulated for inclusion in the indispensable instantial conditions for the application of T. More generally, we do not know at present the detailed physicochemical composition of any living organism, nor the forces that may be acting between the elements on the lowest level of its hierarchical organization. We are therefore currently unable to state in exclusively physicochemical terms the initial and boundary conditions requisite for the application of T to vital systems. Until we can do this, we are in principle precluded from deducing biological laws from mechanistic theory. Accordingly, although it may indeed be true that a living organism is not the sum of its physico-chemical parts, the available evidence does not warrant the assertion either of the truth or of the falsity of this dictum.

Although the point just stressed is elementary, organismic biologists often appear to neglect it. They sometimes argue that, while mechanistic explanations may be possible for traits of organic parts when these parts are studied in "abstraction" (or isolation) from the organism as a whole, such explanations are not possible when the parts function conjointly in mutual dependence as actual constituents of a living thing. But this claim ignores the crucial fact that the initial conditions required for a mechanistic explanation of the traits of organic parts manifested when the parts exist *in vitro* are generally insufficient for accounting mechanistically for the conjoint functioning of the parts in a biological organism. For it is evident that when a part is isolated from the rest of the organism it is placed in an environment which is usually different from its normal environment, where it stands in relations of mutual dependence with other parts of the organism. It therefore follows that the initial conditions for using a given theory to explain the behavior of a part in isolation will also be different from the initial conditions for using that theory to explain behavior in the normal environment. Accordingly, although it may indeed be beyond our actual competence at present or in the foreseeable future to specify the instantial conditions requisite for a mechanistic explanation of the functioning of organic parts *in situ*, there is nothing in the logic of the situation that limits such explanations in principle to the behavior of organic parts *in vitro*.

One final comment must be added. It is important to distinguish the question of whether mechanistic explanations of vital phenomena are possible from the quite different though related problem of whether living organisms can be effectively synthesized in a laboratory out of nonliving materials. Many biologists seem to deny the first possibility because of their skepticism concerning the second. In point of fact, however, the two issues are logically independent. In particular, although it may never become possible to manufacture living organisms artificially, it does not follow that vital phenomena are therefore incapable of being explained mechanistically. A glance at the achievements of the physical sciences will suffice to establish this claim. We do not possess the power to manufacture nebulae or solar systems, despite the fact that we do possess physicochemical theories in terms of which nebulae and planetary systems are tolerably well understood. Moreover, while modern

physics and chemistry provide competent explanations for various properties of chemical elements in terms of the electronic structure of the
atoms, there are no compelling reasons for believing that, for example,
men will some day be capable of manufacturing hydrogen by putting
together artificially the subatomic components of the substance. On the
other hand, the human race possessed skills (e.g., in the construction of
dwellings, in the manufacture of alloys, and in the preparation of foods)
long before adequate explanations for the traits of the artificially constructed articles were available.

Nonetheless, organismic biologists often develop their critique of the
mechanistic program in biology as if its realization were equivalent to
the acquisition of techniques for literally taking apart living things and
then overtly reconstituting the original organisms out of their dismem-
bered and independent parts. However, the conditions for achieving
mechanistic explanations for vital phenomena are quite different from
the requirements for the artificial manufacture of living organisms. The
former task is contingent on the construction of factually warranted
theories of physicochemical substances; the latter task depends on the
availability of suitable physicochemical materials, and on the invention of
effective techniques for combining and controlling them. It is perhaps unlikely that living organisms will ever be synthesized in the laboratory
except with the help of mechanistic theories of vital processes; in the
absence of such theories, the artificial manufacture of living things, were
this ever accomplished, would be the outcome of a fortunate but im-
probable accident. But in any event, the conditions for achieving these
patently different tasks are not identical, and either may some day be
realized without the other. Accordingly, a denial of the possibility of
mechanistic explanations in biology on the tacit supposition that these
conditions do coincide, is not a cogently reasoned thesis.

The main conclusion of this discussion is that organismic biologists
have not established the absolute autonomy of biology or the inherent
impossibility of physicochemical explanations of vital phenomena. Nevertheless, the stress they place on the hierarchical organization of living
things and on the mutual dependence of organic parts is not a misplaced
one. For, although organismic biology has not convincingly secured all
its claims, it has demonstrated the important point that the pursuit of
mechanistic explanations for vital processes is not a *sine qua non* for
valuable and fruitful study of such processes. There is no more reason
for rejecting a biological theory (e.g., the gene theory of heredity) because it is not a mechanistic one (in the sense of "mechanistic" we have
been employing) than there is for discarding some physical theory (e.g.,

modern quantum theory) on the ground that it is not reducible to a theory in another branch of physical science (e.g., to classical mechanics). A wise strategy of research may indeed require that a given discipline be cultivated as a relatively independent branch of science, at least during a certain period of its development, rather than as an appendage to some other discipline, even if the theories of the latter are more inclusive and better established than are the explanatory principles of the former. The protest of organismic biology against the dogmatism often associated with the mechanistic standpoint in biology is salutary.

There is, however, an obverse side to the organismic critique of that dogmatism. Organismic biologists sometimes write as if any analysis of vital processes into the operation of distinguishable parts of living things entails a seriously distorted view of these processes. For example, E. S. Russell has maintained that in analyzing the activities of an organism into elementary processes "something is lost, for the action of the whole has a certain unifiedness and completeness which is left out of account in the process of analysis." [20] Analogously, J. S. Haldane claimed that we cannot apply mathematical reasoning to vital processes, since a mathematical treatment assumes a separability of events in space "which does not exist for life as such. We are dealing with an indivisible whole when we are dealing with life." [21] And H. Wildon Carr, a professional philosopher who subscribed to the organismic standpoint and wrote as one of its exponents, declared that "Life is individual; it exists only in living beings, and each living being is indivisible, a whole not constituted of parts." [22]

Such pronouncements exhibit an intellectual temper that is as much an obstacle to the advancement of biological inquiry as is the dogmatism of intransigent mechanists. In biology as in other branches of science knowledge is acquired only by analysis or the use of the so-called "abstractive method"—by concentrating on a limited set of properties things possess and ignoring (at least for a time) others, and by investigating the traits selected for study under controlled conditions. Organismic biologists also proceed in this way, despite what they may say, for there is no effective alternative to it. For example, although J. S. Haldane formally proclaimed the "indivisible unity" of living things, his studies on respiration and the chemistry of the blood were not conducted by considering the body as an indivisible whole. His researches involved the examination of relations between the behavior of one part of the

[20] E. S. Russell, *The Interpretation of Development and Heredity*, p. 171.
[21] J. S. Haldane, *The Philosophical Basis of Biology*, London, 1931, p. 14.
[22] Quoted in L. Hogben, *The Nature of Living Matter*, London, 1930, p. 226.

body (e.g., the quantity of carbon dioxide taken in by the lungs) and the behavior of another part (the chemical action of the red blood cells). Like everyone else who contributes to the advance of knowledge, organismic biologists must be abstractive and analytical in their research procedures. They must study the operations of various prescinded parts of living organisms under selected and often artificially instituted conditions—on pain of mistaking unenlightening statements liberally studded with locutions like 'wholeness,' 'unifiedness,' and 'indivisible unity' for expressions of genuine knowledge.

Methodological Problems
of the Social Sciences

13 The study of human society and human behavior molded by social institutions has been cultivated for about as long as has the investigation of physical and biological phenomena. However, much of the "social theory" that has emerged from such study, in the past as well as the present, is social and moral philosophy rather than social science, and is made up in large measure of general reflections on the "nature of man," justifications or critiques of various social institutions, or outlines of stages in the progress or decay of civilizations. Although discussions of this type often contain penetrating insights into the functions of various social institutions in the human economy, they rarely pretend to be based on systematic surveys of detailed empirical data concerning the actual operation of societies. If such data are mentioned at all, their function is for the most part anecdotal, serving to illustrate rather than to test critically some general conclusion. Despite the long history of active interest in social phenomena, the experimental production and methodical collection of evidence for assessing beliefs about them are of relatively recent origin.

But in any event, in no area of social inquiry has a body of general laws been established, comparable with outstanding theories in the natural sciences in scope of explanatory power or in capacity to yield precise and reliable predictions. It is of course true that, under the inspiration of the impressive theoretical achievements of natural science, comprehensive systems of "social physics" have been repeatedly constructed, purporting to account for the entire gamut of diverse institutional structures and changes that have appeared throughout human history. However, these ambitious constructions are the products of doubtfully adequate notions of what constitutes sound scientific procedure, and, though some of them continue to have adherents, none of them stands up under careful scrutiny.[1] Most competent students today do not believe that an empirically warranted theory, able to explain in terms of a single set of integrated assumptions the full variety of social phenomena, is likely to be achieved in the foreseeable future. Many social scientists are of the opinion, moreover, that the time is not yet ripe even for theories designed to explain systematically only quite limited ranges of social phenomena. Indeed, when such theoretical constructions with a restricted scope have been attempted, as in economics or on a smaller scale in the study of social mobility, their empirical worth is widely regarded as a still unsettled question. To a considerable extent, the problems investigated in many current centers of empirical social research are admittedly concerned with problems of moderate and often unimpressive dimensions.

It is also generally acknowledged that in the social sciences there is nothing quite like the almost complete unanimity commonly found among competent workers in the natural sciences as to what are matters of established fact, what are the reasonably satisfactory explanations (if any) for the assumed facts, and what are some of the valid procedures in sound inquiry. Disagreement on such questions undoubtedly occurs in the natural sciences as well. But it is usually found at the advancing frontiers of knowledge; and, except in areas of research that impinge intimately upon moral or religious commitments, such disagreement is generally resolved with reasonable dispatch when additional evidence is obtained or when improved techniques of analysis are developed. In contrast, the social sciences often produce the impression that they are a battleground for interminably warring schools of thought, and that even subject matter which has been under intensive and prolonged study remains at the unsettled periphery of research. At any rate, it is a matter of public record that social scientists continue to be divided on central issues in the logic of social inquiry which are implicit in the questions mentioned above. In particular, there is a long-standing divergence in professed scientific aims between those who view the explanatory systems and logical methods of the natural sciences as models to be emulated in social research, and those who think it is fundamentally inappro-

---

[1] Many of these systems are "single factor" or "key cause" theories. They identify some one "variable"—such as geographic environment, biological endowment, economic organization, or religious belief, to mention but a few—in terms of which the institutional arrangements and the development of societies are to be understood.

priate for the social sciences to seek explanatory theories that employ "abstract" distinctions remote from familiar experience and that require publicly accessible (or "intersubjectively" valid) supporting evidence.

In short, the social sciences today possess no wide-ranging systems of explanations judged as adequate by a majority of professionally competent students, and they are characterized by serious disagreements on methodological as well as substantive questions. In consequence, the propriety of designating any extant branch of social inquiry as a "real science" has been repeatedly challenged—commonly on the ground that, although such inquiries have contributed large quantities of frequently reliable information about social matters, these contributions are primarily descriptive studies of special social facts in certain historically situated human groups, and supply no strictly universal laws about social phenomena. It would not be profitable to discuss at any length an issue framed in this manner, particularly since the requirements for being a genuine science tacitly assumed in most of the challenges lead to the unenlightening result that apparently none but a few branches of physical inquiry merit the honorific designation. In any event, it will suffice for present purposes to note that, although descriptive studies of localized social fact mark much social research, this statement does not adequately summarize its achievements. For inquiries into human behavior have also made evident (with the increasing aid, in recent years, of rapidly developing techniques of quantitative analysis) some of the relations of dependence between components in various social processes; and those inquiries have thereby supplied more or less firmly grounded generalized assumptions for explaining many features of social life, as well as for constructing frequently effective social policies. To be sure, the laws or generalizations concerning social phenomena made available by current social inquiry are far more restricted in scope of application, are formulated far less precisely, and are acceptable as factually sound only if understood to be hedged in by a far larger number of tacit qualifications and exceptions, than are most of the commonly cited laws of the physical sciences. In these respects, however, the generalizations of social inquiry do not appear to differ radically from generalizations currently advanced in domains usually regarded as unquestionably respectable subdivisions of natural science—for example, in the study of turbulence phenomena and in embryology.

The important task, surely, is to achieve some clarity in fundamental methodological issues and the structure of explanations in the social sciences, rather than to award or withhold honorific titles. However, attempts at such clarification run into a difficulty that is perhaps distinctive of the social sciences. Enough has already been said about the disagreements flourishing in these disciplines to suggest that almost any product of social research selected for logical analysis runs the risk of being judged by many professional students as unrepresentative of important achievements in its area, even though other students with similar professional competence may judge the question differently. Moreover, the issues raised for analysis by the materials selected, as well as the analysis itself, face the analogous hazard of being condemned either as irrelevant to the significant logical problems of social inquiry or as exhibiting a narrow partisan bias toward some particular school of social thought. Despite these hazards, it is the aim of the present and the succeeding chapters to examine a number of broad logical issues that persistently recur in methodological discussions of the social sciences. The present chapter will first consider various difficulties, alleged to be created by the special subject matter of social inquiry and frequently cited as serious if not fatal obstacles to establishing general laws of social phenomena. The next chapter will then discuss the question whether explanations in the social sciences differ in form as well as in substantive content from those in other branches of inquiry; and certain features of probabilistic explanations will receive fuller treatment than has thus far been given them. The final chapter will deal with problems of historical knowledge, and will discuss further aspects of the probabilistic pattern as well as examine the structure of genetic explanations.[2]

I.   *Forms of Controlled Inquiry*

On the assumption that the paramount aim of theoretical social science is to establish general laws which can serve as instruments for systematic explanation and dependable prediction, many students of social phenomena have tried to account for the relative paucity of reliable laws in their disciplines. We shall examine some of the reasons that have been suggested. The reasons to be discussed call attention to difficulties confronting the social sciences, either because of certain alleged distinctive features inherent in the subject matter studied, or because of certain supposed consequences of the fact that the study of society is part of its own subject matter. These difficulties are generally not mutually independent, so that the issues they raise do not always differ sharply. It is nevertheless convenient to list and examine the problems separately.

Perhaps the most frequently mentioned source of difficulty is the allegedly narrow range of possibilities for controlled experiments on social subject matter. Let us first state the difficulty in the form it receives when a very strict sense is associated with the term "controlled experiment." In a controlled experiment, the experimenter can manipulate at

---

[2] Probabilistic and genetic explanations were identified and illustrated in Chapter 2, and the former was briefly discussed in Chapter 10.

will, even if only within limits, certain features in a situation (often designated as "variables" or "factors") which are assumed to constitute the relevant conditions for the occurrence of the phenomena under study, so that by repeatedly varying some of them (in the ideal case, by varying just one) but keeping the others constant, the observer can study the effects of such changes upon the phenomenon and discover the constant relations of dependence between the phenomenon and the variables. Controlled experiment thus involves not only directed changes in variables that can be reliably identified and distinguished from other variables, but also the reproduction of effects induced by such changes upon the phenomenon under study.

However, experiment in this strict sense can apparently be performed at best only rarely in the social sciences, and perhaps never in connection with any phenomenon which involves the participation of several generations and large numbers of men. For social scientists do not usually possess the power to institute experimentally designed modifications into most social materials that are of scientific interest. Moreover, even if such power were theirs, and moral scruples did not stand in the way of subjecting human beings to various changes with unforeseeable but possibly injurious effects upon the lives of men, two important problems would arise concerning any experiments they might perform. The exercise of power to modify social conditions for experimental purposes is evidently itself a social variable. Accordingly, the manner in which such power is exercised may seriously compromise the cognitive significance of an experiment, if that use of power affects the outcome of the experiment to an unknown degree. Furthermore, since a given change introduced into a social situation may produce (and usually does produce) an irreversible modification in relevant variables, a repetition of the change to determine whether or not its observed effects are constant will be upon variables that are not in the same initial conditions at each of the repeated trials. In consequence, since it may thus be uncertain whether the observed constancies or differences in the effects are to be attributed to differences in the initial states of the variables or to differences in other circumstances of the experiment, it may be impossible to decide by experimental means whether a given alteration in a social phenomenon can be rightly imputed to a given type of change in a certain variable.[8] In addition to all this, the scope of experimentation in the social sciences is severely limited by the circumstance that a con-

[8] This difficulty also arises in sciences dealing with nonhuman materials. It can usually be overcome in these domains by employing a fresh sample in each repeated trial, the new samples being homogeneous in relevant respects with the initial one. In the social sciences the problem cannot be resolved so easily, because even if an adequate supply of samples is available, they may not be sufficiently alike in pertinent features.

trolled experiment can be performed only if it is possible to produce repeatedly observable modifications in the phenomenon studied—a possibility that seems clearly out of the question for those social phenomena which are ostensibly nonrecurrent and historically unique (such as the rise of modern industrial capitalism, or the unionization of American labor during the New Deal).

These claims as to the restricted scope of controlled experiment in the social sciences raise many important issues. However, discussion of them will be confined for the present to the following two, others being reserved for later examination: (1) Is controlled experimentation a *sine qua non* for achieving warranted factual knowledge, and in particular for establishing general laws? and (2) Is there in fact only a negligible possibility in the social sciences for controlled empirical procedure?

1. Inquiries in which controlled experiments can be instituted possess familiar and undeniably great advantages. It is indeed unlikely that various branches of science (e.g., optics, chemistry, or genetics) could have achieved their present state of advanced theoretical development without systematic experimentation. Nevertheless, this conjecture is obviously unsound if extended to all domains of inquiry in which comprehensive systems of explanation have been established. Neither astronomy nor astrophysics is an experimental science, even if each employs many assumptions that are patently based on the experimental findings of other disciplines. Although during the eighteenth and nineteenth centuries astronomy was rightly held to be superior to all other sciences in the stability of its comprehensive theory and in the accuracy of its predictions, it certainly did not achieve this superiority by experimentally manipulating celestial bodies. Moreover, even in branches of inquiry nowhere near the theoretical level of astronomy (e.g., geology, or until relatively recently embryology), lack of opportunity for controlled experiment has not prevented scientists from arriving at well-grounded general laws. It is in consequence beyond dispute that many sciences have contributed, and continue to contribute, to the advancement of generalized knowledge despite severely limited opportunities for instituting controlled experiments.

However, every branch of inquiry aiming at reliable general laws concerning empirical subject matter must employ a procedure that, if it is not strictly controlled experimentation, has the essential logical functions of experiment in inquiry. This procedure (we shall call it "controlled investigation") does not require, as does experimentation, either the reproduction at will of the phenomena under study or the overt manipulation of variables; but it closely resembles experimentation in other respects. Controlled investigation consists in a deliberate search for contrasting occasions in which the phenomenon is either uniformly manifested (whether in identical or differing modes) or manifested in some cases but not in others, and in the subsequent examination of certain factors discriminated in those occasions in order to ascertain whether variations in these factors are related to differences in the phenomena—where these factors as well as the different manifestations of the phenomenon are selected for careful observation because they are assumed to be relevantly related. From the perspective of the logical role which empirical data play in inquiry, it is clearly immaterial whether the observed variations in the assumed determining factors for observed changes in the phenomenon are introduced by the scientist himself, or whether such variations have been produced "naturally" and are simply found by him—provided that in each case the observations have been made with equal care and that the occurrences manifesting the variations in the factors and in the phenomenon are alike in all other relevant respects. It is for this reason that experimentation is often regarded as a limiting form of controlled investigation, and that sometimes the two provisos are not even distinguished. It may indeed be the case that the second of the above two provisos can be satisfied more easily when experiments can be performed than when they cannot; and it may also be the case that, when experiment is feasible, relevant factors can be subjected to variations which are rarely if ever found to occur naturally, but which must nevertheless be obtained if general laws are to be established. These comments direct attention to matters of undoubtedly great importance in the conduct of inquiry, but they do not annul the identity in logical function of controlled experiment and controlled investigation.

In short, although it is possible to make scientific headway without experiment, either controlled experimentation (in the narrow sense we have associated with this phrase) or controlled investigation (in the sense just indicated) appears to be indispensable. We shall say that an inquiry using one or the other of these two procedures is a "controlled empirical inquiry." [4]

2. In consequence, it becomes pertinent to ask whether in the social sciences the scope for procedures that either are strictly experimental or have the same logical role as experiment is as close to the vanishing

[4] It is of some importance not to confuse what is frequently called "controlled (sensory) observation" with controlled empirical inquiry in the above sense. Observations are usually said to be "controlled" if they are not haphazard but are performed with care and are instituted for the sake of resolving some question and in the light of some conception concerning the requirements for reliable observation. Controlled observation in this sense is essential to both controlled experimentation and controlled investigation. However, controlled observation is only a necessary but not a sufficient condition for controlled empirical inquiry.

point as is frequently portrayed. The claim that this scope is quite small commonly rests on some misconceptions which we shall now briefly discuss.

a.   Although John Stuart Mill was a foremost advocate in nineteenth-century England for employing the logical methods of the natural sciences in social inquiry, he was convinced that experimentation directed toward establishing general laws was not feasible in the social sciences. He held this view essentially because he saw no prospects for applying in these latter disciplines either his Method of Agreement or his Method of Difference, the two of his five "Methods of Experimental Inquiry" which were for him definitive of what it is to be an experiment. According to the Method of Agreement, two instances of a phenomenon are required that are unlike in *all* respects but one (which may then be identified as the "cause" or "effect" of the phenomenon); and according to the Method of Difference, two situations are required, the phenomenon being present in one and not in the other, that are otherwise alike in *all* respects but one (which may again be identified as the "cause" or "effect" of the phenomenon). Mill apparently took for granted that theoretically significant social experiments must be performed upon historically given societies in their entirety: and since he believed with obviously good reasons that no two such societies actually conform to the requirements of either of his two Methods, as well as that by no contrivance could they be made to do so, he denied the possibility of social experimentation.[5]

Mill's account of experimental method suffers from the serious defect that he underestimated if he did not ignore the crucial point that, since two situations are *never* either completely alike or completely unlike in *all but one* respect, his Methods are workable only within some framework of assumptions stipulating which features (or respects) of a situation are to count as the relevant ones.

But even if Mill's analysis is corrected on this point, his reasons for denying the possibility of social experimentation remain inconclusive. For his contention is based in part on the supposition that controlled experimentation (or for that matter, even controlled investigation) requires the occurrence of variation in just one (relevant) factor at a time —a notion that is commonly held but is nonetheless an oversimplified view of the conditions for competent empirical analysis. The supposition does indeed state an ideal of experimental procedure, often realized at least approximately. It is well to remember, however, that the question

[5] Mill recommended what he called the "Concrete Deductive Method" as the appropriate one for social inquiry. According to this method, various consequences derived from some set of theoretical assumptions are verified by observation.

whether only a "single" factor is being varied in an experiment, or even what is to count as a "single" factor, are matters that depend on the antecedent assumptions underlying the experiment. It is beyond human power even in the most carefully run laboratory to eliminate completely variations in all but one circumstance of an experiment; and the point has already been emphasized that assumptions concerning the changes to be singled out as relevant are implicit in every inquiry. Furthermore, to illustrate the point that special assumptions may be involved in judging a factor to be a "single" one, although in many experiments changing the quantity (e.g., the number of grams) of chemically pure oxygen counts as a variation in a single factor, in other experiments this is not a satisfactory way of specifying what is a single factor because of the assumption, relevant in this second class of experiments but not in the first, that there are isotopes of oxygen. For, since the proportions in which these isotopes are contained in different quantities of chemically pure oxygen are not constant, varying the quantity of pure oxygen may significantly alter the proportions.

In any event, there are areas of inquiry in the natural sciences in which it is not possible to vary one at a time even the relevant and admittedly "single" factors in an experiment, but in which we are not thereby prevented from establishing laws. For example, in experiments on physicochemical systems in thermodynamical equilibrium it is not generally possible to vary the pressure exerted by a system without varying its temperature. It is nevertheless possible to ascertain what constant relations of dependence hold between these variables and other factors in the system, and what are the effects produced on the system by changes in just one of these variables. Moreover, modern statistical analysis is sufficiently general to enable us to cope with many situations in which variables do not vary one at a time, even in the case of phenomena for which theory is far less advanced than it is in physics or for which only techniques of controlled investigation but not of strict experimentation are available. For example, the size of the crop produced in a given cornfield is affected by both temperature changes and variations in the rainfall, though these factors cannot be varied independently. Nevertheless, statistical analysis of data on their simultaneous variations enables us to isolate the effects of rainfall upon crop yield from the effects of temperature.[6] In short, the injunction that factors are to be varied one at a time represents a frequently desirable but by no means universally indispensable condition for controlled inquiry.

b. Accordingly, the field for controlled empirical inquiry into social phenomena is in principle much larger than unduly narrow conceptions

[6] Analyses of this type will receive further attention later in this chapter.

of what is essential for such inquiry may lead one to suppose. But let us survey briefly the main forms in which controlled empirical study actually occurs in the social sciences.

i. Despite widespread claims that experimentation in the strict sense is not feasible, several types of experiments are in fact employed in the social sciences. One of them is the laboratory experiment, in essentials similar to laboratory experiments in the natural sciences. It consists of constructing an artificial situation that resembles "real" situations in social life in certain respects, but conforms to requirements normally not satisfied by the latter in that some of the variables assumed to be relevant to the occurrence of a social phenomenon can be manipulated in the laboratory situation while other relevant variables can be kept at least approximately constant. For example, a laboratory experiment was devised to determine whether voters are influenced by their knowledge of the religious affiliations of candidates for office. For this purpose, a number of clubs were created, whose members were carefully selected so that none were previously acquainted; each club was asked to elect one of its own members to an office, and information about the religious affiliations of the members was supplied to half the clubs but withheld from the others. The election results indicated that a goodly number of voters who were given this information were influenced by it.

Laboratory experiments have been employed in increasing numbers in many areas of social inquiry. It is evident, however, that an extensive class of social phenomena does not lend itself to such experimental study. Moreover, even when social phenomena can be investigated in this manner, it is generally not possible in a laboratory to produce changes in variables that compare in magnitude with changes sometimes occurring in those variables in natural social situations. For example, the sense of crucial importance frequently generated by issues in political elections cannot easily be provoked in subjects participating in a laboratory vote. It is a misguided criticism of laboratory experiments in social science that, since a laboratory situation is "unreal," its study can throw no light on social behavior in "real" life. On the contrary, many such experiments have been illuminating; for example, a number of experiments have been made on the behavior of children when the conditions under which they engaged in play activities were varied. However, it *is* a sound observation that no generalizations concerning social phenomena based exclusively on laboratory experiments can be safely assumed without further inquiry to hold in natural social environments.

ii. A second type of experiment is the so-called "field experiment." In such experiments, instead of an artificially created miniature social system, some "natural" though limited community is the experimental subject in which certain variables can be manipulated, so that one can ascertain by repeated trials whether or not given changes in those variables generate determinate differences in some social phenomenon. In one such field experiment, for example, changes were made in the way groups of workers in a certain factory were organized, the various types of organization being defined in the inquiry. It turned out that those groups in which more "democratic" forms of organization were introduced were eventually more productive than groups organized less democratically.

Field experimentation has some clear advantages over experimentation in the laboratory, but it is equally clear that in field experiments the difficulty of keeping relevant variables constant is in general greater. For obvious reasons, moreover, the opportunities for instituting field experiments have thus far been relatively meager; indeed, most of the experiments performed have been undertaken in connection with problems that are only of a narrowly practical interest.

iii. But the bulk of controlled empirical inquiry in the social sciences is not experimental in the sense we have associated with this term, even though such inquiries are frequently designated as "natural experiments," "ex post facto experiments," or are qualified in analogous ways. The aim of these investigations is in general to ascertain whether, and if so in what manner, some event, set of events, or complex of traits is causally related to the occurrence of certain social changes or characteristics in a given society. Examples of subjects discussed in such inquiries are human migrations, variations in birth rate, attitudes toward minority groups, the adoption of new forms of communication, changes in interest rates by banks, differences in the distribution of various personality traits in various social groups, and the social effects of legislative enactments.

Inquiries of this type can be subdivided in various ways: those seeking to ascertain the social effects of phenomena, as distinguished from those concerned with their causes; inquiries addressed to individual actions, as distinguished from those investigating group behavior; inquiries directed to relations between traits occurring more or less simultaneously, as distinguished from those dealing with traits manifested in some temporal sequence; and so on. Each of these subdivisions is associated with special methodological problems and techniques of inquiry. But despite such differences, and despite the fact that the variables assumed to be relevant in these investigations cannot be manipulated at will or that variations in such variables may not even have been planned by anyone, the investigations satisfy to a greater or lesser degree the requirements for controlled empirical inquiry. In a fairly representative study of this type, for example, the problem was to ascertain the influence of television upon the church attendance of children. For this purpose, a sample survey obtained answers to questions concerning the church attendance, age, and sex of each child in the sample, whether or not the child was a television viewer, and the church attendance of the child's parent. When the answers were classified according to whether a child who attended church was or was not a television viewer, the proportion of children attending church in the class of those who were viewers was found to be smaller than the proportion of children attending church in the class of those who were not viewers; and these proportions remained substantially unaltered when children with similar sex and age were compared. On the other hand, when the sample answers were further classified according to the church attendance of a child's parents, in the class of children who were viewers and whose parents attended church, the proportion of children attending church was not significantly different from the proportion of children attending church in the class of those who were not viewers but whose parents also attended church. Analysis of the data in the sample thus supplied some evidence that the church attendance of children is not influenced by their watching television.

We shall discuss later the structure of such analyses in greater detail. For the present, let us make explicit what it is in investigations of this type that qualifies them as in some degree controlled empirical inquiries. Since by hypothesis the relevant factors cannot be overtly manipulated in these investigations, the control must be effected in some other manner. As the above example suggests, this control is achieved if sufficient information about these factors can be secured, so that analysis of the information can yield symbolic constructions in which some of the factors are represented to be constant (and hence without influence upon any alterations in the phenomenon under study), in contrast to the correlations (or lack of correlations) between the recorded data on variations in the other factors and the recorded data on the phenomenon. Accordingly, the subjects manipulated in these investigations are the *recorded (or symbolically represented) data of observation* on relevant factors, rather than the factors themselves. These inquiries therefore attempt to obtain information about a phenomenon and the factors assumed to be relevant to its occurrence, so that by subjecting the recorded data to the manipulations of statistical analysis it may be possible either to eliminate some of the factors as causal determinants of the phenomenon or to provide grounds for imputing to some factors a causal influence upon the phenomenon.

However, the difficulties associated with basing causal imputations upon investigations of this type are notorious. There are not only serious and sometimes intractable technical problems generated by various special areas of social research, for example, problems concerning the identification and definition of variables, the choice of relevant variables, the selection of representative sample data, and the finding of enough data to permit reliable inferences to be drawn from comparisons of various classes of data in the sample. There is also the crucial general problem concerning the nature of the evidence required for ascribing validly causal significance to correlations between data. The history of social study amply testifies to the ease with which the familiar *post hoc* fallacy can be committed when data about sequentially manifested events are interpreted as indicating causal connections. This general problem, as well as the rationale for distinguishing spurious from genuinely causal correlations, will receive attention later. We shall conclude for the present with the observation that much empirical research in the social sciences does not even attempt to be controlled inquiry, and that investigations of this type differ considerably among themselves in the completeness with which they satisfy the conditions for such inquiry.

II. *Cultural Relativity and Social Laws*

A second difficulty often cited as an obstacle to establishing general laws in the social sciences, closely related to the difficulty already discussed, is the "historically conditioned" or "culturally determined" character of social phenomena. Although most if not all societies in the past and present have a number of analogous institutions—e.g., all known societies possess some kind of family organization, some form of education for the young, some provision for maintaining order, and so on—in general these institutions have been developed in response to different environments and embody different cultural traditions, so that the internal structures and interrelations of corresponding institutions in different societies are in general also different. Accordingly, since the forms assumed by human social behavior depend not only on the immediate occasions that call forth the behavior but also on the culturally instituted habits and interpretations of events involved in the response to the occasion, the patterns of social behavior will vary with the society in which the behavior occurs and with the character of its institutions at a given historical period. In consequence, conclusions reached by controlled study of sample data drawn from one society are not likely to be valid for a sample obtained from another society. Unlike the laws of physics and chemistry, generalizations in the social sciences therefore have at best only a severely restricted scope, limited to social phenomena occurring during a relatively brief historical epoch within special institutional settings. For example, Snell's law on the refraction of light formulates relations between phenomena that are apparently invariant throughout the universe. On the other hand, the manner in which human birth rate varies with social status in one community at a given time is in general different from the way these items are related in another community, or even in the same community at another time.

The substance of this account of one serious obstacle to establishing comprehensively general social laws is undeniable. Human behavior is undoubtedly modified by the complex of social institutions in which it develops, despite the fact that all human actions involve physical and physiological processes whose laws of operation are invariant in all societies. Even the way members of a social group satisfy basic biological needs—e.g., how they obtain a living or build their shelters—is not uniquely determined either by their biological inheritance or by the physical character of their geographic environment, for the influence of these factors on human action is mediated by existing technologies and traditions. The possibility must certainly be admitted that nontrivial but reliably established laws about social phenomena will always have only a narrowly restricted generality.

However, the facts under discussion have been frequently misinterpreted, and in consequence many students of human affairs have maintained that "transcultural" laws of social phenomena (i.e., social laws valid in different societies) are in principle impossible. We shall therefore examine this issue.

1. A common source of skepticism toward the prospects for transcultural social laws is the tacit assumption that scientific laws must enable us to make precise predictions into the indefinite future, so that astronomy is taken as the paradigm of any science worthy of the name. It has been maintained, for example, that if a science of society

> were a true science, like that of astronomy, it would enable us to predict the essential movements of human affairs for the immediate and the indefinite future, to give pictures of society in the year 2000 or the year 2500 just as astronomers can map the appearance of the heavens at fixed points of time in the future. Such a social science would tell us exactly what is going to happen in the years to come and we should be powerless to change it by any effort of will.[7]

But since "owing to the development of human experience, men and women as individuals, and as groups, races, and nations, are always growing and changing" so that "closed schemes cannot be made out of the data of the social sciences," and since in consequence the social

---

[7] Charles A. Beard, *The Nature of the Social Sciences*, New York, 1934, p. 29.

sciences cannot make such predictions, the conclusion is that there is no "social science in any valid sense of the term as employed in real science." [8]

However, no extended discussion is needed to show that the circumstances which permit long-range predictions in astronomy do not prevail in other branches of natural science, and that in this respect celestial mechanics is not a typical physical science. Such predictions are possible in astronomy because for all practical purposes the solar system is an isolated system which will remain isolated, there is reason to believe, during an indefinitely long future. In most other domains of physical inquiry, on the other hand, the systems under study do not satisfy this requirement for long-range predictions. Moreover, in many cases of physical inquiry we are ignorant of the pertinent initial conditions for employing established theories to make precise forecasts, even though the available theories are otherwise entirely adequate for this purpose. For example, we can predict with great accuracy the motions of a given pendulum as long as it is isolated from the influence of various disturbing factors, because the theory of pendular motion and the requisite factual data concerning that specific system are known; but the predictions cannot be confidently made far into the future, because we have excellent reasons for believing that the system will not remain immune indefinitely from external perturbations. On the other hand, we cannot predict with much accuracy where a leaf just fallen from a tree will be carried by the wind in ten minutes; for, although available physical theory is in principle capable of answering the question provided that relevant factual data are supplied about the wind, the leaf, and the terrain, we rarely if ever have at our command knowledge of such initial conditions. Inability to forecast the indefinite future is thus not unique to the study of human affairs, and is not a certain sign that comprehensive laws have not been established or cannot be established about the phenomena.

Moreover, it is an obvious error to maintain, as the passage quoted above seems to suggest, that theoretical knowledge is possible only in those domains in which effective human control is lacking. Crude ores can be transformed into refined products, not because no theory for such changes is available, but quite frequently just because there is. And conversely, a domain does not cease to be a field for theoretical knowledge if, after suitable techniques are developed, changes that could not be controlled previously become controllable. Will the principles of meteorology lose their validity, should we discover some day how to manufacture the weather? Men are certainly able to alter various fea-

[8] *Ibid.*, pp. 26, 33. 37.

tures of their modes of social organization, but this fact does not establish the impossibility of a "real" science of human affairs.

2. A related misconception is the supposition that wide differences in the specific traits and regularities of behavior manifested in a class of systems excludes the possibility that there is a common pattern of relations underlying these differences, and that the patently dissimilar characteristics of the various systems cannot therefore be understood in terms of a single theory about those systems. This supposition usually originates in a failure to distinguish between the question whether there is a structure of relations invariant in a class of systems and capable of being formulated as a comprehensive theory (even if in highly abstract terms), and the question whether the initial conditions appropriate for applying the theory to any one of the systems are uniformly the same in all the systems.

Consider, for example, the following purely physical phenomena: a lightning storm, the motions of a mariner's compass, the appearance of a rainbow, and the formation of an optical image in the range finder of a camera. These are undeniably quite dissimilar occurrences, incomparable on the basis of their manifest qualities; and it may not seem antecedently likely that they could be illustrations for a single set of integrally related principles. Nevertheless, as is well known, these phenomena can all be understood in terms of modern electromagnetic theory. There are of course different special laws for each of these phenomena; but the theory can explain all the laws, since different laws are obtained from the theory when different initial conditions, corresponding to the evident dissimilarities of the various systems, are supplied.

Accordingly, the fact that social processes vary with their institutional settings, and that the specific uniformities found to hold in one culture are not pervasive in all societies, does not preclude the possibility that these specific uniformities are specializations of relational structures invariant for all cultures. For the recognized differences in the ways different societies are organized and in the modes of behavior occurring in them may be the consequences, not of incommensurably dissimilar patterns of social relations in those societies, but simply of differences in the specific values of some set of variables that constitute the elementary components in a structure of connections common to all the societies. However, it is any man's guess whether a comprehensive social theory of this sort is destined to remain permanently as a logical but unrealized possibility. The present discussion, which is not intended to be an exercise in crystal gazing, seeks merely to note a misconception that arises when this possibility is overlooked.

3. Another reminder is pertinent to the consideration of the limited scope of social laws because of the "historically conditioned" character of social phenomena. It is obvious that, if a law in any domain of inquiry is to cover a wide range of phenomena exhibiting admittedly relevant and important differences, the formulation of the law must ignore these differences, so that the terms employed in the formulation make no explicit mention of traits specific to the phenomena occurring in special circumstances. Such a formulation can sometimes be achieved by using variables (in the usual mathematical sense of this word), the application of the law to particular situations being mediated by assigning to the variables constant values which may vary from situation to situation. For example, although the gravitational "constant" mentioned in Galileo's law for freely falling bodies does not have the same value at all latitudes, in the customary formulation of the law these variations in its value are not cited; and a greater generality in statement is obtained by employing the variable 'g' for the law than would be the case if some particular value were mentioned.[9]

However, this technique for securing generality in statement is not always possible or convenient. A different device commonly employed in the natural sciences is to formulate a law for a so-called "ideal case," so that the law states some relation of dependence which supposedly holds only under certain limiting conditions, even though these conditions may be rarely if ever realized. For example, Galileo's law for freely falling bodies is formulated for bodies moving in a vacuum, although terrestrial bodies normally if not always move through some resisting medium; and the law of the lever is stated for perfectly rigid and homogeneous bars, although actual levers only approximate to this condition. In consequence, when a concretely given situation is analyzed with the help of a law so formulated, additional assumptions or postulates must be introduced to bridge the gap between the ideal case for which the law is stated and the concrete circumstances to which the law is applied. Such additional assumptions are frequently quite complicated; they may be formulated with far less precision than is the law; and they may not even be fully stated, whether because explicit mention of all the assumptions would be too onerous (so that many are simply taken for granted), or because knowledge is lacking of all the factors relevantly differentiating the actual from the ideal case. Accordingly,

---

[9] The statement of the law must therefore be supposed to contain an existential logical quantifier for the variable 'g.' Thus, the familiar formula relating the distance s with the time t of a free fall, $s = gt^2/2$, must be understood to assert that *there is at least one value* of 'g' for which this relation holds, and that this value is constant at places whose distances from the earth's center are equal, although the values of 'g' are different at unequal distances from the earth's center.

although a law in its formal statement may appear to have a comprehensive generality and great simplicity, the formal statement may fail to disclose the restriction in scope and the complication in asserted content to which the statement is often subjected when the actual conditions for applying the law to concrete situations are introduced.

It is therefore clear that the "historically conditioned" character of social phenomena is no inherent obstacle to the formulation of comprehensive transcultural laws. In point of fact, the two logical devices just mentioned have been used in the social sciences with this end in view. For example, these techniques have been repeatedly employed in economics, in particular in the construction of economic theories involving the notion of perfect competition among buyers and sellers, or the notion of economic agents who seek simply to maximize their respective financial profits (or other "utilities"). To be sure, attempts to use these techniques for constructing inclusive laws, in economics and elsewhere in social inquiry, have thus far had only moderate success at best. However, it is a mistake to attribute the failures of these attempts, as is sometimes done, to some basic flaw in the general strategy of formulating social laws in terms of "ideal cases." The scarcity of indisputably successful achievement of this sort is to be credited in part to the specific theoretical notions employed in these attempts, but perhaps in greater measure to difficulties in knowing how statements using "ideal" notions are to be modified in the light of the special circumstances that are present in concrete social situations to which the statements may be applied.

However, analyses of social phenomena directed toward establishing general laws have for the most part been conducted in terms of distinctions men make in their day-by-day social activities. Even when these ordinarily imprecise common-sense notions are made less vague, it is difficult to eliminate from them essential reference to matters specific to some particular society (or particular social tradition). Moreover, the precise conditions under which generalizations stated with the help of such concepts hold are rarely known completely. In consequence, more often than not the generalizations either are statements of statistical correlations rather than of strictly universal relations of dependence, or they are "quasi-general" (in the sense that, though they may be expressed as strictly universal in form, they are in fact asserted without any intent to rule out various exceptions, which are sometimes explicitly covered by the familiar proviso that the relations of dependence mentioned in a generalization hold only if "other things are equal"). In either case, the relevance or the validity of a generalization for social groups belonging to other societies may be quite uncertain. For example, the generalization (based on a study of American soldiers in World War

II) that better-educated men drafted into the armed forces of a nation show fewer psychosomatic symptoms than those with less education, is quasi-general in the above sense. For it is unlikely that the generalization would be rejected as false should some particular group of college-educated draftees display a larger number of such symptoms than a group of draftees with only primary grade schooling, if it should also turn out, for example, that the commanding officer of the two groups had a special animus against college men and enjoyed making life miserable for them. However, although continued adherence to the generalization may be quite reasonable despite this particular exception, it would hardly be feasible to state in exhaustive detail the types of situations that the generalization is not intended to cover, and whose occurrence is therefore not to be taken as a genuine exception to the generalization. It is also obvious that, although the generalization is not invalidated by the fact that differences in formal education do not occur in a number of societies (for example, among the warriors of the Nuer people in northeast Africa), the generalization is inapplicable (because it is irrelevant) to the consideration of human behavior in those social systems.

In short, if social laws or theories are to formulate relations of dependence that are invariant throughout the wide range of cultural differences manifested in human action, the concepts entering into those laws cannot denote characteristics occurring in just one special group of societies. But it is clearly impossible to provide guarantees that satisfactory concepts will eventually be devised which do not designate such parochial characteristics, but which can nevertheless enter into factually warranted statements of culturally invariant social laws. Attempts made thus far to establish comprehensive transcultural laws have employed various kinds of concepts (or "variables") that appear to cut across cultural differences, for example, variables referring to physical factors (such as climate), biological factors (such as organic drives), psychological factors (such as desires or attitudes), and economic factors (such as forms of property relations), as well as more strictly sociological factors (such as social cohesion or social role). The social laws that have been perhaps most frequently proposed in terms of such concepts state orders of supposedly inevitable social changes, and maintain that societies or institutions succeed one another in some fixed sequence of developmental stages. None of these attempts or proposals has been successful; and in the light of past failures, as well as for reasons based upon a general analysis of historical processes, it seems most unlikely that a comprehensive social theory will be a theory of historical development. Moreover, the possibility must also be recognized that in comparison with the variables employed in the past in proposed transcultural laws, the concepts required for this purpose may have to be much more "abstract," may need to be separated by a greater "logical gap" from the familiar notions used in the daily business of social life, and may necessitate a mastery of far more complicated techniques for manipulating the concepts in the analysis of actual social phenomena.

III.   *Knowledge of Social Phenomena as a Social Variable*

A third difficulty confronting the social sciences, sometimes cited as the gravest one they face, has its source in the fact that human beings frequently modify their habitual modes of social behavior as a consequence of acquiring fresh knowledge about the events in which they are participating or the society of which they are members. This difficulty has two facets, one relating to the investigation of social phenomena, the other to the conclusions reached in such investigations.

1.   The point has already been noted that the manner in which experiments on social subject matter are conducted may introduce changes of unknown magnitude into the materials under study, and may therefore vitiate from the outset the conclusion advanced on the basis of some experiment. This observation can be extended to cover more than strictly experimental inquiries. For example, current empirical research into such matters as attitudes toward minority groups, voting behavior, or investment plans in business makes extensive use of questionnaires; and the answers obtained in various types of interviews employed in opinion surveys are the data on which conclusions concerning these matters are eventually based. However, even if we assume that the interviewers are properly trained for this task and do not introduce gross distortions into the data they collect by transparently inept techniques of interviewing, the problem remains whether, because of the fact that the respondents *know* they are being interviewed, their replies do express views or attitudes they held before the interview and will continue to hold after the interview. The circumstance that a respondent is aware of being an object of some interest to the interviewer, the consequences he believes his replies may have for matters of concern to himself, as well as the particular way in which the interview is conducted, may bring into operation influences that can radically affect the responses he gives, either by inducing him to give confident answers to questions upon which he had never reflected, or by inclining him to state opinions unrepresentative of his actual beliefs and unrevealing of his habitual behavior. Accordingly, if the process of gathering evidence for some hypothesis about a given subject matter yields only data whose characteristics, identified as constituting the relevant evidence, are created by the process itself, it is patently unsound to evaluate the hypothesis simply on the basis of such data.

This difficulty is undeniably a serious one, and there is no general formula for outflanking it; but it is not a difficulty that is unique to the social sciences, and it is not insuperable in principle. Thus, students of natural science have long been familiar with the fact that the instruments used for making measurements may produce alterations in the very magnitude being measured; this fact has received much attention particularly in recent years, in connection with the interpretation of the Heisenberg uncertainty relations in quantum mechanics. For example, the temperature recorded by a thermometer immersed in a liquid does not represent the precise temperature of the liquid before the immersion, since before immersion the temperature of the thermometer is usually different from that of the liquid, so that the two initial temperatures will change before the thermometer and the liquid are in thermal equilibrium with one another. However, it clearly makes little sense to claim that the magnitude of a measured property is altered by the very process of measuring it, unless it is possible to adduce independent evidence for the assumption that the measuring instrument employed in the process produces in the property changes of a specified kind. In consequence, to be intelligible such a claim must be accompanied by some notion (even if only a hazy one) of the extent to which the property may be altered because of its interaction with the measuring instrument. Accordingly, one of the following possibilities will present itself: the effects produced by such an interaction are known to be comparatively minute, and may therefore be ignored; the effects can be precisely calculated on the basis of known laws, and allowance made for them when a numerical value is assigned to the magnitude of the measured property; the effects cannot be calculated precisely, but they can be shown on the basis of known laws not to exceed certain limits, so that only an approximate value is assigned to the magnitude of the measured property; or finally, because of ignorance of various special circumstances under which the given type of measurement happens to be made, no estimate of the effects can be formed, so that assignment of a value to the property being measured must wait upon removal of this ignorance or upon the development of measuring instruments whose effects on the property can in fact be estimated.

The logic for handling the difficulty just discussed in relation to the subject matters explored by the natural sciences is not altered when the difficulty is examined in connection with the materials studied by the social sciences. In both groups of disciplines, the difficulty arises because changes are produced in a subject matter by the means used to investigate that subject matter. However, although in the social sciences (but not in the natural sciences) such changes can in part be attributed to the knowledge men possess of the fact that they are the subjects of an inquiry, this difference bears on the particular mechanism involved in bringing about the changes in one domain; and this difference in mechanism by which the changes are produced does not affect the nature of the logical problem created by the changes. Nevertheless, it is in general less easy to discount such changes in the social sciences because there are fewer well-established laws in these disciplines with the help of which the extent of the changes may be estimated. On the other hand, the social sciences can frequently employ techniques of inquiry for which the difficulty either does not occur at all or occurs in less acute form— for example, various devices for observing social behavior, where the participants in that behavior simply do not know they are being observed; or so-called "projective techniques," where the subjects do know they are under study but are unaware of the objectives of the study, so that they can only guess what aspect of their behavior is under scrutiny.[10]

2. The second aspect of the difficulty under discussion concerns the validity of conclusions reached in social inquiry. It has been often noted that, although the forces which keep the stars in their courses or the mechanisms which transmit the hereditary traits of the human organism are not affected by advances in astrophysics or biology, the relations of dependence that are the subjects of study in the social sciences may be profoundly modified as a consequence of developments in these latter disciplines. For even when generalizations about social phenomena and predictions of future social events are the conclusions of indisputably competent inquiries, the conclusions can literally *be made invalid* if they become matters of public knowledge and if, in the light of this knowledge, men alter the patterns of their behavior upon whose study the conclusions are based. In consequence, the claim is frequently made that it is hopeless to look for social laws which are valid for an indefinite future, and that prediction about social behavior is inherently unreliable.

Two types of such predictions are sometimes distinguished, each type illustrating a way in which actions generated by beliefs about human affairs may affect the validity of these beliefs themselves. One type is the so-called "suicidal prediction," consisting of predictions that are soundly grounded at the time they are made, and hence most likely to be confirmed by future events, but which are nevertheless falsified because of actions undertaken as a consequence of announcing the predictions. For example, on the basis of an apparently adequate analysis of the state of the American economy, economists predicted a business "recession" for 1947. However, because of this warning businessmen lowered the prices of a number of products occupying strategic positions in the operations of the economic market, so that the effective demand for these goods was increased and the predicted recession did not take place. The second type is the so-called "self-fulfilling prophecy," consisting of predictions that are false to the actual facts at the time the predictions are made, but that nevertheless turn out to be true because of the actions taken as a consequence of belief in the predictions. For example, although the United States Bank (a private bank in New York City, despite its name) was in no serious financial difficulties in 1928, many of its depositors came to believe that it was in dire trouble and would soon fail. This belief led to a run on the bank, so that the organization was in fact compelled to go into bankruptcy.[11]

The fact to which such predictions direct attention—i.e., that beliefs about human affairs can lead to crucial changes in habits of human behavior that are the very subjects of those beliefs—is sometimes presented as if the difficulty it raises for inquiry were unique to the social sciences because of the alleged "freedom of the human will." However, this ancient question is quite irrelevant to the methodological problems of social inquiry, as is evident from the circumstance that both types of prediction can also be illustrated by examples drawn from the subject matters of the natural sciences. For example, the pointing and firing of an anti-aircraft gun can be effected by means of a purely physical mechanism. Such a mechanism includes, we may assume, a radar for locating the target, an automatic computer for calculating the direction in which the gun should be pointed to hit the target as reported by the radar, an adjusting device for pointing and firing the gun, and some system for transmitting the calculations of the computer as a series of signals to the adjusting apparatus. Let us now suppose that, were the gun fired in accordance with the calculations of the computer on a given occasion, the target would be hit; but let us also suppose that the signals transmitting these calculations have disturbing effects (whether on the adjusting apparatus or on the target) for which the computer has made no allowance. Accordingly, although the gun is set and fired in accordance with calculations that were correct at the time they were made, it nevertheless fails to hit the target because of the changes introduced by the process of transmitting those calculations. The situation as described does not differ in essential respects from a suicidal prediction in social inquiry, despite the fact that only purely physical assumptions enter into the example. A physical analogue of a self-fulfilling prophecy in human affairs can be constructed in similar fashion. Thus, assume in the above example that either the radar equipment or the computer suffers from some "defect," such that if the gun

[11] Cf. Robert K. Merton, *Social Theory and Social Structure*, Rev. ed., Glencoe, Ill., 1957, Chap. 11.

were pointed and fired in accordance with the calculations of the computer on a given occasion, the gun would in fact fail to hit the target. It is nevertheless obviously possible that, though the gun is fired according to calculations that were incorrect at the time they were made, the target is successfully hit because of the perturbations produced by the process of transmitting those calculations.[12]

But however this may be, the frequent occurrence of suicidal and self-fulfilling predictions concerning human affairs is undeniable; and no theory adequate to the subject matter of the social sciences can ignore the fact that actions undertaken in the light of knowledge about some patterns of social behavior can often change those patterns. Nevertheless, as the preceding paragraph suggests, the interpretations sometimes placed upon this fact can be highly dubious. In particular, although this fact undoubtedly complicates the search for warranted generalizations about social phenomena, it does not eliminate, as is commonly alleged, the very possibility of establishing general social laws. Let us make explicit why not.

a. In the first place, those who make this allegation overlook the elementary point that a statement purporting to be a law is *conditional* in logical form, even if the particular wording employed does not reveal this explicitly. Such statements assert simply that *if* certain conditions are satisfied, *then* certain other things are also realized (whether invariably or only with some more or less precisely formulated relative frequency). Accordingly, the *factual validity* of a proposed social law does not depend on whether or not a given instance of the antecedent clause of the conditional is categorically true, even though the applicability of the law to a given situation does depend on whether or not the conditions mentioned in the antecedent are realized in that situation. For example, a simplified version of a familiar economic law asserts that, if the selling price of a commodity is lowered, the effective demand for the commodity is increased. Suppose that in a certain society a steady drop in the prices of various commodities (and in particular, of candy) during a long period is accompanied by a constant increase in the consumption of those commodities, so that the law comes to be asserted as correct. But suppose further that, in order to discourage the consumption of candy (perhaps for reasons dictated by studies on the effects of such consumption on overweight), steps are taken in view of this law to reverse the price trend of this commodity, so that eventually the ef-

[12] The example used in this paragraph is an adaptation of the one employed for identical purposes in Adolf Grünbaum, "Historical Determinism, Social Activism, and Prediction in the Social Sciences," *British Journal for the Philosophy of Science*, Vol. 7 (1956), pp. 236-40.

fective demand for candy is decreased. It is obvious, however, that the law is not invalidated by the circumstance that, because of the action taken in the light of the law, the price of candy is gradually increased—any more than is the fact that men generally avoid exposure to the fumes of hydrocyanic acid when they became familiar with the law that if the gas is inhaled death rapidly follows, a disproof of this law. In sum, if action based on knowledge of a given law is not one of the conditions the law mentions in its antecedent clause and asserts to be accompanied by certain consequences when those conditions are realized, the law is not shown to be erroneous when situations are discovered in which such action does take place but the stated consequences do not occur.

b. In the second place, there are no sound reasons for ruling out once for all the possibility of laws whose antecedent clauses do take cognizance of the presence of actions deliberately instituted on the basis of knowledge concerning social processes. On the contrary, it is in fact sometimes possible to foresee, if only in a general way, what are the likely consequences for established social habits of the acquisition of new knowledge or new skills. For example, the manufacture of equipment required as means of transportation and communication generally increases with the growing industrialization of a society. On the other hand, there is also evidence for the generalization that, when men discover the advantages of more rapid forms of transportation and communication, they tend to use these forms in preference to older but slower forms. In consequence, when knowledge of more rapid forms becomes widespread, the manufacture of equipment for the established forms will tend either to decrease altogether or to increase at a diminished rate, and at the same time the natural resources needed for that manufacture will either be exploited on a smaller scale or be assigned other uses. Even if the effects of newly acquired knowledge upon patterns of social behavior may not be predictable in minute detail, at least a rough account can sometimes be supplied concerning the probable consequences of such innovations. In short, if the knowledge men possess of social processes is a variable that enters into the determination of social phenomena, there are no a priori grounds for maintaining that changes in that variable and the effects they may produce cannot be the subject of social laws.

The point under consideration should not be confused with the quite different question whether the achievement of new knowledge and the form it takes can be predicted. Such prediction is undoubtedly not generally possible, except perhaps in those areas where advance in knowledge happens to depend on the solution of some special class of problems, for whose resolution effective techniques and adequate resources are already available. The point at issue is the question whether, when knowledge of relations of dependence between social phenomena is once acquired, it is in principle possible to establish laws that take into account the consequences which the use of such knowledge may have for those relations. The discussion has tried to make evident why the claim is untenable that laws of this kind are inherently impossible.

c. Finally, although the influence of men's beliefs and aspirations upon human history has been frequently underrated, it is equally easy to exaggerate the controlling role of deliberate choice in the determination of human events, even when the choice is based on considerable knowledge of social processes. It is a common experience of mankind that, despite carefully laid plans for realizing some end, the actions adopted result in entanglements that had not been foreseen and had certainly not been intended. For planned actions rarely if ever take place in a social setting over which men have total mastery. The consequences that follow a deliberate choice of conduct are the products not simply of that conduct; they are also determined by various attendant circumstances, whose relevance to the intended aim of the action may not be always understood, and whose modes of operation are in any case not within complete effective control of those who have made that choice. Eli Whitney did not invent the cotton gin in order to strengthen a social system based on human slavery; Pasteur would have been horrified to learn that his researches on fermentation would become the theoretical basis for bacteriological warfare; and French support of the American revolutionary cause against England did not aim at founding a nation that would eventually make it difficult for France to continue as a colonial power in North America.

This familiar incongruity between the intent and the outcome of social action has considerable bearing on the question whether the role played by knowledge of social processes in modifying those processes precludes the possibility of establishing general social laws. The intended aims of planned social action are undoubtedly subject to wide variation, since such aims generally depend on characteristics more or less distinctive of the individuals who do the planning and acting, as well as upon the knowledge of social processes they happen to possess; and it is indeed often difficult to foresee what those aims will be. On the other hand, as has just been suggested, the actual results attained by such action usually fall into a much more limited range of alternatives, because of the constraints imposed on individual social behavior by the relatively stable institutions within which individuals seek to realize their aims. For although planned effort can certainly transform the character of social institutions, the actions men perform on any given occasion are for the most part not the manifestations of reflective thought directed toward resolving some problem specific to that occasion, but rather of habits of behavior that cannot be changed simultaneously and that can usually be counted on to remain unchanged. Accordingly, the effects produced by efforts on behalf of some intended objective are frequently swamped by effects produced either by conduct conforming to habitual patterns of social behavior or by other events over which the actors have no control whatsoever. Although there is always a genuine possibility that action based on knowledge of social processes will alter the character of those processes, that possibility can therefore often be ignored, since such action does not generally transform radically the over-all pattern of habitual social behavior. For this reason, as well as for the reasons already discussed, that possibility does not constitute a fatal obstacle to the establishment of social laws.[18]

IV. *The Subjective Nature of Social Subject Matter*

A fourth set of related methodological questions is raised by the familiar claim that objectively warranted explanations of social phenomena are difficult if not impossible to achieve, because those phenomena have an essentially "subjective" or "value-impregnated" aspect.

The subject matter of the social sciences is frequently identified as purposive human action, directed to attaining various ends or "values," whether with conscious intent, by force of acquired habit, or because of unwitting involvement. A somewhat more restrictive characterization limits that subject matter to the responses men make to the actions of other men, in the light of expectations and "evaluations" concerning how

[18] In recent years, the question we have been discussing has been the subject of a number of theoretical and empirical investigations. It has been shown, for example, that a pollster can in principle always publish his prediction of an election result in such a form that, despite the reactions of voters to the forecast, the prediction is not falsified by those reactions. Cf. Herbert A. Simon, *Models of Man*, New York, 1957, Chap. 5, which has the title "Bandwagon and Underdog Effects of Election Predictions."

Moreover, a branch of inquiry has also been inaugurated recently whose aim is to specify, given a competitive action directed toward attaining a certain goal, the strategy to be followed which is in a certain sense the "best" strategy, with an outcome unaffected by any information the parties to the competition (the "players in the game") may have concerning each other's plans. This "theory of games" thus provides rules for deciding upon a course of conduct that need not be altered to achieve its goal, even if the other "players" acquire new knowledge in the course of the "game." The basic theory was developed by John von Neumann and Oskar Morgenstern, *The Theory of Games and Economic Behavior*, Princeton, 1944. See also John C. C. McKinsey, *Introduction to the Theory of Games*, New York, 1952; and R. D. Luce and H. Raiffa, *Games and Decisions*, New York, 1957.

these others will respond in turn.[14] On either delimitation of that subject matter, its study is commonly said to presuppose familiarity with the motives and other psychological matters that constitute the springs of purposive human behavior, as well as with the aims and values whose attainment is the explicit or implicit goal of such behavior.

According to many writers, however, motives, dispositions, intended goals, and values are not matters open to sensory inspection, and can be neither made familiar nor identified by way of an exclusive use of procedures that are suitable for exploring the publicly observable subject matters of the "purely behavioral" (or natural) sciences. On the contrary, these are matters with which we can become conversant solely from our "subjective experience." Moreover, the distinctions that are relevant to social science subject matter (whether they are employed to characterize inanimate objects, as in the case of terms such as 'tool' and 'sentence,' or to designate types of human behavior, as in the case of terms such as 'crime' and 'punishment') cannot be defined except by reference to "mental attitudes" and cannot be understood except by those who have had the subjective experience of possessing such attitudes. To say that an object is a tool, for example, is allegedly to say that it is *expected* to produce certain effects by those who so characterize that object. Accordingly, the various "things" that may need to be mentioned in explaining purposive action must be construed in terms of what the human actors *themselves believe* about those things, rather than in terms of what can be discovered about the things by way of the objective methods of the natural sciences. As one proponent of this claim states the case, "A medicine or a cosmetic, e.g., for the purposes of social study, are not what cures an ailment or improves a person's looks, but what people think will have that effect." And he goes on to say that, when the social sciences explain human behavior by invoking men's knowledge of laws of nature, "what is relevant in the study of society is not whether these laws of nature are true in any objective sense, but solely whether they are believed and acted upon by the people." [15]

In short, the categories of description and explanation in the social sciences are held to be radically "subjective," so that these disciplines are forced to rely on "nonobjective" techniques of inquiry. The social scientist must therefore "interpret" the materials of his study by imaginatively identifying himself with the actors in social processes, view-

[14] Max Weber, *The Theory of Social and Economic Organization*, New York, 1947, p. 118. On the more restrictive definition, a farmer tilling the soil merely to provide food for himself is not engaged in social activity. His behavior is social only if he evaluates plans to satisfy his own wants by reference to the assumed wants of other men.

[15] F. A. Hayek, *The Counter-Revolution of Science*, Glencoe, Ill., 1952, p. 30.

ing the situations they face as the actors themselves view them, and constructing "models of motivation" in which springs of action and commitments to various values are imputed to these human agents. The social scientist is able to do these things, only because he is himself an active agent in social processes, and can therefore understand in the light of his own "subjective" experiences the "internal meanings" of social actions. A purely "objective" or "behavioristic" social science is in consequence held to be a vain hope; for to exclude on principle every vestige of subjective, motivational interpretation from the study of human affairs is in effect to eliminate from such study the consideration of every genuine social fact.[16]

This account of social science subject matter raises many issues, but only the following three will receive attention in the present context: (1) Are the distinctions required for exploring that subject matter exclusively "subjective"? (2) Is a "behavioristic" account of social phenomena inadequate? and (3) Do imputations of "subjective" states to human agents fall outside the scope of the logical canons employed in inquiries into "objective" properties?

1. It is beyond dispute that human behavior is frequently purposive; and it is likewise beyond question that when such behavior is described or explained, whether by social scientists or by laymen, various kinds of "subjective" (or psychological) states are commonly assumed to underlie its manifestations. Nevertheless, as is evident from the biological sciences, many aspects of goal-directed activities can frequently be investigated without requiring the postulation of such states. But what is more to the point, even when the behaviors studied by the social sciences are indisputably directed toward some consciously entertained ends, the social sciences do not confine themselves to using only distinctions that refer to psychological states exclusively; nor is it clear, moreover, why these disciplines should place such restrictions upon themselves. For example, in order to account for the adoption of certain rules of conduct by a given community, it may be relevant to inquire into the ways in which members of the community cultivate the soil, construct shelters, or preserve food for future use; and the overt behaviors these individuals exhibit in pursuing these tasks cannot be described in purely "subjective" terms.

Furthermore, even though purposive action is sometimes partly ex-

---

[16] R. M. MacIver, *Social Causation*, New York, 1942, Chap. 14; Max Weber, *op. cit.*, Chap. 1, esp. Sec. 1; Charles H. Cooley, *Sociological Theory and Social Research*, New York, 1930, pp. 290-308; Ludwig von Mises, *Theory and History*, New Haven, Conn., 1957, Chap. 11; Peter Winch, *The Idea of a Social Science*, London, 1958, esp. Chap. 2.

plained with the help of assumptions concerning dispositions, intentions, or beliefs of the actors, other assumptions concerning matters with which the actors are altogether unfamiliar may also contribute to the explanation of their action. Thus, as the passage quoted above makes clear, if we wish to account for the behavior of men who believe in the medicinal properties of a given substance, it is obviously important to distinguish between the question whether that belief has any influence upon the conduct of the believers, and the question whether the substance does in fact have the assumed medicinal properties. On the other hand, there appear to be excellent reasons for rejecting the conclusion, alleged to follow from this distinction, that in explaining purposive behavior the social scientist must use no information available to himself but not available to those manifesting the behavior.[17] For example, southern cotton planters in the United States before the Civil War were certainly unacquainted with the laws of modern soil chemistry, and mistakenly believed that the use of animal manure would preserve indefinitely the fertility of the cotton plantations. Nevertheless, a social scientist's familiarity with those laws can help explain why, under that treatment, the soil upon which cotton was grown gradually deteriorated, and why in consequence there was an increasing need for virgin land to raise cotton if the normal cotton crop was not to decrease. It is certainly not evident why such explanations should be ruled out from the social sciences. But if they are not ruled out, and since they patently involve notions not referring to the "subjective" states of purposive agents, it does become evident that the categories of description and explanation in those sciences are not exclusively "subjective" ones.

2. The standpoint in social science known as "behaviorism" is an adaptation of the program of research first adopted by many psychologists in the second decade of this century. That program was the expression of a widespread revolt against the vagueness and general unreliability of psychological data obtained by introspective analyses of mental states, and its proponents took as their immediate model for psychological inquiry the procedures employed by students of animal behavior. In its initial formulation, behaviorism recommended the wholesale rejection of introspection as a technique of psychological study, and its announced aim was to investigate human behavior in the manner of inquiries into chemical processes or into the behavior of animals, without any appeal

[17] "Any knowledge which we may happen to possess about the true nature of the material thing [i.e., the alleged medicine], but which the people whose actions we want to explain do not possess, is as little relevant to the explanation of their actions as our private disbelief in the efficacy of a magic charm will help us to understand the behavior of the savage who believes in it."—F. A. Hayek, *op. cit.*, p. 30.

or reference to the contents of consciousness. Moreover, some of its advocates advanced distinctive views on substantive psychological issues (for example, on the "conditioning" mechanisms involved in learning or in literary creation), though the simple-minded "mechanistic" theories they adopted were not entailed by their rejection of introspection. It is worth passing notice, however, that even exponents of this radical form of behaviorism did not *deny the existence* of conscious mental states; and their rejection of introspection, in favor of the study of overt behavior, was controlled primarily by a *methodological* concern to base psychology upon publicly observable data.[18]

But in any event, behaviorism has undergone important transformation since its initial formulation, and there are perhaps no psychologists (or for that matter, social scientists) currently calling themselves "behaviorists" who subscribe to the earlier version's unqualified condemnation of introspection. On the contrary, professed behaviorists today generally accept introspective *reports* by experimental subjects, not as statements *about* private psychic states of the subjects, but as observable verbal *responses* the subjects make under given conditions; and accordingly, introspective reports are included among the objective data upon which psychological generalizations are to be founded. Furthermore, contemporary behaviorists operating within this more liberal methodological framework have been investigating many (frequently nonoverlapping) areas of human behavior, both individual (e.g., perceptual discrimination, learning, or problem solving) and social (e.g., communication, group decisions, or group cohesiveness); and they have proposed a number of special mechanisms to account for these various phenomena—mechanisms that for the most part differ among themselves and also differ substantially from the simple mechanisms associated with earlier expounders of the behavioristic standpoint. However, none of these more recently suggested mechanisms is known to be adequate for explaining the entire range of human conduct, so that behaviorism (like most "schools" of contemporary psychology) continues to be a diversified program of research which stresses certain methodological considerations, rather than a school committed to some particular, minutely articulated substantive theory. A similar state of affairs exists at present among social scientists who either profess to be behaviorists or manifest sympathies for a behavioristic approach. In consequence, the term "behaviorism" does not have a precise doctrinal connotation; and students of human conduct who designate themselves as behaviorists do so chiefly

[18] Cf. J. B. Watson, "Psychology as the Behaviorist Views It," *Psychological Review*, Vol. 20 (1913), pp. 158-77, and the same author's *Behaviorism*, New York, 1930.

because of their adherence to a methodology that places a premium on objective (or intersubjectively observable) data.[19]

In the light of this situation, however, it is not easy to assess the claim that a "behavioristic" approach to the study of social phenomena is self-defeating, because it is usually not clear what is the intended target of the criticism. Much of the criticism is certainly directed against what is a caricature of that approach. Thus, when it is asserted that a consistent behaviorist cannot properly talk of "the reactions of people to what our senses tell us are similar objects" (such as red circles) but only of "the reactions to stimuli which are identical in a strictly physical sense" (i.e., of the effects of light waves of a given frequency on a particular part of the retina of the human eye),[20] or when it is said that a behaviorist does not recognize the difference between purely reflex action (such as a knee jerk) and purposive behavior (such as is manifested in the building of a railroad),[21] the attack is in each case leveled against a straw man, created on the model of a biophysicist debauched by dubious epistemology, and not against a position held by any existing behaviorists. To be sure, behaviorists have sometimes shown themselves to be grossly insensitive to important features of human experience; and they have also often proposed explanations of psychological and social processes that turned out to be far too crude to deal adequately with the actual complexities of human conduct. But behaviorists do not have a monopoly in either type of failing; and as has been already indicated, acceptance of behaviorism as a methodological approach in no way necessitates the acceptance of any particular substantive theory.

An assumption underlying much criticism of behaviorism is that a consistent behaviorist must deny the very existence of "subjective" or "private" mental states; and it is therefore pertinent to discuss this contention briefly. In the first place, probably everyone recognizes the distinction between, say, a directly felt pain and the overtly behavioral manifestations of being in pain (such as groans or muscular spasms). In any event, anyone who rejects such distinctions as invalid is controverting facts too well established to be open to significant doubt. But in the second place, a behaviorist to be consistent is compelled neither to re-

[19] Cf. Kenneth W. Spence, "The Postulates of 'Behaviorism,'" *Psychological Review*, Vol. 55 (1948), pp. 67-78; Gardiner Murphy, *Historical Introduction to Modern Psychology*, New York, 1951, Chaps. 18 and 19; *The Science of Man in the World Crisis* (ed. by Ralph Linton), New York, 1945, esp. the chapters by Clyde Kluckhohn and William H. Kelly, "The Concept of Culture," Melville J. Herskovits, "The Processes of Cultural Change," and George P. Murdock, "The Common Denominator of Culture"; and Paul F. Lazarsfeld, "Problems in Methodology," in *Sociology Today* (ed. by Robert K. Merton, Leonard Broom, and Leonard S. Cottrell, Jr.), New York, 1959.
[20] F. A. Hayek, *op. cit.*, p. 45.
[21] Ludwig von Mises, *op. cit.*, p. 246.

nounce such familiar distinctions nor to abandon the central postulates of his methodological position. For he need not be a "reductive materialist," for whom the term 'pain' (or other admittedly "subjective" terms) is *synonymous* with some expression containing only terms belonging unmistakably to the languages of physics, physiology, or general logic. On the contrary, he will be well advised to reject this reductive thesis, since it confounds facts established in physics or physiology with the quite different types of facts established in logical inquiries into relations of meaning—so that it commits the error commonly made in another context when, for example, the meaning of the word 'red' (as used both currently as well as long before the advent of the electromagnetic theory of light, to designate a visible color) is identified with the meaning of, say, 'electromagnetic vibrations with wave lengths of approximately 7100 angstrom units.' [22] A behaviorist who does reject this mistaken thesis can therefore readily acknowledge that men are capable of having emotions, images, ideas, or plans; that these psychic states are "private" to the individual in whose body they occur, in the sense that this individual alone can *directly* experience their occurrence, because of the privileged relation his body has to those states; and that consequently a man can in general attest to being in some psychic state without having to examine first the publicly observable state of his own body (e.g., his own facial expression or his own utterances), although other men can ascertain whether he is in that psychic state only on the basis of such examination.[23]

However, the behaviorist also assumes that psychic states occur only in bodies having certain types of organization; that such states are "adjectival" or "adverbial" of those bodies, rather than substantive agents (or "entities") inhabiting the bodies; that the occurrence of a psychic state in a body is always accompanied by certain overt and publicly observable behaviors (frequently on a "molar" or macroscopic level) of that body; that such overt behaviors (including verbal responses) constitute a sufficient basis for grounding conclusions about the entire range

[22] Cf. the discussion of this issue in Chapter 11.

[23] Just how much confirming evidence for a statement is needed to warrant its acceptance is a difficult problem for which there is no general solution. There are undoubtedly many cases in which a minimum of confirming evidence suffices, so that additional evidence is sometimes regarded as gratuitous. Introspective statements frequently fall into this class, although not all of them are of this sort, since they may indeed be false and are sometimes accepted as true only when elaborate controls are instituted. However, introspective statements are not unique in being accepted on the basis of a bare minimum of supporting evidence. Thus, a chemist who observes that a piece of blue litmus paper turns red when it is immersed in a liquid may assert that the paper has indeed turned red and that the liquid is an acid. Moreover, he may regard it a waste of time to look for further evidence to support these claims, even though additional evidence could be found for his statements.

of human experience; and that observation of such overt behavior is not only the sole source of information anyone has concerning other men's experiences and actions, but also provides generally more reliable data for conclusions even about an individual's own character and capacities than is supplied by introspective analyses of psychic states. Accordingly, a behaviorist can maintain without inconsistency that there are indeed such things as private psychic states, and also that the controlled study of overt behavior is nevertheless the only sound procedure for achieving reliable knowledge concerning individual and social action.

Moreover, although some contemporary behaviorists believe a science of man can be developed which employs only distinctions "definable" in terms of molar human behavior, there is nothing in the methodological orientation of behaviorism to preclude even such behaviorists from adopting psychological theories postulating various kinds of mechanisms that are not open to direct public observation. Many such behaviorists do in fact subscribe to theories of this type. There are, to be sure, some behaviorists who, without denying the *existence* of psychic states, seek to develop theories all of whose terms refer exclusively to states and processes (whether molar or molecular) that are either physical, chemical, or physiological. Behaviorists in this category are therefore hostile to psychological theories purporting to *explain* overt human behavior by reference to various "mentalistic" occurrences—for example, theories invoking "subjective" intentions or ends-in-view to account for men's overt behaviors. However, behaviorism of this variety is clearly a program of theoretical and experimental research, comparable to the program of mechanists in biology, which hopes to achieve a comprehensive system of explanation for human behavior through the "reduction" of psychology to other sciences. The objectives of the program have certainly not been attained, and perhaps never will be. But provided that the program does not dismiss well-attested forms of human conduct as in some sense "unreal"—and there is no reason inherent to the program why it should—it cannot be ruled out on a priori grounds as illegitimate or as intrinsically absurd.

It is therefore difficult to escape the conclusion that behaviorism as a methodological orientation (as distinct from behaviorism as some particular substantive theory of human behavior) is not inherently inadequate to the study of purposive human action, and that in consequence the repeated claims asserting the essential inappropriateness of a behavioristic approach to the subject matter of the social sciences rest on no firm foundations.

3.   But however this may be, let us assume that the distinctive aim of the social sciences is to "understand" social phenomena in terms of

"meaningful" categories, so that the social scientist seeks to explain such phenomena by imputing various "subjective" states to human agents participating in social processes. The crucial question that thus remains to be examined is whether such imputations involve the use of logical canons which are different from those employed in connection with the imputation of "objective" traits to things in other areas of inquiry.

It will be helpful in discussing this issue to have before us some examples of "meaningful" explanations of human actions. Let us begin with a simple one, in which the writer stresses the essential difference

> between a paper flying before the wind and a man flying from a pursuing crowd. The paper knows no fear and the wind no hate, but without fear and hate the man would not fly nor the crowd pursue. If we try to reduce fear to its bodily concomitants we merely substitute the concomitants for the reality expressed as fear. We denude the world of meanings for the sake of a theory, itself a false meaning which deprives us of all the rest. We can interpret experience only on the level of experience.[24]

A more complex illustration is supplied by a historian, who maintains that

> We reject the theory that the intellectual movement of the 18th century was the sole cause of the French Revolution because we know that there participated in that upheaval large masses of peasants and workers, illiterate masses lacking any knowledge of philosophical or political doctrines; and by analogy with our own personal experience we hold that, were we illiterate and ignorant, and were we to revolt against the society in which we live, the cause of our revolutionary activities should be traced not to ideological impulses but to other causes—for instance, to our economic ills. On the other hand, we hold that among the causes of the French Revolution should be numbered the philosophical and political doctrines developed in France during the half century preceding the Revolution, because we have noticed that the cultivated classes continually invoked such doctrines while they were destroying the old regime; and again the analogy with our personal experience leads us to think that none of us when taking part in a revolutionary movement would publicly profess philosophical and political doctrine which did not really form an ingredient in our beliefs. All the reasonings of the historian and the social scientist can be reduced to this common denominator of analogy with our inward experience, whereas the [natural] scientist lacks the help of this analogy.[25]

But the example that has come to serve as the classical model for "meaningful" explanations of social phenomena is Max Weber's carefully worked out account of modern capitalism, in which he attributes the development of this type of economic enterprise at least in part to the spread of the religious beliefs and the precepts of practical conduct

[24] R. M. MacIver, *Society*, New York, 1931, p. 530.
[25] Gaetano Salvemini, *The Historian and Scientist*, Cambridge, Mass., 1939, p. 71.

associated with ascetic forms of Protestantism.[26] Weber's discussion is too detailed to permit brief summary. However, the structure of his argument (and of other "meaningful" explanations) can be represented by the following abstract schema. Suppose a social phenomenon E (e.g., the development of modern capitalistic enterprise) is found to occur under a complex set of social conditions C (e.g., widespread membership in certain religious groups, such as those professing Calvinistic Protestantism), where some of the individuals participating in C generally also participate in E.[27] But individuals who participate in E are assumed to be committed to certain values (or to be in certain "subjective" states) V_E (e.g., they prize honesty, orderliness, and abstemious labor); and individuals who participate in C are assumed to be in the subjective state V_C (e.g., they believe in the sacredness of a worldly calling). However, V_C and V_E are also alleged to be "meaningfully" related, in view of the motivational patterns we find in our own personal experiences—for example, by reflecting on how our own emotions, values, beliefs, and actions hang together, we come to recognize an intimate connection between believing that one's vocation in life is consecrated by divine ordinance, and believing that one's life should not be marked by indolence or self-indulgence. Accordingly, by imputing subjective states to the agents engaged in E and C, we can "understand" why it is that E occurs under conditions C, not simply as a mere conjuncture or succession of phenomena, but as manifestations of subjective states whose interrelations are familiar to us from a consideration of our own affective and cognitive states.

These examples make it clear that such "meaningful" explanations invariably employ two types of assumptions which are of particular relevance to the present discussion: an assumption, singular in form, characterizing specified individuals as being in certain psychological states at indicated times (e.g., the assumption, in the first quotation above, that members of the crowd hated the man they were pursuing); and an assumption, general in form, stating the ways such states are related to one another as well as to certain overt behaviors (e.g., the assumption in the second quotation that men participating in revolutionary movements do not publicly profess a political doctrine unless they believe in it). However, neither of such assumptions is self-certifying, and evidence is required for each of them if the explanation of which they are parts is to be more than an exercise in uncontrolled imagination. Competent evidence for assumptions about the attitudes and actions of other men is often difficult to obtain; but it is certainly not obtained merely by introspecting one's

[26] Max Weber, *The Protestant Ethic and the Spirit of Capitalism,* London, 1930.
[27] Weber tried to show that E did not occur in the absence of C. But this point is not directly relevant to the specific issue under discussion.

own sentiments or by examining one's own beliefs as to how such sentiments are likely to be manifested in overt action—as responsible advocates of "interpretative" explanations have themselves often emphasized
(e.g., with vigor and illumination by Max Weber). We may identify
ourselves in imagination with a trader in wheat, and conjecture what
course of conduct we would adopt were we confronted with some problem requiring decisive action in a fluctuating market for that commodity. But conjecture is not fact. The sentiments or envisioned plans we
may impute to the trader either may not coincide with those he actually possesses, or even if they should so coincide may eventuate in
conduct on his part quite different from the course of action we had
imagined would be the "reasonable" one to adopt under the assumed
circumstances. The history of anthropology amply testifies to the blun-
ders that can be committed when categories appropriate for describing
familiar social processes are extrapolated without further scrutiny to
the study of strange cultures. Nor is the frequent claim well founded
that relations of dependence between psychological processes with which
we have personal experience, or between such processes and the overt
actions in which they may be manifested, can be comprehended with a
clearer "insight" into the reasons for their being what they are than can
any relations of dependence between nonpsychological events and processes. Do we really understand more fully and with greater warranted
certainty why an insult tends to produce anger, than why a rainbow is
produced when the sun's rays strike raindrops at a certain angle?

Moreover, it is by no means obvious that a social scientist cannot account for men's actions unless he has experienced in his own person
the psychic states he imputes to them or unless he can successfully
recreate such states in imagination. Must a psychiatrist be at least partly
demented if he is to be competent for studying the mentally ill? Is a
historian incapable of explaining the careers and social changes effected
by men like Hitler unless he can recapture in imagination the frenzied
hatreds that may have animated such an individual? Are mild-tempered
and emotionally stable social scientists unable to understand the causes
and consequences of mass hysteria, institutionalized sexual orgy, or
manifestations of pathological lusts for power? The factual evidence
certainly lends no support to these and similar suppositions. Indeed, *discoursive* knowledge—i.e., knowledge statable in *propositional form*,
about "common-sense" affairs as well as about the material explored by
the specialized procedures of the natural and social sciences—is not a
matter of *having* sensations, images, or feelings, whether vivid or faint;
and it consists neither in identifying oneself in some ineffable manner
with the objects of knowledge, nor in reproducing in some form of direct experience the subject matter of knowledge. On the other hand,

discoursive knowledge is a *symbolic* representation of only certain se-lected phases of some subject matter; it is the product of a process that deliberately aims at formulating relations between traits of a subject matter, so that one set of traits mentioned in the formulations can be taken as a reliable sign of other traits mentioned; and it involves as a necessary condition for its being warranted, the possibility of verifying these formulations through controlled sensory observation by anyone prepared to make the effort to verify them.

In consequence, we can *know* that a man fleeing from a pursuing crowd that is animated by hatred toward him is in a state of fear, with-out our having experienced such violent fears and hatred or without imaginatively recreating such emotions in ourselves—just as we can *know* that the temperature of a piece of wire is rising because the veloci-ties of its constituent molecules are increasing, without having to imag-ine what it is like to be a rapidly moving molecule. In both instances "internal states" that are not directly observable are imputed to the ob-jects mentioned in explanation of their behaviors. Accordingly, if we can rightly claim to *know* that the individuals do possess the states im-puted to them and that possession of such states tends to produce the specified forms of behavior, we can do so only on the basis of evidence obtained by observation of "objective" occurrences—in one case, by ob-servation of overt human behavior (including men's verbal responses), in the other case, by observation of purely physical changes. To be sure, there are important differences between the specific characters of the states imputed in the two cases: in the case of the human actors the states are psychological or "subjective," and the social scientist making the imputation may indeed have first-hand personal experience of them, but in the case of the wire and other inanimate objects they are not. Nevertheless, despite these differences, the crucial point is that the log-ical canons employed by responsible social scientists in assessing the objective evidence for the imputation of psychological states do not appear to differ essentially (though they may often be applied less rig-orously) from the canons employed for analogous purposes by respon-sible students in other areas of inquiry.

In sum, the fact that the social scientist, unlike the student of inani-mate nature, is able to project himself by sympathetic imagination into the phenomena he is attempting to understand, is pertinent to questions concerning the *origins* of his explanatory hypotheses but not to questions concerning their validity. His ability to enter into relations of empathy with the human actors in some social process may indeed be heuristi-cally important in his efforts to *invent* suitable hypotheses which will ex-plain the process. Nevertheless, his empathic identification with those individuals does not, by itself, constitute *knowledge*. The fact that he achieves such identification does not annul the need for objective evidence, assessed in accordance with logical principles that are common to all controlled inquiries, to support his imputation of subjective states to those human agents.[28]

V.   *The Value-Oriented Bias of Social Inquiry*

We turn, finally, to the difficulties said to confront the social sciences because the social values to which students of social phenomena are committed not only color the contents of their findings but also control their assessment of the evidence on which they base their conclusions. Since social scientists generally differ in their value commitments, the "value neutrality" that seems to be so pervasive in the natural sciences is therefore often held to be impossible in social inquiry. In the judgment of many thinkers, it is accordingly absurd to expect the social sciences to exhibit the unanimity so common among natural scientists concerning what are the established facts and satisfactory explanations for them. Let us examine some of the reasons that have been advanced for these contentions. It will be convenient to distinguish four groups of such reasons, so that our discussion will deal in turn with the alleged role of value judgments in (1) the selection of problems, (2) the determination of the contents of conclusions, (3) the identification of fact, and (4) the assessment of evidence.

1.   The reasons perhaps most frequently cited make much of the fact that the things a social scientist selects for study are determined by his conception of what are the socially important values. According to one influential view, for example, the student of human affairs deals only with materials to which he attributes "cultural significance," so that a "value orientation" is inherent in his choice of material for investigation. Thus, although Max Weber was a vigorous proponent of a "value-free" social science—i.e., he maintained that social scientists must appreciate (or "understand") the values involved in the actions or institutions they are discussing but that it is not their business as objective scientists to approve or disapprove either those values or those actions and institutions —he nevertheles argued that

> The concept of culture is a *value-concept.* Empirical reality becomes "culture" to us because and insofar as we relate it to value ideas. It includes those segments and only those segments of reality which have become significant to us because of this value-relevance. Only a small portion of exist-

[28] The heuristic function of such imaginary identification is discussed by Theodore Abel, "The Operation Called *Verstehen,*" *American Journal of Sociology,* Vol. 54 (1948), pp. 211-18.

ing concrete reality is colored by our value-conditioned interest and it alone is significant to us. It is significant because it reveals relationships which are important to us due to their connection with our values. Only because and to the extent that this is the case is it worthwhile for us to know it in its individual features. We cannot discover, however, what is meaningful to us by means of a "presuppositionless" investigation of empirical data. Rather perception of its meaningfulness to us is the presupposition of its becoming an *object* of investigation.[29]

It is well-nigh truistic to say that students of human affairs, like students in any other area of inquiry, do not investigate everything, but direct their attention to certain selected portions of the inexhaustible content of concrete reality. Moreover, let us accept the claim, if only for the sake of the argument, that a social scientist addresses himself exclusively to matters which he believes are important because of their assumed relevance to his cultural values.[30] It is not clear, however, why the fact that an investigator selects the materials he studies in the light of problems which interest him and which seem to him to bear on matters he regards as important, is of greater moment for the logic of social inquiry than it is for the logic of any other branch of inquiry. For example, a social scientist may believe that a free economic market embodies a cardinal human value, and he may produce evidence to show that certain kinds of human activities are indispensable to the perpetuation of a free market. If he is concerned with processes which maintain this type of economy rather than some other type, how is this fact more pertinent to the question whether he has adequately evaluated the evidence for his conclusion, than is the bearing upon the analogous question of the fact that a physiologist may be concerned with processes which maintain a constant internal temperature in the human body rather than with something else? The things a social scientist *selects for study* with a view to determining the conditions or consequences of their existence may indeed be dependent on the indisputable fact that he is a "cultural being." But similarly, were we not human beings though still capable of conducting scientific inquiry, we might conceivably have an interest neither in the conditions that maintain a free market, nor in the processes involved in the homeostasis of the internal temperature in human bodies, nor for that matter in the mechanisms that regulate the height of tides, the succession of seasons, or the motions of the planets.

In short, there is no difference between any of the sciences with respect to the fact that the interests of the scientist determine what he selects for investigation. But this fact, by itself, represents no obstacle

[29] Max Weber, *The Methodology of the Social Sciences*, Glencoe, Ill., 1947, p. 76.
[30] This question receives some attention below in the discussion of the fourth difficulty.

to the successful pursuit of objectively controlled inquiry in any branch of study.

2. A more substantial reason commonly given for the value-oriented character of social inquiry is that, since the social scientist is himself affected by considerations of right and wrong, his own notions of what constitutes a satisfactory social order and his own standards of personal and social justice do enter, in point of fact, into his analyses of social phenomena. For example, according to one version of this argument, anthropologists must frequently judge whether the means adopted by some society achieves the intended aim (e.g., whether a religious ritual does produce the increased fertility for the sake of which the ritual is performed); and in many cases the adequacy of the means must be judged by admittedly "relative" standards, i.e., in terms of the ends sought or the standards employed by that society, rather than in terms of the anthropologist's own criteria. Nevertheless, so the argument proceeds, there are also situations in which

> we must apply absolute standards of adequacy, that is evaluate the end-results of behavior in terms of purposes we believe in or postulate. This occurs, first, when we speak of the satisfaction of psycho-physical 'needs' offered by any culture; secondly, when we assess the bearing of social facts upon survival; and thirdly, when we pronounce upon social integration and stability. In each case our statements imply judgments as to the worth-whileness of actions, as to 'good' or 'bad' cultural solutions of the problems of life, and as to 'normal' and 'abnormal' states of affairs. These are basic judgments which we cannot do without in social enquiry and which clearly do not express a purely personal philosophy of the enquirer or values arbitrarily assumed. Rather do they grow out of the history of human thought, from which the anthropologist can seclude himself as little as can anyone else. Yet as the history of human thought has led not to one philosophy but to several, so the value attitudes implicit in our ways of thinking will differ and sometimes conflict.[81]

It has often been noted, moreover, that the study of social phenomena receives much of its impetus from a strong moral and reforming zeal, so that many ostensibly "objective" analyses in the social sciences are in fact disguised recommendations of social policy. As one typical but moderately expressed statement of the point puts it, a social scientist

[81] S. F. Nadel, *The Foundations of Social Anthropology*, Glencoe, Ill., 1951, pp. 53-54. The claim is sometimes also made that the exclusion of value judgments from social science is undesirable as well as impossible. "We cannot disregard all questions of what is socially desirable without missing the significance of many social facts; for since the relation of means to ends is a special form of that between parts and wholes, the contemplation of social ends enables us to see the relations of whole groups of facts to each other and to larger systems of which they are parts."—Morris R. Cohen, *Reason and Nature*, New York, 1931, p. 343.

cannot wholly detach the unifying social structure that, as a scientist's theory, guides his detailed investigations of human behavior, from the unifying structure which, as a citizen's ideal, he thinks ought to prevail in human affairs and hopes may sometimes be more fully realized. His social theory is thus essentially a program of action along two lines which are kept in some measure of harmony with each other by that theory—action in assimilating social facts for purposes of systematic understanding, and action aiming at progressively molding the social pattern, so far as he can influence it, into what he thinks it ought to be.[32]

It is surely beyond serious dispute that social scientists do in fact often import their own values into their analyses of social phenomena. It is also undoubtedly true that even thinkers who believe human affairs can be studied with the ethical neutrality characterizing modern inquiries into geometrical or physical relations, and who often pride themselves on the absence of value judgments from their own analyses of social phenomena, do in fact sometimes make such judgments in their social inquiries.[33] Nor is it less evident that students of human affairs often hold conflicting values; that their disagreements on value questions are often the source of disagreements concerning ostensibly factual issues; and that, even if value predications are assumed to be inherently capable of proof or disproof by objective evidence, at least some of the differences between social scientists involving value judgments are not in fact resolved by the procedures of controlled inquiry.

In any event, it is not easy in most areas of inquiry to prevent our likes, aversions, hopes, and fears from coloring our conclusions. It has taken centuries of effort to develop habits and techniques of investigation which help safeguard inquiries in the natural sciences against the intrusion of irrelevant personal factors; and even in these disciplines the protection those procedures give is neither infallible nor complete. The problem is undoubtedly more acute in the study of human affairs, and the difficulties it creates for achieving reliable knowledge in the social sciences must be admitted.

However, the problem is intelligible only on the assumption that there is a relatively clear distinction between factual and value judgments, and that however difficult it may sometimes be to decide whether a given statement has a purely factual content, it is in principle possible to do so. Thus, the claim that social scientists are pursuing the twofold program mentioned in the above quotation makes sense, only if it is possible to distinguish between, on the one hand, contributions to theoretical understanding (whose factual validity presumably does not depend on the

---

[32] Edwin A. Burtt, *Right Thinking*, New York, 1946, p. 522.
[33] For a documented account, see Gunnar Myrdal, *Value in Social Theory*, London, 1958, pp. 134-52.

social ideal to which a social scientist may subscribe), and on the other hand contributions to the dissemination or realization of some social ideal (which may not be accepted by all social scientists). Accordingly, the undeniable difficulties that stand in the way of obtaining reliable knowledge of human affairs because of the fact that social scientists differ in their value orientations are practical difficulties. The difficulties are not necessarily insuperable, for since by hypothesis it is not impossible to distinguish between fact and value, steps can be taken to identify a value bias when it occurs, and to minimize if not to eliminate completely its perturbing effects.

One such countermeasure frequently recommended is that social scientists abandon the pretense that they are free from all bias, and that instead they state their value assumptions as explicitly and fully as they can.[34] The recommendation does not assume that social scientists will come to agree on their social ideals once these ideals are explicitly postulated, or that disagreements over values can be settled by scientific inquiry. Its point is that the question of how a given ideal is to be realized, or the question whether a certain institutional arrangement is an effective way of achieving the ideal, is on the face of it not a value question, but a factual problem—to be resolved by the objective methods of scientific inquiry—concerning the adequacy of proposed means for attaining stipulated ends. Thus, economists may permanently disagree on the desirability of a society in which its members have a guaranteed security against economic want, since the disagreement may have its source in inarbitrable preferences for different social values. But when sufficient evidence is made available by economic inquiry, economists do presumably agree on the factual proposition that, *if* such a society is to be achieved, then a purely competitive economic system will not suffice.

Although the recommendation that social scientists make fully explicit their value commitments is undoubtedly salutary, and can produce excellent fruit, it verges on being a counsel of perfection. For the most part we are unaware of many assumptions that enter into our analyses and actions, so that despite resolute efforts to make our preconceptions explicit some decisive ones may not even occur to us. But in any event, the difficulties generated for scientific inquiry by unconscious bias and tacit value orientations are rarely overcome by devout resolutions to eliminate bias. They are usually overcome, often only gradually, through the self-corrective mechanisms of science as a social enterprise. For modern science encourages the invention, the mutual exchange, and the free but responsible criticisms of ideas; it welcomes competition in the quest for

[34] See, e.g., S. F. Nadel, *op. cit.*, p. 54; also Gunnar Myrdal, *op. cit.*, p. 120, as well as his *Political Element in the Development of Economic Theory*, Cambridge, Mass., 1954, esp. Chap. 8.

knowledge between independent investigators, even when their intellectual orientations are different; and it progressively diminishes the effects of bias by retaining only those proposed conclusions of its inquiries that survive critical examination by an indefinitely large community of students, whatever be their value preferences or doctrinal commitments. It would be absurd to claim that this institutionalized mechanism for sifting warranted beliefs has operated or is likely to operate in social inquiry as effectively as it has in the natural sciences. But it would be no less absurd to conclude that reliable knowledge of human affairs is unattainable merely because social inquiry is frequently value-oriented.

3.   There is a more sophisticated argument for the view that the social sciences cannot be value-free. It maintains that the distinction between fact and value assumed in the preceding discussion is untenable when purposive human behavior is being analyzed, since in this context value judgments enter inextricably into what appear to be "purely descriptive" (or factual) statements. Accordingly, those who subscribe to this thesis claim that an ethically neutral social science is in principle impossible, and not simply that it is difficult to attain. For if fact and value are indeed so fused that they cannot even be distinguished, value judgments cannot be eliminated from the social sciences unless all predications are also eliminated from them, and therefore unless these sciences completely disappear.

For example, it has been argued that the student of human affairs must distinguish between valuable and undesirable forms of social activity, on pain of failing in his "plain duty" to present social phenomena truthfully and faithfully:

> Would one not laugh out of court a man who claimed to have written a sociology of art but who actually had written a sociology of trash? The sociologist of religion must distinguish between phenomena which have a religious character and phenomena which are a-religious. To be able to do this, he must understand what religion is. . . . Such understanding enables and forces him to distinguish between genuine and spurious religion, between higher and lower religions; these religions are higher in which the specifically religious motivations are effective to a higher degree. . . . The sociologist of religion cannot help noting the difference between those who try to gain it by a change of heart. Can he see this difference without seeing at the same time the difference between a mercenary and nonmercenary attitude? . . . The prohibition against value-judgments in social science would lead to the consequence that we are permitted to give a strictly factual description of the overt acts that can be observed in concentration camps, and perhaps an equally factual analysis of the motivations of the actors concerned: we would not be permitted to speak of cruelty. Every reader of such a description who is not completely stupid would, of course,

see that the actions described are cruel. The factual description would, in truth, be a bitter satire. What claimed to be a straightforward report would be an unusually circumlocutory report. . . . Can one say anything relevant on public opinion polls . . . without realizing the fact that many answers to the questionnaires are given by unintelligent, uninformed, deceitful, and irrational people, and that not a few questions are formulated by people of the same caliber—can one say anything relevant about public opinion polls without committing one value-judgment after another?[35]

Moreover, the assumption implicit in the recommendation discussed above for achieving ethical neutrality is often rejected as hopelessly naive —this is the assumption, it will be recalled, that relations of means to ends can be established without commitment to these ends, so that the conclusions of social inquiry concerning such relations are objective statements which make *conditional* rather than categorical assertions about values. This assumption is said by its critics to rest on the supposition that men attach value only to the ends they seek, and not to the means for realizing their aims. However, the supposition is alleged to be grossly mistaken. For the character of the means one employs to secure some goal affects the nature of the total outcome; and the choice men make between alternative means for obtaining a given end depends on the values they ascribe to those alternatives. In consequence, commitments to specific valuations are said to be involved even in what appear to be purely factual statements about means-ends relations.[36]

We shall not attempt a detailed assessment of this complex argument, for a discussion of the numerous issues it raises would take us far afield. However, three claims made in the course of the argument will be admitted without further comment as indisputably correct: that a large number of characterizations sometimes assumed to be purely factual descriptions of social phenomena do indeed formulate a type of value judgment; that it is often difficult, and in any case usually inconvenient in practice, to distinguish between the purely factual and the "evaluative" contents of many terms employed in the social sciences; and that values are commonly attached to means and not only to ends. However, these admissions do not entail the conclusion that, in a manner unique to the study of purposive human behavior, fact and value are fused beyond the possibility of distinguishing between them. On the contrary, as we shall

---

[35] Leo Strauss, "The Social Science of Max Weber," *Measure*, Vol. 2 (1951), pp. 211-14. For a discussion of this issue as it bears upon problems in the philosophy of law, see Lon Fuller, "Human Purpose and Natural Law," *Natural Law Forum*, Vol. 3 (1958), pp. 68-76; Ernest Nagel, "On the Fusion of Fact and Value: A Reply to Professor Fuller," *op. cit.*, pp. 77-82; Lon L. Fuller, "A Rejoinder to Professor Nagel," *op. cit.*, pp. 83-104; Ernest Nagel, "Fact, Value, and Human Purpose," *Natural Law Forum*, Vol. 4 (1959), pp. 26-43.

[36] Cf. Gunnar Myrdal, *Value in Social Theory*, London, 1958, pp. xxii, 211-13.

try to show, the claim that there is such a fusion and that a value-free social science is therefore inherently absurd, confounds two quite different senses of the term "value judgment": the sense in which a value judgment expresses *approval or disapproval* either of some moral (or social) ideal, or of some action (or institution) because of a commitment to such an ideal; and the sense in which a value judgment expresses *an estimate* of the degree to which some commonly recognized (and more or less clearly defined) type of action, object, or institution is embodied in a given instance.

It will be helpful to illustrate these two senses of "value judgment" first with an example from biology. Animals with blood streams sometimes exhibit the condition known as "anemia." An anemic animal has a reduced number of red blood corpuscles, so that, among other things, it is less able to maintain a constant internal temperature than are members of its species with a "normal" supply of such blood cells. However, although the meaning of the term "anemia" can be made quite clear, it is not in fact defined with complete precision; for example, the notion of a "normal" number of red corpuscles that enters into the definition of the term is itself somewhat vague, since this number varies with the individual members of a species as well as with the state of a given individual at different times (such as its age or the altitude of its habitat). But in any case, to decide whether a given animal is anemic, an investigator must judge whether the available evidence *warrants* the conclusion that the specimen is anemic.[37] He may perhaps think of anemia as being of several distinct kinds (as is done in actual medical practice), or he may think of anemia as a condition that is realizable with greater or lesser completeness (just as certain plane curves are sometimes described as better or worse approximations to a circle as defined in geometry); and, depending on which of these conceptions he adopts, he may decide either that his specimen has a certain kind of anemia or that it is anemic only to a certain degree. When the investigator reaches a conclusion, he can therefore be said to be making a "value judgment," in the sense that he has in mind some standardized type of physiological condition designated as "anemia" and that he *assesses* what he knows about his specimen with the measure provided by this assumed standard. For the sake of easy reference, let us call such evaluations of the evidence, which conclude that a given characteristic is in some degree present (or absent) in a given instance, "characterizing value judgments."

[37] The evidence is usually a count of red cells in a sample from the animal's blood. However, it should be noted that "The red cell count gives only an estimate of the *number of cells per unit quantity of blood*," and does not indicate whether the body's total supply of red cells is increased or diminished.—Charles H. Best and Norman B. Taylor, *The Physiological Basis of Medical Practice*, 6th ed., Baltimore, 1955, pp. 11, 17.

On the other hand, the student may also make a quite different sort of value judgment, which asserts that, since an anemic animal has diminished powers of maintaining itself, anemia is an undesirable condition. Moreover, he may apply this general judgment to a particular case, and so come to deplore the fact that a given animal is anemic. Let us label such evaluations, which conclude that some envisaged or actual state of affairs is worthy of approval or disapproval, "appraising value judgments." [38] It is clear, however, that an investigator making a characterizing value judgment is not thereby logically bound to affirm or deny a corresponding appraising evaluation. It is no less evident that he cannot consistently make an appraising value judgment about a given instance (e.g., that it is undesirable for a given animal to continue being anemic), unless he can affirm a characterizing judgment about that instance independently of the appraising one (e.g., that the animal is anemic). Accordingly, although characterizing judgments are necessarily entailed by many appraising judgments, making appraising judgments is not a necessary condition for making characterizing ones.

Let us now apply these distinctions to some of the contentions advanced in the argument quoted above. Consider first the claim that the sociologist of religion must recognize the difference between mercenary and nonmercenary attitudes, and that in consequence he is inevitably committing himself to certain values. It is certainly beyond dispute that these attitudes are commonly distinguished; and it can also be granted that a sociologist of religion needs to understand the difference between them. But the sociologist's obligation is in this respect quite like that of the student of animal physiology, who must also acquaint himself with certain distinctions—even though the physiologist's distinction between, say, anemic and nonanemic may be less familiar to the ordinary layman and is in any case much more precise than is the distinction between mercenary and nonmercenary attitudes. Indeed, because of the vagueness of these latter terms, the scrupulous sociologist may find it extremely difficult to decide whether or not the attitude of some community toward its acknowledged gods is to be characterized as mercenary; and if he should finally decide, he may base his conclusion on some inarticulated "total impression" of that community's manifest behavior, without being able to state exactly the detailed grounds for his decision. But however this may be, the sociologist who claims that a certain attitude manifested by a given religious group is mercenary, just as the physiologist who claims

[38] It is irrelevant to the present discussion what view is adopted concerning the ground upon which such judgments supposedly rest—whether those grounds are simply arbitrary preferences, alleged intuitions of "objective" values, categorical moral imperatives, or anything else that has been proposed in the history of value theory. For the distinction made in the text is independent of any particular assumption about the foundations of appraising value judgments, "ultimate" or otherwise.

that a certain individual is anemic, is making what is primarily a characterizing value judgment. In making these judgments, neither the sociologist nor the physiologist is necessarily committing himself to any values other than the values of scientific probity; and in this respect, therefore, there appears to be no difference between social and biological (or for that matter, physical) inquiry.

On the other hand, it would be absurd to deny that in characterizing various actions as mercenary, cruel, or deceitful, sociologists are frequently (although perhaps not always wittingly) asserting appraising as well as characterizing value judgments. Terms like 'mercenary,' 'cruel,' or 'deceitful' as commonly used have a widely recognized pejorative overtone. Accordingly, anyone who employs such terms to characterize human behavior can normally be assumed to be stating his disapprobation of that behavior (or his approbation, should he use terms like 'nonmercenary,' 'kindly,' or 'truthful'), and not simply characterizing it.

However, although many (but certainly not all) ostensibly characterizing statements asserted by social scientists undoubtedly express commitments to various (not always compatible) values, a number of "purely descriptive" terms as used by natural scientists in certain contexts sometimes also have an unmistakably appraising value connotation. Thus, the claim that a social scientist is making appraising value judgments when he characterizes respondents to questionnaires as uninformed, deceitful, or irrational can be matched by the equally sound claim that a physicist is also making such judgments when he describes a particular chronometer as inaccurate, a pump as inefficient, or a supporting platform as unstable. Like the social scientist in this example, the physicist is characterizing certain objects in his field of research; but, also like the social scientist, he is in addition expressing his disapproval of the characteristics he is ascribing to those objects.

Nevertheless—and this is the main burden of the present discussion—there are no good reasons for thinking that it is inherently impossible to *distinguish* between the characterizing and the appraising judgments implicit in many statements, whether the statements are asserted by students of human affairs or by natural scientists. To be sure, it is not always easy to make the distinction formally explicit in the social sciences—in part because much of the language employed in them is very vague, in part because appraising judgments that may be implicit in a statement tend to be overlooked by us when they are judgments to which we are actually committed though without being aware of our commitments. Nor is it always useful or convenient to perform this task. For many statements implicitly containing both characterizing and appraising evaluations are sometimes sufficiently clear without being reformulated in the manner required by the task; and the reformulations would frequently be too unwieldy for effective communication between members of a large and unequally prepared group of students. But these are essentially practical rather than theoretical problems. The difficulties they raise provide no compelling reasons for the claim that an ethically neutral social science is inherently impossible.

Nor is there any force in the argument that, since values are commonly attached to means and not only to ends, statements about means-ends relations are not value-free. Let us test the argument with a simple example. Suppose that a man with an urgent need for a car but without sufficient funds to buy one can achieve his aim by borrowing a sum either from a commercial bank or from friends who waive payment of any interest. Suppose further that he dislikes becoming beholden to his friends for financial favors, and prefers the impersonality of a commercial loan. Accordingly, the comparative values this individual places upon the alternative means available to him for realizing his aim obviously control the choice he makes between them. Now the *total* outcome that would result from his adoption of one of the alternatives is admittedly different from the *total* outcome that would result from his adoption of the other alternative. Nevertheless, irrespective of the values he may attach to these alternative means, each of them would achieve a result—namely, his purchase of the needed car—that is common to both the total outcomes. In consequence, the validity of the statement that he could buy the car by borrowing money from a bank, as well as of the statement that he could realize this aim by borrowing from friends, is unaffected by the valuations placed upon the means, so that neither statement involves any special appraising evaluations. In short, the statements about means-ends relations are value-free.

4. There remains for consideration the claim that a value-free social science is impossible, because value commitments enter into the very *assessment of evidence* by social scientists, and not simply into the content of the conclusions they advance. This version of the claim itself has a large number of variant forms, but we shall examine only three of them.

The least radical form of the claim maintains that the conceptions held by a social scientist of what constitute cogent evidence or sound intellectual workmanship are the products of his education and his place in society, and are affected by the social values transmitted by this training and associated with this social position; accordingly, the values to which the social scientist is thereby committed determine which statements he *accepts* as well-grounded conclusions about human affairs. In this form, the claim is a *factual* thesis, and must be supported by detailed empirical evidence concerning the influences exerted by a man's moral and social values upon what he is ready to acknowledge as sound social analysis.

In many instances such evidence is indeed available; and differences between social scientists in respect to what they accept as credible can sometimes be attributed to the influence of national, religious, economic, and other kinds of bias. However, this variant of the claim excludes neither the possibility of recognizing assessments of evidence that are prejudiced by special value commitments, nor the possibility of correcting for such prejudice. It therefore raises no issue that has not already been discussed when we examined the second reason for the alleged value-oriented character of social inquiry (pages 487-90).

Another but different form of the claim is based on recent work in theoretical statistics dealing with the assessment of evidence for so-called "statistical hypotheses"—hypotheses concerning the probabilities of random events, such as the hypothesis that the probability of a male human birth is one-half. The central idea relevant to the present question that underlies these developments can be sketched in terms of an example. Suppose that, before a fresh batch of medicine is put on sale, tests are performed on experimental animals for its possible toxic effects because of impurities that have not been eliminated in its manufacture, for example, by introducing small quantities of the drug into the diet of one hundred guinea pigs. If no more than a few of the animals show serious after-effects, the medicine is to be regarded as safe, and will be marketed; but if a contrary result is obtained the drug will be destroyed. Suppose now that three of the animals do in fact become gravely ill. Is this outcome significant (i.e., does it indicate that the drug has toxic effects), or is it perhaps an "accident" that happened because of some peculiarity in the affected animals? To answer the question, the experimenter must *decide* on the basis of the evidence between the hypothesis H_1: the drug is toxic, and the hypothesis H_2: the drug is not toxic. But how is he to decide, if he aims to be "reasonable" rather than arbitrary? Current statistical theory offers him a rule for making a reasonable decision, and bases the rule on the following analysis.

Whatever decision the experimenter may make, he runs the risk of committing either one of two types of errors: he may reject a hypothesis though in fact it is true (i.e., despite the fact that H_1 is actually true, he mistakenly decides against it in the light of the evidence available to him); or he may accept a hypothesis though in fact it is false. His decision would therefore be eminently reasonable, were it based on a rule guaranteeing that no decision ever made in accordance with the rule would commit either type of error. Unhappily, there are no rules of this sort. The next suggestion is to find a rule such that, when decisions are made in accordance with it, the relative frequency of each type of error is quite small. But unfortunately, the risks of committing each type of error are not independent; for example, it is in general logically impossible to find a rule so that decisions based on it will commit each type of error with a relative frequency not greater than one in a thousand. In consequence, before a reasonable rule can be proposed, the experimenter must compare the relative importance to himself of the two types of error, and state what risk he is willing to take of committing the type of error he judges to be the more important one. Thus, were he to reject H_1 though it is true (i.e., were he to commit an error of the first type), all the medicine under consideration would be put on sale, and the lives of those using it would be endangered; on the other hand, were he to commit an error of the second type with respect to H_1, the entire batch of medicine would be scrapped, and the manufacturer would incur a financial loss. However, the preservation of human life may be of greater moment to the experimenter than financial gain; and he may perhaps stipulate that he is unwilling to base his decision on a rule for which the risk of committing an error of the first type is greater than one such error in a hundred decisions. If this is assumed, statistical theory can specify a rule satisfying the experimenter's requirement, though how this is done, and how the risk of committing an error of the second type is calculated, are technical questions of no concern to us. The main point to be noted in this analysis is that the rule presupposes certain appraising judgments of value. In short, if this result is generalized, statistical theory appears to support the thesis that value commitments enter decisively into the rules for assessing evidence for statistical hypotheses.[39]

However, the theoretical analysis upon which this thesis rests does not entail the conclusion that the rules actually employed in every social inquiry for assessing evidence necessarily involve some *special* commitments, i.e., commitments such as those mentioned in the above example, as distinct from those generally implicit in science as an enterprise aiming to achieve reliable knowledge. Indeed, the above example illustrating the reasoning in current statistical theory can be misleading, insofar as it suggests that alternative decisions between statistical hypotheses must invariably lead to alternative actions having immediate practical consequences upon which different special values are placed. For example, a theoretical physicist may have to decide between two statistical hypotheses concerning the probability of certain energy exchanges in atoms; and a theoretical sociologist may similarly have to choose between two statistical hypotheses concerning the relative frequency of childless marriages under certain social arrangements. But neither of these men may

[39] The above example is borrowed from the discussion in J. Neymann, *First Course in Probability and Statistics*, New York, 1950, Chap. 5, where an elementary technical account of recent developments in statistical theory is presented. For a nontechnical account, see Irwin D. J. Bross, *Design for Decision*, New York, 1953, also R. B. Braithwaite, *Scientific Explanation*, Cambridge, Eng., 1953, Chap. 7.

have any *special* values at stake associated with the alternatives between which he must decide, other than the values, to which he is committed as a member of a scientific community, to conduct his inquiries with probity and responsibility. Accordingly, the question whether any special value commitments enter into assessments of evidence in either the natural or social sciences is not settled one way or the other by theoretical statistics; and the question can be answered only by examining actual inquiries in the various scientific disciplines.

Moreover, nothing in the reasoning of theoretical statistics depends on what particular subject matter is under discussion when a decision between alternative statistical hypotheses is to be made. For the reasoning is entirely general; and reference to some special subject matter becomes relevant only when a definite numerical value is to be assigned to the risk some investigator is prepared to take of making an erroneous decision concerning a given hypothesis. Accordingly, if current statistical theory is used to support the claim that value commitments enter into the assessment of evidence for statistical hypotheses in social inquiry, statistical theory can be used with equal justification to support analogous claims for all other inquiries as well. In short, the claim we have been discussing establishes no difficulty that supposedly occurs in the search for reliable knowledge in the study of human affairs which is not also encountered in the natural sciences.

A third form of this claim is the most radical of all. It differs from the first variant mentioned above in maintaining that there is a necessary *logical* connection, and not merely a contingent or causal one, between the "social perspective" of a student of human affairs and his standards of competent social inquiry, and in consequence the influence of the special values to which he is committed because of his own social involvements is not eliminable. This version of the claim is implicit in Hegel's account of the "dialectical" nature of human history and is integral to much Marxist as well as non-Marxist philosophy that stresses the "historically relative" character of social thought. In any event, it is commonly based on the assumption that, since social institutions and their cultural products are constantly changing, the intellectual apparatus required for understanding them must also change; and every idea employed for this purpose is therefore adequate only for some particular stage in the development of human affairs. Accordingly, neither the substantive concepts adopted for classifying and interpreting social phenomena, nor the logical canons used for estimating the worth of such concepts, have a "timeless validity"; there is no analysis of social phenomena which is not the expression of some special social standpoint, or which does not reflect the interests and values dominant in some sector of the human scene at a certain stage of its history. In consequence, although a sound distinction can be made in the natural sciences between the origin of a man's views and their factual validity, such a distinction allegedly cannot be made in social inquiry; and prominent exponents of "historical relativism" have therefore challenged the universal adequacy of the thesis that "the genesis of a proposition is under all circumstances irrelevant to its truth." As one influential proponent of this position puts the matter,

> The historical and social genesis of an idea would only be irrelevant to its ultimate validity if the temporal and social conditions of its emergence had no effect on its content and form. If this were the case, any two periods in the history of human knowledge would only be distinguished from one another by the fact that in the earlier period certain things were still unknown and certain errors still existed which, through later knowledge were completely corrected. This simple relationship between an earlier incomplete and a later complete period of knowledge may to a large extent be appropriate for the exact sciences. . . . For the history of the cultural sciences, however, the earlier stages are not quite so simply superseded by the later stages, and it is not so easily demonstrable that early errors have subsequently been corrected. Every epoch has its fundamentally new approach and its characteristic point of view, and consequently sees the "same" object from a new perspective. . . . The very principles, in the light of which knowledge is to be criticized, are themselves found to be socially and historically conditioned. Hence their application appears to be limited to given historical periods and the particular types of knowledge then prevalent.[40]

Historical research into the influence of society upon the beliefs men hold is of undoubted importance for understanding the complex nature of the scientific enterprise; and the sociology of knowledge—as such investigations have come to be called—has produced many clarifying contributions to such an understanding. However, these admittedly valuable services of the sociology of knowledge do not establish the radical claim we have been stating. In the first place, there is no competent evidence to show that the principles employed in social inquiry for assessing the intellectual products are *necessarily* determined by the social perspective of the inquirer. On the contrary, the "facts" usually cited in support of this contention establish at best only a contingent causal relation between a man's social commitments and his canons of cognitive validity. For

[40] Karl Mannheim, *Ideology and Utopia*, New York, 1959, pp. 271, 288, 292. The essay from which the above excerpts are quoted was first published in 1931, and Mannheim subsequently modified some of the views expressed in it. However, he reaffirmed the thesis stated in the quoted passages as late as 1946, the year before his death. See his letter to Kurt H. Wolff, dated April 15, 1946, quoted in the latter's "Sociology of Knowledge and Sociological Theory," in *Symposium on Sociological Theory* (ed. by Llewellyn Gross), Evanston, Ill., 1959, p. 571.

example, the once fashionable view that the "mentality" or logical operations of primitive societies differ from those typical in Western civilization—a discrepancy that was attributed to differences in the institutions of the societies under comparison—is now generally recognized to be erroneous, because it seriously misinterprets the intellectual processes of primitive peoples. Moreover, even extreme exponents of the sociology of knowledge admit that most conclusions asserted in mathematics and natural science are neutral to differences in social perspective of those asserting them, so that the genesis of these propositions is irrelevant to their validity. Why cannot propositions about human affairs exhibit a similar neutrality, at least in some cases? Sociologists of knowledge do not appear to doubt that the truth of the statement that two horses can in general pull a heavier load than can either horse alone, is logically independent of the social status of the individual who happens to affirm the statement. But they have not made clear just what are the inescapable considerations that allegedly make such independence inherently impossible for the analogous statement about human behavior, that two laborers can in general dig a ditch of given dimensions more quickly than can either laborer working alone.

In the second place, the claim faces a serious and frequently noted dialectical difficulty—a difficulty that proponents of the claim have succeeded in meeting only by abandoning the substance of the claim. For let us ask what is the cognitive status of the thesis that a social perspective enters essentially into the content as well as the validation of every assertion about human affairs. Is this thesis meaningful and valid only for those who maintain it and who thus subscribe to certain values because of their distinctive social commitments? If so, no one with a different social perspective can properly understand it; its acceptance as valid is strictly limited to those who can do so, and social scientists who subscribe to a different set of social values ought therefore dismiss it as empty talk. Or is the thesis singularly exempt from the class of assertions to which it applies, so that its meaning and truth are not inherently related to the social perspectives of those who assert it? If so, it is not evident why the thesis is so exempt; but in any case, the thesis is then a conclusion of inquiry into human affairs that is presumably "objectively valid" in the usual sense of this phrase—and, if there is one such conclusion, it is not clear why there cannot be others as well.

To meet this difficulty, and to escape the self-defeating skeptical relativism to which the thesis is thus shown to lead, the thesis is sometimes interpreted to say that, though "absolutely objective" knowledge of human affairs is unattainable, a "relational" form of objectivity called "relationism" can nevertheless be achieved. On this interpretation, a social scientist can discover just what his social perspective is; and if he then formulates the conclusions of his inquiries "relationally," so as to indicate that his findings conform to the canons of validity implicit in his perspective, his conclusions will have achieved a "relational" objectivity. Social scientists sharing the same perspective can be expected to agree in their answers to a given problem when the canons of validity characteristic of their common perspective are correctly applied. On the other hand, students of social phenomena who operate within different but incongruous social perspectives can also achieve objectivity, if in no other way than by a "relational" formulation of what must otherwise be incompatible results obtained in their several inquiries. However, they can also achieve it in "a more roundabout fashion," by undertaking "to find a formula for translating the results of one into those of the other and to discover a common denominator for these varying perspectivistic insights." [41]

But it is difficult to see in what way "relational objectivity" differs from "objectivity" without the qualifying adjective and in the customary sense of the word. For example, a physicist who terminates an investigation with the conclusion that the velocity of light in water has a certain numerical value when measured in terms of a stated system of units, by a stated procedure, and under stated experimental conditions, is formulating his conclusion in a manner that is "relational" in the sense intended; and his conclusion is marked by "objectivity," presumably because it mentions the "relational" factors upon which the assigned numerical value of the velocity depends. However, it is fairly standard practice in the natural sciences to formulate certain types of conclusions in this fashion. Accordingly, the proposal that the social sciences formulate their findings in an analogous manner carries with it the admission that it is not in principle impossible for these disciplines to establish conclusions having the objectivity of conclusions reached in other domains of inquiry. Moreover, if the difficulty we are considering is to be resolved by the suggested translation formulas for rendering the "common denominators" of conclusions stemming from divergent social perspectives, those formulas cannot in turn be "situationally determined" in the sense of this phrase under discussion. For if those formulas were so determined, the same difficulty would crop up anew in connection with them. On the other hand, a search for such formulas is a phase in the search for invariant relations in a subject matter, so that formulations of these relations are valid irrespective of the particular perspective one may select from some class of perspectives on that subject matter. In consequence, in acknowledging that the search for such invariants in the social sciences is not inherently

41 Karl Mannheim, *op. cit.*, pp. 300-01.

bound to fail, proponents of the claim we have been considering abandon what at the outset was its most radical thesis.

In brief, the various reasons we have been examining for the intrinsic impossibility of securing objective (i.e., value-free and unbiased) conclusions in the social sciences do not establish what they purport to establish, even though in some instances they direct attention to undoubtedly important practical difficulties frequently encountered in these disciplines.

Explanation and Understanding
in the Social Sciences

14

The net outcome of the discussion in the previous chapter is that none of the methodological difficulties often alleged to confront the search for systematic explanations of social phenomena is unique to the social sciences or is inherently insuperable. On the other hand, problems are not resolved merely by showing that they are not necessarily insoluble; and the present state of social inquiry clearly indicates that some of the difficulties we have been considering are indeed serious. Despite these difficulties, social scientists are able to advance explanations for a large variety of social phenomena, even if the comprehensiveness of proposed explanatory premises is often small and their merit is frequently in dispute. We shall not survey these proposed explanations nor discuss any of them in detail, for it is not our aim to deal with the substantive content of special areas of social study; but in conformity with our aim we will examine several structural (or formal) characteristics that are exhibited by various types of explanations prominent in current social inquiry.

I. Statistical Generalizations and Their Explanations

As has already been noted, most if not all the generalizations empirical social research has succeeded in establishing are formulated in terms of familiar "common-sense" distinctions, and possess a comparatively narrow scope of valid application (or low-order generality). More-

over, most if not all of these generalizations assert relations of dependence that hold between stated phenomena only in a (more or less precisely specified) fraction of the instances of those phenomena, rather
than invariably or with strict universality—for example, generalizations
like "Most rural Americans belong to some religious organization" or
"The annual suicide rate among Protestants is in general higher than
among Catholics." For convenience, we shall refer to such generalizations
as "statistical" or "probabilistic," even though (as in these examples)
no numerical values for statistical or probability coefficients are mentioned in them. Statistical laws are not unique to the social sciences; statistical assumptions are contained in several physical and biological
theories, and statistical experimental laws are common in various
branches of natural science, such as meteorology, physiology, and animal behavior. It is nevertheless noteworthy that experimental laws in
social science are perhaps exclusively statistical; and we shall therefore
first consider why this is so, and whether it is inevitable. We shall then
examine the structure of explanations for statistical generalizations, with
primary attention directed to such explanations in the social sciences;
but we shall postpone until the next chapter discussing the role of statistical and other laws in explanations of particular historical occurrences.

1.   Two main reasons, not altogether unrelated, are commonly offered
for the pervasively statistical nature of generalizations obtained in empirical social study. The first attributes this fact to the inherent complexity of social science subject matter, so that, because of our consequent inability to identify individually all the pertinent variables, we
are unable to state the precise conditions upon which different types of
human conduct invariably depend. The second reason stresses the element of volition that enters into the determination of human conduct.
This account is sometimes based on the claim that the human will is
"free" and its manifestations in overt action are therefore not completely
predictable, so that there can be no invariable regularities in social phenomena. However, writers who do not subscribe to the free-will doctrine
state this second reason somewhat differently. This alternative account
maintains that men's actions are governed by their *interpretations* of
external stimuli, rather than by such stimuli *directly*. Accordingly, since
the responses men make to social situations vary because their interpretations differ, whether because of differences in their personal developments or in their native endowments, we cannot establish strictly
universal generalizations relating external stimuli and human reactions
to them.

These reasons undoubtedly have some merit, especially if the issues
raised by the free-will doctrine are dismissed as irrelevant to the ques-

tion under discussion. However, the complexity of a subject matter is at best not a precise notion, and problems that appear to be hopelessly complex before effective ways for dealing with them are invented often lose this appearance after the inventions have been made. Before the introduction of Arabic numerical notation, only unusually gifted men were able to perform arithmetical computations that a normal teen-ager today can take in his stride; and after Newtonian mechanics was developed, properly trained students were able to analyze motions of bodies that some of the best minds of preceding generations found too complex for human understanding. In any event, though social phenomena may indeed be complex, it is by no means certain that they are in general more complex than physical and biological phenomena for which strictly universal laws have been established. Moreover, while it is true that responses to a given social situation are mediated by the variable interpretations men place upon it, this fact by itself does not explain why there are no universal laws relating each of the several interpretations placed upon a given type of social stimulus to a particular form of human response.

At any rate, two further points should be noted, since they are methodologically more fundamental in the present context than are the considerations thus far mentioned. Both points are related to the discussion earlier in Chapter 13 of transcultural social laws. The first point directs attention to the nature of the terms or distinctions employed in formulating the generalizations of empirical social research; the second recalls a logical device commonly adopted in many branches of empirical inquiry in order to make possible the assertion of strictly universal laws.

a. In the first place, we must remind ourselves that the terms used in the universal laws of many branches of science usually have a quite precise connotation, and frequently signify traits that are more or less "idealized" versions of actually observed properties. Each such term is in consequence intended to designate some class of items that are highly homogeneous in certain indicated respects; and a law containing such terms is neither expected to be, nor is it actually, in strict agreement with observed data, if the items to which the terms are in fact applied fail to possess the required degree of homogeneity. For example, the term 'silver' as employed in universal laws in physics and chemistry designates a class of objects that satisfy, among other precisely specified conditions, certain requirements of chemical purity. Accordingly, universal laws such as, for example, that at a given temperature the ratio of the mass of silver to its volume (i.e., its density) is constant, cannot be expected to be in more than approximate agreement with experimental data if the samples of silver upon which experiments are performed do not fully meet these requirements of chemical homogeneity; and analogous comments are in order concerning the other terms mentioned in the laws.

On the other hand, the terms current in empirical social study are for the most part adaptations of distinctions employed in everyday discussions of social questions, and are often used in formulating empirical generalizations with little redefinition of their vague common-sense meanings. Examples of such terms in empirical social research are 'sense of deprivation,' 'morale,' and 'role.' Moreover, even when the meaning of a term is made relatively precise, the precision is frequently achieved by way of some essentially statistical procedure, so that the items falling into its intended designation can possess different specific forms of the property connoted by the term. Thus, the definition of the term 'authoritarian family structure' adopted in recent empirical research includes, among other things, reference to the frequencies with which parents use corporal punishment and interfere with various activities of children. Furthermore, many terms in current use that are quite precise without having a statistical connotation (e.g., 'foreign-born' or 'voting at the last election') nevertheless designate classes of individuals who often vary widely in other characteristics which may be highly relevant to the problem under inquiry. In short, the terms employed in empirical social research frequently possess an indeterminate connotation; they codify less refined or detailed distinctions than do the terms occurring in the laws of the natural sciences; and the items subsumed under them are in consequence usually far less homogeneous in pertinent respects than are these latter terms.

Under the circumstances, it is perhaps inevitable that the generalizations of present empirical social research should be statements of statistical rather than strictly invariable relations of dependence. An analogy will help make this point. Suppose that, after recognizing a gross distinction between metals and nonmetals, we investigated the electrical conductivity of metals, but without introducing further distinctions between different kinds of metals. In the light of what we now know, would it be surprising if the generalizations we succeeded in establishing concerning the variation of electrical conductivity with, say, temperature, were statistical in form? A competent physicist would certainly tell us that on the level of analysis we had adopted nothing else could reasonably be expected, and that if we wished to obtain strictly universal relations of dependence we would have to refine our distinctions, basing them eventually on assumptions concerning the microscopic structures of metallic substances.

The obvious moral of this analogy is that social scientists should likewise develop more discriminating classifications of social phenomena, if strictly universal social laws are to be established. Whether this seemingly plausible suggestion has any merit can of course be settled only in the light of the results of actual trial, and not by any quantity of a priori reasoning. However, there are some grounds for doubting that the social sciences are likely to refine their current distinctions beyond a certain point—a point fixed by the general character of the problems they investigate and the level of analysis appropriate for dealing with those problems—unless indeed these disciplines as presently constituted are transformed almost beyond recognition. For suppose that in order to obtain universal social laws, it would be necessary to classify social phenomena in part by reference to minutely differentiated physical and physiological traits of human participants in those phenomena, and in part on the basis of detailed data concerning the culturally acquired habits and beliefs each participant possesses. To give some flesh to this conjecture, let us consider a grossly oversimple and fanciful example. Suppose that the determining factors of authoritarian parental attitudes were known, and that an adequate categorization of such attitudes required, among other things, the use of variables referring to details of bone structure of each parent, quantity of calcium deposit in their joints, differences in chemical composition of their blood, and variations in the spatial distribution of their nerve filaments; and suppose further that, if subdivisions of authoritarian parents were instituted in terms of these variables, universal laws could be established concerning the attitudes which children reared by such parents exhibit toward minority groups. Nevertheless, even on the assumption that without the proposed classification of the phenomena only a statistical generalization about them is obtainable, it may not be advantageous to abandon such a generalization in favor of the strictly universal one based on the highly discriminating set of categories mentioned in the example.

For the variables listed refer to characteristics which do not fall into the special province of current social inquiry, since they are not specifically social traits; and, in view of the training social scientists normally receive, few if any of them would be competent to analyze social phenomena in terms of those variables. This circumstance alone would probably militate against the introduction into social study of a highly refined system of distinctions such as the one proposed in the example. Moreover, on the assumption that empirical social research will continue to remain focused on problems concerning relations of dependence between commonly recognized and practically important forms of social conduct, such refined distinctions may require that phenomena be discriminated far beyond the needs of the problems under inquiry, so that their adoption may not effectively augment our knowledge of the connections between the phenomena in which we are actually interested;

and, in consequence, universal laws formulated in terms of distinctions more subtle than is necessary for achieving the objectives of empirical research, may be just so much dead lumber. A high-powered microscope is not an improvement on a simple magnifying glass as an instrument for reading small print. Similarly, social scientists may find it more advantageous to establish statistical generalizations rather than strictly universal ones, if the former are more effective means than the latter for answering the sort of questions we normally ask about social phenomena. Accordingly, if the essentially "practical" nature of our current interests in social phenomena is not radically altered, then, although strictly universal social laws are not inherently impossible, the prospects for establishing such laws in the foreseeable future on the basis of empirical research do not appear to be bright.

b. The second point which helps to account for the pervasively statistical character of empirical generalizations in social science can be made quite briefly, since its substance has been stated earlier. We begin by noting the familiar fact that the experimental evidence for the universal laws of physical science is rarely if ever in perfect agreement with them. Accordingly, if physicists were to formulate their laws in strict adherence to what observation establishes about physical phenomena, those laws would have a statistical rather than a universal form. For example, had Galileo sought to establish the laws for freely falling bodies simply by correlating observed data, he would certainly have found that the velocity of falling bodies varies with their weight and shape; and he would have also found that there is only a high correlation rather than an invariable proportionality between the distances bodies fall and the squares of the lapsed times of their fall, so that a generalization based entirely on these findings would have been statistical in form.

The universal form which physical laws nevertheless possess is the fruit of a successful logical strategy. As has been explained previously, it is possible in many branches of natural science to state laws as universally valid under certain "ideal" conditions and for "pure cases" of the phenomena investigated, and to account systematically for any discrepancies between what the laws assert and what observation reveals in terms of more or less well-authenticated discrepancies between those ideal conditions and the actual ones under which observations are made.

However, this strategy is not customary in the social sciences, certainly not in inquiries which seek to establish relations of dependence between phenomena by correlating raw empirical data. Perhaps the chief reason for this is that adequate theoretical notions have not been developed in most of these disciplines, to suggest how laws universally valid for "pure cases" of social phenomena might be fruitfully formulated. The strategy has indeed been attempted in economics. However, the discrepancy between the assumed ideal conditions for which economic laws have been stated and the actual circumstances of the economic market are so great, and the problem of supplying the supplementary assumptions needed for bridging this gap is so difficult, that the merits of the strategy in this domain continue to be disputed. But whatever may be the reasons for the fact that the strategy is not commonly employed in the social sciences, this fact helps to explain why generalizations in these disciplines are for the most part statistical in form. As the history of science as well as common experience amply testify, correlations between empirical data are rarely perfect, and generalizations based exclusively on such correlations are almost inevitably bound to be statistical.[1]

2. The social sciences succeed not only in establishing statistical generalizations, but sometimes also in explaining them; and we shall therefore examine how such explanations are effected. It will speed the discussion, however, to recall the structure of explanations (briefly noted in Chapters 10 and 11) in which physical theories containing statistical assumptions are the explanatory premises for various physical laws. Most of the laws explained in this way are themselves statistical, although contrary to what is sometimes claimed many of such laws are strictly universal; but the pattern of the explanations for both types of laws is uniformly deductive in structure. Furthermore, there are apparently no instances of explanations in the natural sciences in which statistical laws are explained with the help of premises that are exclusively universal (or nonstatistical). The expectation is therefore not unreasonable that the formal structure of explanations of statistical generalizations in the social sciences is also deductive, and that the premises in such explanations similarly contain statistical assumptions. This expectation is indeed fully confirmed; and in consequence, nothing further needs to be said concerning the over-all pattern exhibited by explanations for statistical social generalizations. Nevertheless, in large measure

[1] It should not be overlooked in this connection that our practical interests determine which generalizations are explicitly formulated in the social sciences. It is not too difficult to state well-founded universal generalizations about social phenomena. However, such generalizations would frequently be regarded as trivial, either because they assert what is "obvious" or because they fail to make distinctions that are held to be "important." For example, there appear to be no exceptions to the generalization that every religion has some form of collective ritual for renewing the common sentiments of its adherents, nor to the generalization that all delinquent children are found in societies in which there is a socially structured tension between cultural goals and institutionalized means for achieving them. The first of these is perhaps a candidate for the class of "obvious" trivia, the second for the class of "unimportant" ones (since it does not distinguish between types of tensions or between kinds of goals that are commonly regarded as of the greatest practical moment).

because of the present state of empirical research and of the relatively primitive character of current social theory, such explanations have important variant forms in social inquiry. We shall therefore briefly outline a schematism that codifies in an illuminating way the major types of interpretation social scientists frequently advance when they account for empirically established relations of statistical dependence.[2]

Let us begin with a typical (though for purposes of exposition a considerably simplified) example of empirical social research. Suppose that the problem under study is absenteeism among women working in factories; and suppose that in a sample of 205 such women, 100 of whom are married and the rest single, 25 of the former are absentees (defined as absent from work three or more days each month) but only 10 of the latter. This information is conveniently presented in the table, where M is the class of married women, $\overline{M}$ the class of those unmarried or single, A the class of absentees and $\overline{A}$ the class of those who are not absentees.

	A	$\overline{A}$
M	25	75
$\overline{M}$	10	95

We shall assume throughout this discussion that the samples mentioned are representative of the populations from which they are drawn, and that the relative frequencies with which various attributes occur in the samples can be extrapolated from them to yield warranted generalizations about relative frequencies, or relations between relative frequencies, in the corresponding populations. Two such generalizations implicit in the present example are: "In the population of women factory employees, the relative frequency of absentees among those married is 25/100 or 0.25," and "In the population of women factory employees, the relative frequency of absentees among those single is 10/105 or 0.09+," each of these being of the form: 'In the population K, the relative frequency with which the attribute X occurs in the class of those having the attribute Y is $f_{X,Y}$.' Since the first of these relative frequen-

[2] This schematism was first proposed by Paul F. Lazarsfeld, and the discussion in the text is based almost entirely on his work. Cf. Paul F. Lazarsfeld, "Interpretations of Statistical Relations as a Research Operation," in *The Language of Social Research* (ed. by Paul F. Lazarsfeld and Morris Rosenberg), Glencoe, Ill., 1955, pp. 115-25; and also Patricia L. Kendall and Paul F. Lazarsfeld, "Problems of Survey Analysis," in *Continuities in Social Research* (ed. by Robert K. Merton and Paul F. Lazarsfeld), Glencoe, Ill., 1950, pp. 133-96. Lazarsfeld's schematism makes clear the relevance to the study of scientific explanation of Yule's calculus of association developed in G. Udny Yule, *Introduction to the Theory of Statistics*, London, 1929, Chaps. 3 and 4. An analysis of statistical explanations, analogous in some respects to Lazarsfeld's, is also given in Herbert A. Simon, *Models of Man*, New York, 1957, Chaps. 1, 2, and 3.

cies is significantly greater than the second, there appears to be a definite connection between the marital state of women and absenteeism. (Let us also note in passing but for future use two other generalizations implicit in the example, since they illustrate statistical generalizations with forms slightly different from the one just mentioned. One such generalization is "In the population of women factory workers, the relative frequency of absentee factory workers is 35/205 or 0.17+," which has the form 'In population K, the relative frequency of the attribute X is f_X'; the other generalization is "In the population of women factory workers, the relative frequency of married absentees is 25/205 or 0.12+," which has the form 'In the population K, the relative frequency of individuals having both attributes X and Y is f_{XY}.')

However, the fact that only a fraction of married women are absentees and that absenteeism also occurs among those who are single suggests that it is not marital state *as such* which is responsible for absenteeism. Suppose that an attempt is therefore made to account for the statistical generalizations already established, by exhibiting the dependence of absenteeism upon some third (or "test") variable. Let this test variable be the number of hours a woman devotes to housework, calling this quantity "a great deal" if it is six or more hours per week, and "little or none" in the contrary case. Suppose further that when the sample is analyzed (or "stratified") in terms of this third variable, the following is found: 76 of the women do a great deal of housework (we shall say they have the attribute H) and 129 do not ($\bar{H}$); in the former group, of those who are married (M) 24 are absentees (A) but 33 are not ($\bar{A}$), while of those who are single ($\bar{M}$) 8 are absentees but 11 are not; in the latter group of women performing little or no housework, 1 married woman is an absentee but 42 are not, while of those who are single 2 are absentees and 84 are not. These data are more clearly presented in tabular form.

| | H | | | $\bar{H}$ | |
	A	$\bar{A}$		A	$\bar{A}$
M	24	33	M	1	42
$\bar{M}$	8	11	$\bar{M}$	2	84

It will be evident from these tables that in the subpopulation H the relative frequency of absentees among those who are married is 24/57, and is equal to the relative frequency of absentees among those who are single; and in the subpopulation $\bar{H}$ the corresponding relative frequencies are also equal. Accordingly, within each stratified part of the sample (and in consonance with our assumption, within each stratified part of the entire population of women factory workers) marital state and absenteeism are statistically independent. The statistical dependence between these attributes asserted by the various generalizations for the unstratified population is thus entirely accounted for in terms of the statistical dependence between each of these attributes (or variables) and the test variable.

The main point illustrated by this example is that a statistical generalization about relations of dependence between two variables X and Y is accounted for by showing that, if the population is stratified with respect to a third variable T, there is no significant statistical relation between those two variables in either part of the stratified population. However, the discussion thus far has not carried us much beyond the analysis essentially contained in John Stuart Mill's Canons of Experimental Inquiry. But that analysis can be made more general, so as to provide a basis for a systematic classification of types of situations in which interpretations are proposed for statistical regularities.

To this end, let us assume that statistical generalizations like those in the example are available concerning the relative frequencies with which individuals in a given population K possess the attributes X and Y, where f_X, f_Y, and f_{XY} are the relative frequencies with which individuals in K have the attributes X, Y, and both X and Y, respectively. It is easily seen that the attributes X and Y are not significantly related in K when $f_{XY} = f_X \times f_Y$, and that there is some degree of statistical dependence (or "association") between the variables when this equality does not hold.[3] Various measures for this degree of association can be constructed, differing in their advantages. But for our purposes it will suffice to take one of the simplest of such measures—the difference between the left- and right-hand terms in the equation just stated for statistical independence, or $(f_{XY} - f_X \times f_Y)$, which will be denoted by the index 'd_{XY}.' The generalization that in the population K, the degree of statistical dependence between the variables X and Y is d_{XY}, will be represented by 'S_{XY}' (or simply by 'S,' when it is clear which are the variables intended).[4]

---

[3] For suppose that K contains n members in all, n_X with the attribute X, n_Y with the attribute Y, and n_{XY} with both of these attributes. Then n_{XY}/n_X is the relative frequency of individuals with both attributes X and Y in the class of individuals with the attribute X; and similarly, n_Y/n is the relative frequency with which the attribute Y occurs in the entire population. But if these relative frequencies are equal (i.e., if $n_{XY}/n_X = n_Y/n$) it is clear that X and Y are statistically independent. However, this equation is equivalent to $n_{XY}/n = (n_X/n)(n_Y/n)$, which is simply $f_{XY} = f_X \times f_Y$, as stated in the text. It will also be evident that $f_{XY,X}$ (i.e., the relative frequency of individuals with the attribute X who also possess the attribute Y) is equal to f_{XY}/f_X.

This discussion is based on the supposition that K is a finite class. If K is not finite, the various f's mentioned in this footnote as well as in the text must be understood as the *limits* of relative frequencies, on the assumption that such limits exist.

[4] The measure d_{XY} is equal to zero when the variables are statistically independent; it is positive if the relative frequency of individuals in K possessing both attributes X and Y exceeds that required for statistical independence; and it is negative if this relative frequency is less than that required for such independence. The

Suppose next that a test variable T is introduced, to test the hypothesis that the degree of association between X and Y is affected by the presence or absence of the attribute T. The population K is therefore divided into two mutually exclusive and exhaustive subpopulations T and $\bar{T}$, and the degree of association between X and Y in each of these subpopulations is then determined. Let this degree in T (called the degree of "partial association" of the two variables in this subpopulation) be measured by the difference $(f_{XYT} \times f_T - f_{XT} \times f_{YT})$ and denoted by the index '$d_{XY,T}$'; and let this degree in the subpopulation $\bar{T}$ be measured by the difference $(f_{XY\bar{T}} \times f_{\bar{T}} - f_{X\bar{T}} \times f_{Y\bar{T}})$ and denoted by the index '$d_{XY,\bar{T}}$'—where, for example, f_{XYT} is the relative frequency with which individuals in K possess all three attributes X, Y, and T; $f_{Y\bar{T}}$ is the relative frequency with which individuals in K possess the two attributes Y and $\bar{T}$; and so on. The statistical generalization that in the population K, the degree of partial association between X and Y in the subpopulation T is $d_{XY,T}$ will be represented by '$S_{XY,T}$'; and '$S_{XY,\bar{T}}$' will be used in an analogous manner.

It can be shown, however, that $d_{XY} = (d_{XY,T}/f_T) + (d_{XY,\bar{T}}/f_{\bar{T}}) + (d_{XT} \times d_{YT}/f_T \times f_{\bar{T}})$.[5] But the content of this mathematical identity becomes clearer if the denominators in the equation are ignored; and the relations which the identity asserts to hold between the various measures of statistical association are adequately represented for our purposes by the schematic formula:

$$d_{XY} = D_{XY,T} + D_{XY,\bar{T}} + (D_{XT} \times D_{YT}) \qquad (1)$$

in which the capital letters with subscripts replace the corresponding indices with denominators in the exact equation.

This formula expresses the degree of association of X with Y in K as the sum of three terms: the first two refer to the degrees of partial association of those variables when K is stratified with respect to a test variable T (i.e., except for the coefficients we have decided to ignore in the denominators of the mathematical identity, they state the degrees of association of X and Y in the subpopulations T and $\bar{T}$, respectively); the last term is a product, whose factors (commonly called the degrees of "marginal association") refer to the degrees of association in K of X with T and of Y with T, respectively. The possible numerical values of these terms are endless; but if we consider only certain critical ones, we obtain

maximum value of d_{XY} is 0.25, that will sometimes occur when there is a perfect positive association between the variables, that is, when either all X's are Y or all Y's are X; its minimum value is -0.25, possible only for a perfect negative association, that is, when either all X's are non-Y or all non-Y's are X.

[5] A proof of this identity is given in Yule, *op. cit.*, Chap. 4; also in M. G. Kendall, *The Advanced Theory of Statistics*, New York, 1952, Vol. 1, Chap. 13.

the following two major types of analyses, which account in some fashion for the statistical generalization S_{XY}.

Type 1. Each of the first two terms in (1) vanishes, so that $d_{XY} = (D_{XT} \times D_{YT})$.

In this case, when the population is stratified with respect to T, the variables X and Y are statistically *independent* within each stratified part of K; and the statistical dependence between the variables asserted by S is shown to be the consequence of a statistical association between each of the variables and T. The above example belongs to this type, as do perhaps most of the examples usually cited to illustrate the "explanation" of generalizations. Indeed, this type includes as a limiting case the deductive explanation of universal laws by way of what is traditionally known as the "introduction of a middle term." [6] Since for this type the *partial* associations between X and Y are both zero, the degree of association asserted by S between these variables is expressible as a product. However, no degree of association can have a value exceeding one, so that each of the factors in this product must refer to a degree of marginal association whose absolute magnitude (i.e., irrespective of its being positive or negative) also cannot exceed one. It therefore follows that in any "explanation" of this type for S, at least one statistical premise must assume that the degree of statistical dependence between one of the variables mentioned in S and the test variable is greater in absolute value than the degree of dependence asserted by S.

Two important variants of this type can be distinguished, on the basis of the temporal order that sometimes exists between the test variable and the two mentioned in S. (a) In the first variant, the time at which individuals acquire the attribute T is later than the time they acquire one of the others (for example, X), and precedes the time at which they

---

[6] Thus, suppose a universal law of the form 'All X is Y' (e.g., 'All ice floats in water') is explained by deducing it from two other universal laws of the form 'All T is Y' (e.g., 'All objects whose density is less than that of water float in water') and 'All X is T' (e.g., 'All ice has a density less than that of water'), where 'T' (objects whose density is less than that of water) is the middle term. Viewed formally, and for the purposes of the present discussion, this explanation consists in showing that when the population K is stratified with respect to T, X is statistically independent of Y in each of the subpopulations T and $\bar{T}$. To fix our ideas, assume that in a sample of 200 objects, 10 are pieces of ice, all of which float in water; 70 are pieces of wood, all of which also float in water; and the remaining 120 objects are pieces of metal, none of which floats in water. Suppose further that, when this sample is stratified with respect to the attribute of having a density less than that of water, the 10 pieces of ice and 70 pieces of wood have this property, while the remaining 120 objects do not. It is immediately evident that in this case the attributes X (ice) and Y (floating in water) are statistically independent in each of the two subpopulations T (objects with density less than that of water) and $\bar{T}$ (objects with density not less than that of water).

acquire Y. Under these circumstances, T will be called the "intermediary" variable, and X the "antecedent" one. (b) In the second variant, the time at which individuals acquire T precedes the time at which both of the others are acquired, so that now T is called the "antecedent" variable. There are many situations in which neither of these temporal relations exists, or in which a temporal order can be specified only arbitrarily. Nevertheless, each of the two possibilities mentioned occurs often enough to merit brief consideration.

a. The above example dealing with absenteeism among women factory workers can be construed as belonging to Type 1a if we assume that marital state precedes doing (or not doing) a great deal of housework, and also that involvement in housework precedes being (or not being) an absentee. As the example makes clear, the net import of an analysis of this subtype is that the antecedent condition X (marriage) upon which, according to the generalization S, the occurrence of Y (absenteeism) in some way depends, is replaced by a quite different condition T (performance of a great deal of housework) that may be related in some manner to X but is in any case not identical with it. The intermediary variable does not, in general, identify a necessary and sufficient condition for the occurrence of Y, though it may do so sometimes; for as the example indicates, some married women are *not* absentees despite their doing a great deal of housework, and some married women *are* absentees despite their doing little or no housework. What the intermediary variable does do is to specify a condition under which there is an increase or decrease in the relative frequency of Y, in comparison with its relative frequency on which S is based. In the above example, there is an increase in the relative frequency of absentees from 0.25 in the class of married workers to 0.42+ in the more narrowly restricted class of married workers doing a great deal of housework, and a decrease in the relative frequency of absentees from 0.10 in the class of unmarried workers to 0.02+ in the class of unmarried workers doing little or no housework.

b. A numerical illustration for Type 1b analysis can be constructed from the materials presented in the above example. As has just been observed, the attribute of doing a great deal of housework (H) is neither a sufficient nor a necessary condition for the occurrence of absenteeism (A). The question may therefore arise whether the facts stated in the example may not be accounted for more completely in terms of some additional variable. As a first step in such further inquiry, let us look into the relations of statistical dependence between A and H, as revealed by those facts. A simple calculation enables us to state these relations in a table.

	H	H̄
A	32	3
Ā	44	126

This table shows that while the relative frequency of absenteeism among women workers doing a great deal of housework is 32/76 (or 0.42+), it is only 3/129 (or 0.02+) among those doing little or none. The relative frequency of A thus certainly seems to be dependent on the presence or absence of H.

Nevertheless, the relative frequency with which A fails to occur is 44/76 (or 0.57+), despite the presence of H, and this may give us reason for pause. In any event, is the appearance of a significant connection between A and H trustworthy, so that, for example, absenteeism among women factory workers would be reduced were arrangements instituted to relieve them of most household chores? Or is the seeming dependence only spurious, masking the operation of some hitherto unnoted factor, so that the relative frequency of absenteeism would not be affected by such measures? To test these conjectures, a test variable is therefore introduced. Suppose, rather unrealistically, that the variable is the genetically determined physique of human beings; and assume that on the basis of some criterion, related to various kinds of employment in which women engage, a woman can be characterized as possessing either a satisfactory physique (P) or an unsatisfactory one ($\bar{P}$). And suppose, finally, that, when the sample is stratified with respect to this variable, relations of dependence between A and H are found as stated in the table, where as before these relations are assumed to hold in the entire population of women factory workers.

| | P | | | | P̄ | |
	H	H̄			H	H̄
A	0	0		A	32	3
Ā	44	126		Ā	0	0

Accordingly, although there is a significant degree of statistical association between A and H in the unstratified population, these variables are statistically independent in each of the subpopulations P and $\bar{P}$; e.g., in the group of workers with unsatisfactory physiques, absenteeism occurs with the same relative frequency among those doing a great deal of housework as it does among those doing little or none. Since P is obviously an antecedent variable, the example illustrates an analysis belonging to Type 1b. Incidentally, the example has been so constructed that $\bar{P}$ (unsatisfactory physique) is a necessary and sufficient condition for A (absenteeism), for as the preceding table indicates, a woman factory worker is an absentee if, and only if, she has an unsatisfactory physique (and so presumably suffers from recurrent ill health), irrespective of the amount of housework she does or of her marital state.

As the example suggests, at least one function of analyses of this type is to correct mistaken (or "spurious") causal imputations. Such an analysis "explains" a statistical generalization S_{XY}, in the sense that it provides grounds for rejecting the supposition that X and Y are therefore causally related, by demonstrating S to be the consequence of assumptions concerning the statistical association of each of these variables with some antecedent variable T, which may therefore be a common factor in the "causal" conditions for the occurrence of the attributes X and Y.[7] Thus, it would be clearly absurd to ascribe the death of patients to the ministrations of their physicians, on the strength of the conceivably well-founded generalization that the relative frequency with which patients die varies directly with the relative frequency with which physicians visit them. For this generalization can most likely be explained by showing that the frequencies mentioned are statistically independent in each of the subpopulations of patients, classified in terms of the antecedent variable of the gravity of their ailment, since gravity of illness is presumably related causally both to frequency of death and to frequency of physicians' visits.[8]

But let us turn to the second major type of analysis for statistical generalizations in empirical social science.

Type 2. The third term in the schematic formula (1) vanishes, so that $d_{XY} = D_{XY,T} + D_{XY,\bar{T}}$.

Since the third term is the product of two factors, at least one of them must therefore vanish in this case; and it is obvious that in consequence the test variable must be statistically independent of at least one of the variables mentioned in S. It can be easily proved, moreover, that one of the degrees of partial association between X and Y (i.e., the degree of association between these variables in one of the stratified parts of the population) must be greater in absolute magnitude than the degree of

---

[7] In the above example, an unsatisfactory physique may not only be a factor in the conditions producing ill health and therefore absenteeism, as already suggested, but also a factor in the obstacles to marriage.

[8] On the other hand, although S may thus be shown to be "spurious" with the help of a given antecedent variable T_0, it does not follow that T_0 is therefore causally related to any of the variables mentioned in S, say the variable Y. Despite the disagreements noted in Chapter 4 over the precise conditions two variables must satisfy to be causally related, it would generally be admitted that it is not sufficient for T_0 to be temporally prior to Y and correlated with Y only statistically. According to one suggestion, to show that T_0 is causally related to Y, it is necessary to show that, for *every* antecedent variable T, the degrees of partial correlation between T_0 and Y do not vanish when the population is stratified with respect to T. Cf. Paul F. Lazarsfeld, *op. cit*, p. 125.

association between them in the unstratified population. Accordingly, in analyses of this type the test variable specifies a subpopulation in which the relation of dependence between X and Y more closely approaches a strictly universal connection than obtains between them in the entire population. As in the case of the first major type, two variant forms of this second type can be distinguished, according to whether the test variable is intermediary or antecedent. However, the difference between these forms is not of great moment; and a numerical example in which the test variable is intermediary will suffice to make clear what is distinctive of analyses belonging to either variant of this major type.

Suppose that a study is undertaken to ascertain whether there is any connection between the annual income of men at least 30 years old, and the annual income of their parents. Men of this age are therefore classi-fied as well-to-do (W) if they have an annual "real" income (estimated on the basis of some agreed-upon measure of the purchasing power of money at different times and places) of at least $25,000, and as not well-to-do ($\overline{W}$) in the contrary case; and their parents are similarly classified as well-to-do if the "real" family income was at least $25,000 annually ($F$), and as not well-to-do ($\overline{F}$) otherwise. Suppose that a sample of 200 men in the indicated age group yields the information contained in the table below, with the assumption of the previous examples still in force, that the data suffice to establish generalizations concerning the relations of these variables in the entire population.

	F	$\overline{F}$
W	80	30
$\overline{W}$	20	70

Since the relative frequency with which men are well-to-do in the class of those with well-to-do parents is 0.80, but only 0.30 in the class of those whose parents are not, the two variables are statistically dependent.

However, since it is evident from the data that the sons of well-to-do parents are not invariably well-to-do, we shall suppose that an attempt is made to discover whether there is not some attribute favorable to achieving financial success that characterizes some sons of such parents but not other sons. The population of men who are 30 or over is there-fore stratified into those with a college education (E) and those without one ($\overline{E}$), with the assumed results shown in a new table.

	E				$\overline{E}$	
	F	$\overline{F}$			F	$\overline{F}$
W	50	20		W	30	10
$\overline{W}$	10	40		$\overline{W}$	10	30

The test variable is evidently an intermediary one. It is also clear that there is a statistical dependence between W and F in each of the sub-populations E and $\bar{E}$, but that this dependence is greater in E than it is in $\bar{E}$—for the relative frequency in E of being well-to-do among these with well-to-do parents is 50/60 or 0.83+, but only 30/40 or 0.75 in the subpopulation $\bar{E}$. Moreover, if we inquire into the statistical dependencies between the test variable and each of the other two, we first obtain by simple calculation from the above data the information contained in a third table.

	E	$\bar{E}$		E	$\bar{E}$
W	70	40	F	60	40
$\bar{W}$	50	40	$\bar{F}$	60	40

We therefore conclude that the proportion of well-to-do men among those with a college education is greater than it is among those without one, but that F and E are statistically independent.

This example thus illustrates an analysis belonging to Type 2, with the test variable as an intermediary one. A non-numerical example of this major type with an antecedent test variable is the following. Assume that in the adult population the suicide rate among those married is less than it is among those single, and also that adults can be distinguished according to whether their childhood was happy or not. Suppose that this latter attribute is statistically independent of marital state, and is adopted as a test variable. Assume, finally, that in the subpopulation of adults who had an unhappy childhood, the suicide rate among those married is still less than it is among those single, but is in each case greater than the corresponding rate in the unstratified population. This example is again of Type 2, but the test variable is an antecedent one. Both examples show that in analyses of this type, in contrast to those of Type 1, the dependence of Y upon X which S asserts is not only confirmed, but its degree is even *increased* in the more restricted group of individuals who satisfy a certain condition T in addition to the condition X; i.e., it is increased in that special subclass of individuals with attribute X who also have attribute T. Accordingly, an analysis belonging to this second major type "explains" a statistical generalization S only in the sense of deducing S from other statistical assumptions which assert, in effect, that the dependence of Y upon X formulated by S holds in an even greater measure under explicitly stated circumstances, which are identified by "refining" or qualifying the description of the condition X given in S.

We cannot in this book examine further details in the structure of explanations for statistical laws. However, the general characterizations of such explanations which the discussion has sought to establish would require little if any emendation were the subject pursued further. In any event, the main points that have emerged in the discussion are the following: explanations of statistical laws are uniformly deductive in pattern; at least one premise in such explanations must be statistical in form; and the degree of statistical dependence assumed in at least one premise must be greater than the degree of dependence stated in the generalization for which the explanation is proposed.

II.   *Functionalism in Social Science*

In the judgment of many students, a comprehensive theory of social phenomena is most likely to be achieved within the framework of systematically "functional" analyses of social phenomena. Indeed, such a theory is sometimes claimed to be already available in various current formulations of the standpoint in social inquiry known as "functionalism." In any event, although functionalism has been the subject of much critical debate, many of its ideas enjoy a wide influence in contemporary social science, particularly but not exclusively in anthropology and sociology. We shall therefore examine a number of central issues raised by functional interpretations of social phenomena.

Under various names, most of the major ideas commonly associated with functionalism have had a long past, in some cases reaching back to Greek antiquity. In its modern versions, however, the functional approach in social science was developed partly in reaction to the preoccupation of much nineteenth-century social inquiry with questions concerning the origins of social institutions, and to the largely speculative reconstructions of their genesis and evolution that were frequently the chief products of this concern. As stated by some of its advocates, moreover, functionalism represents an often explicitly avowed attempt to account for social phenomena in a manner modeled on the pattern (as distinct from the substantive concepts) of functional (or "teleological") explanations in physiology. For these reasons, its proponents frequently contrast historico-causal accounts of social facts in terms of their historical antecedents, with their own *prima facie* very different functional analyses. More generally, functionalism is a standpoint in the social sciences which, not unlike the standpoint of organismic biology in its relation to mechanistic approaches in biological science, insists upon the "autonomous" character of these disciplines, and is opposed to "reductionist" interpretations of social fact in terms of nonhuman traits or forms of behavior.

However, it is not easy to give a brief statement of the positive tenets of functionalism. For, although its proponents are usually agreed as to its great promise, they do not generally agree on what are the essential features of a functional analysis; and the following statements by two of its major exponents are not fully representative of its many varieties. Nevertheless, these statements will serve to introduce some of the issues that certain forms of the functional approach generate.

In a general account of functionalism in anthropology, Malinowski declared that a functional analysis of culture

> aims at the explanation of anthropological facts at all levels of development by their function, by the part which they play within the integral system of culture, by the manner in which they are related to each other within the system, and by the manner in which this system is related to the physical surroundings. . . . The functional view of culture insists therefore upon the principle that in every type of civilization, every custom, material object, idea and belief fulfills some vital function, has some task to accomplish, represents an indispensable part within a working whole.[9]

Moreover, according to Malinowski the major institutions of society correspond to the primary biological needs of human beings, and like the latter can survive only if a certain set of conditions is fulfilled. A functional explanation of a social fact must therefore exhibit the survival value of that fact, by exhibiting its function in satisfying those conditions and thereby also the primary needs of human beings.[10]

A somewhat different statement of functionalism by Radcliffe-Brown stresses the analogy between functional analysis in physiology and in social science:

> If we consider any recurrent part of the life-process [of an organism], such as respiration, digestion, etc., its *function* is the part it plays in, the contribution it makes to, the life of the organism as a whole. As the terms are here being used a cell or an organ has an *activity* and that activity has a *function*. It is true that we commonly speak of the secretion of gastric fluid as a 'function' of the stomach. As the words are here used we should say that this is an 'activity' of the stomach, the 'function' of which is to change the proteins of food into a form in which these are absorbed and distributed by the blood to the tissues. We may note that the function of a recurrent physiological process is thus a correspondence between it and the needs (i.e., necessary conditions of existence) of the organism. . . .

> To turn from organic life to social life, if we examine such a community as an African or Australian tribe we can recognize the existence of a social structure. Individual human beings, the essential units in this instance, are connected by a definite set of social relations into an integrated whole. The continuity of the social structure, like that of an organic structure, is not

[9] Bronislaw Malinowski, "Anthropology," *Encyclopaedia Britannica*, Suppl. Vol. 1, New York and London, 1936, pp. 132-33.
[10] Bronislaw Malinowski, "The Functional Theory," in *A Scientific Theory of Culture*, Chapel Hill, N. C., 1944, pp. 147-76; and see also Malinowski's article "Culture," in the *Encyclopedia of the Social Sciences*, New York. 1935. Vol. 4.

destroyed by changes in the units. Individuals may leave the society, by death or otherwise; others may enter it. The continuity of structure is maintained by the process of social life, which consists of the activities and interactions of the individual human beings and of the organized groups into which they are united. The social life of the community is here defined as the *functioning* of the social structure. The *function* of any recurrent activity, such as the punishment of a crime, or a funeral ceremony, is the part it plays in the social life as a whole and therefore the contribution it makes to the maintenance of the structural continuity.[11]

According to Radcliffe-Brown, the two central tasks of a science of society are to ascertain how social systems perpetuate themselves by maintaining the form of their structure, and how social systems change their type by altering their structural form. In each case, the task therefore requires analysis of the functions of various standardized modes of behavior or belief in relation to the total social system to which the modes belong.[12]

Although these formulations are in some respects similar, there are also important differences between them; and they suggest that the label "functionalism" covers a variety of distinct (though in some cases closely related) conceptions. In point of fact, even professed functionalists, who explicitly adopt as the paradigm for explanations in social science the functional analyses in biology, do not construe the character of these analyses in identical ways, and sometimes employ in a single discussion different notions of what constitutes a functional explanation. We shall therefore (1) first examine several senses in which the expression 'functional analysis' is commonly employed, and (2) then consider some conceptual and other difficulties implicit in proposed functional analyses of social processes.

1. The word 'function' is highly ambiguous, and an exhaustive list of its many meanings would be very long. However, only six of them will be mentioned, because they are not always clearly distinguished in discussions of functionalism. In the first place, the word is widely used to signify relations of dependence or interdependence between two or

[11] A. R. Radcliffe-Brown, *Structure and Function in Primitive Society,* London, 1952, pp. 179-80.
[12] A. R. Radcliffe-Brown, *A Natural Science of Society,* Glencoe, Ill., 1957, Part 2. The literature on functionalism is quite extensive. Among the many other statements and discussions of this standpoint, the following are especially useful: Robert K. Merton, *Social Theory and Social Structure,* Rev. ed., Glencoe, Ill., 1957, Chap. 1; Talcott Parsons, *Essays in Sociological Theory,* Glencoe, Ill., 1949, Part 1, as well as the same writer's *Social System,* Glencoe, Ill., 1951, Chap. 2; Raymond Firth, "Functionalism," in *Yearbook of Anthropology* (ed. by William L. Thomas), New York, 1955; and Gregory Bateson, *Naven,* 2nd ed., Stanford, Calif., 1958, esp. Chaps. 3 and 17.

more variable factors (or "variables"), whether or not these factors are measurable. Thus, a physicist who states that the pressure exerted upon the walls of a container by a mixture of alcohol and water vapors is a function of the temperature and the concentration of each vapor is employing the word in this sense, as is the sociologist who declares that the suicide rate in a community is a function of the degree of social cohesion in that society. This is also the sense of the word that is exploited in pure mathematics, where the term 'function' is abstractly defined as any relation between two classes of elements such that for every member of one of the classes there is a uniquely determined member of the other class.[13] Such "functional" relations of dependence or interdependence are often established by functionalists in their analyses of social processes. However, if functional analysis means no more than this, it does not differ in aim or logical character from analyses undertaken in any other domain with the objective of discovering uniformities in some subject matter; and, if 'functional analysis' is so construed, functionalism cannot rightly be regarded as a distinctive approach to social inquiry.

In the second place, the word is sometimes employed to denote a more or less inclusive set of processes or operations within (or manifested by) a given entity, without indication of the various effects these activities produce either upon that entity or any other. It is in this sense that biologists sometimes talk about "the functioning of the stomach," when they refer to the muscular contractions, secretions of acids, absorption of liquids, etc., which take place in that organ. It is also in this sense of the word that Radcliffe-Brown appears to be using the phrase 'the functioning of the social structure' when he defines it in the passage quoted above as "the social life of the community"; and it is in this sense that some other anthropologist would be speaking of "the functioning of the postal system" in our society were he to designate by this expression the class of miscellaneous activities such as the sale of stamps, the collection of mail, or the purchase by postal officials of mailbags. However, analyses undertaken for the sake of discovering what processes are going on in stated objects are not distinctive of social inquiry; and 'functional analysis' so understood cannot plausibly be supposed to constitute an especially promising direction for studying human affairs.

In the third place, the word is commonly used by biologists (as well as by others in an analogous sense) in the phrase 'the vital functions' to refer to certain inclusive types of organic processes occurring in living organisms, such as reproduction, assimilation, and respiration. These

[13] For example, if $y = x^2$, y is said to be a function of x, since for every value of the "independent" variable 'x' there is one and only one value for the "dependent" variable 'y.'

processes are frequently regarded as being carried on by the organism "as a whole" rather than by just some of its parts, even though some of the processes are intimately connected with the operations of certain special parts of the organism. Moreover, these functions are uniquely characteristic of living things, and are usually said to be indispensable for the continued life of an organism (or for the continuance of its species). In consequence, the vital functions can be taken to be the *defining* attributes of *living* organisms (or possibly of some particular type of living organism), so that if an organic body lacks one of these attributes it does not count as a living organism (or a living organism of some stated type). Accordingly, if respiration, for example, is such a defining attribute, to say that respiration is essential or indispensable for the survival of a living thing is to utter an obvious tautology. As we shall see, this point is pertinent in assessing certain assumptions made by functionalists, among others by Malinowski when in the passage quoted above he declares that every cultural object "fulfills some vital function."

Fourth, the word is widely employed to signify some generally recognized use or utility of a thing, or some normally expected effect of an action, as in the statements "The function of an axe is to cut wood," and "The function of spreading manure on a field is to fertilize the soil." The things or actions said to have functions in this sense are often deliberately contrived to yield the utilities or consequences imputed to them. But this is not invariably the case; and things are commonly said to have functions even when they are not human artifacts, or when they have been produced with different utilities as the intended objectives than the ones designated in ascribing certain functions to them. Thus, it is not uncommon to speak of the function of the Pole Star in determining directions when navigating at night; and, although books may not have been designed with a view to being decorative furniture, it is correct to say that in many homes this is their main function. This sense of the word is not the only one in which Malinowski employs it in the above quotation; but it seems to be the sense intended when he insists that according to the functionalist view every cultural object has some task to accomplish, or when he states elsewhere that "function means always the satisfaction of a need."[14] When 'function' is used with this meaning, functional analyses are in the main confined to inquiries dealing with phenomena associated with living organisms, nonhuman as well as human. However, in this sense of the word a functional "explanation" consists in stating either the utility some entity (or type of entity) has for a certain class of living creatures, or the consequences of possessing

[14] Bronislaw Malinowski, *A Scientific Theory of Culture*, p. 159.

such utility that usually follow from some type of action. But an "explanation" of this sort consists of only a single statement (in some cases universal, in other cases not), which simply asserts a factual connection between several items but does not explicitly relate this fact with any other so as to show why that particular connection happens to occur. Accordingly, to the extent that 'functionalism' connotes no more than a search for such "explanations," functionalism is neither a theory of social phenomena nor a distinctive theoretical approach to their study.

In the fifth place, the word 'function' is frequently employed (in a sense close to the one just discussed) to designate a more or less inclusive set of consequences that a given thing or activity has either for "the system as a whole" to which the thing or activity supposedly belongs, or for various other things belonging to the system. It is in this sense that the word is being used in statements such as "One function of the liver is to store sugar in the body, but this is not its only function"; "The historical function of the doctrine of the divine right of kings was to weaken the power of the feudal nobility and to make possible the development of strong national states"; and "The publication of research findings has the function not only of making those findings generally available, but also of submitting them to the criticism of the scientific community as well as of establishing priorities in discovery." In this meaning of the word, the functions of an activity need not be *intended* consequences of the activity, nor need they have any utility for some living creature; and the functions may be either favorable or unfavorable to the continuance of the system (or of some part of the system) upon which the activity produces various effects. It seems likely that functionalists associate this meaning with the word when they stress the importance of recognizing the "multiple functionings" of various sociological items. But, except for the language used to describe what is being done, it is not clear how functional analysis which is directed to discovering the various effects some social item has upon other such items, differs from the analysis of a physicist which is directed to discovering what consequences follow from, say, the radiation of energy from the sun which affect the constitution of the sun itself or of the various planets.

Finally, the word 'function' is used in the sense that occupied us in Chapter 12 and that is illustrated by such expressions as 'the function of the chattering of teeth when a human being is exposed to cold' or 'the function of the governor in a standing engine.' On this meaning of the term, the function of some item signifies the contributions it makes (or is capable of making under appropriate circumstances) toward the *maintenance* of some stated characteristic or condition in a given system to which that item is assumed to belong. This is plainly the meaning that the term has for Radcliffe-Brown in the passage cited above, and sometimes but by no means uniformly perhaps also for Malinowski. It is perhaps something like the meaning which other functionalists associate with the word, when they recommend functionalism as the most promising approach to the study of social phenomena. In any event, it is only if the word is understood in approximately this sense that functional explanations in biology can be taken as paradigms for functional explanations in social inquiry. Nevertheless, even functionalists who formally subscribe to this interpretation of functionalism do not consistently distinguish between the sense of 'function' now under discussion and some of its other senses. For example, among other illustrations of functional analysis cited by one such functionalist is a study according to which the hostile attitude a community manifests toward lawbreakers possesses the "unique advantage" of uniting the community in a common emotion of aggression. On the face of it, however, this account simply states that the hostile attitude has certain consequences which are not intended by those manifesting the hostility. But the study does not even attempt to show that the emotional solidarity of the community is maintained because of the presence of that attitude or of compensating variations of it, despite various changes within or outside the community that would otherwise prevent the continuance or the realization of the solidarity. Accordingly, the study appears at best to establish either a relation of dependence or interdependence between two variables (i.e., a "functional dependence" in the first sense of the word 'function') or the utility specified by one of the variables which the other variable possesses (a 'function' in the fourth sense of the word), rather than the function (in the sixth sense of the word) of one of the variables in maintaining a given system in a state specified by the other variable.

2. In the light of this discussion, functionalism is not a unitary and clearly articulated perspective in social inquiry. We must next examine some important problems that face attempts at functional explanations of social processes.

a. The structure of functional (or "teleological") analyses in biology has already been discussed at length in Chapter 12. Accordingly, to the extent that functional explanations in social science are indeed similar in form to functional explanations in biology, nothing further needs to be said about their over-all pattern. However, the construction of such explanations for social phenomena requires the resolution of serious conceptual problems. For to achieve such explanations, a number of concepts must be defined, in terms of notions applicable to the subject matter of social inquiry, which will correspond to the basic formal distinctions in the pattern of teleological explanations. The moot question is whether the definitions generally proposed are satisfactory.

It will be recalled from the discussion in Chapter 12 that two basic concepts employed in teleological explanations are: the notion of a *system S*, and the notion of some *state* or *condition G* which is maintained in the system. In the functional explanations of biology, the systems commonly examined are individual organisms; and the states of systems that are considered include among others the survival of the organism (i.e., the condition of being a *living* organism), certain characteristic activities of some organ, the internal temperature of the organism, and the chemical state of some internal fluid such as the blood. There is usually no difficulty in biology in specifying what is an individual organism. Moreover, a number of easily identifiable activities carried on by organisms (namely, the "vital functions," such as respiration and assimilation), are generally recognized to constitute the defining attributes of being alive; and similarly, adequate definitions for particular organs, for their characteristic activities, and for other states of organisms that may be investigated can normally be supplied without much trouble. In consequence, since the system S and the state G can be clearly specified in biology, it is intelligible to ask, and to seek an answer by way of experimental inquiry, whether, and if so by what mechanisms, S is maintained in the state G.

The situation in these respects is notoriously different in the social sciences. The systems frequently discussed by functionalists are individual societies or communities; and the states of these systems that have been of particular concern to them include the survival of a society, its social structure, the patterns of various institutional activities, and the roles (or norms of social behavior) prescribed or overtly manifested in a society. As in biology, it is often possible in social inquiry to designate unambiguously the systems to be investigated—with relative ease in primitive societies, though with increasing difficulty in more industrialized ones. On the other hand, in regard to the condition of survival by a society, there is nothing comparable in this domain to the generally acknowledged "vital functions" of biology as defining attributes of living organisms. Societies do not literally die, though to be sure a society may disappear because all the human beings who constitute it die without leaving heirs or are permanently dispersed. It is therefore not easy to fix upon a criterion of social survival that can have fruitful uses and not be purely arbitrary.

For example, were physical nonextinction in the manner just indicated adopted as the criterion for survival, only a relatively few societies in the history of mankind would fail to satisfy it; and, since on that criterion survival would be compatible with any form of organization characterizing the various societies that have appeared in human history, every proposed functional explanation of social survival in terms of social organization would be simply an empty tautology. This outcome is even more obvious if, with Malinowski, everything present in any type of civilization is taken to fulfill a "vital function." For, since on such a stipulation one type of civilization cannot be distinguished from any other except in terms of these vital functions, the logically trivial character of Malinowski's "thesis" is transparent. On the other hand, a criterion that includes apparently more restrictive requirements than mere physical nonextinction is faced with the difficulty either of showing that the restrictions do actually exclude from the class of recognized societies some group of men living together, or of justifying the restrictions as not entirely arbitrary. For example, suppose that the criterion which is adopted stipulates that, to count as a society, a group of human beings living within the confines of a territory must exhibit a political organization. However, if the term 'political organization' is being used so broadly that it covers any form of social control and arrangement for the distribution of authority or power, all human groups living together satisfy this requirement almost as a matter of definition; and, since the continuance of a society so defined is thus compatible with any form of political organization, statements concerning political organization as an indispensable prerequisite for social survival are once again nothing but truisms. But if the term 'political organization' is employed less broadly, to signify certain special forms in which power relations are exhibited, there are primitive peoples who in the judgment of some anthropologists possess no political organization in this narrower sense; and it is not clear how the exclusion of such peoples from the class of societies is to be justified except as the outcome of an arbitrary decision.

Analogous difficulties arise in connection with many other states that functionalists attempt to show are maintained in social systems. We shall consider only the notion of social structure. As was mentioned above, Radcliffe-Brown believed that one major task of social science is to discover how social systems preserve their "structural" form. According to him, the *social structure* of a given system "consists of the sum total of all the social relationships of all individuals at a given moment of time," where by a 'social system' he understood any aggregation of human beings conceptually isolated from the rest of the universe who are jointly adjusting their interests, and by a 'social relationship' any behavior of men which involves such an adjustment; moreover, the *structural form* of a system consists of the various "kinds" of social relations manifested in an actual social structure. It is that which "remains the same" in a social structure, even though different individuals may be participating in those relations.[15]

However, structural form so defined cannot become the subject of *empirical* study, unless much more is said about just what is to be understood by a "kind" of social relation or by "remaining the same." For if no restrictions whatever are placed upon the meaning of "kind," the question of whether a social system preserves its structural form is not a *factual* one, to be decided by empirical inquiry; on the contrary, the question can be settled by a purely *logical* analysis, since it is demonstrable that every society must necessarily possess *some* pattern of social relations invariant under any given class of changes in that society. It was argued in Chapter 10 that the notion of *absolute* disorder is self-contradictory, because every conceivable state of affairs exhibits *some* order, albeit a complex and unfamiliar one, and that a given situation can be said to be "unordered" only in the relative sense of not illustrating a particular class of patterns. But the notion of a *complete* change in the structural form of a society is similarly incoherent. For although *one* kind of social relationship in "the sum total of all the social relationships" may be totally modified, some *other* kind of social relationship must remain unaltered, albeit this latter kind may happen to be one that is of no interest to us normally. In short, a social system can be said to change its structural form also only in the relative sense of an alteration in some particular kinds of social relationships.

A simple example will help to fix this point. Suppose that in a given society private ownership of all industries is replaced by public ownership in a certain year, so that the society is said to have changed its structural form. Nevertheless, quite apart from other social relationships that may not be affected by this change (such as that the social relations in which most members of the society stand to those actually directing industrial operations may be the same after the change as they were before it), the very statement of the example requires us to conclude that in this society *men will continue to engage in industrial activity,* despite the indicated change. In consequence, there *must* be at least one kind of social relationship which is not altered; and it is conceivable that in certain social inquiries this invariance is the paramount one, so that for the purpose of those inquiries the change may not be classified as one in the structural form of the society. It is worth passing notice, moreover, that we are not always prepared to say whether some of the social changes actually occurring in our society are to count as changes in its structural form. Do such changes occur when, for example, new sales taxes are introduced, income tax rates are increased, or

[15] A. R. Radcliffe-Brown, *A Natural Science of Society,* pp. 43, 55, 84, and *Structure and Function in Primitive Society,* p. 192.

public funds come to be used for providing lunches for children in private schools? It is clear that the answer to any one such question will depend on the specific objectives of the inquiry within which the question may arise, and in particular on the degree of refinement in classifying kinds of social relationships which is required by those objectives.

It follows that proposed explanations aiming to exhibit the functions of various items in a social system in either maintaining or altering the system have no substantive content, unless the *state* that is allegedly maintained or altered is formulated more precisely than has been customary. It also follows that the claims functionalists sometimes advance (whether in the form of "axioms" or of hypotheses to be investigated) concerning the "integral" character or "functional unity" of social systems produced by the "working together" of their parts with a "sufficient degree of harmony" and "internal consistency," or concerning the "vital function" and "indispensable part" every element in a society plays in the "working whole," cannot be properly judged as either sound or dubious or even mistaken. For in the absence of descriptions precise enough to identify unambiguously the states which are supposedly maintained in a social system, those claims cannot be subjected to empirical control, since they are compatible with every conceivable matter of fact and with every outcome of empirical inquiries into actual societies.

The difficulty we have been discussing, and in particular the point just made, are often overlooked or insufficiently stressed even by careful students of functionalism who have contributed much to its clarification and development. Some of them, for example, while in basic sympathy with the functional approach, have criticized the large claims mentioned in the preceding paragraph as having doubtful empirical warrant, and as being at best hypotheses to be explored rather than essential postulates of functionalism. However, they have paid little attention to the problem that has been occupying us, even though it is a problem that must first be resolved before questions concerning the factual merits of those large claims can be significantly raised.[16] A resolution of this problem has certainly not been measurably furthered by Talcott Parsons, in his attempt to construct a comprehensive conceptual framework for what he calls a "structural-functional" social theory. Limitations of space forbid a discussion of Parsons' architectonic system of distinctions, and only a brief mention can be made of his account of some of the "functional prerequisites" for a social system that has a "persistent order" or that undergoes "an orderly process of developmental change." One of the "most general" prerequisites he cites is that a social system cannot

[16] Cf. Robert K. Merton, *Social Theory and Social Structure*, pp. 25-37, and Kingsley Davis, "The Myth of Functional Analysis," *American Sociological Review*, Vol. 24 (1959), p. 763.

be "radically incompatible with the conditions of functioning of its component individual actors as biological organisms and as personalities" [i.e., as motivated in their behaviors], "or of the relatively stable integration of a cultural system" [i.e., the "symbolic" system of language and other artifacts, expressing ideas, beliefs, standards, values, etc., serving as media of communication]. Another such prerequisite is that a social system must have "a sufficient proportion of its component actors adequately motivated to act in accordance with the requirements of its role system" [i.e., the various types of participation by actors in interactive processes with other actors, the processes being viewed from the perspective of their functional significance for the social system]. Except in the rare and trivial case of the physical extinction of a social system because of the death of all its members, it would certainly be difficult to determine whether or not a given system satisfies prerequisites formulated in such vague and indefinite terms. The difficulty is not diminished by the further indefiniteness that is introduced when Parsons adds that "in the present state of knowledge, it is not possible to define precisely what are the minimum needs of individual actors," and that "it is not the needs of all the participant actors which must be met, nor all the needs of any one, but only a sufficient proportion for a sufficient fraction of the population."[17]

b. Functional (or teleological) explanations having the objective of exhibiting the contributions some item makes to the maintenance of a society *as a whole* are thus confronted with a still incompletely resolved difficulty. In point of fact, however, functionalists have not been exclusively engaged in developing explanations of this ambitious sort. They have achieved greatest success either in providing explanations of this general type for the maintenance of some state in a system that is far less inclusive than an *entire* society (e.g., a tribal clan, the system of relations known as the family as it occurs in certain human communities, or a professional or political organization in modern society), or in merely showing what analogous or differing uses and consequences (or 'functions,' in the fourth and fifth senses of this word which were distinguished above) various standardized forms of behavior have in given societies (e.g., the functions of punishment or of certain forms of ritual activity). The problem of unambiguously specifying the states supposedly maintained in relatively small groups is understandably much more tractable than it is in an entire society, and can often be solved reasonably well. Nevertheless, there are other difficulties that face those more modest as well as the more ambitious teleological explanations; and we shall briefly remind ourselves of some of them.

[17] Talcott Parsons, *The Social System*, pp. 26-28.

As the formal pattern of functional explanation developed in Chapter 12 suggests, once a system S and a state G supposedly maintained in it are adequately specified, the task of the functionalist is to identify a set of state variables whose operations maintain S in the state G, and to discover just how these variables are related to each other and to other variables in the system or in its environment. In the actual conduct of social inquiry, however, this sequence is usually reversed: some variable (e.g., a religious ritual) is first identified; and inquiry is then directed toward ascertaining what functions it has (perhaps only in the fourth and fifth senses of the word), and whether it does in fact contribute to the maintenance of some state G (e.g., emotional solidarity) which is suspected of being fairly stable. It is therefore quite easy to overlook the requirement that the system S and the state G with which the analysis presumably deals must be carefully delimited, and in consequence to omit explicit mention, in the teleological explanation finally proposed, of the specific system within which the variable allegedly maintains a specific state. It is then also easy to forget that even if the variable does have the function attributed to it of preserving G in S (e.g., the performance of a religious ritual having the function of maintaining the state of emotional solidarity of *each* primitive tribe in which the ritual takes place), it may not have this role in some *other* system S' (e.g., in a confederation of tribes, where the ritual may have a divisive force) to which the variable may also belong; or that it may not have the function of maintaining in the same system S some *other* state G' (e.g., an adequate food supply), with respect to which it may perhaps be *dysfunctional* by obstructing the maintenance of G' in S.

But however this may be, it is hardly possible to overestimate the importance for the social sciences of recognizing that the imputation of a teleological function to a given variable must always be *relative* to some particular state in some particular system, and that, although a given form of social behavior may be functional for certain social attributes, it may also be dysfunctional (or even nonfunctional, in the sense of being causally irrelevant) for many others. Failure to recognize this point, obvious though it is when stated formally, is undoubtedly a major source for the not uncommon confusion of questions of fact with questions of desirable social policy, as well as for the frequent accusation that a functional approach in social science is *necessarily* committed to the values embodied in the social *status quo*. With this point in mind, however, even if individual functionalists are so committed, it will be evident that the accusation that such commitment is inherent to functionalism is baseless.[18]

[18] The related claim that functionalism is *necessarily* restricted to the study of social equilibrium and the conditions which maintain some state in a system, and that

It is also crucial in this connection to distinguish between the function or type of activity exercised by a particular variable in a system, and the variable that exercises this function. Thus, one of the functions of the thyroid glands in the human body is to help preserve the internal temperature of the organism. However, this is also one of the functions of the adrenal glands, so that in this respect there are at least two organs in the body that perform (or are capable of performing) a similar function. Accordingly, although the maintenance of a steady internal temperature may be indispensable for the survival of human organisms, it would be an obvious blunder to conclude that since the thyroid glands contribute to this maintenance they are for this reason indispensable for the continuance of human life. Indeed, there are human beings who, as a consequence of surgical intervention, do not have thyroid glands, but nevertheless remain alive. An identical point requires to be made in the context of social inquiry. Let us assume that one of the functions of a church organization in a given society is to foster religious sentiments and religious activities. However, this function may also be exercised by other institutionalized groups in that society, for example, by individual families or by schools. Moreover, even if these other organizations did not actually perform this function at a given time, they might acquire it at some later time under appropriate circumstances. In consequence, even if it were beyond dispute that religious attitudes and activities are essential for the welfare of human societies, it would not follow that church organizations are indispensable for that welfare.

This point has not been consistently recognized by functionalists. For example, Malinowski argued that because the function of myth is to strengthen tradition by attributing to it a supernatural origin, "Myth is, therefore, an indispensable ingredient of all culture."[19] However, although one may grant, if only for the sake of the argument, the role Malinowski ascribes to myth in strengthening traditions, as well as his tacit claim as to the indispensability of tradition in all societies for the persistence of their cultures, his conclusion is nevertheless a *non sequitur.* For he transfers without warrant the admitted indispensability of *tradition,* to a *particular means or instrument* that happens to be employed in certain societies for sustaining tradition.

functional analysis is therefore inherently concerned with "social statics" rather than "social dynamics," is similarly unfounded, despite the undeniable fact that many functionalists have given little attention to factors which produce disequilibrium and structural change in social systems. For a discussion of this and several other related issues, see Robert K. Merton, *op. cit.,* and also the attempt to state Merton's analysis within the framework of the formal pattern of teleological explanation in Ernest Nagel, "A Formalization of Functionalism," *Logic Without Metaphysics,* Glencoe, Ill., 1956, pp. 247-83.

[19] Bronislaw Malinowski, *Magic, Science and Religion,* New York, 1948, p. 146.

Indeed, it is in general even more difficult in the study of social behavior to establish the indispensability of some particular institutional device for performing a given function (or type of activity) than it is to establish the indispensability of a given function for maintaining a specified state. Thus, to show that a given form of social behavior is essential for the preservation of a certain state in a system—for example, to show that punishment by society of those violating accepted norms of conduct is indispensable for the maintenance of reasonably orderly public behavior—it must be possible to find a number of societies in which, for example, the gravity of punishment and the certainty with which it is inflicted upon violators vary, so as to ascertain whether there is any significant correlation between these differences and variations in deviant behavior in those societies. To be sure, the available data concerning such matters are often insufficient to support reliable conclusions, and there may even be a total lack of pertinent data for assessing the alleged indispensability of a function which is so pervasive that societies do not vary in respect to manifesting it; nevertheless, it is frequently possible to arrive at conclusions about these matters that are not entirely unfounded. On the other hand, to show that a given type of social function can be performed only by a particular social organization (e.g., to show that privately subsidized universities are indispensable for the continuance of unregimented scientific research), it would be necessary to show not only that the given function is not in fact performed by any organization other than the stated one, but also that no other organization (whether already in existence or only envisaged) *could* perform that function. However, in view of the varying functions that the same or similar organizations have exercised in the past, and in view also of human capacity for creating new institutional forms, such a task is almost hopeless.

The situation in biological inquiry is in this respect markedly different. For although one organ in a living body can sometimes assume a function which a different organ in that body normally exercises, and although even the same vital function is sometimes performed by quite different mechanisms in different organisms, such alternative mechanisms for the exercise of the same or similar functions are not as unlimited in biology as they appear to be in the social sciences. At any rate, it is not customary to find that, although in one class of men the lungs are organs of respiration and the stomach the organ of digestion, in some other class of human beings these organs have interchanged their familiar functions. But something not radically unlike this fantasy is a commonplace reality in the study of human societies.

As a consequence of this and the other points that have emerged in this discussion of functionalism in social science, the cognitive worth of functional explanations modeled on teleological explanations in physiology is therefore and in the main very dubious.[20] Certainly the least questionable and most illuminating functional accounts that have been proposed thus far are those which analyze the functions of forms of social behavior in either the fourth or the fifth sense of the word 'function' distinguished earlier—the great many accounts which exhibit relations of interdependence between patterns of standardized conduct in primitive societies, between economic and legal institutions, between religious, social, and economic ideals, between architectural style, social norm, and philosophical doctrine, between social stratification and type of personality, and much else besides. However, though these functional analyses are illuminating and valuable, they cannot be rightly regarded as illustrating a distinctive theoretical approach in the study of human affairs.

III. *Methodological Individualism and Interpretative Social Science*

Even a cursory inspection of generalizations and explanations in the social sciences reveals many differences between the formal characteristics as well as between the substantive content of the various concepts employed by these disciplines. The terms occurring in the statements of these sciences can therefore be classified in a number of alternative ways. However, one familiar way of doing so needs explicit mention. For the distinction upon which this classification rests is also the focus of a long-standing debate among social scientists over an influential view that satisfactory explanations of social phenomena must possess a feature uniquely characteristic of those explanations—a view that requires at least a brief discussion.

If we limit ourselves to notions descriptive of human beings and their behaviors, two classes of terms employed in social inquiry are commonly distinguished. Although the distinction is not sharp and is not free from difficulties, we shall for the moment ignore them. The first class contains only terms referring to *individual human beings* or to *attributes of such individuals*—terms like 'the current President of the United States,' 'ambitious,' 'tolerant in religious matters,' 'absent from work,' or 'student in

---

[20] The explanations are in any event not full-fledged teleological ones, in the sense stated in Chapter 12. To have such explanations it would be necessary to show that the variables (e.g., performance of religious rituals and performance of military duties) maintaining a state in some system are *state variables;* that is, one of the conditions the variables must fulfill is that of being independent of each other, in the sense that the performance of religious rituals at any given time does not depend on whether military duties are performed at that time, and conversely. In point of fact, it is questionable whether any of the proposed functional explanations in social science satisfy this requirement.

medical school.' The second class contains only terms designating *groups of human individuals, attributes characterizing such groups collectively, or forms of organization and activity manifested by such groups*—terms like 'the delegates to the last Democratic presidential convention,' 'the French Enlightenment,' 'corporation,' 'mob hysteria,' or 'degree of social cohesion.' For convenient reference, let us call terms in the first class "individual terms" and those in the second "collective terms." Both classes of terms are employed freely in social inquiry. Nevertheless, (1) many collective terms are a frequent source of perplexity when questions are raised concerning what if anything is designated by them (i.e., what their "extensions" are) and whether collective terms are in general "definable" by way of individual terms; in consequence, (2) social scientists continue to be divided on how social phenomena formulated with the help of collective terms are to be explained. We shall examine a few facets of these two related issues.

1.   No conceptual difficulties are normally encountered in specifying what is denominated by many collective terms. For example, the term 'the delegates to the last Democratic presidential convention' is generally understood to designate a certain readily identifiable class of human beings; and most men would probably find it obviously absurd to postulate as the extension of that term some "entity" other than those human individuals. Accordingly, a statement such as 'The last Democratic presidential convention chose its standard-bearer in a hopeful mood' would be commonly taken to assert, approximately, that each delegate to the convention (or each individual in a very large subclass of the delegates) manifested in some fashion a hopeful attitude on a certain occasion; and few would be tempted to construe the statement to mean that some superhuman person, capable of exercising causal power, looked forward with hope to the outcome of an election.

However, there is much less general agreement or clarity concerning the extensions of such collective terms as 'the French Enlightenment' or 'corporation.' To be sure, perhaps no one questions that individual human beings are in some sense "parts" of what is designated by these and similar expressions. But it is certainly not possible to enumerate exhaustively the individuals who constitute those "parts" or even to say how many such individuals there are; nor can we state with much precision either the actions performed by the individuals who can be enumerated, the philosophic beliefs that must have been advocated by individuals if the latter are to count as "parts," or the distinctive relations in which human beings have stood to each other and to other things in order to be so counted. In short, although the term 'the French Enlightenment' is an undeniably useful one, it is also a term that is highly vague and whose extension cannot be articulated with unlimited detail. This inability to "spell out" in full and with precision the extension of the term is perhaps one reason why some students have conceived the French Enlightenment to be some sort of "unitary whole," and have endowed this "super-organic individual" with powers to direct the course of individual human action.

But however this may be, such a hypostatic transformation of a complex system of relations between individual human beings into a self-subsisting entity capable of exercising causal influence is the analogue of vitalistic doctrines in biology, and is a recurrent theme in the history of social thought. Thus, political theorists have argued that a people possesses a "general will" that is distinct from the wills of its individual members and that may not even be an object of explicit awareness for the latter; psychologists have postulated "group minds" to account for ethnic and racial differences; sociologists have attributed a "psyche" to mobs in order to explain mass hysteria; judges and lawyers have maintained that corporations are "persons," not only in the technical legal sense of being organizations that can bring suit and be sued in courts of law, but in the sense of being substantial entities distinct from and yet comparable with the human beings constituting those organizations in having their own personalities, places of residence, capacities for movement, and inherent rights independent of all legal enactments; and, as we shall indicate more fully in the next chapter, historians have denied the efficacy of individual human effort to alter the course of events because of the overwhelming power they ascribe to supposedly self-subsisting "forces" that determine the direction of historical changes. These hypostatic interpretations of what is denoted by collective terms have frequently been exercises in irresponsible intellectual constructions, and have served as instruments for justifying social iniquities. However, it is generally impossible to assess their validity, since they are usually formulated far too unclearly to permit an unambiguous determination of what if anything follows from them. In any event, like the vitalistic assumptions in biology, such hypostatic interpretations have been useless as guides in inquiry and sterile as premises in explanations. Their introduction into social science is therefore entirely gratuitous; for the methodological assumption that all collective terms designate either groups of human individuals or patterns of human behavior leads to a more fruitful way of identifying the extensions of such terms than does the perplexing hypostasis of mysterious super-individuals.

Nevertheless, it does not necessarily follow from this methodological assumption that all collective terms are *explicitly definable* exclusively by way of individual terms, "in principle" if not in actual practice. For in the first place, the supposition that collective terms are so translatable is obviously indeterminate, unless the class of individual terms is precisely delimited. But, as has already been noted, the distinction between collective and individual terms is not a strict one, and there are no decisive reasons for subsuming some terms under one category rather than under the other. Thus, to illustrate just one variant of this difficulty, the term 'law-abiding' signifies an attribute a human being may possess because of his behavior in relation to norms of conduct adopted and enforced in certain ways by an organized community. Accordingly, the term could be counted as an individual one, since it is predicated of individual human beings; it could nevertheless also be counted as a collective term, on the ground that it involves reference to forms of activity characterizing the behavior of groups of human individuals. However, there are no firm principles for deciding between these alternatives, nor is there much prospect for developing such rules. But if the supposition that all collective terms are definable in the stated manner is thus indeterminate, the claim that this supposition necessarily follows from the above methodological assumption is equally undecidable.

Let us waive this difficulty, and turn to a second. We had occasion to suggest that the notorious vagueness of many collective terms need not stand in the way of their having important uses. However, their vagueness can be an insuperable obstacle to translating them into formulations employing only individual terms. As the brief discussion of the term 'the French Enlightenment' indicated, it is a marked feature of our use of many collective terms that most of the individual details in their extensions cannot be specified, and that certain parts of these extensions can be described only with the help of other collective terms. To cite another example, when we characterize a nation as "warlike," we may be able to state in a general way some of the organized activities of various groups of individuals in that nation, in virtue of which we ascribe a warlike character to it, e.g., large-scale military expenditures and military training programs, influential political and social positions occupied by military officers, "saber-rattling" conduct of foreign affairs, and the like; but we are unable to render the meanings of these essentially collective ascriptions by way of exclusively individual terms. It is therefore not evident just how one is to proceed in order to provide explicit definitions of the required sort for collective terms which possess the feature just noted.

But even if both of these difficulties are waived, and in particular a precise distinction between collective and individual terms is postulated, a final point should be noted. There is no *formal* incompatibility between the above methodological assumption and the possibility that collective as well as individual terms occur among the primitive (or undefined) terms which may be required to define explicitly various other collective terms. An incompatibility would exist only if it were impossible to understand the meaning of any collective term and to learn how to apply it, except by way of the meanings of individual terms. The claim that there is indeed such an impossibility has been made by some writers. It has been argued, for example, that, unlike the natural sciences, the social sciences can never observe directly "collective entities" or their attributes; for in these latter disciplines the relevant data that are directly accessible to us are the attitudes and beliefs of individuals, out of which the various "wholes" of social inquiry are eventually compounded. More specifically, the natural sciences are said to initiate their investigations with direct observations of complex wholes, such as rocks, lightning flashes, or plants; and they then proceed to "explain" these wholes by analyzing them into relations between the theoretically defined but inferred individuals, such as atoms or electrons, which constitute those wholes. In contrast to this, the alleged point of departure of the social sciences is the observation of individual human conduct; and the various collective terms employed in these disciplines (such as 'society,' 'economic system,' or 'imperialistic policy') are therefore theoretical constructions, which are defined exclusively with the help of individual terms.[21]

However, this argument is not only mistaken in its contention that "collective entities" in the natural sciences have an observational status radically different from the one they occupy in the social sciences; it also fails to establish the claim that the meanings of collective terms in the social sciences can be acquired only via the meanings of individual terms. Thus, astronomy is admittedly an exception to the alleged thesis that in the natural sciences inquiry proceeds from observation of "wholes" to their explanation in terms of the individual components reached by analyzing those "wholes." The solar system, for example, is not an observational datum, and our conception of that system is a theoretical construction based on observations of individual constituents of the system. But astronomy is not the sole exception in this respect, for the galaxies discussed in astrophysics, the magnetic field of the earth studied with the help of electromagnetic theory, the earth's atmospheric envelope investigated with the aid of thermodynamics and physical chemistry, the continental masses and the oceans whose motions are analyzed by means of mechanical and hydrodynamical principles, and the numerous species of plants and animals explored in biology are similarly "wholes" which belong to the domain of natural science without being objects of direct observation.

Let us turn next to the social sciences, but ignore as not centrally rele-

[21] F. A. Hayek, *The Counter-Revolution of Science*, Glencoe, Ill., 1952, Chap. 4.

vant to the issue before us the dubious contention that human attitudes are directly observable. However, the crucial claim that "collective entities" are never directly observed in social inquiry appears to be no less dubious. Do we really never directly observe such collective wholes or their attributes as parades, elaborate ceremonies sometimes involving large groups of human actors (as in the enthronement of a monarch, the performance of a religious dance, or the taking of marriage vows in public), the hostility of a crowd, or the orderly procedure in a court of law? The question is obviously rhetorical; and a uniformly negative answer to it must surely be the result of some misunderstanding as to what is being asked—unless indeed we are prepared to construe the fact that most men do profess to observe such things directly, as the consequence of a universally perverse and mysterious misuse of language. To be sure, in claiming that some collective wholes and attributes are directly observable, we cannot rightly mean that these observations are instantaneous, or that they occur without selective attention and without interpretation in the light of various controlling ideas. In these respects, however, what is generally characterized as direct observation in the natural sciences is no different from direct observation of collective wholes in social inquiry. To deny that such wholes are ever directly observed is in effect like denying that we can ever observe a forest, on the ground that when we say we do so the only things we really see are individual trees.

We therefore conclude that, although it is a sound methodological assumption to interpret collective terms in social science as designations for groups of human beings or their modes of behavior, these terms are not in fact invariably *defined* by way of individual terms, nor does the assumption necessitate that collective terms must in principle be so definable.

2.   With this conclusion in mind, let us finally examine the claim that since "group activities are essentially and necessarily activities of the individuals who form groups to attain their ends," the distinctive aim of the social sciences is to "understand" social phenomena by explaining them in terms of the "motivationally meaningful" (or "subjective") categories of human experience.[22] This view has for many years been referred to as "interpretative social science" (or as "verstehende Soziologie," to mention a widely used German label), and more recently it has been frequently advocated under the name of "methodological individualism," and is contrasted with "methodological collectivism" or "holism." [23] Thus, on one version of the principle of methodological in-

---

[22] Ludwig von Mises, *Theory and History*, New Haven, Conn., 1957, p. 258. See also the other references mentioned in footnote 16 in Chapter 13.

[23] F. A. Hayek, *op. cit.*, Chaps. 4 and 6.

dividualism, the social scientist is to "continue searching for explanations of a social phenomenon until he has reduced it to psychological terms"; [24] and, according to a more explicit formulation of the principle,

> the ultimate constituents of the social world are individual people who act more or less appropriately in the light of their dispositions and understanding of their situation. Every complex social situation, institution or event is the result of a particular configuration of individuals, their dispositions, situations, beliefs, and physical resources and environment. There may be unfinished or half-way explanations of large-scale social phenomena (say, inflation) in terms of other large-scale phenomena (say, full employment); but we shall not have arrived at rock-bottom explanations of such large-scale phenomena until we have deduced an account of them from statements about the dispositions, beliefs, resources and inter-relations of individuals. (The individuals may remain anonymous and only typical dispositions, etc., may be attributed to them.) And just as mechanism is contrasted with the organicist idea of physical fields, so methodological individualism is contrasted with sociological holism or organicism. On this latter view, social systems constitute "wholes" at least in the sense that some of their large-scale behavior is governed by macro-laws which are essentially *sociological* in the sense that they are *sui-generis* and not to be explained as mere regularities or tendencies resulting from the behaviour of interacting individuals. On the contrary, the behaviour of individuals should (according to sociological holism) be explained at least partly in terms of such laws (perhaps in conjunction with an account, first of individuals' roles within institutions, and secondly of the functions of institutions within the whole social system). If methodological individualism means that human beings are supposed to be the only moving agents in history, and if sociological holism means that some superhuman agents or factors are supposed to be at work in history, then these two alternatives are exhaustive.[25]

Some of the issues raised by interpretative social science were discussed earlier, in Chapter 13, in connection with the alleged "subjective" character of social phenomena. Our present concern, however, is with just that part of the over-all claims of methodological individualism which asserts in effect that all the descriptive terms occurring in satisfactory explanations of social phenomena must belong to a special subclass of individual terms, namely, terms denoting "subjective" or "psychological" states of individual human beings. Methodological individualism thus subscribes to what is often advanced as a factual thesis (although it is perhaps best regarded as a program of research) concerning the *reducibility* of all statements about social phenomena to a

[24] J. W. N. Watkins, "Ideal Types and Historical Explanation," *British Journal for the Philosophy of Science*, Vol. 3 (1952), p. 29.
[25] J. W. N. Watkins, "Historical Explanation in the Social Sciences," *British Journal for the Philosophy of Science*, Vol. 8 (1957), p. 106.

special class of ("psychological") statements about individual human conduct; and we shall therefore be able to assess this thesis in the light of the general requirements for reduction that were stated in Chapter 11.

However, as is evident from the just-quoted statement of methodological individualism, some of its proponents fail to distinguish between what may be called the ontological thesis that "the ultimate constituents of the social world are individual people" (which corresponds to the methodological assumption discussed above in connection with the hypostatization of collective terms), and the reductive thesis that statements about social phenomena are deducible from psychological statements about human individuals. There is little doubt that many students who subscribe to methodological individualism do so because they believe that to reject a hypostatic interpretation of collective terms and to deny that "superhuman agents" are causally operative in human affairs is logically equivalent to the reductive thesis. But in any event, and in the light of our earlier discussion of this issue, that belief is mistaken, so that a commitment to the ontological thesis does not logically require a commitment to the reductive one.

Moreover, we have also found reasons for doubting whether all the collective terms of social science are explicitly definable via individual terms when *no* restrictions are placed upon the individual terms that may appear in the definitions. The likelihood of such definitions is not increased if the stronger requirement is introduced that the primitive terms must be "psychological" ones. Accordingly, should it not in fact be possible to construct definitions meeting this strong requirement, the condition of connectability for reduction will not be satisfied unless (as the discussion of this point in Chapter 11 indicates) collective terms that are not so definable are linked with individual terms either by way of suitable correspondence rules or of various empirical hypotheses. However, neither of these alternatives would apparently contribute anything toward achieving the objectives of methodological individualism. For, although either alternative would make it possible to *verify* statements containing collective terms by observing the conduct of individuals, neither alternative would permit the *elimination* of collective terms from such statements in favor of individual terms exclusively; and in consequence, neither alternative would serve to realize the professed aim of methodological individualism: to deduce statements about large-scale social phenomena from statements "about the dispositions, beliefs, resources and inter-relations of individuals."

It is also obvious, furthermore, that, even if the condition of connectability could be fulfilled in some fashion, the second formal condition for reduction—the condition of derivability—would not thereby be necessarily realized. Among other reasons for this, a simple one, is that no set of premises about the conduct of *individual* human beings might suffice for deducing some given statement about the actions of a *group* of men, and that at least one assumption of the latter kind might be required in any set of premises from which the given statement is deducible. A clear example of just this possibility has already been mentioned in Chapter 11, where the reduction of thermodynamics to mechanics was outlined. In that example, it will be recalled, the condition of connectivity was satisfied by establishing a linkage between the temperature of a gas and the mean kinetic energy of its molecules. But the reduction was not effected simply by satisfying this condition, since it also required the deduction of a certain relation between that energy and the pressure and volume of the gas—namely, $pV = 2E/3$. However, this relation cannot be deduced from premises stating only various mechanical attributes of the molecules *individually;* and the usual Newtonian assumptions about individual molecules were therefore supplemented by a special assumption concerning a statistical property of the *ensemble* of molecules.

A comparable situation apparently exists in several areas of social science, notably in economics. Thus the economic theory currently designated as "microeconomics" (also known as "marginal utility" theory, whose classic formulation in English is contained in Alfred Marshall's influential *Principles of Economics*) analyzes economic phenomena in terms of assumptions concerning the economic preferences of individual producers and consumers of economic goods. A major aim of the theory is to explain the operations of the total economy of a society, by deducing propositions characterizing those operations from premises dealing with dispositions, beliefs, and resources of individual economic agents. The objectives of microeconomics are therefore in complete accord with the program of methodological individualism; indeed, some of the chief advocates of the latter (e.g., F. A. Hayek and L. von Mises) are also prominent exponents of marginal utility analysis. However, in the judgment of many economists, microeconomics fails to account for various important features that frequently mark the total economies of nations (such as recurrent unemployment crises) and does not provide effective tools for controlling the course of large-scale economic events. In consequence, without rejecting marginalist theory *in toto* in favor of an "institutional" (or "historical") approach to economic problems, many students believe that the classical assumptions of the theory do not suffice to achieve the objectives for which the theory was devised, and require supplementation by additional ones.[26]

[26] The "institutional" or "historical" approach to economic problems was in many cases a negative reaction to the "abstract" character of classical economic theory, and to some of the dubious psychological assumptions that earlier forms of the

An important step in this direction was taken in the 1930's by J. M. Keynes, when he proposed an influential "macroeconomic" theory, although similar but less influential proposals had been made before that time by other economists. The only feature of this theory that needs to be noted in the present context, since it is the directly relevant feature in assessing the merits of methodological individualism, is that the basic postulates of the theory are not exclusively "psychological" ones about *individual* economic agents, but include assumptions concerning relations between large-scale *statistical aggregates* (such as national income, total national consumption, and total national savings). To be sure, no proof is available that these macroeconomic assumptions cannot be deduced from microeconomic ones. But there is also no proof that the deduction can be effected, and there is at least a presumption that it cannot be. In any event, despite the absence of such a deduction, economists do not hesitate to employ those macroeconomic postulates in their analyses; for, as one student has put the matter, "one may disagree on particular assumptions, institutional or psychological, regarding the saving patterns of individuals or groups of individuals, and yet may find the concept of aggregate savings useful in describing the actual or probable behavior of national income." [27] But if this is so, and macroeconomic assumptions enable economists to account for aggregative phenomena no less adequately than do microeconomic postulates, the reduction of macroeconomic to microeconomic explanations appears to offer no substantial scientific advantages. In short, there are nonformal considerations as well as purely formal ones for questioning the merits of the reductive thesis of methodological individualism.

We shall therefore conclude this discussion of methodological individualism and interpretative social science with some brief comments on the alleged superiority of "understanding" social phenomena in terms of the "evaluative schema" of human actors, over what has been called the

marginalist theory took for granted (such as the hedonistic interpretation of economic utility, and the possibility of interpersonal comparison of pleasures and pains). These assumptions were eventually modified under the pressure of criticism. However, even some of the modified postulates of more recent versions of marginalism (such as the assumption that all consumers choose economic goods in accordance with definite preference scales, or that changes in preferences of different consumers occur independently of each other) continue to be targets of criticism. For an account of the controversy between institutionalists and abstract theorists, as well as of the critique directed against the "psychological preconceptions" of marginal utility analysis, see, for example, J. N. Keynes, *Scope and Method of Political Economy*, London, 1890; Joseph Schumpeter, *Economic Doctrine and Method*, New York, 1954; T. W. Hutchison, *A Review of Economic Doctrines, 1870-1929*, Oxford, 1953; Allan G. Gruchy, *Modern Economic Thought*, New York, 1947; I. M. D. Little, *A Critique of Welfare Economics*, Oxford, 1950, Chaps. 2 and 3.

[27] Kenneth K. Kurihara, *Introduction to Keynesian Dynamics*, London, 1956, p. 20. For a survey of the central role played by collective terms in sociological inquiry, cf. Robert K. Merton, *Social Theory and Social Structure*, Chap. 9.

'causal-functional" approach of natural science. It has been argued, for example, that if we wish to understand why there has been a marked increase in the divorce rate in the United States during the past half century, a satisfactory explanation must be of the type which reveals

> a change in valuation affecting the status of the family. The general indication is that *divorce is more prevalent in those areas where the continuity of the family through several generations has less significance in the schema of cultural values than formerly or elsewhere.*

And more generally, it is claimed that "Insofar as we are able to discover the changes of the evaluative schema of social groups we can attain, and thus only, a unified explanation of social change." [28]

If knowledge of changes in evaluative schema is indeed a necessary and sufficient condition for unified explanations of social change, it would certainly be folly not to give highest priority to the pursuit of that knowledge. There are nevertheless two related questions which generate legitimate doubt concerning the promise claimed for such knowledge. In the first place, how can alterations in cultural values be ascertained? One obvious source of information consists of explicit statements by individuals concerning their values, whether given in the form of unsolicited private confessions, public addresses, replies to interviewers, or otherwise. However, such information is in general not abundant, so that knowledge of men's values can be obtained in this way in only a comparatively few cases. Moreover, the reliability of such information is frequently questionable. For there is a familiar disparity between what men verbally profess on special occasions, and what they habitually believe and practice; and anthropologists recognize "the existence of 'pretend rules': standards that are honored in spoken word, but breached in customary behavior." [29] Nor is it uncommon for individuals as well as entire societies to continue without conscious hypocrisy a verbal allegiance to some set of values long after an earlier mode of life that once illustrated their active commitment to those values has been radically transformed. Well-trained students of human affairs have therefore made it standard practice to base their conclusions concerning the values operative in a given community not simply on verbal professions, but in large measure on the evidence of other overt activities—of courtship customs, of conduct in domestic life, of business practices, and the like.

In consequence, explanations of social change in terms of alterations in evaluative schema run the risk of being empirically empty tautologies. For example, if a change C in the divorce rate in a community during a period of years is part of the evidence E for the conclusion that there

[28] R. M. MacIver, *Social Causation*, New York, 1942, pp. 338, 374.
[29] E. Adamson Hoebel, "The Nature of Culture," in *Man, Culture and Society* (ed. by Harry L. Shapiro), New York, 1956, p. 175.

has been a change V in the cultural value associated with the continuity of the family, it would not be illuminating to "explain" the former change by the latter one. The point being argued is not that every proposed explanation of social change in terms of a change in evaluative schemas is necessarily sterile. The point is that, if such sterility is to be avoided, the evidence E for the alleged change V in cultural values must be reliable, and must be different from the social change C that the change V is intended to explain. But if the evidence E is different from C, and since E is itself a social change, there is patently a determinate relation between E and C, so that conceivably E itself might serve to explain C. Accordingly, if the introduction of the alteration V as an explanatory link between E and C is not to be purely *ad hoc*, it must be justified by showing that V serves to establish not only the relation of dependence between E and C but also *other* relations of dependence between *other* sets of social changes.

But this brings us to the second question: Do explanations formulated in terms of changes in cultural values (or other "subjective" dispositions) generally possess greater capacities for systematically organizing relations of dependence between social phenomena than do explanations that employ different substantive concepts? However, no straightforward answer to this question can be given, if only because no careful studies have yet been made of the comparative merits of these two types of explanation. It would be absurd to deny that explanations in terms of "meaningful" categories have frequently illuminated social changes, as in the obvious case of Max Weber's discussion of the rise of modern capitalism. But it would be no less absurd to claim that such explanations fully account for all social change, or to deny that explanations in terms of other variables (such as physical environment, state of technology, population density, or form of economic organization) often have at least as much predictive and systematizing power as do explanations in terms of "subjective" dispositions, as in the equally obvious case of many Marxist interpretations of modern legal systems. In any event, the available evidence does not support the claim that the answer to the question is unmistakably and uniformly affirmative; and there are therefore no compelling reasons for supposing that only by discovering the changes in the evaluative schema of social groups can we attain a unified explanation of social changes.

Accordingly, although methodological individualism and interpretative social science rightly emphasize that social phenomena are constituted out of interactions between purposive human agents, neither of these essentially similar approaches to social inquiry possesses the unqualifiedly pre-eminent merits that are claimed for it.

Problems in the Logic
of Historical Inquiry

15 Although it will be convenient to employ the term 'history' in the comprehensive sense of signifying the study of sequential changes that have occurred in any subject matter and not only in human affairs, the present chapter nevertheless deals with questions that relate primarily to explanations of past human actions. After a brief discussion of the general character of historical inquiry, we will analyze some of the forms historical explanations usually take; we will subsequently consider some problems recurrent in historical study, and finally examine a number of issues raised by the doctrine known as "historical inevitability."

I. The Central Focus of Historical Study

According to Aristotle, poetry, like theoretical science, is "more philosophic and of graver import" than history, because poetry is concerned with what is pervasive and universal but history is addressed to what is special and singular. Aristotle's remark is a possible source of a widely accepted distinction between two allegedly different types of sciences: the *nomothetic*, which seek to establish abstract general laws for indefinitely repeatable events and processes; and the *ideographic*, which aim to understand the unique and the nonrecurrent. It is often maintained that the natural sciences and some of the social ones are nomothetic, whereas history (in the sense of an account of human events, as distinct from the events themselves) is pre-eminently ideographic.[1] In consequence, it is frequently claimed that the logical structure of the concepts and explanations required in human history is fundamentally different from the logical structure of concepts and explanations in the natural (and other "generalizing") sciences. Let us examine the basis for these contrasts.

Even a hasty inspection of treatises in theoretical natural and social science on the one hand (such as optics and economics), and of books on history on the other hand, suffices to reveal a striking difference between them. For by and large the statements occurring in the former are general in form and contain few if any references to specific objects, dates, or places, whereas the statements in the latter are almost without exception singular in form and are replete with proper names, designations for particular times or periods, and geographic specifications. To this extent, at any rate, the contrast between the natural and some social sciences as nomothetic, and human history as ideographic, appears to be well founded.

It would be a gross error, however, to conclude that singular statements play no role in the theoretical sciences or that historical inquiry makes no use of universal ones. As previous chapters have repeatedly noted, no conclusions concerning the actual character of specific things and processes can be derived from general statements alone; for theories and laws must be supplemented by initial conditions (i.e., by statements singular or instantial in form) if those general assumptions are to serve for explaining or predicting any particular occurrence. Nor does the familiar and often useful distinction between "pure" and "applied" natural science impair the relevance of this point—on the supposed ground that the pure sciences (such as theoretical electrodynamics or genetics) are concerned with establishing only general statements, and that only applied sciences (such as electrical engineering or agronomy) need to concern themselves with particular cases. For even the pure natural sciences can assert their general statements as empirically warranted only on the basis of concrete factual evidence, and therefore only by making use of singular statements. Moreover, many statements commonly acknowledged to be laws of "pure" science have a generality that is at least geographically restricted—for example, the familiar law that a freely falling body at sea level in latitudes between 38° and 39° on the earth's surface undergoes an acceleration of 980 centimeters per second. If laws of this kind, which are specializations of laws not similarly restricted, are excluded from theoretical treatises, the exclusion is at best only a

---

[1] The distinction was first stated in this terminology by Wilhelm Windelband in his essay "Geschichte und Naturwissenschaft," reprinted in the collection of his essays *Präludien*, 5th ed., Tübingen, 1915, Vol. 2, pp. 136-60.

question of convenience rather than of principle. Furthermore, some branches of natural science, such as geophysics or animal ecology, are primarily concerned with spatiotemporal distributions and the development of specific individual systems, and are therefore engaged to a large extent in establishing statements singular in form. In short, neither the natural sciences in their entirety nor any of their purely theoretical subdivisions are exclusively nomothetic.

But neither can historical study dispense with at least a tacit acceptance of general statements of the kind cited in theoretical treatises. Thus, although the historian may be concerned with the nonrecurrent and the unique, he must obviously select and abstract from the concrete occurrences he is engaged in studying, and his discourse about what is unquestionably individual requires the use of common names or general descriptive terms. Accordingly, the historian's characterizations of individual things assume that there are various *kinds* of occurrences, and in consequence that there are more or less determinate empirical regularities which are associated with each kind and which differentiate one kind from other kinds. For example, Greek colonial expansion during the sixth century B.C. has been attributed by one historian to the needs of Greek commercial interests combined with the adventurous spirit of the Greeks; [2] and he obviously took for granted that human beings possess several types of needs, that each type is generally manifested in certain characteristic modes of behavior, that some of these modes frequently result in the founding of colonies, and so on. Furthermore, one phase of historical inquiry consists of so-called "external and internal criticism"— of efforts directed to ascertaining the authenticity of documents or other remains from the past, the precise meanings of recorded statements, and the reliability of testimony concerning past events. But to accomplish these tasks, historians must be armed with a wide assortment of general laws, some of which are undoubtedly accepted tacitly as "common-sense knowledge" while others are adopted because they are endorsed by some natural or social science.

Moreover, historians are rarely mere chroniclers of the past; and they do not always terminate their investigations of some special group of events, even when they have settled the sequential order in which those events actually happened—for example, when they have established that Antony fell in love with Cleopatra before he fled from the battle of Actium. On the contrary, historians usually seek to understand and explain the events they record in terms of causes and consequences, and to find relations of causal dependence between some of the sequentially ordered events—for example, by showing that Antony fled from the

---

[2] J. B. Bury, *History of Greece*, New York, 1937, Chap. 2.

battle of Actium *because* of his love for Cleopatra. To be sure, a historian's claim that two such events are causally related can be mistaken; but the historian making the claim presumably believes he has competent grounds for doing so. However, historians do not as a rule profess an ability to apprehend causal connections between individual occurrences by way of some direct, infallible intuition of such connections; and in any case, a given pair of past events can be shown to be causally related only with the help of causal generalizations (whether strictly universal or statistical in form) which are the products of inquiries designated as "controlled investigations" in a preceding chapter. Accordingly, the causal imputations historians make in explanations of human actions in the past are based on assumed laws of causal dependence. In brief, history is therefore not a purely ideographic discipline.

Nonetheless, there is an important asymmetry between theoretical (or "generalizing") science and history. A theoretical discipline like physics seeks to establish both general and singular statements, and in order to do so physicists employ previously assumed statements of both types. Historians, on the other hand, aim to assert warranted singular statements about the occurrence and the interrelations of specific actions and other particular occurrences. However, although this task can be achieved only by assuming and using general laws, historians do not regard it as part of their aim to *establish* such laws. It is unlikely that anyone would find something radically wrong with a treatise on theoretical thermodynamics which did not contain a single proper name or a single reference to any particular date. But it is even more unlikely that anyone using the word 'history' in its customary meaning would classify a book as a *history* if it mentioned no particular individuals, times, and places but stated only generalizations about human behavior. The distinction between history and theoretical science is thus fairly analogous to the difference between geology and physics, or between medical diagnosis and physiology. A geologist seeks to ascertain, for example, the sequential order of geologic formations, and he is able to do so in part by applying various physical laws to his materials of study; but it is not the geologist's task, *qua* geologist, to establish the laws of mechanics or of radioactive disintegration which he employs in his investigations.

However, this discussion must not be construed as an attempt to rule out by a priori reasoning the possibility of "historical laws" of developmental change. There have been many attempts, in recent years by Oswald Spengler and Arnold Toynbee among others, to show that every society or civilization manifests a uniform pattern of successive change, so that, for example, each society allegedly passes through a fixed series of evolutionary stages, in a manner comparable to the birth, adolescence, maturation, and decay of individual biological organisms. Although none of these purported "laws" has found acceptance among competent students, their validity can be assessed only in the light of actual historical evidence and cannot be settled by examining only the formal structure of statements contained in the writings of historians. It is nevertheless pertinent to note that, irrespective of the factual worth professional historians ascribe to these alleged "historical laws," they tend to regard attempts at discovering such laws as contributions to sociology (or to some other branch of the "generalizing" or theoretical sciences) rather than to "history proper." [3] Accordingly, despite the fact that some historians undoubtedly use the evidence of the human past to establish laws of developmental change, they do not do so *qua* historians, in the opinion of most of their professional colleagues as well as on the evidence of the great bulk of historical writing.

II. *Probabilistic and Genetic Explanations*

In any event, we shall assume that historical inquiry is concerned primarily with particular occurrences, and we shall therefore be concerned with the explanations that are typically offered for them. However, historians sometimes undertake to account for some action of just one human being, sometimes for an aggregative occurrence involving the actions of many men. Since there are important differences between explanations for these two kinds of happenings, it will be convenient to discuss separately the two corresponding major types of historical explanations. Moreover, as was suggested by the discussion in Chapter 10 of various generic differences between explanatory variables, historical explanations also differ in the temporal magnitudes of the events they mention. In particular, individual actions as well as the circumstances under which they occur are sometimes described as if they had no temporal dimensions, while the circumstances under which another action occurs, but not the action itself, are characterized in terms of their temporal spread. In consequence, some historical accounts of individual actions in effect assume that the conditions which supposedly explain the actions can be viewed for the purposes at hand as practically instantaneous; but other accounts of individual actions provide developmental or genetic explanations for them. We will therefore begin the discussion with an example of a historical explanation belonging to a subdivision of the first major type, i.e., one that attempts to account for some action of a single individual by stating for its occurrence a condition whose duration

[3] See the various critical studies of Toynbee's attempt to establish such laws in *Toynbee and History* (ed. by M. F. Ashley Montagu), Boston, 1956. The comment by A. J. P. Taylor on Toynbee's work, that "this is not history" (p. 115), is characteristic.

is ignored. On the supposition that the explanation is representative of the type, we will then (a) discuss the question whether in explanations of this sort the premises contain general laws, (b) give some reasons for maintaining that the logical structure of such explanations is probabilistic rather than strictly deductive, and (c) consider briefly some senses in which this characterization of the structure is to be understood.

1. The example that will serve to illustrate historical explanations of the first type deals with a circumstance connected with Elizabeth's ascension to the British throne in the sixteenth century. As a result of Henry VIII's quarrel with the Roman Catholic Church, his official title at the time of his death read essentially as follows: "By the Grace of God, King of England, Ireland, and France, Defender of the Faith and Only Supreme Head on Earth of the Church of England called *Anglicana Ecclesia*." It will be recalled, however, that when his daughter Mary ("Bloody Mary") became queen in 1553 after the death of her brother Edward VI, she repealed the acts establishing the ecclesiastical sovereignty of the British monarch and reaffirmed the supremacy of the Pope in Rome. But when Elizabeth succeeded to the throne five years later, she proclaimed herself "Elizabeth, by the Grace of God Queene of Englande Fraunce and Ireland defendour of the fayth. &c."; and in doing so she became the first British sovereign to "etceterate" herself in an unabbreviated official title. Why did she do so? The legal historian F. W. Maitland proposed the following explanation. He first showed that the '&c.' in the proclamation was not there by inadvertence but had been introduced deliberately. He also pointed out that Elizabeth was confronted with the alternatives either of acknowledging with the late Mary the ecclesiastical supremacy of the Pope or of voiding the Marian statutes and breaking with Rome as her father had done—a decision for either alternative being fraught with grave perils, because the alignment of political and military forces both at home and abroad which favored each alternative was unsettled. Maitland therefore argued that in order to avoid committing herself to either alternative for the moment, Elizabeth employed an ambiguous formulation in the proclamation of her title—a formulation which could be made compatible with any decision she might eventually make. In consequence, according to his own succinct summary statement of the explanation, "So we might expand the symbol thus:— '&c.' = and (if future events shall so decide, but no further or otherwise) of the Church of England and also of Ireland upon earth the Supreme Head." [4]

---

[4] F. W. Maitland, "Elizabethan Gleanings," in *Collected Papers*, London, 1911, Vol. 3, pp. 157-65. Maitland was mistaken in assuming that Elizabeth is the first British sovereign to have "etceterated" herself in the manner described. She had

a. We shall suppose that Maitland's explanation is fully warranted, since our aim is to analyze its structure rather than to discuss its factual validity. However, such an analysis cannot be performed unless the assumptions underlying the explanation are fully articulated; and, as is normal practice in historical writings, Maitland did not make explicit all the assumptions implicit in his account, whether or not he was aware that he was making them. For example, he took for granted, though without formal mention, the fact (crucial for his argument) that Elizabeth did not believe a patently ambiguous proclamation of her stand on the Roman question would by itself provoke the Pope to initiate armed hostilities against England. But in addition to such singular statements, Maitland also took for granted without mention generalizations about human behavior no less crucial for his argument. For example, in maintaining that Elizabeth adopted a noncommittal wording of her ecclesiastical claims in order to permit herself to postpone a major decision until it became safer to make it, Maitland tacitly assumed some sort of generalization about human conduct—a generalization whose precise form is unclear but which in any case must assert a relation between (a) public statements men are expected to issue concerning their ostensible commitment to some policy at a time when definitive commitment is hazardous, and (b) the use of ambiguous language in such statements for the sake of avoiding a premature commitment. Without some such assumption there would be no ground for holding that the formulation Elizabeth adopted in announcing her sovereign claims had anything to do with the quandary with which she was faced.

However, such generalizations are indispensable in all historical explanations of individual actions (even if, as is usually the case, the generalizations are not stated explicitly). This large claim cannot be established beyond every shadow of doubt, except perhaps by a systematic survey of historical writings; nevertheless, the contention can be shown to be highly credible. For historical explanations of this type aim to state the reason (or *a* reason) why some given individual *x* decided more or less deliberately to act in the manner *y* under circumstances *z*. But the possible reasons for individual actions can be subsumed under a small number of broad categories, where each category can in turn be divided into suitable subordinate classes; and by considering how the possible reasons in each category could be shown to be actually determining (or "causal") conditions for an individual's action, we can persuade ourselves that the above contention is sound. Indeed, three such major categories of reasons seem to be sufficient, and we shall briefly describe them.

been anticipated in this by her sister Mary, as was noted by the English historian A. F. Pollard.

Since by hypothesis the actions to be explained are more or less purposeful, they are presumably consequences of decisions between alternative courses of action available to the actors. Accordingly, one category consists of reasons asserting that certain characteristics of the alternatives are among the determining conditions of the action. Thus, in the Maitland example, Elizabeth appears to have recognized three alternative ways of acting (though other alternatives may in fact have been open to her, but did not occur to her): to proclaim herself immediately as the sovereign head of the English Church, to acknowledge at once the supremacy of the Pope, or to temporize. She seems also to have attributed certain characteristics to these alternatives—for example, she ascribed to each of the first two the "latent" (i.e., not actually realized or overtly manifest) capacity for generating serious disturbances in England or abroad, and to the third the absence of this capacity. The second category contains reasons ascribing a causal role to characteristics of the actor. Thus Elizabeth possessed various native impulses and propensities (e.g., a quick intelligence), as well as a large number of acquired personal aims and dispositions (e.g., a capacity for compromise); and she also had objectives, values, and obligations as a member of the aristocracy and as the reigning monarch (e.g., to prevent civil war in Britain). The third category of reasons refers to circumstances surrounding the action which exert pressure on the actor in connection with the alternatives. For example, Elizabeth was exposed to advice from her official councilors and she was aware of the general unpopularity of Mary's religious fanaticism, but she also was receiving intelligences about the inclinations of Philip II of Spain to use force against England on behalf of the Pope.[5]

[5] Historical explanations of individual actions are answers to questions similar to those investigated in current social research under the label of "reason analysis." "Reason analysis" signifies a methodologically self-conscious use of recently developed techniques of research for determining why people behave as they do on given occasions—for example, why immigrants from a particular country left their native land, why people who had expressed their intention to vote at a certain election failed to do so, or why men join one book club rather than another. Answers to such queries are obtained chiefly from interviews with the people whose actions are being studied; and the value of the answers obtained in this way depends in large measure on whether the questions asked are constructed in conformity with an adequate "accounting scheme" (or system of categories of questions). The importance of such accounting schemes for the analysis of historical explanations has been stressed by Paul F. Lazarsfeld, and the use made of them in the text has been suggested by his work. For an account of reason analysis, see *The Language of Social Research* (ed. by Paul F. Lazarsfeld and Morris Rosenberg), Glencoe, Ill., 1955, Sec. 5, esp. pp. 387-91; also Hans Zeisel, *Say It With Figures*, 4th ed., New York, 1957, Chaps. 6 and 7. The procedures employed in reason analysis provides additional support for the general position taken in Chapter 13 that human dispositions and intentions can be successfully investigated by behavioristic methods. Moreover, the products of reason analysis confirm the view (and effectively challenge recent claims to the contrary) that historians' explanations of individual actions in terms of the reasons mo-

Now one important step in historical inquiries of the type under discussion evidently consists in showing that the factors in each category which are conjectured to be determining conditions for a given action have in fact been present on the occasion of the act. However, though this step is essential, it clearly does not establish *which* of these factors (or for that matter, whether any of them) was the actual reason for the actor's conduct, since a factor may indeed have been present on the occasion of the act and yet may have been causally inoperative. Thus the fact that a person on trial for murder is known to have hated the victim does not suffice to show that he committed the murder, or that even if he killed the victim he did so *because* of his hatred—for, although the accused individual may be guilty of manslaughter, he may have killed the deceased by accident, because he was paid to do so, or for a number of other reasons. Similarly, even though a historian produces incontrovertible evidence that Antony was madly in love with Cleopatra, this does not establish the claim that Antony fled from the battle of Actium *because* of his love. For Antony possessed other dispositions and objectives than those associated with his love for Cleopatra, so that his action may have been the result, for example, of his ambition to make Egypt a granary for Rome.

But if mere proof that a given factor was one circumstance under which an individual performed some particular act does not establish the claim that the factor was a *reason* why the individual acted as he did, how then can a historian warrant his imputation of a causal role to the factor? If we dismiss as not responsive the suggestion sometimes made that historians have a special faculty for recognizing the reasons for human actions, there seems to be but one viable answer. The historian can justify his causal imputation by the assumption that, when the given factor is a circumstance under which men act, they *generally* conduct themselves in a manner similar to the particular action described in the imputation, so that the individual discussed by the historian presumably also acted the way he did because the given factor was present. In short, generalizations of some sort are required in historical explanations of individual actions.

b. The logical structure of an explanation can be determined only if all the tacitly assumed premises are made explicit; and, since historians do not as a rule state all the assumptions they make in explaining individual actions (indeed, they are often unaware of many crucial assumptions they take for granted), the pattern into which their accounts

tivating the actors do not differ from explanations of individual happenings in other domains of science, either in the tacit use of generalizations in their premises or in the logic required for warranting causal imputations.

fall is not immediately obvious. It would certainly be unwarranted to conclude that explanations of this type are not deductive in form, simply on the ground that such explanations as actually presented by historians do not exhibit this structure, just as it would be unjustified to conclude that a man is not reasoning deductively when his argument happens to be enthymematic—for example, when a man argues that Mars shines by reflected light, because Mars is a planet, but does not explicitly mention the premise that all planets shine by reflected light. Accordingly, although it would be an onerous (and perhaps even practically hopeless) task to state in full the assumptions taken for granted in historical explanations, this task must be supposed to have been completed when the structure of such explanations is being characterized.

Let us first examine the view that historical explanations of this type exhibit a deductive pattern. It is evident that such explanations can indeed be cast into this form, provided that the premises can be freely chosen. For example, one portion of Maitland's explanation for the unusual way in which Elizabeth proclaimed her title can be given a rigorously deductive form, if a strictly universal assumption is introduced as an additional premise, as follows: Whenever an individual is compelled to announce publicly which one of several alternative policies he is ostensibly adopting, the circumstances under which the announcement is made being such that he believes that the proclamation of a definitive commitment to any one of these policies at that time is fraught with grave perils for himself, the individual will formulate his announcement in ambiguous language; Elizabeth was required to proclaim her stand on the Roman question at a time when she regarded as hazardous an announcement of her decision either way; therefore Elizabeth formulated her proclamation in ambiguous language.

However, although the second premise in this formally valid argument repeats in substance what Maitland explicitly maintained, the first or strictly universal premise does not; nor is it likely that he would have accepted it or that anyone else would do so. For it is simply not the case that all men use ambiguous language under the conditions specified in the first premise, since there are human beings (whether uncompromising and courageous in their probity, merely foolhardy, or just stupid) who announce their adoption of some course of action in a forthright manner, even when it would be advantageous for them to temporize. Accordingly, the first premise in the above argument is a false generalization about the matters mentioned in it, so that the argument as it stands is not a satisfactory explanation of Elizabeth's conduct. On the other hand, a credible assumption about those matters will have what is at best only a statistical rather than a strictly universal form; it will assert, for example, that *most* men, or that a certain *percentage* of men, behave in the indicated manner. But if the first premise is replaced by some acceptable statistical generalization, the resulting argument is not a formally valid deductive one, and its premises entail the conclusion not with necessity but only with some "degree of probability."

Accordingly, on the assumption that, if the tacit generalization in Maitland's explanation is to be credible, the generalization is statistical in form, the explanation must be characterized as probabilistic rather than deductive in its structure. But is this assumption sound? The possibility of establishing strictly universal generalizations pertinent to the matters at hand cannot, of course, be excluded on principle. However, no such generalizations seem to be currently available. Moreover, in the light of the discussion in the preceding chapter on the statistical nature of social science generalizations,[6] if well-founded universal laws about human conduct are ever established, they are likely to be formulated in terms of highly refined distinctions which fall outside the customary range of interest of historians. Thus, suppose that Elizabeth's proclamation of her ambiguously worded title could be strictly deduced from premises containing, among others, assumptions formulated in quantum theoretical terms about the state of her glands, the condition of her neural synapses, the organization of her brain cells, and the intensities of various physical stimuli to which she was exposed. It is not an implausible conjecture that most historians as well as most readers of historical literature would turn away from such an explanation, on the ground that it is not the sort of historical explanation to which they have become accustomed or in which they have much interest.[7] In this respect, Maitland's account of the reason why Elizabeth proclaimed her title in ambiguous language can be safely taken as representative of historical explanations of individual actions. In consequence, there is a firm basis for the contention that, in general, historical explanations of this type have a probabilistic structure.

This contention is strengthened by a further consideration, which can also be illustrated by the Maitland example. We have been discussing that example as if Maitland's sole intent were to account for the ambiguous language of Elizabeth's proclamation. It will be recalled, however, that he undertook to explain why Elizabeth "etceterated" herself in announcing her sovereign claims, and not simply why she employed *an*

[6] See pages 505-09 above.

[7] The point has been made by a professional anthropologist as follows: "The human sciences do not press their analyses very far beyond the 'apparent' reality into an order of things where waves of probability undulating in nothingness offer the ultimate common denominators. For better or worse, the human sciences are concerned with the phenomenal surface of reality; if they discarded it entirely, they would destroy their subject matter."—S. F. Nadel, *The Foundations of Social Anthropology*, Glencoe, Ill., 1951, p. 195.

ambiguous formulation in doing so. But even if it were possible to account for the latter fact by deducing it strictly from some set of credible premises, the statement that Elizabeth used the particular phrase '&c.' (rather than some other expression with which she might also have achieved her aim, such as 'and so forth') would not thereby be explained in a deductive fashion. On the contrary, under the assumptions made, the fact that she employed the phrase '&c.' would have been shown to be only probable. On the other hand, if the distinctions are retained which are appropriate to the level of analysis normally practiced by historians, the likelihood is negligible that strictly universal laws can be established concerning the use of the particular locution '&c.' by individuals who are undecided upon a course of action but wish to conceal their indecision. There is therefore no genuine prospect that an explanation with a deductive structure can be given for Elizabeth's use of this locution in her proclamation.

The point just made in terms of an example can be stated more generally. Let A_1 be a specific action performed by an individual x on some occasion t in order to achieve some objective O. However, historians do not attempt to explain the performance of the act A_1 in all its concrete details, but only the performance by x of a *type* of action A whose specific forms are $A_1, A_2, \ldots, A_n$. Let us suppose further that x could have achieved the objective O had he performed on occasion t any one of the actions in the subset $A_1, A_2, \ldots, A_k$ of the class of specific forms of A. Accordingly, even if a historian were to succeed in giving a deductive explanation for the fact that x performed the type of action A on occasion t, he would not thereby have succeeded in explaining deductively that x performed the specific action A_1 on that occasion. In consequence, and at best, the historian's explanation shows only that, under the assumptions stated, x's performance of A_1 on occasion t is probable.

c. Historical explanations of individual actions are thus probabilistic in structure, their form being the outcome of the essentially statistical character of the available generalizations about human conduct that enter into the explanatory assumptions. But in what sense is the characterization "probabilistic" to be construed? An adequate discussion of this much-debated and still unsettled question would require detailed analyses of the logic of probable reasoning and of inductive inference, neither of which can be attempted here. Nevertheless, something must be said, though only in brief, to explicate the meaning of that characterization.

Undoubtedly the distinctive feature of probabilistic arguments in general, and of historical explanations in particular, is that their conclusions are not logically necessary consequences of their premises, even when all the empirically warranted but only tacitly employed assumptions are explicitly formulated. Accordingly, the actions historians succeed in explaining could not have been predicted (in the sense of being strictly deduced) from the information contained in the premises of the explanations by anyone who had access to just that information prior to those happenings; that is, the truth of the premises in a historical explanation is entirely compatible with the falsity of its conclusion. Thus, suppose that 'The individual x behaved in manner A on occasion t' is a schematic representation of some action for which a historian attempts to find a reason, and that the explanation finally given has the form 'The individual x was in circumstances C on the occasion t,' with the additional tacit premise 'For the most part, individuals under circumstances C behave in the manner A.' It is obvious that, since the premises do not suffice to establish the conclusion logically, the premises could be incontrovertibly sound even if the individual x had not acted in the manner A on the occasion t.

In point of fact, historians are rarely if ever in a position to state the *sufficient* conditions for the occurrence of the events they investigate. Most if not all historical explanations, like explanations of human conduct in general—and indeed, like many explanations of concrete events in the natural sciences—mention only some of the *indispensable* (or, as is commonly also said, *necessary*) conditions for those occurrences. We must, however, make explicit the sense in which this claim is to be understood, by indicating what is intended by the contrast between "sufficient" and "necessary" conditions in the present context and throughout this chapter. Suppose that an event A occurs when a certain set of conditions C is realized, so that the statement S_1, 'If C is realized, then A occurs,' is assumed to be true; but we shall not assume that the converse of S_1 (i.e., 'If A occurs, then C is realized') is true, in order to allow for the possibility that A will occur when some set of conditions C' different from C is satisfied. Suppose further that the condition C consists in the conjuncture of a number of factors, one of which is C_1 while the remaining ones are C_2; and assume that A does not occur when either C_1 alone or C_2 alone is realized, but that the statement S_2, 'If C_2 is realized, then A occurs if and only if C_1 is also realized,' is true. In virtue of statement S_1, and in consonance with standard usage in formal logic, the condition C is said to be a "sufficient condition" for the event A, and A to be a "necessary condition" for C; but in view of the further assumptions we made, it is evident that C_1 is *not* a necessary condition for A in this sense of 'necessary.' Nevertheless, since according to statement S_2, the event A will not occur when C_2 is realized but C_1 is not, although A will occur when both C_1 and C_2 are realized, C_1 is an indispensable condition for the occurrence of A if we assume that the condition C_2 is already satisfied. Relative to C_2, the condition C_1 can therefore be said to be a "contingently necessary condition" for A, in order to distinguish this sense of 'necessary condition' from the one specified in formal logic; and for purposes of easy reference, we shall use the label "absolutely necessary condition" for this latter standard sense. With the help of these distinctions, we can now state more clearly how the claim is to be construed that historical explanations do not mention the sufficient conditions for events, but only some of the "necessary conditions" for them: the quoted expression is to be understood in the sense of 'contingently necessary condition' rather than of 'absolutely necessary condition.' However, since there will be little occasion in the present chapter to refer to anything other than to contingently necessary conditions for events, we shall for the most part dispense with this longer phrase, and use the briefer 'necessary condition' instead.

Let us illustrate the contention that historical explanations in the main cite only some of the contingently necessary conditions for occurrences. In Maitland's account of why Elizabeth evaded an immediate decision on the Roman question, he described a complex background of events which he believed was the reason (and hence *one* necessary condition) for Elizabeth's conduct. However, the events Maitland explicitly cited were clearly not sufficient to bring about her action, so that many other circumstances, no less indispensable for her acting the way she did (such as that Elizabeth was sane in mind, or that she wanted to avoid civil war), are not mentioned in his account. Some of these additionally necessary conditions were perhaps not cited by Maitland because they seemed to him too obvious to need formal statement. But if there are less obvious ones that he failed to mention, it is undoubtedly because he did not know *all* the conditions in the absence of which the action he sought to explain would not have taken place. In consequence, explanations of particular occurrences (in the natural sciences as well as in the study of the human past) are frequently accepted only with various qualifications, a common one being the notorious *ceteris paribus* reservation that the conditions explicitly mentioned in an explanation can be assumed to account for some happening, provided that "other things are equal," where these "other things" are often unknown or are only hazarded.[8]

---

[8] The *ceteris paribus* clause is often tacitly employed even in highly developed branches of physics. For example, the path traversed by a bullet on a given occasion can be explained with the help of Newtonian theory, supplemented by instantial data concerning a number of items. The explanation of the bullet's trajectory mentions explicitly the latitude at which the gun is fired, the direction in which the gun is pointed, the muzzle velocity of the projectile, and the resistance of the air; but it is not likely to mention the position of the earth with respect to its own and other galactic systems. The explanation ignores these latter items because of the assumption, built into the Newtonian theory, that the mass of the projectile is constant and is independent not only of its velocity but also of its distance from other bodies. Until Ernst Mach's critique of Newtonian mechanics, it apparently did not occur to

The incompleteness of the premises when measured by the standards of valid deductive reasoning, and their formulation of necessary rather than sufficient conditions for the occurrence of events, are two generally acknowledged traits that explicate in part the sense in which historical explanations are "probabilistic." This partial explication of the characterization is all that is essential for the purposes of the present discussion. However, in order to suggest where some of the still unsettled issues lie when a more complete explication is attempted, we select for brief mention two interpretations out of the great many that have been proposed for the word 'probability' and its derivatives.

According to one of the oldest conceptions, the probability of a given statement h (or "hypothesis," to use the customary designation) relative to given premises or "evidence" e measures the *confidence* (or on some formulations the *intensity of belief*) that an individual x has in the truth of h when x has the information e, on the assumption that x is in some sense "rational" or "reasonable" in the credence he gives to beliefs. Since this interpretation is compatible with the assignment of different degrees of probability by different individuals to the same hypothesis relative to the same evidence, probability on this view is not a completely "objective" relational property of statements and has an uneliminable "subjective" component. This "subjective" interpretation was the dominant one for some two centuries, but it eventually lost favor with the majority of students of the subject because of various apparently inherent difficulties in it—among others, the technical difficulty of defining a quantitative measure for differences in subjective states of confidence.

More recently, however, this interpretation has been revived in improved versions, and under the label of "personalistic probability" it plays a prominent role in current developments of statistical decision theory. According to one of these versions, for example, the probability which an individual with evidence e in his possession should assign to the hypothesis h is defined in terms of the *odds* he would be willing to accept were he to bet on the truth of h and against its falsity, so that the probability will measure the risk the individual is prepared to take in adopting h and rejecting its denial. Thus suppose that in the light of given evidence an individual offers odds of 9 to 1 in favor of the truth of the statement 'There will be no snowfall in New York City next April'; that is, he agrees to pay nine dollars, should he be wrong in believing

physicists that the inertia of a body might be a function of its distance from all other bodies in the universe. (The assumption that this is so has come to be called "Mach's principle" and has received serious consideration in current physical cosmology.) Accordingly, although a projectile's distance from all other bodies obviously varies, the variation is normally not mentioned in explanations of a bullet's trajectory, and is tacitly subsumed under the *ceteris paribus* reservation. Cf. Olaf Helmer and Nicholas Rescher, "On the Epistemology of the Inexact Sciences," *Management Science*, Vol. 6 (1959), especially pp. 25-33.

the statement to be true, provided that he will receive one dollar should he be right. Then for this individual the probability of that statement relative to the evidence in his possession is 9/10.

On the assumption that this example is a paradigm for interpreting probabilistic arguments in general, the probabilistic structure of historical explanations can supposedly be explicated in an essentially similar manner. To illustrate such an explication, let us ignore various complications, and assume that a certain individual x does not know that Elizabeth "etceterated" herself when she proclaimed her title, but that he is given the information stated in Maitland's explanatory premises for her action. Assume further that x is then asked to bet whether, under the stated circumstances, Elizabeth would "etceterate" herself, and that he accepts odds of 7 to 3 in favor of her doing so. Then relative to the evidence available to him, the probability for x that Elizabeth would act in that manner is 7/10.

However, it is a safe estimate that most students today who employ statistical assumptions and statistical analyses in their inquiries subscribe in variant forms to a patently very different interpretation of the term 'probability.' On this alternative view, the term can be used significantly only in connection with classes containing repeated instances of given attributes (such as the attribute of being a male in the class of human births); and, according to a commonly held version of this interpretation, the probability of a given attribute P in a given class R is the relative frequency with which instances of P occur in R. For example, if in the first 100 cases in a series of human births 52 are male children, and if the ratio of males to the total number of births does not appreciably alter as the series becomes progressively longer, then the probability of a male child in this class of human births is 52/100. A degree of probability on this interpretation, in contrast to the personalistic one, is therefore a measure of a completely "objective" property, since this property is entirely independent of any human belief about the classes possessing the property.

But this interpretation requires a slight reformulation to make evident its possible relevance for explicating the sense in which historical explanations are probabilistic. The reformulation relies on a logical device that has already been discussed in connection with the instrumentalist account of scientific theories.[9] Given an argument with an instantial conclusion and a general statement as one of the premises, the procedure consists in eliminating the general statement from the premises, replacing it by a leading principle (or rule of inference), and then deducing the instantial conclusion *in accordance with* the leading principle from premises that are exclusively instantial. Thus, suppose that a historical

[9] Cf. Chapter 6 above, especially pages 138-39.

explanation can be represented by the simple schematic form: Most individuals under circumstances C behave in the manner A. The individual i on occasion t_0 was in the circumstances C; therefore (probably) the individual i on occasion t_0 behaved in the manner A, where 'C' and 'A' are constant predicates, 'i' designates a particular individual, and 't_0' a particular occasion. Let '$C_{x,t}$' be an abbreviation for the statement-form "The individual x on occasion t was in circumstances C,' '$A_{x,t}$,' for the statement-form "The individual x on occasion t behaved in the manner A,' and 'L' for the leading principle postulating that a statement of the form '$A_{x,t}$' is derivable from a statement of the form '$C_{x,t}$.' Now define a class R of arguments as follows: Each argument in R has a conclusion of the form '$A_{x,t}$' which is derived in accordance with L from the single premise of the form '$C_{x,t}$.' Assume that there is a definite relative frequency with which a *true* conclusion is derived from a *true* premise in R in accordance with L. This relative "truth frequency" is then by definition the probability that in R an argument with a true premise has a true conclusion; moreover, the above historical explanation is equivalent to an argument belonging to R. On this interpretation, accordingly, a historical explanation is probabilistic, in the sense that it corresponds to an argument belonging to a class of arguments in which the relative truth frequency is less than 1.

However, although this explication as well as the one based on the personalistic interpretation of 'probability' has a considerable following, neither has won general acceptance because each has features that many regard as grave faults. For example, according to most critics of personalistic probability, its chief weakness is that on this interpretation judgments of probability rest ultimately upon the variable idiosyncracies of human beings, and that in consequence no firm reasons can be given for preferring statements with high rather than low probabilities as conclusions of reliable explanations or as grounds for dependable predictions. On the other hand, the central difficulty commonly noted in the truth-frequency interpretation is that, since a frequency cannot be significantly ascribed to a single statement, it is strictly nonsensical to talk about the probability of a given hypothesis relative to given evidence. Accordingly, that interpretation is often held to be inherently unsuitable for explicating the sense in which a historical explanation of some particular action shows the action to have been probable. But, although this criticism of the truth-frequency conception is by no means fatal to the claims of its proponents, and the force of the criticism can be effectively turned, the discussion of this and related questions cannot be pursued here any further.[10]

[10] The literature on the foundations of probability is vast. For a general survey of alternative positions that have been taken, cf. Ernest Nagel, "Principles of the

2. We have thus far been discussing explanations for individual actions in terms of conditions whose durations are ignored. We must now examine the pattern exhibited by historical accounts of actions in terms of temporally extended circumstances.

Explanations of the latter sort usually take the form of narrations. Although the relations of dependence between the events they describe are often not explicitly formulated, the selection of events for sequential mention is undoubtedly based on the tacit assumption that some of these events are necessary conditions for some of the others. An example of such explanations will help make evident their structure. Let us therefore consider the account given by the historian G. M. Trevelyan of why the first Earl of Buckingham finally opposed the marriage of the young Prince Charles (afterwards Charles I) to the Spanish Infanta Maria, despite the fact that until early in 1623 Buckingham had been an enthusiastic supporter of the project. After noting that Buckingham had become impatient over the delay in the marriage plans and that he received James I's permission to go with Charles to Spain in order to fetch the princess to England, Trevelyan continues:

> They took ship secretly, galloped in disguise across France, and presented themselves in the astonished streets of Madrid. Charles, though he was not permitted by Spanish ideas of decorum to speak to the poor Princess, imagined that he had fallen in love at first sight. Without a thought for the public welfare, he offered to make every concession to English Catholicism, to repeal the Penal Laws, and to allow the education of his children in their mother's faith. The Spaniards, however, still lacked the guarantee that these promises would really be fulfilled, and still refused to evacuate the Palatinate [one of the objectives James hoped to achieve through the projected marriage]. Meanwhile a personal quarrel arose between Buckingham and the Spanish nation. The favorite [i.e., Buckingham] . . . observed neither Spanish etiquette nor common decency. The lordly hidalgos could not endure the liberties he took. . . . The English gentlemen, who soon came out to join their runaway rulers [i.e., Buckingham and Charles], laughed at the barren lands, the beggarly populations and the bad inns through which they

Theory of Probability," in *International Encyclopedia of Unified Science*, Chicago, 1939, Vol. 1, No. 6. For modern versions of the subjectivist interpretation, see Frank Ramsey, "Truth and Probability," in *The Foundations of Mathematics*, London, 1931; Bruno de Finetti, "La prévision: ses lois logiques, ses sources subjectives," *Annales de l'Institute Henri Poincaré*, Vol. 7 (1937); and, more recently, Leonard J. Savage, *The Foundations of Statistics*, New York, 1954, esp. Chaps. 3 and 4. For the relative frequency conception, see Richard von Mises, *Probability, Statistics and Truth*, New York, 1939; Hans Reichenbach, *Experience and Prediction*, Chicago, 1938; and for the truth-frequency version, Charles S. Peirce, *Collected Papers* (ed. by Charles Hartshorne and Paul Weiss), Cambridge, Mass., 1932, Vol. 2, pp. 415-77. For a still different approach to the subject, see John M. Keynes, *A Treatise on Probability*, London, 1921; and Rudolf Carnap, *Logical Foundations of Probability*, Chicago, 1950.

passed, and boasted of their England. They were not made welcome to Madrid, and fancied themselves pestered by the priests. . . . They began to hate the Spaniards and to dread the match. Buckingham was sensitive to the emotions of those immediately around him, and he soon imparted the change of his own feelings about Spain to the silent and sullen boy, whom he could always carry with him on every flood of short-lived passion.[11]

This narration can be viewed as describing a finite number of events or states of affairs, some of them occurring in consecutive order, the others more or less simultaneously with several of the latter; and each circumstance mentioned is ostensibly a contingently necessary condition for some later happening in the series. Thus, let 'c_i' be an abbreviation for a description of an occurrence, with subscripts indicating events in their temporal order and superscripts indicating approximately concurrent happenings. The following sequence is one way of listing the occurrences described in the narration, c_0 (Buckingham desired the marriage between Charles and the Infanta); c_1 (various circumstances, not mentioned in the above quotation, frustrated the early realization of his desire); c_2 (Buckingham became impatient over the delay); c_3 (Buckingham decided to achieve his aim by fetching the Infanta from Spain); c_4 (he obtained the consent of James I to go to Spain with Charles); c_5 (Buckingham sailed to France with Charles); c_6 (he rode across France to Spain with Charles); c_7 (Charles made promises concerning the treatment of Catholics in England); c_7' (Spain declined to move out of the Palatinate); c_7'' (Buckingham took questionable liberties in his behavior); c_8 (the Spaniards manifested antagonism toward Buckingham); c_8' (a number of Englishmen arrived in Spain to be with Buckingham and Charles); c_9 (the behavior of these Englishmen antagonized the Spaniards still further); c_{10} (these Englishmen began to hate the Spaniards and the prospect of the marriage between Charles and the Infanta); c_{11} (Buckingham was made aware of this hatred as well as of the Spanish antagonism); c_{12} (Buckingham changed his mind about the desirability of the marriage, and became an opponent of the plan).

Let us now make explicit, as a preliminary to characterizing the structure of Trevelyan's explanation, how some of the twelve items we have extracted from this narrative account are related (or fail to be related). In the first place, there appears to be no connection between c_0 and c_{12} (the action for which an explanation is being proposed), other than that the latter is the "opposite" of the former. It is difficult to imagine a reasonable generalization that would permit us, given c_0, to conclude that c_{12} would probably occur; and in any case no such generalization is

[11] George M. Trevelyan, *England Under the Stuarts*, New York and London, 1906, pp. 128-29.

known. Nevertheless, assuming for the sake of the argument that Trevelyan's narrative is factually sound, his account does show why the transition from the earlier to the later state did take place. He succeeded in showing this by intercalating a number of happenings between c_0 and c_{12} and thereby "filling in" the temporal gap between the initial and terminal states in a sequence. The problem is to determine in what way the introduction of these additional items into the account contributes to the explanation of the change in Buckingham's attitude.

In the second place, there is no generalization that would enable us to infer from c_0 the likelihood of c_1. On the contrary, on the basis of the information contained in Trevelyan's account, c_1 can only be viewed as an occurrence entirely extraneous to c_0 and must be accepted as just a "brute fact"—as an unexplained initial condition like c_0, although coming into existence later than c_0. The identical point can also be made about some of the other events mentioned in the narrative, e.g., about c_7 in its relation to the occurrences preceding it in the above series.

But thirdly, once c_0 and c_1 are given, we possess reliable grounds for expecting that c_2 will also take place if we accept as sound the assumption (let us designate it for future reference by '$L_{01,2}$') that in general men become impatient when they believe their plans are repeatedly frustrated by persons whom they dislike. Accordingly, we are able to account for c_2. Moreover, since the condition c_0 continues to be present throughout the existence of c_1, the two conditions can be regarded as simultaneously existing initial conditions for $L_{01,2}$, so that the explanation has the probabilistic structure with which we are already familiar. Analogous explanations can be given for other events in the above sequence, once the successively realized necessary conditions for them are assumed to have occurred, and if acceptable generalizations concerning those conditions and events are available. In particular, c_{12} can be explained in this manner if we suppose that c_{11} has already occurred as well as that there is some well-founded generalization $L_{11,12}$ about men's responses to conditions such as those present in c_{11}—for example, the generalization that most men, believing themselves and their own kind to be disliked by a foreign group, and sensing also the hatred with which members of their own kind reciprocate that dislike, themselves develop strongly antagonistic attitudes toward that group.

It should be noted, finally, that the number of happenings Trevelyan's account interpolates between c_0 and c_{12} could be enlarged as well as diminished. How large this number is depends on various considerations, among others, on the amount of detail a historian finds it appropriate to include in his story, perhaps for reasons of artistry or because of wanting to mention only what is "most important"; on the information about that past which is actually at his disposal; on the scope of his investigation and the level of analysis he adopts in pursuing it; on the tacit generalizations he assumes for explaining interpolated events; and on his conception of the evidence required for showing convincingly that some supposed relation of dependence does hold between the events he is examining. Some writers have argued that the appropriate model for satisfactory explanations of particular events in every domain of inquiry, and not only in human history, is that of a "continuous series of happenings." [12] However, if a temporally ordered series of happenings is to be mentioned at all, only a finite number of them can be mentioned; and in consequence, no actual explanations can illustrate the "continuous series" model, whether in human history or anywhere else. It is nevertheless true that, in historical explanations of the type we are now considering, historians attempt to "fill in" temporal gaps in their accounts of events by intercalating other happenings. But this fact does not show, as is sometimes claimed, that such historical explanations dispense with general assumptions in the manner we have been indicating. On the contrary, that fact is entirely congruous with the use of general assumptions by historians; and we have just canvassed some of the reasons why historians attempt such interpolations, as well as some of the considerations that control their selection of the happenings they interpolate.

Let us now characterize the logical structure of Trevelyan's explanation of why, by the autumn of 1623, Buckingham opposed the marriage of Prince Charles to the Infanta. His account is an example of what is commonly called a "genetic explanation" of some particular event or state of affairs—a type of explanation mentioned in Chapter 2 and frequent in biology (in ontogenetic analyses), historical geology, and other branches of natural sciences, and not only in human history. We shall therefore formulate the pattern of genetic explanations in general terms, without special reference to Trevelyan's narrative, although we shall continue to use with slight modifications the notation introduced in discussing that example.

A genetic explanation of a particular event or state of affairs c_t occurring at time t shows c_t to be the result of a series of occurrences whose initial term is some occurrence or state of affairs c_0 that existed before c_t.[13] Accordingly, the explanation involves reference to a series of events $c_0, c_1, \ldots, c_t, \ldots, c_k, c_k', c_k'', \ldots, c_t$. Some of the events may have come into existence more or less simultaneously (these are indicated by

[12] William Dray, *Laws and Explanation in History*, New York, 1957, Chap. 3, esp. pp. 66-72, 79-81.
[13] The reasons for selecting c_0 rather than some other occurrence in the past as the initial term of the series are usually mixed. They may depend on such things as the subject under inquiry, the level of analysis adopted, the information already in possession of the audience for whom the explanation is intended, or even the need for some convenient place at which to begin the explanation.

letters with the same subscripts but different superscripts) and may have overlapping durations; but most of them have come into existence at different times. Moreover, an event is presumably included for mention in the series only if it is an indispensable condition for the occurrence of some later event in the series.

The logical structure of a genetic explanation for a particular occurrence can therefore be characterized as follows: (a) Its premises belong to one or the other of two classes C and G. Each statement S_i in C is singular in form, and asserts that the event (or "condition") c_i has occurred. Although the statements in G are rarely formulated explicitly in genetic explanations, they are general in form and are usually statistical (or quasi-statistical) rather than strictly universal. These generalizations assert relations of dependence between various traits of the events c_i; for example, the generalization $L_{ij,k}$ might state that events analogous in certain respects to c_i and c_j are for the most part followed by events analogous to c_k. (b) The instantial statements (i.e., those in C) fall into two subclasses C_1 and C_2. For each statement in C_1 an explanation can be given which is either probabilistic or (more rarely) deductive, some of whose premises belong to C and the others to G, with the understanding that an instantial statement S_i is a premise in an explanation neither for c_i nor for any occurrence earlier than c_i. The instantial statement S_t (asserting the occurrence of the event for which the genetic explanation is being proposed) must obviously belong to C_1, since otherwise c_t would not be explained genetically. On the other hand, the subclass C_2 contains all statements of C which do not belong to C_1 and which cannot therefore be explained in the way those in C_1 are explained. Accordingly, statements in C_2 are those that formulate the initial conditions in the genetic explanation which must be accepted simply as *data*. C_2 must contain at least one statement, namely S_0, and in general it will contain many more. Indeed, it is a distinctive feature of genetic explanations that the statements of initial conditions in C_2 are fairly numerous, that the conditions they specify do not all come into existence simultaneously, and that for the most part those conditions cannot be stated in advance of their occurrence.

In short, a genetic explanation of a particular event is in general analyzable into a sequence of probabilistic explanations whose instantial premises refer to events that happen at different times rather than concurrently, and that are at best only some of the necessary conditions rather than a full complement of sufficient ones for the occurrences which those premises help to explain.

3. We turn finally to explanations of aggregative events, constituted out of the actions of many men, such as the emergence of some new

social institution, the increase in the population of a given country during a stated period, or the outbreak of a particular war. Occurrences of this sort, especially when they involve large numbers of human beings or have a considerable temporal spread, are commonly not the outcome of deliberate plan or concerted action; and they are frequently not even ends-in-view envisaged by any individual participant in them. In consequence, proposed explanations for them are far more controversial among historians than are accounts of individual actions. For the "factors" or "social forces" to which such collective events are attributed by different historians often vary considerably, and the sharp disagreements that continue to exist among unquestionably competent students over the adequacy of many of these accounts testify to the absence of well-established and generally accredited theories of social change. However, we shall not be concerned with substantive issues raised by explanations of this type, nor with the validity of any of the numerous "theories of historical causation" (i.e., the more or less clearly articulated assumptions about the determinants of social change) that frequently underlie such accounts. We shall be concerned exclusively with the abstract pattern exhibited by explanations of this type. But since limitations of space preclude the extended and detailed examination of several examples of such explanations, which alone could do justice to the intricacies of that pattern, we will be able to note only a few of its prominent features.

It is rarely possible to account for a collective event which possesses an appreciable degree of complexity by regarding it as an instance of some recurring *type* of event, and then exhibiting its dependence upon antecedently existing conditions in the light of some (tacit or explicit) generalization about events of that type. For example, historians do not attempt to explain the Protestant Reformation by thinking of it simply as a case of reformations-in-general (or even of the narrower class of religious-reformations-in-general) and arguing, in effect, that since reformations take place under certain conditions and since those conditions were realized in Germany in the sixteenth century, the Protestant Reformation was their outcome. For in the first place, on the assumption that such large-scale occurrences can be usefully classified as instances of various suitably described types, the number of known cases of a given type is usually quite small; and, since evidence is scarce for generalizations about the conditions under which events of a given type occur, reliable generalizations about events belonging to the various types are at best quite rare. But secondly, on the further assumption that several instances of a given type of large-scale collective event have in fact been realized, there are bound to be important differences between them; and, even if reliable generalizations about events of that type are available, the generalizations are not likely to be very useful in explaining the occurrence of a given instance of the type. For example, political revolutions have occurred repeatedly in the past, and the phenomenon has been extensively studied.[14] Although the American Revolution of 1775, the Chinese Revolution of 1911, and the Russian Revolution of 1917 are all instances of this phenomenon, they differ in significant ways in the circumstances under which they happened and in the course of their development; and the available generalizations about revolutions as a class of collective events are of little help in explaining why a revolution occurred in Russia in 1917.

Accordingly, critics of what has come to be called the "covering law" model of explanation (i.e., the view that satisfactory explanations of particular events are deductive in form, so that for an event to be explained it must be subsumed under a strictly universal law serving as a premise in the explanation) are undoubtedly correct in claiming that historical explanations of aggregative events do not exhibit this pattern.[15] Indeed, this claim remains sound if it is broadened to maintain that the form of such explanations is commonly not even probabilistic, in the sense that the event to be explained can be simply subsumed under an appropriate statistical generalization. Nevertheless, these admissions do not establish the further contention of critics of the covering law model that *no* general assumptions (whether explicit or implicit) enter into explanations of collective events. For if the admitted facts are assumed to establish this contention, a similar argument would require us to accept the conclusion that no physical laws are needed to explain the behavior of a particular steam locomotive, for example, since there are in fact no physical laws which are *specifically* about steam locomotives and under which the designated locomotive might be subsumed as a single unit. But this conclusion would be patently erroneous. To explain the behavior of a locomotive, it must be recognized to be a more or less integrated *system of component parts* (e.g., firebox, boiler, driving wheels, headlight, etc.), whose several operations can be accounted for only in terms of various laws about certain physical phenomena manifested by the components (e.g., various laws of mechanics, thermodynamics, optics, etc.), so that characteristics of the entire system (e.g., tractive force, speed of motion, efficiency of operation, etc.) may finally be exhibited as products of interactions between some of the parts.

It is indeed in a quite analogous way that historians explain an aggregative occurrence which is marked by considerable complexity. Historians cannot deal with such an event as a single whole, but must first

---

[14] Cf. Alfred Meusel, "Revolution and Counter-Revolution," in *Encyclopedia of the Social Sciences*, New York, 1934, Vol. 13, pp. 367-75; and Crane Brinton, *The Anatomy of Revolution*, Rev. ed., New York, 1952.

[15] William Dray, *op. cit.*, Chap. 2.

analyze it into a number of *constituent "parts"* or *"aspects."* The analysis is frequently undertaken in order to exhibit certain "global" characteristics of the inclusive event as the outcome of the particular combination of components which the analysis seeks to specify. The primary objective of the historian's task, however, is to show why those components were actually present; and he can achieve this aim only in the light of (usually tacit) general assumptions concerning some of the conditions under which those components presumably occur. In point of fact, even the analysis of a collective event is controlled in large measure by such general assumptions. First of all, the delimitation of the event itself—the selection of some of its features rather than others to describe it and thereby also to contrast it with earlier states of affairs out of which it presumably developed, and the adoption of some particular time or circumstance for fixing its supposed beginnings—depends in part on the historian's general conception of the "basic" variables in terms of which the event is to be understood. Secondly, the components a historian distinguishes in an event when he seeks to account in a piecemeal fashion for its occurrence are usually those whose "most important" determining conditions are specified by the generalizations he normally assumes about those components, so that these determinants are frequently the ones he tries to discover in some actual configuration of happenings that took place antecedently or concurrently with the collective event he is investigating. In short, generalizations of some sort appear as essentially in the premises of explanations for aggregative occurrences as they do in explanations of individual actions.

The specific manner in which a collective event is conceptually broken up into manageable "parts" or "aspects" thus varies with the preconceptions a historian brings to its study, as well as with the magnitude of the event and the circumstances present in it. It is nevertheless convenient to distinguish between two broad classes of explanations for events that are analyzed into components: those which deal with events that are taken to have "abrupt" beginnings, such as the Protestant Reformation, the American Civil War, or the fall of the German Weimar Republic; and those which are concerned with events that supposedly have no sharply definable beginnings but are "continuous" with preceding states of affairs, such as the feudalization of Europe, the development of modern capitalism, or the Industrial Revolution. However, our purposes will be served if we discuss only explanations of the first kind.

In explanations of this sort, historians often distinguish between the "immediate" (or "precipitating") cause of an event, and its "underlying" (or "basic") causes. An immediate cause is usually some occurrence of relatively short duration that initiates the collective one; it may be a "natural event" (e.g., a cataclysmic earthquake), an individual action (the deed of an assassin), or a collective happening (a military defeat). The underlying causes to which historians frequently refer are commonly designated, in obviously metaphorical terms, as "social forces" and are constituted out of relatively enduring modes of action as well as of less normal forms of behavior which are manifested by various groups of anonymous individuals. The social forces often mentioned in historical explanations are such things as the constraints imposed by political structures, the influence of economic interests and institutions, the controls exercised by organized religions, the coercions stemming from military activities and establishments, and the operation of various beliefs, ideas, and aspirations as manifested in the attitudes and activities of those who entertain them.

Let us examine briefly a typical example of explanations of this sort. In his account of the fall of the Weimar Republic, the British historian Barraclough finds that several social classes, identified partly in terms of their divergent aspirations, played capital roles in that event: the military officer class and the Junkers, devoted to the ideals of the Prussian landed aristocracy; the economic groups represented by the industrial and financial magnates; the industrial workers inspired by socialistic objectives; the middle classes of tradesmen and white-collar workers, and the peasants, separated from organized labor by traditionally acquired political and religious attitudes; and the counterparts of German industrialists and financiers in the victorious Allied countries, opponents of socialism abroad as well as at home.

Barraclough describes the alignment of these groups as it existed before the collapse of the Weimar Republic, in some cases tracing the reasons for the alignment back to the Thirty Years' War; but the immediately relevant part of his account begins with the close of World War I. According to him, as far as the eventual fate of the Weimar Republic was concerned, "the die was, as early as 1919, already cast" because the new Constitution did not supplement its liberal political forms with indispensable redistributions of economic and political power. Thus, despite the Reichswehr's defeat in the war, the class it represented continued to have a controlling voice in German politics, though it was overtly hostile to the liberal institutions of the new order. Organized labor did not give the Weimar Constitution its undivided allegiance because it saw no hope of realizing its own basic aims within the framework of that system. The power of the great industrial interests was left uncurbed, partly out of fear of foreign intervention if socialistic measures were adopted. The Weimar Republic did have the support of the middle classes; but they exercised no effective power and saw no reasons for allying themselves with left-wing movements. This unstable equilibrium of social forces was finally disturbed by the economic crisis of

1929. Already impoverished by the inflation that had occurred seven years earlier, the middle classes lost confidence in the Weimar Republic and looked to National Socialism to improve their lot. But the army, the Junkers, and the industrialists, counting on the compliance of their counterparts in the Allied countries, also saw in Hitler an opportunity to eliminate decisively the threat of Communism in Germany, and with Hitler as their instrument to establish their own unchallenged rule. Hitler's rise to power, Barraclough concludes, "was the work of Hindenburg representing the army, of Papen representing the aristocracy, of Hugenberg, the press-lord, and of Thyssen representing the industrialists of the Ruhr." [16]

With this example in mind, the pattern of this type of explanation can be represented schematically as follows: Let E_t be a collective event whose beginning is fixed at some time t; and suppose that, when E_t is analyzed, it is found to have had as components a set of social forces $F_1, F_2, \ldots, F_i, \ldots, F_n$ interacting in a manner R_t. Suppose further that analysis reveals those forces to have been related at some time s shortly before t in the manner R_s (which we shall designate as a state of "equilibrium"). The task, as usually conceived, of explaining why E_t occurred thus has two parts: Why was the alignment of forces changed from R_s to R_t? And why were those forces in the alignment R_s at the time s? The first question is usually answered in terms of the occurrence at time t of a precipitating event e_t which produced some effect upon one or more of the component forces F_i, and thereby upset the equilibrium R_s.[17] It will be clear, however, that this answer takes for granted

[16] G. Barraclough, *The Origins of Modern Germany*, New York, 1957, p. 450.

[17] Some historians tend to minimize the role of precipitating events. According to one recent writer, for example, ". . . the 'immediate cause' is not really a cause; it is merely the point in a chain of events, trends, influences, and forces at which the effect begins to become visible. It is the precipitating event that serves as the dropping of a match in a combustible pile or the tripping of a hammer on an explosive. As such it is a good lead toward the antecedents that may be more satisfactorily described as 'causes.' The more satisfying line of inquiry is not: What might have happened had this 'accident' not occurred? It is rather: How did circumstances get to such a pass? How could a mere accident like the late delivery of a message or the taking of a wrong turn in a parade [the references here are to Louis XVI's delay in notifying the National Assembly that no meeting was to be held on June 20, 1789, and to the wrong turning taken by Franz Ferdinand's chauffeur at Sarajevo, respectively] lead to a world revolution or a world war? When that line is followed, the answer to 'what might have been' usually becomes easy; one often becomes satisfied that, if not this accident now, some other later would have had the same effect, for the determining trends, influences and factors were still operating. . . ."— Louis Gottschalk, *Understanding History*, New York, 1950, pp. 210-11. Although this view is understandable as a reaction against historical explanations which attempt to account for the past exclusively in terms of precipitating events, it in effect throws out the baby with the bath water. If it is assumed that an event E would not have happened at time t unless some precipitating event e had happened at that time, it is patently absurd to claim that in the absence of e the event E

some generalization about the likely effects of events like e_t upon circumstances like those subsumed under the social forces F_i. For example, according to Barraclough, the economic crisis of 1929 destroyed the allegiance of the German middle classes to the Weimar Republic. The tacit assumption underlying this claim seems to be that, when hard-pressed by economic disasters not of their own doing but imputed to the operation of a social system, men generally become disaffected with that system especially if it holds out no promise of early improvement. The second part of the task of explaining the occurrence of E_t requires that an account be given for the development of each of the social forces from some earlier stage in the past to its state at the time s. Such accounts commonly contain several steps similar to the one required in completing the first part of the task, but, despite this complication, each of these accounts has the form of a genetic explanation. In short, explanations of aggregative events are constituted out of strands of subordinate accounts whose patterns are those of probabilistic and genetic explanations.

Collective events that are appreciably complex are thus not usually explained by subsuming them as single units under abstract concepts appearing in generalizations. In consequence, it is often claimed not only that historical explanations of such events (especially in human history) differ basically in logical pattern from explanations in the generalizing sciences, but also that the very *concepts* employed in human history have a logical structure radically unlike the structure of "general concepts" in the generalizing sciences. In particular, general concepts conform to the familiar logical principle according to which the extensions of terms vary inversely with their intensions. For example, the general terms 'living organisms,' 'animal,' and 'man' are arranged in order of decreasing extension, so that the class of things designated by a term includes the class designated by a term following it; but the intensions of these terms are increasing, so that the attributes connoted by a term include the attributes connoted by a preceding term; thus, although the class of men is included in the class of animals, the defining attributes of the term 'animal' are only a part of the defining attributes of 'man.' On the other hand, this principle is allegedly not satisfied by the "individual concepts"

would nevertheless have happened at time t. To be sure, if E had not happened at time t, it might have happened at some *other* time; but since without e the "still operating" conditions are not sufficient to produce E, then for E to happen at some other time some other precipitating event e' must happen—and it is quite possible that none does happen. Were the quoted claim sound, it would be folly to prevent an individual with a lighted match from dropping it into a combustible pile, since according to the reasoning on which the claim is based a combustion will take place anyway. What the writer apparently has in mind is a distinction between the "more important" and the "less important" causes of events, a distinction that will occupy us in the next section of this chapter.

employed in historical study, since the more inclusive the event designated by such a term, the "richer" and "fuller" is its meaning. Thus, the term 'French Enlightenment' has been said to have a wider extension than the term 'life of Voltaire' but also to possess a fuller intension.[18]

However, neither of these claims stands up under examination. We have already noted that applied natural sciences, such as the engineering sciences concerned with automotive mechanisms, commonly account for the operations of complex systems by first analyzing such systems into components, so that the pattern of these explanations is not entirely unlike the pattern of many historical explanations for complex collective events. Moreover, since explanations in physical cosmogony and evolutionary biology (to mention but two theoretical sciences) are partly genetic in character, the over-all structure of these explanations is not distinguishable from the pattern of the historical accounts we have been discussing. Accordingly, although this pattern may occur more frequently in history than in other branches of inquiry, it is not a total stranger to the generalizing sciences.

But in any event, it is demonstrably erroneous to claim that the principle of the inverse variation of extension with intension does not hold for the "individual concepts" of historical study. For the claim confuses two quite different relations: the relation of *inclusion* between the extensions of two terms with the relation of *whole-to-part* between an instance of a term and some component of that instance. Thus the extension of the term 'carburetor' is *not* included in the extension of the term 'vehicle' (since carburetors are not vehicles), although a carburetor may be a *part* of an automobile which *is* a vehicle (and in consequence a *part* of an *instance* of the extension of the term 'vehicle'); and, although the term 'vehicle' may have a "richer meaning" than the term 'carburetor,' this greater richness does no violence to the above logical principle. Similarly, the extension of the term 'life of Voltaire' is *not* included in the extension of the term 'French Enlightenment,' despite the fact that the life of Voltaire is *part* of the French Enlightenment (so that the term 'French Enlightenment' is doubtless "richer in meaning" than the term 'life of Voltaire'). Accordingly, the logical principle under discussion is simply not applicable to these terms, any more than it is to the terms 'carburetor' and 'vehicle'; and, in consequence, neither pair of terms can count as a possible exception to the principle. In short, there appears to be no foundation for the contention that historical inquiry into the human past differs radically from the generalizing natural or social sciences, in respect to either the logical patterns of its explanations or the logical structures of its concepts.

[18] Heinrich Rickert, *Die Grenzen der naturwissenschaftlichen Begriffsbildung*, 4th ed., Tübingen, 1921, pp. 294-95.

III.   *Recurrent Issues in Historical Inquiry*

However, even if historical explanations possess no absolutely unique logical features, and even though the methodological problems of history have their counterparts in other branches of inquiry, some of these problems generate difficulties and disagreements that are particularly acute in the search for reliable explanations of past human events. We shall examine three such recurrent problems: the import of the selective character of historical inquiry for the achievement of historical objectivity; the rationale for assigning orders of relative importance to causal factors; and the role and basis of contrary-to-fact judgments about the past.

1.   It is a platitude that research in history as in other areas of science selects and abstracts from the concrete subject matter of inquiry, and that however detailed a historical discourse may be it is never an exhaustive account of what actually happened. Curiously enough, although natural scientists have rarely been agitated by parallels in their own branches of study to these obvious features of historical inquiry, the selective character of historical research continues to be a major reason historians give for the sharp contrast they frequently draw between other disciplines and the study of the human past, as well as the chief support for the skepticism many of them profess concerning the possibility of achieving "objective" historical explanations. We have already canvassed most of the issues raised by these skeptical claims, when we discussed (in Chapter 13) the obstacles to a value-free social science. We will not traverse the same ground, and will examine briefly only some alleged difficulties in the way of establishing well-founded explanations which have been cited primarily in connection with historical inquiry.

a.   Historians are sometimes greatly troubled by the circumstance that they cannot hope to render the "full reality" of what has transpired or to state the total set of causal conditions for what has happened, because a historical account of an event can consider only a few of its aspects and must stop at some point in tracing its antecedents. According to Charles A. Beard, for example,

> The over-arching question is: What can we know about this all-embracing totality called history? Millions, billions, of historical facts have been established beyond debate by the researches of competent scholars. Libraries are crowded with them. . . . But can we grasp this totality including all relations, know it, formulate its laws, reduce it to an exact science, or any kind of science? If every particular theme of human affairs deals with only an aspect and that aspect is conditioned by other aspects, we must ask our-

selves this question, unless we deliberately decide to be dogmatic, to fix arbitrary boundaries to discussion, to be false to our own knowledge.

But since the human mind cannot encompass the seamless continuum of the past, Beard maintained that "picking 'events' and 'causes' out of the totality is an act of human will, for some purpose arising in human conceptions of values and interests," so that every proposed explanation of a past occurrence bears the mark of arbitrariness and subjectivity.[19]

The basic question raised by this claim is whether an account of some past event becomes inevitably distorted and erroneous by the mere fact that the historian addresses himself to a limited problem and attempts to solve it by investigations that do not deal with the entire past. However, the contention that the answer to the quesion is affirmative entails the view that we cannot have competent knowledge of anything unless we know everything; it is a corollary to the philosophical doctrine of the "internality" of all relations.[20] Were this doctrine sound, every historical account that could be constructed by a finite intelligence would have to be considered a necessarily mutilated version of what actually happened; indeed, all science and all analytical discourse would have to be condemned in an identical manner. But the claim that all historical explanations are inherently arbitrary and subjective is intelligible only on the assumption that *knowledge* of a subject matter must be *identical with* that subject matter or must *reproduce it* in some fashion; and this assumption, as well as the claim accompanying it, must be rejected as absurd. Thus, a map cannot be sensibly characterized as a distorted version of the region it represents, merely because the map does not coincide with the region or does not mention every item that may actually exist in that region; on the contrary, a "map" which was drawn to scale and which omitted nothing would be a monstrosity utterly without purpose.

Similarly, knowledge as the outcome of historical inquiry is not inadequate merely because it is not about everything in the past, or because it answers only the specific question about the past that initiated the inquiry but fails to answer every other problem concerning what happened. However, *all* discursive knowledge is the product of research instituted for the sake of resolving determinate (and hence delimited) questions. It is therefore not only an unrealizable but also an absurd

---

[19] From *The Discussion of Human Affairs* by C. A. Beard, The Macmillan Co., 1936, with the permission of Wm. Beard and Mrs. Miriam B. Vagts. Pp. 79-81.

[20] This is the doctrine that the attributes or relations a thing has (e.g., in the case of a human being, of being married, or of having five dollars in his pocket on a certain day) are logically necessitated by the attributes or relations of all other things, so that everything is relevant in some degree to everything else. For a critique of the doctrine, cf. Ernest Nagel, *Sovereign Reason*, Glencoe, Ill., 1954, Chap. 15.

ideal of objectivity that requires a historical explanation to be characterized as "subjective" if the explanation fails to state "all that has been
said, done, and thought by human beings on the planet since humanity
began its long career." [21] Accordingly, the bare fact that historical inquiries deal with selected aspects of the past, or that historical explanations do not regard everything as causally relevant to everything else, is
not a cogent reason for skepticism concerning the possibility of objectively warranted human history.

b.   A somewhat different though related misconception also underlies these skeptical doubts. It is sometimes tacitly assumed that, since
every causal condition for an event has its own causal conditions, the
event is never properly explained unless the terms in the entire regressive
(and theoretically endless) series of causal conditions are also explained.
For example, it has been argued that

> A Baptist sermon in Atlanta, if we seek to explain it, takes us back through
> the Protestant Reformation to Galilee—and far beyond in the dim origins of
> civilization. We can, if we choose, stop at any point along the line of rela-
> tions, but that is an arbitrary act of will and does violence to the quest for
> truth in the matter.[22]

But is violence being done to the truth by stopping at some arbitrary
point in the regressive series? Is B not a cause of A, merely because C
is a cause of B? When the position of a planet at a stated time is explained in terms of gravitational theory and information about the initial
condition of the solar system at some previous time, is the explanation
unsatisfactory on the ground that these initial conditions have not themselves been explained and that they are the outcome of still earlier configurations of the solar system? Is there a fault in explaining Boyle's law
in terms of the kinetic theory of gases, because this theory is not itself
explained? Is a proof of the Pythagorean theorem suspect because the

[21] Charles A. Beard, *op. cit.*, p. 69.

[22] Charles A. Beard, *op. cit.*, p. 68. There is irony in the fact that Beard's theoretical skepticism did not prevent him as a practicing historian from offering vigorously stated explanations for numerous historical occurrences. His confident explanation of the American Civil War as the culmination of a struggle between two incompatible economic systems is well known. He was equally outspoken in claiming that
the Dred Scott Decision by the United States Supreme Court in 1857 was the product of the political ambitions of Justice McLean, one of the anti-slavery judges, who
planned to issue a dissenting opinion favoring the restriction of slavery with a view
to winning the Republican nomination for the presidency. "Beyond doubt," Beard
concludes his account of the matter, "the stiff insistence of Justice McLean upon the
promulgation of his views at all costs was a leading factor in forcing the pro-slavery
judges to come out against the validity of the Missouri Compromise."—Charles A.
and Mary R. Beard, *The Rise of American Civilization*, New York, 1930, Vol. 2,
p. 19, quoted by kind permission of the publishers, The Macmillan Company.

starting point of the demonstration is a set of assumptions that are not proved in turn? These are rhetorical questions, and the answers to all of them are obviously and uniformly in the negative. To suppose that no explanation is ultimately satisfactory unless all the elements out of which it is constructed are also explained, is to subscribe to the confusion underlying romantic philosophies of irrationalism, which despair of the capacity of discursive human intelligence to discover the "real" nature of things because scientific inquiry cannot answer the question why something exists rather than nothing at all.

Moreover, precisely what is it in connection with the Baptist sermon in Atlanta for which an explanation is to be sought? Is it why a given individual delivered a sermon at a stated time and occasion, or why he chose a particular text and theme, or why that occasion happened to arise, or why Baptists flourish in Atlanta, or why the Baptists developed as a Protestant sect, or why the Protestant Reformation occurred, or why Christianity arose in antiquity, or why civilized life originated? These are all distinct questions; and an adequate answer to one of them does not answer any of the others, and is not even remotely relevant to the problems raised by some of the others. Accordingly, once the event to be accounted for is made reasonably definite, it is self-contradictory to maintain that a historian's explanation of the event is objectively warranted only if he first completes a series of explanations each term of which is an explanation for the data assumed in the previous one. On the other hand, the fact that one problem may suggest another, and so lead to a possibly endless series of new inquiries and further explanations, testifies simply to the vast complexity of a given subject matter and to the progressive character of the scientific enterprise. That fact does not support the contention that, unless such a series is completed, every proposed solution to a given problem is necessarily a mutilation of the truth.

c. One further facet of the skeptical argument for the inherent subjectivity (or "relativity") of human history requires brief notice. According to one influential form of this argument, history is but an artificial extension of memory and has many of the defects of that human faculty. In particular, the things an individual remembers are not only a fragment of the things he has lived through, but are indelibly colored by his own self-image and his shifting concerns at different times. Similarly, so the claim continues, history as social memory is radically affected by the needs society has for preserving its traditions and ideals as envisaged in the light of current problems, and for anticipating what the future may bring in view of the remembered past. In consequence, so Carl Becker maintained, living history, the ideal series of events that we affirm and hold in memory, since it is so intimately associated with what we are doing and with what we hope to do, cannot be precisely the same for all at any given time, or the same for one generation as for another.

For, although historians may achieve objectively grounded knowledge concerning relatively "simple facts," the establishment of such facts is only a small part of their task.[23] Indeed, "generally speaking, the more simple a historical fact is, the more clear and definite and provable it is, the less use it is to us in and for itself." On the contrary, historians seek to "interpret" such facts; and they thereby continue to perform the function exercised by bards and storytellers in former days

> to enlarge and enrich the specious present common to us all to the end that "society" (the tribe, the nation, or all mankind) may judge of what it is doing in the light of what it has done and what it hopes to do.

Accordingly, so the argument concludes, since facts do not proclaim their own meanings but have meanings imposed on them by historians, and since in the nature of the case a historian's imputation of significance to a particular occurrence in the past cannot be tested by examining repeated "enactments" of it under varying conditions, an uneliminable personal or subjective element enters into every historical reconstruction.[24]

However, although it is beyond dispute that biases generated by various commitments (social, religious, ideological, moral, or ethnic) often color the reconstructions even of historians whose competence and personal integrity are unquestionable,[25] the argument hardly warrants a

[23] When modern techniques of internal and external criticism can be employed for assessing the authenticity and reliability of various kinds of testimony, there is usually almost complete agreement among historians on such "simple facts" as *whether* some alleged event actually did take place (e.g., whether a chiliastic panic occurred in Europe on the eve of the year 1000 A.D.); *when* it really happened (e.g., whether the Declaration of Independence was signed on August 2, 1776, rather than on July 4 of that year); *who* were the individuals participating in it (e.g., whether King George IV of England was present at the battle of Waterloo), and so on.

[24] Carl Becker, "Everyman His Own Historian," *American Historical Review*, Vol. 37 (1931-32), pp. 227-32; see also the same author's "What Are Historical Facts?" *Western Political Quarterly*, Vol. 8, (1955), pp. 327-40. These skeptical doubts did not prevent Becker from concluding his argument in the latter essay with the observation that, unlike historical research, developments in the natural sciences have had a profound influence upon social life. "A hundred years of scientific research has transformed the conditions of life. How it has done this is known to all." It is obviously difficult for a professional historian to practice the skepticism about historical knowledge that he may formally profess.

[25] Deliberate falsification in the interest of some favored cause is today rare among professional historians in nonauthoritarian countries; and uncritical acceptance of demonstrably erroneous or inadequately supported statements of alleged fact is not the most frequent way in which historians reveal their partisan allegiances. Indeed, it is not inconceivable that each of two historical accounts of the same

wholesale skepticism toward the possibility of historical objectivity. There is no evidence, in the first place, to support the claim that the current problems of a society invariably determine the character of a historian's inquiry into specific questions about the past. Indeed, historians sometimes propose quite similar accounts for a given occurrence, despite their being members of different social groups or having different personal commitments; and inversely they sometimes advance quite dissimilar explanations, despite sharing common preconceptions.[26] Moreover, even if the social climate in which historians work did have a decisive influence upon their investigations, the prospects for objectively based conclusions in historical research would not therefore be necessarily hopeless, for the pursuit of objective historical inquiry might very well be one of the ideals prized and fostered by a society and controlling a historian's researches.

In the second place, although explanations historians propose for a given event frequently differ, they are not necessarily incompatible. As was noted earlier in the present chapter, historical explanations do not state the sufficient conditions for a given happening. Alternative accounts of some past occurrence may in consequence differ (and often do differ) only in mentioning different necessary conditions for that event, so that the alternative explanations supplement rather than contradict one another. To be sure, historians frequently disagree on the relative "importance" they assign to the various factors they may cite as necessary conditions for an occurrence; but, as we shall see, although serious difficulties exist in connection with such judgments, the difficulties are not insuperable in principle.

But thirdly, the obvious *logical* impossibility of re-enacting a given happening in the past does not prove that historical explanations for it are not testable, and are therefore incapable of being objectively grounded. Were this argument sound, a strictly analogous argument would prove that no decisions by courts of law concerning the guilt of defendants accused of some deed could possibly be based on objective evidence. However, although legal trials sometimes do result in mis-

period could contain only indisputably correct statements about matters of particular (or "simple") fact but that each would nevertheless be marked by a distinctive bias. For two such accounts might differ in what they mention or fail to mention, in the way they juxtapose the events both of them report, or in the emphases they place upon various factors both admit were operative; and, in consequence, one of the accounts might in effect be an argument for a conception of the goals and limits of human endeavor in opposition to the conception defended by the other account.

[26] For a discussion of different approaches that historians have followed in dealing with the American Civil War, see Howard K. Beale, "What Historians Have Said About the Causes of the Civil War," in *Theory and Practice in Historical Study,* Social Science Research Council Bulletin No. 54 (1946), pp. 55-102.

taken decisions, it would be an absurd exaggeration to claim that every litigation terminates in a miscarriage of justice, or even that the correctness of a court's conclusion is just a matter of chance. As has already been indicated (in Chapter 13), there are other techniques than that of overt experimental manipulation for obtaining reliable factual knowledge.

It must be admitted, finally, that history is often practiced as a fine art, comparable in some respects to poetry, and that historical reconstructions are frequently designed not simply to communicate knowledge but to portray in dramatic form men's past actions in order to arouse and fortify active sympathy for certain human qualities and aspirations. Nevertheless, the deliberately intended moral flavor of a historical essay is not inherently incompatible with its being an adequate objective account of the events it discusses. Natural scientists are sometimes similarly motivated by moral and aesthetic aims, and the moral passion and literary artistry with which some of them write about achievements in their domains of inquiry (e.g., Galileo in physics or D'Arcy Thompson in recent years in biology) do not automatically impair the objectively warranted content of their expositions.

In short, none of the considerations that have been mentioned justifies an unqualified skepticism concerning the possibility of reliable historical knowledge.

2. Although the goal of scientific explanation is sometimes defined as the discovery of the necessary and sufficient conditions for the occurrence of phenomena, we have had repeated occasions for noting that this ideal is rarely achieved even in the most highly developed branches of natural science. Moreover, historical inquiry is perhaps not even tacitly directed toward this goal, and is in any case farther removed from it than are the physical and biological sciences. In their normal research activities as distinct from what they sometimes say, historians appear to be unperturbed by the patent fact that their explanations never state anything but some of the indispensable conditions for the happenings they investigate. They may acknowledge their ignorance of the sufficient conditions by prefixing the *ceteris paribus* clause to their explanations, but their efforts go into specifying a partial rather than the complete set of determinants of some occurrence and into identifying in this partial set those they judge to be the "most important," "main," "primary," "chief," or "principal" factors. For example, according to one historian the "main cause" of America's entrance into World War I was Germany's adoption of unrestricted submarine warfare; and, although other "contributing" factors are also mentioned as having played important parts, it is not assumed that the factors cited exhaust the determinants of the event.

Such "weighting" of causal factors in respect to their "relative importance" is frequently dismissed as essentially "arbitrary" or even as "meaningless," either on the ground that there is no warrant for selecting one occurrence as the cause of a given event rather than some happening before that occurrence (for example, since unrestricted submarine warfare was Germany's response to the British blockade, this latter occurrence is sometimes said to have been as much the cause of America's entrance into the war as was the former), or on the ground that no verifiable sense can be attached to such characterizations as "most important" or "principal" in connection with causal factors. It must be admitted that the natural sciences do not seem to have any need for ascribing degrees of relative importance to the causal variables entering into their explanations; and it is tempting to deny peremptorily the possibility that such grading of variables has an objective basis, on the ground that if a phenomenon occurs only when certain conditions are realized then all these conditions are equally essential, so that no one of the conditions can meaningfully be said to be "more basic" than the others. Moreover, it must also be acknowledged that most historians do not appear to associate any definite sense with their statements about the relative importance of various causal factors, and that such statements often have only a rhetorical force but no empirically verifiable content.

Nevertheless, such statements are found not only in the writings of historians but also in the publications of other students of human affairs as well as in the language men employ about ordinary affairs of life. For example, social scientists claim that broken homes are a more important cause of juvenile delinquency than is poverty, or that the lack of a trained labor force is a more fundamental reason for the backward state of an economy than is the lack of natural resources; and parents will sometimes argue that overcrowded classrooms are the chief cause of their children's poor performance in school. Something is apparently intended by statements of this sort, even if it is usually not clear what is meant by them. Although most individuals who make such statements might perhaps agree that the *truth* of their statements is often debatable, they would probably reject the suggestion that the statements are *meaningless* and that whenever they assert a statement of this sort they are invariably uttering nonsense.

We must therefore attempt to make explicit what is intended by such statements. However, statements ascribing an order of relative importance to the determining factors of social phenomena appear to be associated with a variety of meanings, so that several distinct senses of "more important" must be distinguished. To this end, let us assume that A and B are two such factors, each specified with reasonable care and clarity, upon which a phenomenon C depends in some manner; and let us consider a few of the possible meanings that seem to be frequently expressed by statements having the form 'A is a more important (or basic, or fundamental) determinant of C than is B.'

a.   Suppose first that A and B are both contingently necessary conditions for the occurrence of C, leaving it open whether or not their joint presence is sufficient for producing C. Let us assume further that when "other things are equal" variations in A (with consequent variations in C) take place quite frequently (and are perhaps beyond actual control), but that variations in B occur so infrequently (or can be so effectively controlled) that they can be ignored for all practical purposes. This specifies one sense in which A is sometimes said to be more important than B as a determinant of C. Thus, suppose that a strong dislike of foreigners and an acute need for additional economic markets are both necessary conditions for the adoption of an imperialist foreign policy by an industrial nation; and suppose also that xenophobia in the country varies little if at all during relatively short periods, whereas the need for foreign markets steadily increases. In this first sense of 'more important,' the need for additional economic markets is a more important cause of imperialism than is dislike of foreigners. Accordingly, if a certain country is found to have embarked upon a policy of imperialist aggression at a specified time, and if investigation shows that before this event there had been no marked change in xenophobic attitudes of its citizens but that recurrent overproduction in a number of its industries had produced a growing demand for new markets, a historian might claim that of these two factors the latter had been the more important (in this first sense) in bringing about the adoption of the imperialist policy.

b.   A second sense of 'more important' is a bit more involved. Assume once more that A and B are both necessary for the occurrence of C. But suppose that there is some way of "measuring" the variations in each of the variables A, B, and C—at least in the limited sense that, although the magnitudes of changes in one variable may not be comparable with the magnitudes of changes in the other variables, the changes in any one of the variables can be compared. Let us assume further that a greater proportional change in C is produced by any given proportional change in A than by an equal proportional change in B. In consequence, A might be assigned a higher degree of importance as a determinant of C than would B. For example, suppose that an adequate supply of coal as well as a trained labor force is indispensable for industrial productivity; but suppose that a 10 per cent increase in the trained labor force yields a considerably larger volume of goods produced (as measured by some convenient index) than is obtained by a 10 per cent increase in the coal supply. Accordingly, in this second sense of 'more important,' the availability of a trained labor force would be a more important determinant of industrial productivity than the availability of coal.

c. Suppose next that *A* is a contingently necessary condition for *C*, and that, although *B* is not, it nevertheless belongs to a set *K* of mutually independent factors (*B, B₁, . . . , Bₙ*) such that the presence of some one member of *K* is a necessary condition for *C*. Suppose also that the various members of *K* occur with about the same frequency, the frequency of occurrence being in each case considerably greater than the frequency with which *A* is present. Accordingly, since the frequency with which some member of *K* occurs is greater than the frequency of *B*'s occurrence, then, even if *B* should not be present when *A* is present, the necessary conditions for *C* may nonetheless be realized because of the presence of some other member of *K*. These stipulations specify a sense of 'more important' that is perhaps most frequently intended when *A* is said to be more important than *B*. Thus, suppose that a necessary condition *A* for emigration from one country to another is discontent with political or economic conditions in the mother country, and that a further necessary condition is the occurrence of some one "precipitating" event (such as loss of employment, receiving promising news about brighter prospects in the foreign country, acquisition of funds to defray the cost of the journey, and the like), where the likelihood that one or the other of these alternative precipitating events occurs is high and is greater than the likelihood that a particular one of them occurs, for example, that a prospective emigrant acquires unexpected funds. On these assumptions, political or economic discontent would be a more important cause of emigration than acquisition of funds for travel. It is perhaps in the present sense that Germany's adoption of unrestricted submarine warfare is alleged to be the "main cause" of America's entrance into World War I.

d. There is a fourth sense of 'more important,' analogous to the first sense listed but distinct from it. Suppose that the joint presence of *A* and *B* is not a necessary condition for the occurrence of *C*, but that *C* occurs either when *A* is present conjointly with *X* or when *B* is present conjointly with *Y*, where *X* and *Y* are otherwise unspecified determining factors; and suppose also that *A* in conjunction with *X* occurs much more frequently than does *B* in conjunction with *Y*. In that case, *A* might again be said to be a more important determinant of *C* than is *B*. Thus, suppose that automobile accidents happen either because of the negligence of motorists or because of mechanical failures in relevant parts of automobiles, and also that the frequency with which such mechanical failures cause accidents is very much less than the frequency with which negligence is responsible for them. In that case, negligence would be a more important cause of automobile accidents (in this third sense of the phrase) than would mechanical failures. Using analogous assumptions, a historian might similarly conclude that Austria and Germany's fear of Pan-Slavism was a more fundamental reason for the outbreak of World War I in 1914 than was the assassination of the Archduke Franz Ferdinand at Sarajevo.

e. Suppose once more that the joint presence of A and B is not necessary for the occurrence of C; but now assume that the relative frequency with which C occurs when A is realized but B is not, is greater than the relative frequency of C's occurrence when B is realized but A is not. When assumptions somewhat like these are taken for granted, the factor A is frequently said to be more important than the factor B as a determinant of C. For example, a statement such as that broken homes are a more fundamental cause of juvenile delinquency than is poverty can perhaps be best interpreted to mean that the relative frequency of delinquents among juveniles coming from broken homes which are not poverty-ridden is much greater than among children whose parents are poor but live together amicably with their offspring. A historian who studies a rise in juvenile delinquency in a certain community during a given period and who attributes the rise to an increase both in the number of broken homes and in the number of poverty-stricken ones may in consequence assign greater importance to the first of these assumed causal factors than to the second.

f. One final sense of 'more important' requires explicit formulation. Suppose that A is one of the basic (primitive or defined) theoretical notions in a theory T but that B is not, and also that T can account for a large class of phenomena (including C when it is supplemented by various suitably specialized assumptions); and suppose further that in order to explain C a specialized assumption referring to B must be introduced, although most of the other phenomena in the range of application of T can be accounted for without such specialized assumptions that refer to B. In consequence, the set of phenomena that can be explained by T when the premises contain a reference to A but not to B is much more numerous than is the set of phenomena whose explanations require a reference to both A and B. The factor A is thus a determinant not only of phenomena for which B is also a determinant (such as the phenomenon C), but also of phenomena for which B is not a determining factor; and A may therefore be said to be more important (or basic) than B as a determinant of C. For example, the notion of inertial force is central to Newtonian mechanics, but the notion of frictional force is not; and an extensive class of phenomena can be explained with the help of Newtonian theory without reference to friction. On the other hand, the motions of marbles rolling down an inclined plane can be properly accounted for only if the assumptions of Newtonian theory are supplemented by some special assumptions about frictional forces. In the present sense of 'more important,' inertial forces are thus more important than frictional ones as determinants of the motions of bodies on inclined planes. Something like this sense of the phrase appears to be intended by those who claim that the relations of production and distribution of wealth operative in a society constitute more basic determinants of its legal institutions than do its religious practices and beliefs. Proponents of this claim usually maintain that a large class of social phenomena can be accounted for with the help of a theory formulated exclusively in terms of economic relations. However, advocates of the claim generally recognize the influence of organized religion upon society. But they appear to hold that the theory requires supplementation by special assumptions about religious institutions only when the theory is used to explain certain limited areas or aspects of social behavior, such as the enactment of certain laws or the appointment of certain individuals to judicial office.

Other senses of 'more important' or 'more basic' could be distinguished but the six mentioned seem to be those most frequently intended in discussions of human affairs. It is pertinent to note, however, that, although a fairly definite meaning can often be found for statements using these and related expressions, the statements may not be factually warranted. Indeed, even when a historian's statement concerning the relative importance of various determining factors for an event has an undeniably clear and verifiable content, it is doubtful in most cases whether the statement is supported by competent evidence. There are next to no statistical data available on the relative frequency with which most of the phenomena that are of special concern to historians take place. Students of human history are therefore compelled, willy-nilly, to rely on guesses and vague impressions in assigning weights to causal factors. In consequence, there are often wide divergences in judgment as to what are the main causes of a given event, and one historian's opinion may be no better grounded than another's. Whether this defect in establishing causal imputations in current historical research will eventually be remedied is an open question; but the prospects for substantial improvements in this respect do not seem bright, since the probable cost of remedial measures in terms of labor as well as money is staggering. But for the present, at any rate, a judicious skepticism toward most if not all judgments concerning the relative importance of causal determinants for events appears to be in order.

3. However, no mention has thus far been made of a familiar special form in which historians frequently assign an order of relative importance to events, namely, when they assert contrary-to-fact conditionals about the past. This form requires brief comment. Contrary-to-fact judgments are often explicitly introduced into historical analyses, usually to support some claim that a certain event had consequences crucial for subsequent developments. To cite a famous example, many historians believe that the battle of Marathon in 490 B.C. was one of the decisive military conflicts in human history; and they support this belief by the contrary-to-fact judgment that, had the Persians been victorious, an Oriental theocratic-religious culture would have been established in Athens, with the consequence that Greek science and philosophy, in which Western civilization has its roots, would not have been developed.

It is sometimes argued that contrary-to-fact judgments have no proper place in historical analyses, either because it is not the business of the historian to introduce such judgments or because the task of finding adequate grounds for them is hopeless. According to one influential school of thought, for example, the historian's job is to discover what actually happened and to ascertain by what continuous transformations one period of human life grew out of a preceding one; in consequence, although it is fitting that a poet or a moralist should concern himself with what might have been, it is inappropriate for a serious student of the past to do so. Moreover, so a number of historians have declared, contrary-to-fact judgments are based on the assumption that the past can be analyzed into a set of simple, isolable, self-contained, and externally related events, so that the occurrence or nonoccurrence of any one of these events allegedly does not radically modify the relations between the remaining ones. In the view of these writers, however, such an assumption is untenable for the subject matter of human history, even if it should be valid for the subject matter of natural science. According to them, the events of the past are so interrelated that the hypothetical nonoccurrence of a single event entails a fundamental transformation of all the others; and they therefore conclude that it is impossible to establish just what would have happened had some particular event not taken place.[27]

[27] Cf. Charles A. Beard, *The Discussion of Human Affairs*, pp. 42-46. The following states concisely the case against the possibility of contrary-to-fact judgments in history: "It is absolutely futile to ask oneself what the world would have become if some particular hypothetical event had been realized, or if some particular real event had not been realized. What would have happened if Hannibal had destroyed Rome? If Louis XVI had been able to escape abroad? If Napoleon had not been born? The course of history would have been transformed by either of these hypotheses:

But assumptions as to what might have been cannot be excised from history by such arguments. Contrary-to-fact judgments are unavoidable except by eschewing all judgments of relevance and all attempts at explaining what has happened. We had occasion to note much earlier (in Chapter 4) the intimate connection between scientific laws and counterfactual statements; and, since historical explanations require at least the tacit use of general assumptions, such explanations thereby assert at least by implication contrary-to-fact conditionals. Thus, a historian who finds that the spread of Arabic culture throughout northern Africa and southern Spain was one of the factors responsible for the revival of learning in western Europe during the eleventh century, is in effect maintaining that, had Mohammedan arms not been victorious in Africa and Spain, the subsequent cultural development of Europe would have had a different issue. Otherwise his apparent imputation of a causal role to the spread of Arabic culture would be nothing but a chronological listing of the events he is discussing. Accordingly, those who reject the possibility of contrary-to-fact judgments in human history must in consistency also deny the possibility of explaining any event in the human past.

Nevertheless, it is in general by no means an easy task to provide reasonably firm grounds for contrary-to-fact judgments in human history. The task is undoubtedly more difficult than the analogous task in many other disciplines, partly because (as has so often been noted) it is impossible to perform experiments on nonrecurrent events, but in large measure because of the paucity of relevant data on most of the questions about which historians make such judgments. Despite these disadvantages, the task is not quite so hopeless as is frequently claimed. An example of how contrary-to-fact judgments are actually warranted will help make clear the difficulties that must be faced in such an undertaking as well as the considerations historians introduce in attempting to resolve them. With this end in view, let us examine the grounds upon which some historians base their contention that the battle of Marathon was decisive for Western civilization.

According to one account, it was the general practice of the ancient Persians to use the extant religious institutions of a country they had subjugated (e.g., Judea) as instruments for governing that territory; and in consequence strict compliance with religious dogma was enforced upon the defeated population. Such potential instruments of political

but in what way? After a definitive victory of Carthage, the play of history would have offered a multitude of diverse possibilities among which chance would have decided, as it did among the immeasurable possibilities which were opened up to humanity the day following the victory of Rome. This incessant interaction of chance renders purely fanciful any reconstruction of the present and the future upon an unrealized hypothesis."—Pierre de Tourtoulon, *Philosophy in the Development of Law,* New York, 1922, p. 631.

control certainly existed in Athens, in the form of various mystery cults that discouraged a questioning attitude and may even have been of Oriental origin. However, rigid conformity to the demands of religious creed and observance is compatible neither with political democracy nor with an unfettered pursuit of art, science, and philosophy. In consequence, it is likely that a Persian victory at Marathon would have given to those mystery cults a commanding place in Athens, and would thereby have made dominant all those elements which were inimical to the use of reason and to the rational organization of Athenian society.[28]

The conclusion of this probabilistic argument is obviously not established beyond every conceivable doubt. Nevertheless, the argument does provide reasonably good support for the conclusion; and it serves to illustrate the essential irrelevance to actual historical analyses of wholesale objections, such as those mentioned above, to the possibility of contrary-to-fact judgments in human history. However, legitimate doubts can be raised about the factual validity as well as the applicability to the problem at issue of some of the premises taken for granted in the argument. How sound is the assumption, for example, that art, science, and philosophy do not flourish in authoritarian societies? Despite its wide acceptance by democratically inclined thinkers, there are notable exceptions to the assumption, so that it appears to be warranted only if it is asserted with various qualifications—although it is not clear what qualifications are needed, nor whether a qualified form of the assumption would support the conclusion we are discussing. Consider next the assumption concerning the administrative practices of ancient Persia in conquered countries. Even if this assumption is sound, its applicability to the problem at issue cannot be taken for granted, and we must first make sure that Persia really did aim at *subjugating* Athens. Perhaps Darius had no intention of making Athens a Persian satrapy, but merely planned to chastise offending Greek cities and to restore his friend Hippias to power in Athens. Unless this possibility can be eliminated, Persia's administrative practice in subjugated countries has only doubtful relevance as evidence for Persia's likely behavior in a defeated Athens. Moreover, even if this possibility is eliminated, it is debatable whether that administrative practice would have been followed in Greece. To be sure, Persia ruled her conquests in Asia Minor with an iron hand. But Athens was at a more remote distance from Persia than was Ionia; and perhaps Persian rule in Athens would have been less harsh in consequence.[29] In short, though the conclusion of the argument is not mere

---

[28] Cf. Eduard Meyer, "Zur Theorie und Methodik der Geschichte," in *Kleine Schriften*, Halle, 1902; and Max Weber, "Critical Studies in the Logic of the Cultural Sciences," in *The Methodology of the Social Sciences*, Glencoe, Ill., 1949.

[29] Cf. the discussion of these questions in J. B. Bury, *History of Greece*, New York, 1937, pp. 243-44.

baseless speculation but has some firm factual support, parts of the supporting argument involve important uncertainties which our present knowledge of the laws of human behavior and our available evidence about the past are insufficient to resolve.

However, perhaps a major difficulty in warranting contrary-to-fact judgments in human history—as it is in all reasoning that is not exclusively deductive or purely formal—is the *logical* problem of assessing the weight of often conflicting evidence for a given hypothesis and of comparing the degrees of support alternative hypotheses receive from the available evidence for each. For in sharp contrast to what has been accomplished in codifying the principles of demonstrative inference, there is at present no generally accepted, explicitly formulated, and fully comprehensive system of logical rules for performing these crucially important tasks. It is true that the mathematical calculus of probability is frequently claimed to be such a codification of the logical canons of nondemonstrative inference; but the relevance of the probability calculus for evaluating the strength of evidence is doubtful, and is in any case widely disputed. Accordingly, even conclusions concerning matters of supreme practical importance are often accepted on the basis of evidence whose probative force is estimated in an idiosyncratic fashion and differently by different individuals.

This lack of clearly articulated uniform standards for measuring and compounding the weight of evidential premises is undoubtedly responsible for many of the disagreements among historians when they assess the comparative importance of various causal factors, and in consequence when they make contrary-to-fact judgments about the human past. For example, even though they base their judgments on what appears to be identical evidence, equally competent historians often give incompatible evaluations of the roles various individuals played in some given episode; and they therefore offer divergent estimates for the likelihood of suppositions highly pertinent to their analyses—such as the likelihood that, had Lincoln lived, he would have been more successful than was Andrew Johnson in persuading Congress to modify its vindictive attitude toward the South.

But the absence of common logical standards is especially conspicuous when there is a need to combine several evidential premises and to estimate their resultant probative force. Such a need is illustrated by the discussion of the significance of the battle of Marathon. Let us indicate this schematically by supposing the following: (a) relative to the evidence *e* that an Oriental despotism subjugated a state with democratic institutions, historians agree in assigning a *high* probability to the hypothesis *h* that the victorious state will not tolerate the continuance of those institutions in the vanquished country; and (b), relative to the evidence e' that two countries with primitive means of communication and transportation are separated by a considerable distance, historians also agree in assigning a *low* probability to the hypothesis h' that one of the countries will exert a controlling influence on the other's institutions. Let us now combine the two items of evidence e and e', and assume the augmented evidence e'' to state that an Oriental despotism, possessing only primitive means of communication and separated by a considerable distance from a state with democratic institutions, subjugates the latter. A problem typical of many inquiries is then raised by the query: What is the probability, relative to e'', of the hypothesis h'' that the Oriental despotism will seriously impair or destroy the democratic institutions of the vanquished society? If there were generally recognized and well-founded rules for evaluating this probability from those assumed in (a) and (b), the question whether the events at Marathon were as decisive for Western civilization as is often claimed could be settled with reasonable unanimity. As it is, however, historians, like other men, fall back upon their intuitive judgments when they attempt such evaluations, and in consequence the estimates they render of the required probabilities sometimes vary within wide limits.

On the other hand, though differences concerning the probative force of given evidence are sometimes dishearteningly large, there is fortu- nately also frequent and substantial agreement on the probabilities men assign to many hypotheses concerning matters with which they have had considerable experience. Such agreement indicates that, despite the lack of an explicitly formulated logic of nondemonstrative inference, men have acquired through trial and error many unformulated habits of thought which embody sound principles of nondemonstrative reasoning. For example, a physician may develop habits of analysis after many years of practice that make him a superbly competent diagnostician, without his ever being aware of the logical principles he tacitly employs in drawing his inferences; and similarly, a student of human affairs may acquire by repeated trials an ability for assessing the weight and significance of evidence, though he never formulates the logic of his procedure. Accordingly, although contrary-to-fact judgments in human history are never established beyond the reach of every doubt, and although such judgments in specific cases are often doubtful if not undoubtedly erroneous, such specific doubts and errors are not adequate grounds for rejecting in principle the very possibility of contrary-to-fact judgments.

IV.   *Determinism in History*

In the 1920's a historian of note examined the apparently decisive influence exercised by a number of famous persons upon such important historical events as the Protestant Reformation in England, the American Revolution, and the development of parliamentary government. He then assessed the supposedly critical role which the decisions and actions of these men played in bringing about those events, generalized his findings, and concluded as follows:

> These great changes seem to have come about with a certain inevitableness; there seems to have been an independent trend of events, some inexorable necessity controlling the progress of human affairs. . . . Examined closely, weighed and measured carefully, set in true perspective, the personal, the casual, the individual influences in history sink in significance and great cyclical forces loom up. Events come of themselves, so to speak; that is, they come so consistently and unavoidably as to rule out as causes not only physical phenomena but voluntary human action.
>
> So arises the conception of *law in history*. History, the great course of human affairs, has not been the result of voluntary efforts on the part of individuals or group of individuals, much less chance; but has been subject to law.[80]

The view expressed in this quotation is a variant of a conception of human affairs that is familiar and continues to be widely held. It is a conception that has sometimes been advanced as ancillary to a theodicy; sometimes to a romantic philosophy of cosmic organicism; sometimes to an ostensibly "scientific" theory of civilization which finds the causes of human progress or decline in the operations of impersonal factors such as geography, race, or economic organization. Despite important differences between them, these various doctrines of historical inevitability share a common premise: the impotence of deliberate human action, whether individual or collective, to alter the course of human history, since historical changes are allegedly the products of deep-lying forces which conform to fixed, though perhaps not always known, patterns of development.

The doctrine of historical inevitability has been repeatedly shown to be untenable by historians as well as philosophers; and it is not the aim of the present discussion to exhibit once more its many deficiencies. It will suffice to note that in some of its variant forms the doctrine has no empirical content, since no conceivable empirical evidence can ever be relevant for testing those versions of the doctrine for their truth or falsity. Moreover, when the doctrine is formulated so that it can be empirically tested, the available evidence supports neither the thesis that all human events illustrate a unitary, transculturally invariant law of development, nor the thesis that individual or collective human effort never operates as a decisive factor in the transformations of society. However, the re-

[80] Edward P. Cheney, *Law in History and Other Essays*, New York, 1927, p. 7.

jection of these claims is not to be construed as denying that in many historical situations individual choice and effort may count for little or nothing, nor that there frequently are ascertainable limits to human power for directing the course of social changes—limits that may be set by facts of physics and geography, by biological endowment, by modes of economic production and available technological skills, by tradition and political organization, by human stupidity and ignorance, as well as by various antecedent actions of men.

On the other hand, many recent critics of historical inevitability have gone much beyond denying the manifestly exaggerated claims of the doctrine. They have proceeded to challenge what they believe is the basic premise upon which it rests—the view, namely, that human events generally occur only under determinate and determining conditions. Accordingly, these critics have tried to show that a thoroughgoing determinism is incompatible with the established facts of human history as well as with the assumption, underlying all discussions of moral problems, that human beings are genuinely responsible for their deliberate choices and actions. Moreover, many thinkers who reject the doctrine of historical inevitability are also severe critics of current trends in psychological and social research; they maintain that, since the behavioristic (or "naturalistic") methodology adopted in these inquiries allegedly rests ultimately on the deterministic premise, contemporary social science is destroying the belief in human freedom and is therefore undermining the foundations of moral effort.

It is with some of these criticisms of determinism that the remainder of this chapter is concerned. However, these criticisms seldom make explicit how 'determinism' is to be construed as a general thesis; and, although critics of the thesis sometimes identify it with the doctrine of historical inevitability,[31] a much more comprehensive notion is usually intended. We must therefore briefly recall the account given in Chapter 10 of the sense in which 'determinism' is commonly understood in the natural sciences, since this appears also to be the sense in which determinism is frequently said to be the premise underlying the doctrine of historical inevitability.

It will be convenient to summarize our earlier discussion of determinism with the help of an example of a physicochemical system generally acknowledged to be deterministic.[32] The system consists of a mixture of

---

[31] According to one historian, for example, determinism is the doctrine "according to which we are helplessly caught in the grip of a movement proceeding from all that has gone before."—Pieter Geyl, *Debates with Historians*, New York, 1956, p. 236. But this is not a representative view.

[32] The example is borrowed from Lawrence J. Henderson, *Pareto's General Sociology*, Cambridge, Mass., 1935, Chap. 3, where it is used to illustrate the sense in which, according to Pareto, a social system is deterministic.

soda water, whiskey, and ice, contained in a sealed vacuum bottle. It is assumed that no air is present in the bottle, and that the mixture is completely isolated from everything else, such as sources of heat in the environment. Moreover, the only characteristics of the system that enter into the discussion are the "thermodynamical variables," such as the following: the number of components of the system (the components in the example are water, alcohol, and carbon dioxide), the phases or types of aggregation in which the components occur (in the example, water occurs in a solid, liquid, and gaseous phase), the concentrations of the components in each phase, the temperature of the mixture, and the pressure on the walls of the container. It is well known that for a given temperature and pressure, each component in the system will occur in the various phases with definite concentrations, and conversely. Thus, if the pressure of the mixture is increased (e.g., by pressing down the stopper of the bottle), the concentration of water in the gaseous phase is reduced, but its concentration in the liquid phase is increased; and analogously for a change in temperature. The variables of the system thus stand to each other in definite relations of interdependence, so that the value of a variable at any given time may be said to be "determined" by the values of the other variables at that time.

Now suppose that at some initial time the system is in a definite "state" (i.e., the variables have certain specific values), and that because of some change induced in one or more of the variables at that time the system is in some other definite state after an interval of time t; but suppose also that the system is brought back in some way to its initial state, that exactly the same changes are induced in the variables as before, and that after the same interval t the system is again in the second state. If the system behaves in this manner no matter what state is taken to be the initial state, and no matter how large the interval t, the system is said to be "deterministic" with respect to the specified set of thermodynamical variables.

If the reference to the physicochemical example is dropped, 'determinism' may therefore be defined quite generally as the thesis that for *every* set of attributes (or "variables") there is *some* system which is deterministic with respect to that set. Accordingly, 'determinism in history' is the thesis that, for every set of human actions, individual or collective characteristics, or social changes that may be of concern to the historian, there is some system which is deterministic with respect to those items—where the state variables of the system are, however, not specified. We can now turn to the task we proposed for the remainder of this chapter, to discuss various criticisms of determinism in history. The objections to the deterministic thesis to be examined fall under the following heads: (1) the argument from the falsity of the doctrine of historical inevitability and from the nonexistence of "laws of necessary development" in human affairs; (2) the argument from the unpredictability of human events; and (3) the argument from the incompatibility of determinism with the reality of human freedom; but we will conclude the examination with (4) some reflections on the validity of the thesis itself.

1. The first argument can be briefly stated and also quickly dismissed. It is directed primarily against those grandiose philosophies of history, whether religious or secular in orientation, which claim to find a fixed pattern of development in the manifold succession of events that have taken place since the beginnings of the human race, or at the very least to detect an invariable order of sequential change repeatedly manifested in different societies or civilizations. From the perspective of some of these philosophies, every human act appears to have a definite place in an unalterable structure of changes, and each society must necessarily pass through a fixed series of antecedent stages before it can achieve a subsequent stage. Moreover, though human individuals are the ostensible agents that bring about the movements of history, human actions are seen in many of these philosophies as at best only the "instruments" through which certain "forces," operating and evolving in conformity with timeless laws, become manifest.

Philosophies of history of this type often possess the fascination of great dramatic literature; and few of their readers would be willing to deny the remarkable imaginative powers and amazing erudition that frequently go into their construction. But as has already been mentioned, when the evidence of what has actually happened is at all relevant for judging such philosophies, it is overwhelmingly negative. Critics of these philosophies are on safe ground in rejecting them as false.

Does it nevertheless follow from the falsity of the doctrine of historical inevitability that there are no causal connections in human affairs, and that determinism in respect to the events discussed by historians is a myth? Those recent critics of the doctrine who believe that this does follow offer no explicit grounds for their claim, and appear to base their contention on an extraordinarily narrow conception of what a deterministic system must be like. They seem to think that, since the human past does not exhibit anything like the regular periodicities of a well-constructed chronometer, the events in that past cannot possibly be elements in a deterministic system. However, although a given system may fail to illustrate some relatively simple schema of changes, it may nevertheless manifest a more complex and unfamiliar pattern of relations of dependence. Moreover, even if it should be the case that a particular system is not deterministic in respect to a specified set of characteristics, the system may not be sufficiently well isolated from external influences (as in the case of a clock whose motions show "irregularities" because of the influence of a fluctuating magnetic field); and there may therefore be some other system (perhaps the system which includes these external influences together with the initial system) that is deterministic in respect to the given set of characteristics. In any event, granting that the doctrine of historical inevitability is false and that there are no laws of necessary development in human history, there is competent evidence to show that, for example, the decline of Spanish power in the seventeenth century was in part the consequence of Spanish economic and colonial policy, or that a necessary condition for the success of the Bolshevik Revolution was the leadership supplied by Lenin. In short, the first argument against determinism does not achieve its objective.

2. Critics of determinism tend to place great weight on the objection that human events are in considerable measure unpredictable. Two senses of the latter word are usually distinguished in this connection. An event is "technologically" unpredictable if, because of limitations in the knowledge or technology men possess at a given time, they lack the effective ability either to foretell the event at all, or to foretell it with more than a certain degree of precision. However, it is obviously not a serious objection to determinism that events may be unpredictable in this sense; and no critic of the deterministic thesis is likely to argue that earthquakes, for example, do not have necessary and sufficient conditions for their occurrence, on the ground that we are unable at present to anticipate when the next earthquake will take place.

On the other hand, an event is "theoretically" unpredictable if the assumption that its occurrence can be calculated in advance with unlimited precision is incompatible with the "laws of nature," that is, with the corpus of scientific knowledge, and in particular with established scientific theory. The stock example used to illustrate this sense of the word is the limited precision with which, according to current quantum theory, subatomic processes can be predicted. It will be evident, however, that, even if human events are assumed to be theoretically unpredictable, this supposition has force as an objection to the deterministic thesis only if the thesis is identified with the claim that in principle events can be predicted with absolute precision.[33] To be sure, a connotation can un-

[33] Such an identification was made by Moritz Schlick: " 'A determines B' can mean nothing else than: B can be calculated from A. And this means: there is a universal formula, which attests to the occurrence of B, when certain values are substituted for the initial conditions A and in addition definite values are assigned to such variables as the time t. . . . The word 'determined' thus means exactly what is meant by 'predictable' or 'calculable in advance.' "—Moritz Schlick, "Die Kausalität in der gegenwärtigen Physik," *Gesammelte Aufsätze*, Vienna, 1938, pp. 73-74.

doubtedly be *assigned* to 'determinism' so that in consequence its meaning will coincide with that of 'predictable.' But the equivalence that would thus be established between the connotations of the words would be the result of an arbitrary *fiat*, since the words are not in fact generally employed as synonyms. If they were, it would be absurd to suppose that something which is admitted to be theoretically unpredictable might nevertheless be determined. However, despite the circumstance that quantum mechanics belongs to the current corpus of scientific theory, it is not self-contradictory (though it may be mistaken) to hold, as Planck, Einstein, and others have held, that subatomic processes have determining conditions for their occurrence, and that an alternative to quantum theory is desirable which would set no upper bounds, as does the latter, upon the precision with which certain of those processes are predictable.

But in any event, nothing comparable to quantum mechanics is available in the social sciences upon which to rest the assumption that human events are theoretically unpredictable. Nor does the actual evidence establish the claim that human actions are utterly unpredictable in fact. It would be ridiculous to maintain that every detail in man's future can be predicted, or even to pretend that every event in the human past can be inferred from the available data. On the other hand, it is no less ridiculous to hold that we are completely incompetent to predict anything about the human future with any assurance of being correct. It is almost truistic to note that our personal relations with other men, our political arrangements and social institutions, our transportation schedules and our administration of justice, could not be what they are unless fairly safe inferences were possible about the human past and future. To be sure, we cannot predict with any certainty who will be the next president of the United States. But if we take for granted current American attitudes toward domestic and foreign issues, and also take into account the present alignment of the world powers, we do have good grounds for confidence that there will be a presidential election during the next leap year, that neither major political party will nominate a Communist, and that the successful candidate will be neither a woman nor a Negro. These various predictions are indefinite in certain ways, for they do not foretell the future in a manner to exclude all but one conceivable alternative. Nevertheless, the predictions *do* exclude an enormous number of logical possibilities; and they do point up the fact that, though the human beings who will participate in the coming events may have a considerable range of free choice in their actions, their actual choices and actions will fall within fairly definite limits. The obvious import of all this is that not everything which is logically possible is also historically possible during a given period and for a given society of men; and the equally obvious interpretation of this fact is that there are determining conditions for both what has happened and what will happen in human affairs.

On the other hand, even our subsequent historical explanations of past events, as well as our forecasts of future ones, are almost invariably imprecise and incomplete. For our accounts of past occurrences, whether these be individual or collective acts, rarely if ever explain the exact details of what did happen; and, as we have seen, they succeed in exhibiting only the grounds which make probable the occurrence of more or less vaguely formulated characteristics. But we have already examined the reasons for the probabilistic structure of historical explanations, and none of them is a ground for rejecting determinism.

3. The remaining argument we will consider is that a thoroughgoing determinism is incompatible with the fundamental axiom of moral theory that human beings can properly be said to be responsible for their decisions and deliberate actions. This objection to determinism has been a theme of philosophical and theological debate since antiquity, but has been revived in current discussions of human history and social science. We shall examine some of the issues raised by it in the form they are presented in a book by Isaiah Berlin.[34] The book is primarily a devastating critique of philosophies of history which view the human scene as the unfolding of an inevitable destiny that cannot be altered by human effort; and it also maintains that these philosophies are simply corollaries to the assumption that human affairs are strictly determined. We shall ignore this reason Berlin advances for rejecting determinism, since we have already shown that the deterministic thesis does not entail the doctrine of historical inevitability; but we must discuss two further arguments he directs against the thesis.

a. Berlin's point of departure in the first of these arguments is the generally acknowledged commonplace that an individual cannot appropriately be held morally responsible for some action if he was coerced into performing it and if he did not elect to do it of his own free volition. Accordingly, if a person is genuinely responsible for an action, he *could* have acted differently had his choice been different. But Berlin also believes that on the deterministic thesis (understood by him to deny that there is any area of human life not exhaustively determined by law), the person could *not* have chosen differently from the way in which he did *in fact* choose, apparently because the individual's decision at the time it was made was determined by circumstances over which he had

[34] Isaiah Berlin, *Historical Inevitability*, London and New York, 1954.

no control, such as his biological heritage or his character as formed by prior actions. In consequence, to anyone who accepts the deterministic thesis, the supposition that a man could have decided otherwise than he did decide must ultimately be an illusion which rests on our ignorance of the determining facts of his choice. Berlin therefore concludes that determinism entails the elimination of individual responsibility, since it is not a man's *free* choice, but rather the conditions determining his choice, that according to determinism explain the man's action. He declares, for example,

> Nobody denies that it would be stupid as well as cruel to blame me for not being taller than I am, or to regard the color of my hair or the qualities of my intellect or my heart as being due principally to my own free choice; these attributes are as they are through no decision of mine. If I extend this category without limit, then whatever is, is necessary and inevitable. . . . To blame and praise, consider possible alternative courses of action, damn or congratulate historical figures for acting as they do or did, becomes an absurd activity.

And he adds:

> If I were convinced that although choices did affect what occurred, yet they were themselves wholly determined by factors not within the individual's control (including his own motives and springs of action), I should certainly not regard him as morally praiseworthy or blameworthy.[35]

Two comments are in order.

i. In the first place, it is far from clear what is the conception of the "human self" with which Berlin operates. For on his view, the human self must apparently be distinguished not only from the human body but also from any of the choices, no longer within a man's control, which determine at least in part the choice he is about to make, as well as from his springs of action, his disposition, and his motives insofar as these latter are also beyond his control. It is therefore difficult to know what does remain of the self, when all the things are eliminated that in the slightest way influence a man's conduct during the knife-edge instant of the immediate present.

The difficulty is not diminished when we try to understand Berlin's conception of the self whose decisions are "free" in his sense of the term, in the context of imagining some person deliberating over a course of action he ought to adopt, and finally deciding between several alternatives he has been contemplating. The individual is usually unaware that the decision he finally makes may be the expression of a set of more or less stable habits, transient impulses, the careful attention he gave to some of the alternatives but not to the others, and so on—any more than he is normally aware of his own heartbeat or of the organ that produces it. It seems certainly unlikely that, when the individual recovers from his perplexed surprise on being asked whether the choice he finally made was really his own, he would hesitate to say that of course it was. But should the individual become aware of these things about himself, as he sometimes may become aware, would he regard his choice as any less his *own* choice? This too seems unlikely, just as it is unlikely that he would disclaim the pulse beat in his temples as his own, when he discovers that it is being produced by the rhythmic contractions of his heart.

According to Berlin, however, the answer to the question as to whether the decision was the individual's own must in either case apparently be negative. Berlin is therefore faced with a puzzle that is nothing short of being irresolvable—the puzzle of finding some activity or trait that is an intrinsic attribute of the human self, but with the proviso that anything which is causally dependent on something else is thereby automatically disqualified from being a genuine part of the self. His problem is like the one with which he would be confronted were he to set himself the task of describing a moving baseball, for example, without mentioning any attributes which owe their presence in the ball to any agency (such as the manufacturer who made it, the batter who struck it, or the sun which shines upon it)—for the reason that, since familiar attributes of the ball like size, shape, color, and state of motion have been determined by external factors, they are not genuinely intrinsic to the ball itself.

Just.how and where the boundaries of the individual human self are to be drawn are undoubtedly not easy questions to decide; and the answers to them may vary with different contexts of self-identification, and may even depend on cultural differences in the ways the human self is conceived. But however they are drawn, the lines should not be so drawn that in the end nothing can be identified as the self. Certainly an artificial and insoluble puzzle should not be made of the fact that we are frequently conscious of acting of our own free volition and without external constraints, even if we recognize that some of our choices are the products of our dispositions, past actions, and present impulses.

ii. A second comment must be made on Berlin's argument. On the face of it, his discussion of the conditions under which human beings can be properly regarded as responsible agents closely resembles the reasoning often used to show that in the light of the findings of modern physics the common-sense view of the world is an illusion. For example, it has been argued that since, according to physics, common-sense objects like tables are complex systems of rapidly moving minute particles separated by relatively great distances, it is illusory to suppose that tables are "really" hard solids possessing continuous surfaces. But, as has been frequently noted, such an argument is a nest of fallacies. It commits the fundamental error of supposing that, since common-sense terms such as 'solid,' 'hard,' and 'continuous' are admittedly not applicable, when used in their ordinary meanings, to things like congeries of molecules, the terms are therefore not correctly applicable to macroscopic objects like tables.[36]

Berlin's discussion suffers from a similar flaw, for it argues in an analogous way that, if there are determining biological and psychological conditions under which responsible behavior occurs, men cannot be genuinely responsible for any of their acts, for the reason that responsibility (in the same sense of the term) cannot be properly ascribed to those *conditions*. It is nevertheless an empirical fact, as well attested as any, that men often do deliberate and decide between alternatives; and nothing we have discovered or will discover about the physiological and psychological *conditions* that make deliberation and choice possible can be used as evidence (except on pain of a fatal incoherence) for denying that such deliberative choices *do* occur.

It is pertinent to note, on the other hand, that it is an empirical question whether a given individual is correctly held responsible for some action, and that we may be badly mistaken in supposing that he is. We may discover, for example, that an individual continues to be a petty thief, despite our best efforts to educate him by way of rewards and punishments, and despite his own apparently serious attempts to mend his ways. We may then conclude that he suffers from a mild derangement and cannot control certain of his actions, so that it would be an error to continue holding him responsible for them. The fact nevertheless remains that the distinction between actions over which a human being does have control and those over which he does not, is not thereby impugned—nor does the distinction become spurious should we discover the conditions under which the capacity for such control is acquired and manifested. In short, a person is correctly characterized as a responsible moral agent if he behaves in the manner in which a normal moral agent behaves; and the characterization remains correct even if the organic and psychological conditions that make it possible for him to function as a moral agent are not within his control on any of those occasions when he is acting as a responsible person.

[36] For an influential statement of the argument, cf. Arthur S. Eddington, *The Nature of the Physical World*, New York, 1929, esp. pp. xi–xiv; for a vigorous criticism of Eddington, see L. Susan Stebbing, *Philosophy and the Physicists*, London, 1937.

b. There is a second argument Berlin directs against determinism. He maintains that, irrespective of the truth of the deterministic thesis, the ordinary thoughts of the majority of men are not in fact colored by belief in the thesis. If they were, so he claims, the language men employ in making moral distinctions and in expressing moral suasions would not be what it actually is. For the customary use of this language tacitly assumes that men are free to choose and to act differently from the way they in fact do choose and act. However, if we really believed in determinism, Berlin concludes, our ordinary moral distinctions would not be applicable to anything, and our moral experience would be unintelligible.[87]

But let us examine this claim that a consistent determinist cannot employ ordinary moral discourse in its customary meanings.

i. Is this claim to be assessed on the basis of straightforward empirical evidence? If so then, although relevant data have not been systematically gathered and the available information is undoubtedly inconclusive, much of the evidence we do possess certainly does not support the contention. The language of many devout religious believers, to say nothing of philosophers like Spinoza, provides some ground for maintaining that, despite explicit and wholehearted adherence to a thoroughgoing determinism, many men have found no psychological obstacles to their making normal moral appraisals. To cite but one instance, Bishop Bossuet composed his *Discourse on Universal History* with the intent to offer guidance to the Dauphin on the proper conduct of a royal prince, but in the course of it declared that

> the long concatenation of particular causes which make and undo empires depends on the decrees of Divine Providence. High up in His heavens God holds the reins of all kingdoms. He has every heart in His hand. Sometimes He restrains passions, sometimes He leaves them free, and thus agitates mankind. By this means God carries out His redoubtable judgments according to ever infallible rules. He is it who prepares results through the most distant causes, and who strikes vast blows whose repercussion is so wide-spread. Thus it is that God reigns over all nations.[88]

[87] "If the determinist hypothesis were true and adequately accounted for the actual world, there is a clear sense in which (despite all the extraordinary casuistry which has been employed to avoid this conclusion) the notion of human responsibility, as ordinarily understood, would no longer apply to any actual, but only to imaginary or conceivable, states of affairs. . . . To speak, as some theorists of history (and scientists with a philosophical bent) tend to do, as if one might accept the determinist hypothesis, and yet to continue to think and speak much as we do at present, is to breed intellectual confusion."—Isaiah Berlin, *op. cit.*, pp. 32-33.

[88] Jacques Bossuet, *Discourse on Universal History*, Part 3, Chap. 8, published in 1681 and quoted in G. J. Renier, *History: Its Purpose and Method*, London, 1950, p. 264.

Bossuet believed that the reconciliation of Divine Omnipotence with the reality of human freedom is a transcendent mystery. But however this may be, he appears to have had no difficulty in subscribing to a Providential (and therefore deterministic) conception of human history, and also employing (in contradiction to Berlin's contention) ordinary moral language to express familiar moral distinctions.

ii. Let us suppose, however, that Berlin is right in thinking that, if we really did come to believe in a thoroughgoing determinism, the meanings of our moral discourse would be altered. What would this assumed fact establish? It is pertinent to recall comparable situations in other domains of thought, where the meanings associated with various linguistic expressions were modified as a consequence of the adoption of new beliefs. Thus, most educated men today accept the heliocentric theory of planetary motions, but although they continue to employ such terms as 'sunrise' and 'sunset,' they do not use them with the same meanings those terms had when the Ptolemaic theory was dominant. Nevertheless, some of the distinctions these terms codified when they were associated with geocentric ideas are not without foundation even today, since in many contexts of observation and analysis it is not incorrect to describe the facts by saying that the sun rises in the east and sinks in the west. We have evidently learned to use such language to express distinctions that are still sound, without committing ourselves to other distinctions that depend entirely on accepting the geocentric theory.

Accordingly, and by parity of reasoning, if in agreement with Berlin's supposition we really came to believe in determinism, we would not therefore have to ignore the distinction between those acts described in current language as "freely chosen" and those acts which are not, or between those traits of character and personality over which an individual has effective control and those over which he does not. In any event, moreover, when as a result of the assumed change in belief the shifts in the meanings of currently used expressions are completed, it would still be the case that certain types of conduct are affected by praise and blame and other types are not, that men are able to control and modify by suitable discipline some of their impulses but not others, that some men by making an effort can improve the quality of their performances while other men are unable to do so, and so on. In short, our ordinary moral language with its associated customary meanings, as well as our differential capacities for various kinds of actions, would survive in considerable measure a general acceptance of the deterministic thesis. To deny this is to subscribe to the hardly credible assumption that merely by adopting a belief in determinism men would be transformed into creatures almost unrecognizably different from what they were before this change in their theoretical convictions.

Belief in determinism is therefore not incompatible, either logically or psychologically, with the normal use of moral discourse or with the imputation of moral responsibility to human beings. The alleged incompatibility can be established, so it appears, only if the question-begging premise is introduced that the very making of moral distinctions entails disbelief in determinism.

4. Nevertheless, although none of the arguments directed against determinism is cogent, if the deterministic thesis is construed as a statement about a categorial feature of everything whatsoever, it has neither been conclusively established nor can it be conclusively refuted. The thesis so construed has not been conclusively established, since there are perhaps an endless set of events for which we do not know the determining conditions; and, as an earlier discussion (in Chapter 10) of the notion of "absolute chance" indicated, it is at least logically possible that for some of these events no determining conditions in fact exist. On the other hand, the thesis cannot be definitively disproved, since failure to discover the determining conditions for some event does not prove that there are in fact no such conditions. Accordingly, the thesis asserted in strictly universal form cannot be successfully defended as a well-established generalization about the world as we actually know it.

However, the operative role in inquiry of the deterministic thesis, as of the principle of causality, is seen most clearly when it is construed as a regulative principle that formulates in a comprehensive way one of the major objectives of positive science, namely, the discovery of the determinants for the occurrence of events. Determinism as a regulative principle is undoubtedly most fruitful when it is given a more specialized form than the highly generalized version we have been considering, so that it mentions the particular substantive variables for which search is to be instituted in the effort to find the determining conditions of certain types of events. Thus, the Laplacian notion of determinism we discussed earlier is one such specialized instance of the general principle, in which the substantive variables mentioned are positions, momenta, and forces; and for a time it served as the guiding principle in all physical investigations, although eventually it was replaced even in physics by a different specialized form of the deterministic principle. Similarly, special versions of the general principle are employed to fruitful ends in the psychological and social sciences—for example, regulative principles which stipulate as the determinants of various phenomena such factors as heredity, conditioning by training, modes of economic production, or social stratification.

But although such specialized guiding principles are fruitful only within limited ranges, it will now be evident that the limited value of any one of them is no reason for condemning determinism as a general regulative principle. A dogmatic insistence upon the use of some special form of the deterministic principle has undoubtedly often hindered the advance of knowledge; and it is also undeniable that particular versions of the principle have been frequently used to defend iniquitous social practices. Nevertheless, to abandon the deterministic principle itself is to withdraw from the enterprise of science. However acute our awareness may be of the rich variety of human experience, and however great our concern over the dangers of using the fruits of science to obstruct the development of human individuality, it is not likely that our best interests would be served by stopping objective inquiry into the various conditions determining the existence of human traits and actions, and thus shutting the door to the progressive liberation from illusion that comes from the knowledge achieved by such inquiry.

Index